Essential Human Virology

Essential Human Virology

Jennifer Louten
Kennesaw State University, Kennesaw, GA, United States

Illustrations by

Niki Reynolds

AMSTERDAM • BOSTON • HEIDELBERG • LONDON • NEW YORK • OXFORD • PARIS
SAN DIEGO • SAN FRANCISCO • SINGAPORE • SYDNEY • TOKYO

Academic Press is an imprint of Elsevier

Academic Press is an imprint of Elsevier
125 London Wall, London EC2Y 5AS, UK
525 B Street, Suite 1800, San Diego, CA 92101-4495, USA
50 Hampshire Street, 5th Floor, Cambridge, MA 02139, USA
The Boulevard, Langford Lane, Kidlington, Oxford OX5 1GB, UK

British Library Cataloguing-in-Publication Data
A catalogue record for this book is available from the British Library

Library of Congress Cataloging-in-Publication Data
A catalog record for this book is available from the Library of Congress

ISBN: 978-0-12-800947-5

For information on all Academic Press publications
visit our website at https://www.elsevier.com/

Publisher: Sara Tenney
Acquisition Editor: Jill Leonard
Editorial Project Manager: Fenton Coulthurst
Production Project Manager: Edward Taylor
Designer: Matthew Limbert

Typeset by TNQ Books and Journals
www.tnq.co.in

Contents

Preface

Essential Human Virology is intended to be an approachable and concise introduction to the field of virology. It focuses on the conceptual framework that is needed for students to understand the replication and transmission of viruses and the diseases they cause in humans. Its goal is to provide the essential foundations necessary to understand more advanced molecular virology and scientific research articles.

This textbook incorporates several techniques to facilitate student learning. The text purposefully uses increasingly complex scientific language as the chapters progress in an effort to gradually introduce students to the virological terms and principles of the field. It incorporates over 250 figures that contain dynamic, full-color illustrations and images to further engage visual learners in the material. Integrated "Study Break" questions allow students to gauge their understanding as they read the chapter, and special "In-Depth Look" sections provide additional, detailed material on related topics of high interest. At the conclusion of each chapter, a "Summary of Key Concepts" reminds students of the entirety of the material covered in the chapter. The "Chapter Review Questions" give students the opportunity to test their recall of the important details and apply them to bigger-picture scenarios. A list of "Flash Card Vocabulary" is also included at the end of each chapter, because a solid grasp of new terminology is paramount for comprehension of virological principles. Two appendices at the end of the book include a glossary and a list of abbreviations, allowing students to easily look up terms whenever necessary.

In order to understand aspects of virology and viral replication, students must already possess an understanding of basic cellular and molecular processes. These topics are usually covered in an introductory biology class and further detailed in advanced cellular and molecular biology classes. For students who may need a refresher on these fundamentals, chapter "Features of Host Cells: Cellular and Molecular Biology Review" reviews the central dogma of molecular biology and notable cellular features that are important for understanding viral replication strategies. Because this information is interwoven throughout many upper-level biology courses, most students—if not all—benefit from a review of these principles.

This book covers several themes related to viruses. Chapters "The World of Viruses," "Virus Structure and Classification," "Features of Host Cells: Cellular and Molecular Biology Review," and "Virus Replication" describe the characteristics of viruses and their detailed replication strategies. Chapters "Virus Transmission and Epidemiology," and "The Immune Response to Viruses" examine the interactions of viruses with individuals and populations, including how viruses are combatted by the host immune system, spread between individuals, and disseminate within a population. Chapters "Detection and Diagnosis of Viral Infections," and "Vaccines, Antivirals, and the Beneficial Uses of Viruses" introduce students to traditional and newer methods of viral diagnosis, outline current and experimental vaccines and antivirals, and discuss the beneficial uses of viruses for gene therapy and anticancer therapeutics. Chapter "Viruses and Cancer" examines those viruses that are associated with the development of cancer. The next six chapters provide an in-depth look into human viruses of clinical significance. These chapters cover the replication strategy, pathobiology, and epidemiology of influenza viruses, human immunodeficiency viruses, the hepatitis viruses, the herpesviruses, poliovirus, and poxviruses. The final chapter highlights examples of emerging diseases and their origins. Actual case studies are incorporated into most chapters in order to give students an opportunity to comprehensively integrate the clinical, diagnostic, and epidemiological aspects of viral infections.

About the Author

Jennifer Louten received her doctoral degree from Brown University Medical School, where she investigated the cellular targets of infection and the induction of type 1 interferons following infection with lymphocytic choriomeningitis virus. Dr. Louten is currently an associate professor of biology at Kennesaw State University, where she has served as a Teaching Fellow and developed courses in virology, biotechnology, immunology, and cell culture techniques. She is presently the biotechnology track coordinator and the director of a scholarship program sponsored by a National Science Foundation S-STEM grant. She is the recipient of a Kennesaw State University Outstanding Early Career Faculty Award and the Student Government Association's Faculty of the Year Award. Before becoming a professor, Dr. Louten performed research in drug discovery at Schering-Plough Biopharma (currently Merck Research Laboratories). She received her Bachelor of Science in biotechnology from the Rochester Institute of Technology.

Instructor Companion Website

The online instructor companion website includes several ideas and examples of activities that can be used to further elaborate upon the topics covered in the textbook. For each chapter, a *Teacher Resources* document provides a summary of the chapter and its student learning objectives, using Bloom's taxonomy action verbs. This document includes a study guide that can be provided to students, as well as many examples of active learning instructional activities, classroom discussion topics, presentation ideas, and related videos and websites. Handouts, hands-on worksheets, and a vocabulary crossword puzzle are also included, as well as links to peer-reviewed research articles that correspond to the material. The instructor can use these activities as a supplement to the textbook for more advanced students or in a "flipped classroom" or blended learning instructional format. In addition, the companion website includes a lecture presentation, image bank, and test bank for each chapter. The website can be found here: www.booksite.elsevier.com/9780128009475.

Chapter 1

The World of Viruses

Consider the following cases:

- In Los Angeles, California, five men ranging in age from 29 to 36 years old are hospitalized with pneumonia (inflammation of the lungs) caused by *Pneumocystis carinii*, a fungus. This sort of pneumonia in previously healthy individuals is extremely rare and most often seen in people with severely suppressed immune systems. All five men die of this condition.
- In Hong Kong, a previously healthy three-year old boy develops a fever, sore throat, and cough. He is hospitalized, and less than a week later he dies of acute respiratory distress syndrome, a condition that prevents sufficient oxygen from getting into the lungs and bloodstream. This was believed to be caused by inflammation in his lungs. Sick poultry are also reported in the area.
- A cruise ship departs on a 21-day trip from Washington to Florida. During the trip, 399 of the 1281 passengers come down with acute gastroenteritis, a sudden stomach illness characterized by nausea, vomiting, and watery diarrhea. The cruise ship is cleaned, but a total of 305 people come down with a similar illness during the next three voyages of the ship.

What do these cases have in common? They are all real cases of illnesses that were caused by viruses. The first case above describes the first documentation of acquired immune deficiency syndrome, or AIDS. Although it sounds like the sick men were suffering from a fungal infection, it was revealed 2 years later, in 1983, that AIDS is caused by a virus, now known as the human immunodeficiency virus (HIV). HIV infects cells of the immune system and causes the immune system to slowly decline until it can no longer fight off pathogens, like the *P. carinii* fungus that infected these men. Worldwide, 36.9 million people are living with HIV, over 1 million people in the United States alone.

In the second case, laboratory tests found that the young boy was infected with the flu, which is caused by the influenza virus. The particular subtype was found to be H5N1 influenza, which had previously been observed only in birds. It can be dangerous when a virus jumps to a new species, because as a population, the new species has never been exposed to the virus and so no individuals will have built up immunity against it. As such, the virus has the potential to spread quickly and cause severe effects. In this case, 18 individuals in Hong Kong ended up with the H5N1 influenza infection, which they acquired through direct interaction with chickens at open-air markets. It was soon discovered that 20% of the chicken population in Hong Kong was infected with this subtype of influenza, and considering the 30% death rate they had so far observed, the government decided to prevent further human exposure to the virus by slaughtering the 1.5 million live chickens found in its markets and poultry farms. This effectively stopped the spread of the virus into any additional humans, but the threat always exists that a similar situation could occur again.

In the final case, passenger stool samples tested positive for Norwalk virus. This type of virus causes diarrhea and vomiting and is very easily transmitted from person to person through the ingestion of contaminated food and water or aerosolized particles, like those that might be generated by flushing a toilet. The virus is difficult to inactivate; alcohol-based hand sanitizers are not completely effective against the virus, so cleaning with a bleach solution is the preferred method. In addition, only a few virus particles need to be ingested to cause illness, which may explain why passengers on subsequent cruises also came down with the illness, even though the ship had been disinfected before their trip. Norwalk virus outbreaks do not occur solely on cruise ships; each year, there are an estimated 20 million cases in the United States of gastroenteritis caused by this virus.

These three cases are just a few examples of the diseases that viruses can cause. In the following chapters, we will explore how viruses replicate, are spread, and cause a variety of diseases in humans.

In-Depth Look: The Difference Between "Virus" and "Virion"

Many people use the word "virus" and "virion" interchangeably, but the two words have subtle but important differences. The word "virion" is used to describe the infectious virus package that is assembled. It is the extracellular form of the virus, also referred to as a virus particle, that is released from one cell and binds to the surface of another cell.

On the other hand, the word "virus" refers to the biological entity in all its stages and the general characteristics that differentiate it from another infectious entry. Rhinovirus and Epstein–Barr virus are two different entities with different properties. These viruses have different genes, structures, and methods of infection—they have different characteristics that differentiate them from each other and from other viruses.

You may be infected with several different viruses at one time, and you likely have millions of virions present in your body from each one.

An analogy to describe the difference between "virus" and "virion" is FedEx and UPS are shipping companies with different properties. They have different revenues, different administrations, different logos, and different business models. When you see a UPS or FedEx truck passing by, however, you are not seeing the company (the virus)—you are seeing how the company's goods get transported from one location to another (the virions). Each company has thousands of trucks delivering goods at any one time, just as each viral infection will generate millions of virions that spread the virus.

1.1 THE IMPORTANCE OF STUDYING VIRUSES

We study viruses and the diseases they cause for a variety of reasons. First, viruses are everywhere. They are found in all of our surroundings: the air, the ocean, the soil, and in rivers, streams, and ponds. They are present wherever life occurs, and it is thought that every living thing has a virus that infects it.

So how many viruses are there? There are around 3000 documented species of viruses that infect a range of living organisms, although there are thousands of different strains and isolates within these species, and thousands more viruses that remain to be discovered.

As described in the *In-Depth Look*, an infectious virus particle is called a **virion**, and when a cell is infected with a virus, millions of these infectious particles are created and released from the cell to infect other cells. How does the number of individual virions compare to other things found in high abundance around us? There are an estimated 10^{18} grains of sand on Earth and 10^{23} stars in the Universe, yet neither of these numbers compare to the number of virions found on Earth. If you multiplied the number of stars in the Universe 100 million times, you would have the number of infectious virus particles in the world. With an estimated 10^{31} total virions, viruses are the most abundant biological entities on our planet. We know that bacteria are abundant and everywhere, but there are 10 times more virions on Earth than bacteria! It is important that we continue trying to understand the viruses that are constantly around us.

All living organisms have a relationship with viruses. Viruses have been around since the beginning of life on Earth and have shaped the course of human evolution and history. The stela shown in Fig. 1.1A dates back to 1580–1350 BC and shows an Egyptian with a walking stick and foot drop, a condition that prevents dorsiflexion of the foot (lifting of the foot at the ankle). This is a common occurrence in people with poliomyelitis, caused by the poliovirus, which causes this condition by infecting and damaging motor neurons. The mummy of Ramses V, shown in Fig.

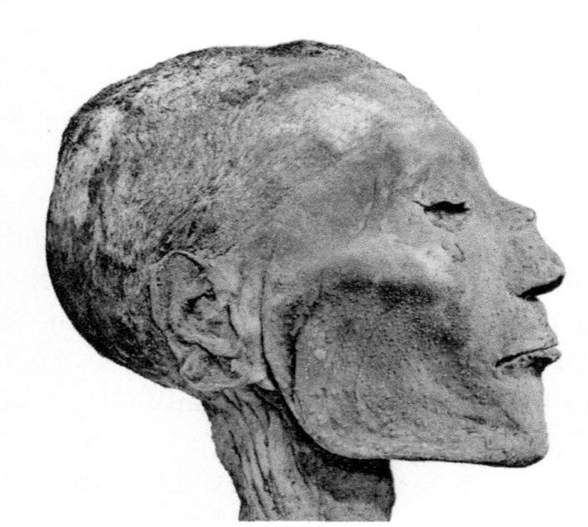

FIGURE 1.1 **Viruses have existed as long as humans have.** (A) An Egyptian stela dating back to 1580–1350 BC depicts a priest with a walking cane and foot-drop deformity, attributed to poliomyelitis. *Photograph by Ole Haupt, courtesy of Ny Carlsberg Glyptotek, Copenhagen.* (B) The mummy of Ramses V, who died in 1157 BC, exhibits a rash on the lower face and neck that is characteristic of a smallpox rash. *Photo courtesy of the World Health Organization.*

1.1B, shows evidence of smallpox-like lesions (seen on the lower face and neck in the photograph), leading scientists to believe he may have died of this poxvirus in 1157 BC. Viruses have been present as long as life has existed, and there is persuasive evidence that viruses may even have existed before life arose.

One of the most compelling reasons we are interested in viruses, however, is because viruses cause diseases. They cause conditions as simple as the common cold or as complex as cancer. A virus can also cause an **epidemic**, an outbreak where the virus infects many more individuals than normal and spreads throughout an area. Some of the world's worst epidemics have been caused by viruses. As mentioned above, the effects of poliovirus have been noted throughout history for thousands of years, but the growth and urbanization of cities provided the conditions that fueled epidemics of the virus (Fig. 1.2A and B). One of the first major epidemics occurred in New York City in 1916 and resulted in over 9000 cases and 2343 deaths. In the early 1950s, over 20,000 cases of paralytic polio (causing temporary or permanent paralysis) occurred each year until, in 1955, the first polio vaccine was introduced and cases dropped precipitously.

A **pandemic** ensues when a virus spreads throughout a much larger area, such as several countries, a continent, or the entire world. Reports and warnings from the Centers for Disease Control and Prevention (CDC) concerning influenza may sometimes seem alarmist, but in late 1918, a strain of influenza originating in the United States spread throughout the entire world, killing 20–50 million people (Fig. 1.2C). In the month of October alone, 195,000 Americans died of this influenza, and when the pandemic was over, over 675,000 Americans had died from the effects of the virus, more Americans than in all the wars of the century combined. Most subtypes of influenza cause mild or moderate respiratory symptoms, but some, like the 1918 influenza, can cause devastating effects. Although medicine has advanced greatly in the last 100 years, surveillance and vaccination is by far the most effective way to prevent another pandemic of this magnitude. Scientific research has led to the development of vaccines that have greatly reduced the burden and death toll of many viral diseases.

Although viruses can be harmful, research using viruses has resulted in a wealth of information revealing how systems work in living organisms. One very important experiment involving viruses helped to verify that deoxyribonucleic acid (DNA) is the molecule that encodes genetic information. In the earlier half of the 20th century, scientists were uncertain as to whether protein (composed of amino acids) or DNA (composed of nucleotides) was the hereditary instructions for cell development and function. There are 20 different amino acids but only 4 different nucleotides, so it was not unreasonable to think that amino acids could encode more possible information than nucleotides could. In 1944, Oswald Avery, Colin MacLeod, and

(A)

QUARANTINE

POLIOMYELITIS

All persons are forbidden to enter or leave these premises without the permission of the HEALTH OFFICER under PENALTY OF THE LAW.

This notice is posted in compliance with the SANITARY CODE OF CONNECTICUT and must not be removed without permission of the HEALTH OFFICER.

Form D-1-Po. ________________ Health Officer.

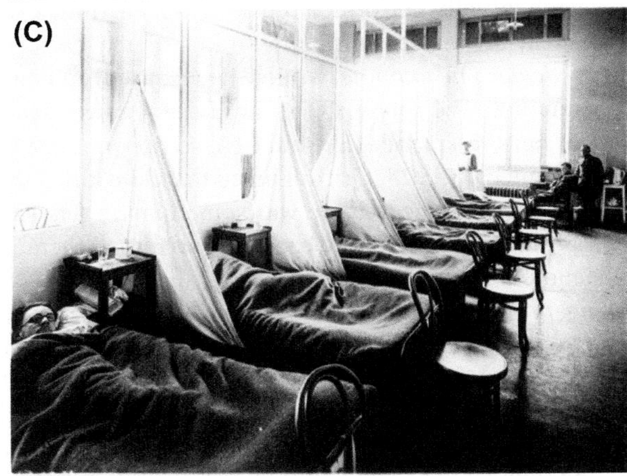

FIGURE 1.2 **Viruses can cause serious epidemics and pandemics.** Public health officials placed quarantine signs, like the one shown in (A), on the houses of people with polio during epidemics. *(Courtesy of the U.S. National Library of Medicine.)* (B) Polio can lead to temporary or permanent paralysis. Muscle weakness or paralysis still present after 12 months is usually permanent. *(Photo courtesy of the CDC/Charles Farmer.)* (C) An influenza ward in a U.S. Army camp hospital in France in 1918, during World War I. *(Courtesy of the U.S. National Library of Medicine.)*

Refresher: Important Biological Molecules

Deoxyribonucleic acid (DNA) is made of many **nucleotides** bonded together. A nucleotide of DNA (Fig. 1.3A) is composed of a sugar, called deoxyribose, with a phosphate group attached at one end of the sugar and a base attached at the other end. There are four different bases in a DNA nucleotide: adenine, guanine, cytosine, and thymine (Fig. 1.3B). DNA is double-stranded in all living organisms, and the bases in one strand bond to the bases in the other strand (Fig. 1.3C). This is known as a **base pair**. Adenine always bonds with thymine, and guanine and cytosine always bond to each other.

Ribonucleic acid (RNA) is also made of nucleotides bonded together, but the sugar used is ribose, rather than deoxyribose. Uracil replaces the base thymine, and RNA is single-stranded, not double-stranded (Fig. 1.3D). In living things, DNA is used as a template for making RNA. DNA and RNA are **nucleic acids**.

Protein is a different biological molecule that is made of **amino acids** bonded together (Fig. 1.3E). There are 20 slightly different amino acids, composed of carbon, hydrogen, oxygen, and nitrogen. Two amino acids, methionine and cysteine, also contain the element sulfur. Enzymes are a very important class of proteins.

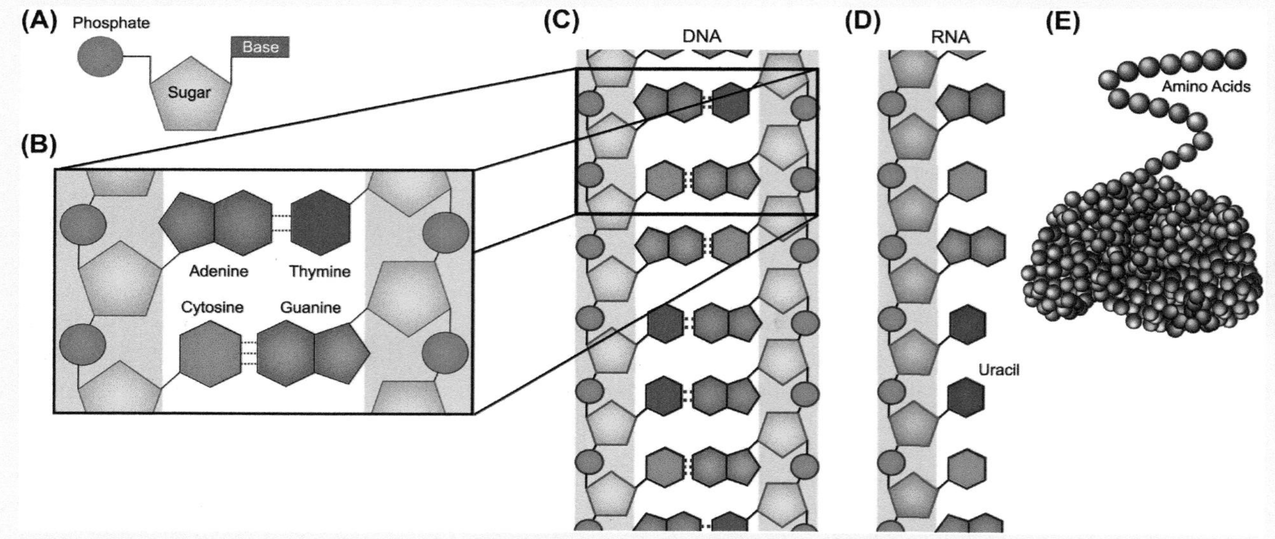

FIGURE 1.3 **Three important biological molecules: (A–C) DNA, (D) RNA, and (E) Protein.**

Maclyn McCarty demonstrated that DNA is the molecule that encodes inheritable traits. They were pursuing a topic of research initially performed by Frederick Griffith, a British medical officer studying two strains of the bacterium *Streptococcus pneumoniae*, which had caused epidemics of pneumonia. The rough (R) strain did not cause disease in mice, but the smooth (S) strain caused pneumonia and killed the mice. In 1928, Griffith noted that if he killed the S bacteria using high heat, there was something in that strain that was taken in by living R bacteria that turned them into the pneumonia-causing S bacteria. He called this the *transforming principle*, although he did not know specifically what molecule was taken in and transformed the R strain.

Avery, MacLeod, and McCarty followed up on Griffith's experiments in an attempt to identify this transforming principle. They isolated the molecule that transformed the bacteria and found it to be DNA. While some scientists believed their results, many were skeptical, arguing that the purified DNA preparations may still have contained some proteins, which they believed to be the true transforming principle. As such, the scientific community was resistant to accepting DNA as the genetic code.

Over the next decade, the field slowly warmed to the idea, and an experiment utilizing viruses conducted in 1952 provided the evidence that was needed to finally convince the world. At Cold Spring Harbor Laboratory on Long Island, New York, Alfred D. Hershey and his laboratory technician Martha Chase were performing research with **bacteriophages** (or **phages**), viruses that infect bacteria (Fig. 1.4). They were working with a bacteriophage called T2 that infects *Escherichia coli*, a bacterium found in the gastrointestinal tract of mammals. T2 attaches to the *E. coli* cell and injects into the chemical instructions to make more T2 bacteriophages.

To test whether these chemical instructions were composed of DNA or protein, Hershey and Chase used radioactive phosphorus (^{32}P) and radioactive sulfur (^{35}S) isotopes. Phosphorus is found in DNA but not in proteins, and sulfur is found in proteins but not in DNA. Hershey and Chase grew two cultures of bacteriophages, one in ^{32}P and one in ^{35}S (Fig. 1.5). Bacteriophages are composed of DNA surrounded by a protein coat, so the bacteriophages grown in ^{32}P incorporated it into their DNA, while the bacteriophages grown in ^{35}S incorporated it into their protein coats.

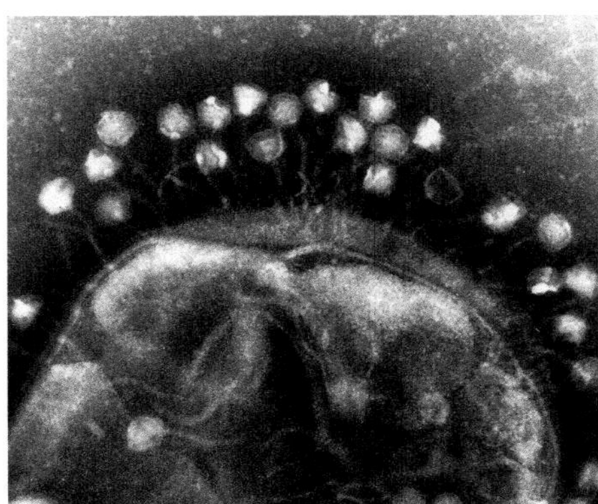

FIGURE 1.4 **Bacteriophages.** An electron micrograph of bacteriophages attached to a bacterial cell. *Courtesy of Dr. Graham Beards CC-BY-3.0.*

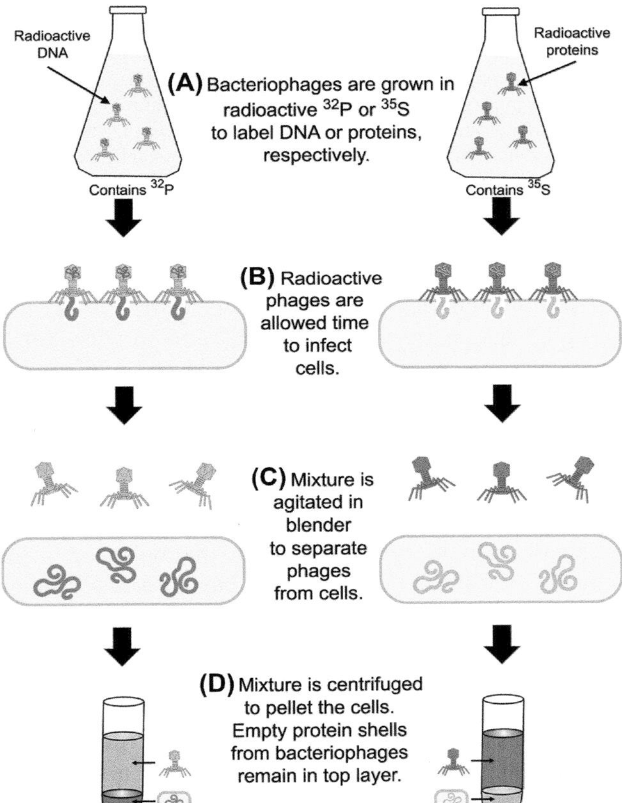

FIGURE 1.5 **The Hershey–Chase experiment.** (A) Bacteriophages were grown in cultures containing radioactive phosphorus (to label DNA) or radioactive sulfur (to label proteins), and the two phage cultures were allowed time to infect separate bacterial cells (B). After infection, a blender was used to agitate the mixture and separate the phage shells from the cells (C). The radioactive DNA remained in the cells while the radioactive proteins were found with the bacteriophage shells (D), indicating that DNA was the hereditary material.

After creating their two sets of bacteriophages, Hershey and Chase infected the bacteria with the ^{32}P-labeled or ^{35}S-labeled bacteriophages. They allowed time for the bacteriophages to infect and then used a blender to violently agitate the cells, shearing off any of the phage still attached to the cell surface. When they used a centrifuge to separate the cells from the empty phages, they found that the ^{35}S-labeled proteins remained outside the cells, while the ^{32}P-labeled DNA entered the cells. In addition, they noted that some of the ^{32}P was also incorporated into the next set of bacteriophages produced in the cell. This experiment showed that when the bacteriophages attached to the cells, the DNA entered the cell while the protein coat remained outside. The Hershey–Chase experiment used viruses to confirm the Avery, MacLeod, and McCarty's findings that DNA, and not protein, is the genetic material. A year later, in 1953, James Watson and Francis Crick presented their double-helix model of DNA structure. Since that time, numerous important scientific discoveries have been elucidated through research with viruses.

Viruses have also been investigated for their use as therapeutics. Nearly 100 years ago, Felix d'Herelle, the man who coined the term bacteriophages, meaning "bacteria eaters," used these viruses for the treatment of bacterial infections in a time when antibiotics did not yet exist. **Phage therapy** declined after antibiotics were introduced but has undergone a renaissance, as of late, with the evolution of certain antibiotic-resistant strains of bacteria. Viruses are also being used for **gene therapy**, the delivery of DNA into cells to compensate for defective genes. Viruses have evolved ways to deliver their genes into cells; in gene therapy, viruses are engineered to deliver a normal copy of the defective human gene. Gene therapy has great potential to cure many genetic diseases, although there are currently procedural obstacles that must be overcome before it can become a mainstream treatment. These therapies will be discussed in detail in Chapter 8, "Vaccines, Antivirals, and the Beneficial Uses of Viruses."

Study Break
Explain how bacteriophages were used to verify that DNA, and not protein, encodes genetic information.

1.2 VIRUSES ARE NOT ALIVE

Virology is the study of viruses, how they replicate, and how they cause disease. It may seem bizarre that virology is a subset of biology—the study of life—because viruses are not considered to be alive. They are, however, intricately tied to the web of life here on Earth.

In order to understand why viruses are not alive, we must revisit the characteristics of living things. To be considered alive, an organism must satisfy several criteria:

1. It must have a genome, or genetic material.
2. It has to be able to engage in metabolic activities, meaning that it can obtain and use energy and raw materials from the environment.

3. It has to be able to reproduce and grow.
4. It must be able to compensate for changes in the external environment to maintain homeostasis.
5. Populations of living organisms are also able to adapt to their environments through evolution.

There is no question that viruses share some of these characteristics. Every virus has genetic material, or a **genome**, although viruses are a bit different because, unlike living organisms that only have DNA genomes, viruses can have genomes composed of DNA or RNA, depending upon the virus. Many viruses also have high mutation rates that lead to the evolution of the virus. For instance, if a person with HIV is treated with one antiviral drug, the virus quickly evolves into a strain that is no longer affected by the drug. The influenza virus continuously acquires small mutations, which is why the flu vaccine you received last year may not protect you from this year's flu. In fact, because viruses mutate so quickly, they function as a great model for studying and observing evolutionary change, which takes much longer in living organisms.

Viruses do not, however, engage in their own metabolic activities. **Metabolism** refers to the collective set of biochemical reactions that takes place within a cell. Biochemical reactions that break down substances to generate energy are constantly taking place within each cell of a living organism. This energy is used in other parts of the cell for thousands of other reactions that are necessary for the survival of the cell. Viruses, however, are unable to perform these metabolic reactions while outside a cell. In essence, they are inert particles that do not have the ability to generate their own energy. They use the cell's energy and machinery to synthesize new virus particles.

This brings us to the next characteristic of living things: the ability to reproduce independently. To reproduce, a cell makes a copy of its DNA, expands in size, and divides the DNA and cell in two. This is known as **binary fission** in prokaryotes and **mitosis** in eukaryotes. Viruses, however, do not reproduce in this way (Fig. 1.6). When a virus particle enters the cell, it completely disassembles. The viral nucleic acid encodes the instructions, and the cell's machinery will be used to make new infectious virus particles (virions). The replication cycle of viruses functions in the same way that a manufacturing factory (the cell) receives a package with instructions (the virus) on how to mass produce a new product, entirely from scratch, that is then shipped to other locations after manufacturing. All cells of living organisms arise from the growth and division of a previously existing cell, and viruses do not reproduce in this manner.

Homeostasis is a steady internal condition that is exhibited by living organisms. For example, if energy levels fall too low within a cell, it will compensate by breaking down stored molecules to generate energy. If you go outside in the winter with a T-shirt on, you start shivering so your muscles generate heat to warm you up, and goose bumps on your skin raise the hairs on your arm in an attempt to keep that heat near your skin. Living organisms have mechanisms to regulate internal highs and lows to maintain homeostasis, but viruses do not. As inert particles, they are unable to compensate for changes in their external environment.

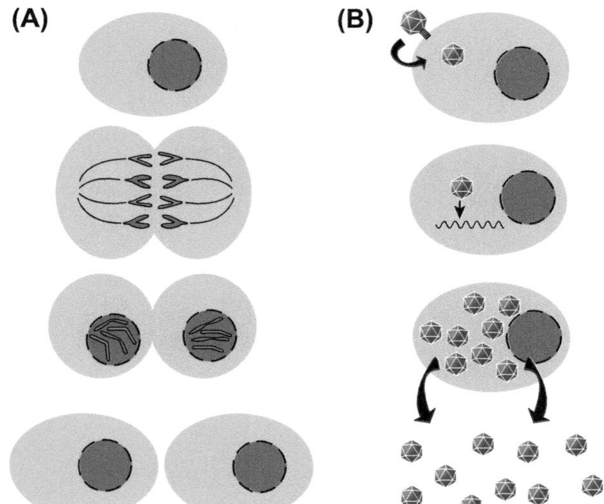

FIGURE 1.6 **Cell division versus viral replication.** (A) Eukaryotic cells make a copy of their genetic information and divide into two cells through the process of mitosis. All cells arise from the growth and division of previously existing cells. (B) Viruses, on the other hand, attach to cells and disassemble within the cell. New virions are assembled from newly made components and are released from the cell.

Despite not being alive, viruses still share many similarities with living organisms. They are composed of the same biological substances, such as nucleic acids or amino acids, and their proteins are translated by ribosomes much in the same way as living organisms. They play a role in the cycling of energy and matter within ecosystems, and some can quickly evolve, as described above. Within the cell, viruses are far from inert: although they use the cell's energy, raw materials, and organelles, their nucleic acid genome encodes the instructions to assemble new infectious virions that are able to carry on the "life cycle" of a virus.

It is interesting to note that viruses are thought to have been around since the beginning of life itself. In fact, one current hypothesis states that viruses were some of the precursors to life on Earth as we know it.

Study Break
What characteristics do viruses share with living organisms? Why are they not considered to be alive?

1.3 THE ORIGIN OF VIRUSES

The question of how viruses arose is a difficult and much debated issue. It is generally accepted that viruses appeared around the same time that life began, and when new information is discovered concerning the origin of life, the hypotheses about the origin of viruses are also revisited. The Earth was a very different environment when life is thought to have originated, around 3.5 billion years ago, and it is a formidable challenge to uncover evidence from that time period. More often, scientists reveal current or past biological processes that help to explain how viruses may have evolved over time. Based upon the evidence that we currently have, there are three viable hypotheses on how viruses originated:

1. The precellular hypothesis (or "virus-first" hypothesis)
2. The escape hypothesis
3. The regressive hypothesis

The **Precellular Hypothesis**, also known as the "virus-first hypothesis," proposes that viruses existed before or alongside cells and possibly contributed to the development of life as we know it. It is now thought that life may have developed in an "RNA World" where RNA, instead of DNA, was the first genetic material. RNA is easier to create than DNA from the precursor chemicals that are thought to have existed on the early Earth, and in present-day cells, the sugar found in DNA, deoxyribose, is made from ribose, the sugar found in RNA. To replicate, DNA also requires complex protein enzymes, but RNA has the unique property that it can encode genetic material and in some cases, like RNA ribozymes, catalyse reactions much like a protein enzyme does. In this way, RNA could have functioned as an enzyme that copied an early RNA genome.

Many important molecules in the cell include RNA or parts of it, such as ATP (used for energy) or the ribosome, which assembles proteins and is composed of RNA. It is thought that these might be conserved remnants of the creation of cells in an RNA World. Similarly, it is conceivable that RNA viruses also originated in this RNA world, either before or alongside RNA-based cells (Fig. 1.7A). All known viruses require a host cell to replicate, however, so it seems more likely that they developed alongside these primitive cells, rather than as precursors of them.

It is presumed that DNA, being a more stable molecule, was selected for and eventually replaced RNA. An interesting thought related to the precellular hypothesis is that DNA first originated in RNA viruses, thereby giving rise to DNA viruses, and that cells with DNA originated from the infection of an RNA cell with a DNA virus (Fig. 1.7B). In this scenario, a DNA virus might have infected an RNA cell and the genome of the virus continued to persist within it, in the same way that the genomes of some currently existing DNA viruses can. Eventually, the viral DNA might have picked up some of the host cell's genes and became the cell's chromosome, resulting in a cell with a DNA genome. Alternatively, the mechanisms used to create DNA within the virus could have been adopted by the cell.

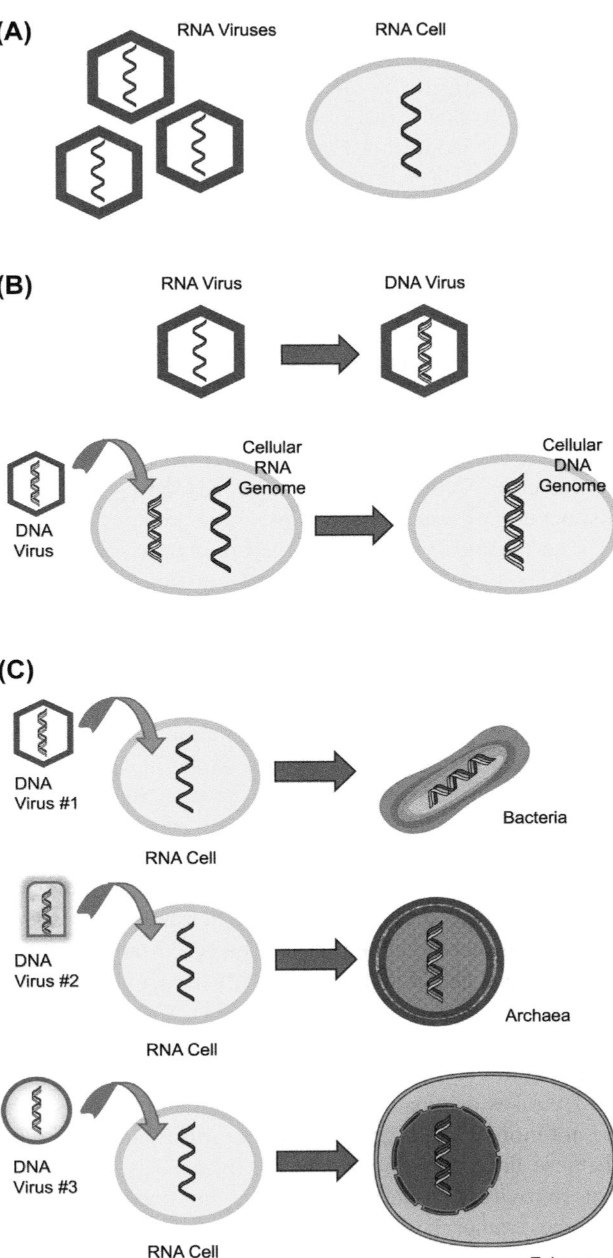

FIGURE 1.7 **Variations of the virus-first hypothesis.** The precellular (or virus-first) hypothesis proposes that viruses initially developed before or alongside cells in an RNA-based world (A). A variation of this hypothesis (B) proposes that RNA viruses evolved into DNA viruses that infected RNA cells, which eventually gained a DNA genome by using the viral DNA or viral DNA-generating mechanisms. (C) A related version speculates that the three Domains of life (*Bacteria*, *Archaea*, and *Eukarya*) may have arisen from infection of cells with three distinct DNA viruses.

Support for this idea came when the DNA polymerase from an algae-infecting virus was found to be related to a DNA polymerase found in eukaryotic cells. Taking this one step further, it has been suggested that the three domains of life—*Bacteria, Archaea,* and *Eukarya*—each arose independently from the infection of cells with three distinct DNA viruses (Fig. 1.7C).

Critics of the precellular hypothesis point out that all viruses are parasitic and require a cellular host. Therefore, it is unlikely that viruses could have existed before cells because they would not have had a reliable source of the materials they need to replicate. In addition, the majority of viral genes are not found in cells, and one should expect to see more similarities between cells and viruses if a DNA virus was the origin of a cell's genetic material.

The other two hypotheses for the origin of viruses presume that cells existed before viruses. The first of these is called the **Escape Hypothesis**. This hypothesis proposes that viruses are pieces of cells that broke away at one point in time (hence they "escaped" from the cell) and gained the ability to travel from cell to cell. By extension, the viruses of Bacteria, Archaea, and Eukarya may have arisen from distinct escape events within those three domains (Fig. 1.8).

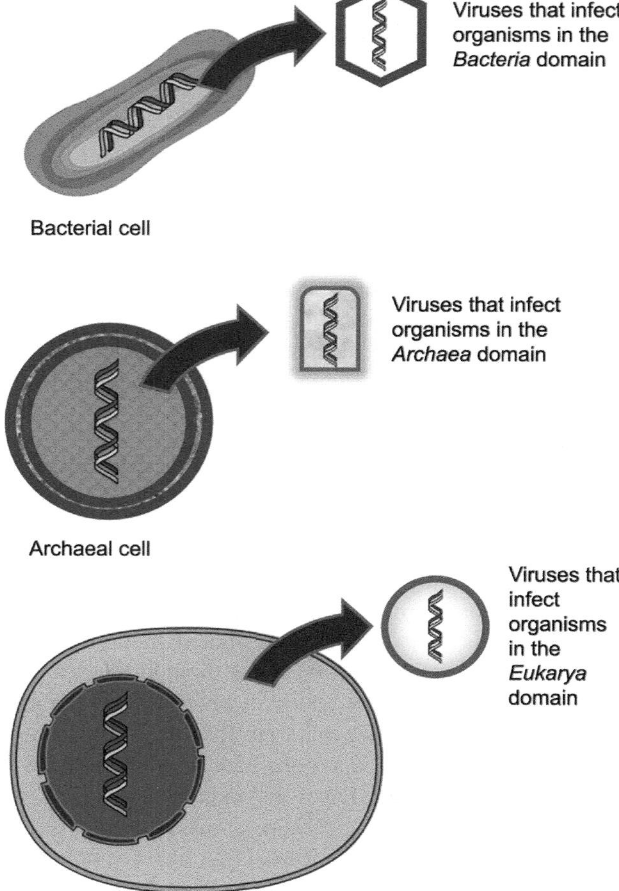

FIGURE 1.8 **The escape hypothesis.** The escape hypothesis proposes that viruses arose from portions of cells that gained the ability to travel from cell to cell. Viruses that infect cells within the three Domains of life (*Bacteria, Archaea,* and *Eukarya*) may have arisen from cells within each of those domains.

The Escape Hypothesis gained popularity when transposable elements were discovered. **Transposable elements**, or transposons, are pieces of DNA that can physically move from one location to another in the genome of a living organism. Some are only a few hundred nucleotides long, while others span thousands of nucleotides. Initially thought to be "junk DNA" with no apparent function, these transposable elements make up nearly half of the human genome, although many are no longer functional. Some of these transposable elements have similarity to **retroviruses**, such as HIV, which incorporate into the host's DNA upon entering a cell. Supporters of the escape hypothesis point out that retroviruses may have originated from the escape of these transposable elements from the cell. Many retrovirus genomes are also found permanently integrated into cellular genomes as relics of past periods. Critics, however, emphasize that the great majority of viral genes have no **homologous** (evolutionarily similar) cellular counterpart, so if viruses originated from escaped cellular genes, why are not more cellular genes found in viruses, and where did all these unique viral genes come from? It is more likely that retroviruses infected cells and integrated into their genomes, rather than retroviruses being derived from them.

The third current hypothesis to explain the origin of viruses is the **Regressive Hypothesis**, which suggests that viruses were once independent intracellular organisms that *regressed* back to a less-advanced state where they were unable to replicate independently. Two organelles currently found within cells, namely the mitochondrion and the chloroplast, are thought to have originated in this manner. Precedent for this idea also comes from the world of bacteria, where certain bacteria such as *Chlamydia* and *Rickettsia* require the intracellular environment of the cell to replicate. Similarly, perhaps viruses were once living intracellular organisms that dissolved their membranes to facilitate easier access to cellular equipment and materials.

The discovery of a giant amoeba-infecting virus, in 2003, lended support to the regressive hypothesis. Mimivirus, short for "microbe-mimicking virus," is so named because of its size: at approximately 750 nm in diameter, the virus was initially thought to be a small bacterium (Fig. 1.9A). It was the largest virus discovered at the time, and a handful of larger viruses have since been characterized. Mimivirus has one of the largest known viral genomes, at over a million base pairs; by comparison, the average virus has a genome composed of thousands or tens of thousands of nucleotides. Its physical and genome size are strikingly large, and several of the genes in the mimivirus genome

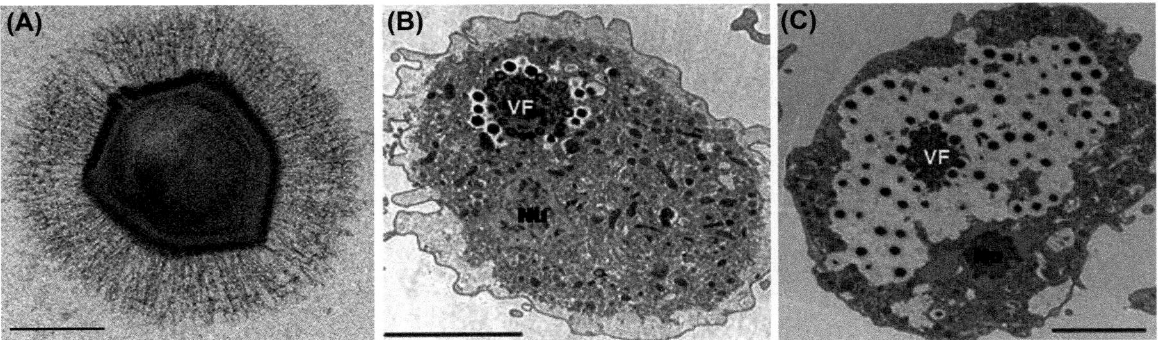

FIGURE 1.9 **The giant mimivirus.** Mimivirus was first discovered in 1992 and initially thought to be a bacterium, due to its large size. Including the protein filaments that project from its surface, mimivirus is about 750 nm in diameter (A). Scale bar = 200 nm. *(Image courtesy of Ghigo, E., Kartenbeck, J., Lien, P., et al., 2008. Ameobal pathogen mimivirus infects macrophages through phagocytosis. PLoS Pathog. 4 (6), e1000087,* http://dx.doi.org/10.1371/journal.ppat.1000087.*)* (B and C) Mimivirus, like other large complex DNA viruses, sets up virus factories (VF) within the cytoplasm of a cell to facilitate the replication of the virus. Scale bar = 5 μm (B) and 3 μm (C). *(Image courtesy of Suzan-Monti, M., La Scola, B., Barrassi L., Espinosa L., Raoult D., 2007. Ultrastructural characterization of the giant volcano-like virus factory of Acanthamoeba polyphaga Mimivirus. PLoS One 2 (3), e328.)*

resemble genes for creating proteins, suggesting that the virus may have been able to create its own proteins at one point in evolutionary time.

After infecting a cell, mimivirus and many large complex DNA viruses set up so-called **virus factories** made of cellular membranes, where the replication and assembly of virions takes place (Fig. 1.9B and C). These factories contain the enzymes necessary to copy the viral genome and either contain or are in close proximity to ribosomes (for making proteins) and mitochondria (for supplying energy in the form of ATP). This scenario is reminiscent of the *Chlamydia* reticulate body, a structure that the bacteria form within a cell that is used as a factory to develop new bacteria. The new bacteria infect new cells, and again form reticulate body factories within the cell, similar to the viral factories observed with infection by large complex DNA viruses. It is possible that our characterization of viruses as inert biochemical packages is too simplistic, and when we observe inert virions outside the cell, we may actually only be observing the infectious portion of the viral life cycle, much in the same way that *Chlamydia* have an inert infectious phase but new bacteria are produced in the intracellular reticulate body.

Critics of the regressive hypothesis point out that although a few mimivirus genes resemble genes in cells, the majority are unlike any genes found in bacteria or eukaryotic cells. If viruses were once parasitic cells, then more viral genes should show similarity to the genes of currently existing cells because they would have shared a common ancestor at one point in time. Perhaps the few viral genes that resemble cellular genes are not artifacts of a free-living organism that evolved into a virus, but were instead stolen from the cell's DNA at one point in time. Known as **horizontal gene transfer**, there are many examples of viral genes that are thought to have originated in this way.

Another criticism of the regressive hypothesis deals with the manner in which viruses are replicated. As described above and in Fig. 1.6, viruses are assembled completely from scratch, rather than splitting in two like cells do. Even the *Chlamydia* reticulate body divides repeatedly to generate the new bacteria that are released from the cell. If viruses were once free-living parasites, what situation could have caused such a major modification, different from every other living thing?

It is understandably much easier to brainstorm than test new hypotheses concerning the origin of viruses. These hypotheses will continue to be refined and modified as more information is revealed from the characterization of known viruses and the discovery of new viruses. The first viruses that existed may have only remotely resembled the plentitude of highly evolved present-day viruses. In any case, it is a distinct possibility that viruses have been continuously evolving alongside life since it began, over three billion years ago.

1.4 THE DISCOVERY OF VIRUSES

Several important discoveries have contributed to the identification of viruses as novel biological entities (Fig. 1.10). In the mid-1800s, the French chemist Louis Pasteur (Fig. 1.11A) performed some simple yet elegant experiments to show that life does not arise through spontaneous generation, the belief at the time that living organisms could spontaneously arise from nonliving matter. In one experiment, Pasteur sterilized beef broth by boiling it in a swan-neck flask, so named for the similarity of the flask's neck to that of a swan (Fig. 1.11B). The curved neck of the flask allowed air to enter but trapped any dust or particles that might have contained microorganisms, and so the broth remained clear and free of any microbes, even after prolonged incubation. In contrast, the broth became contaminated if the top of the neck of the flask was broken off

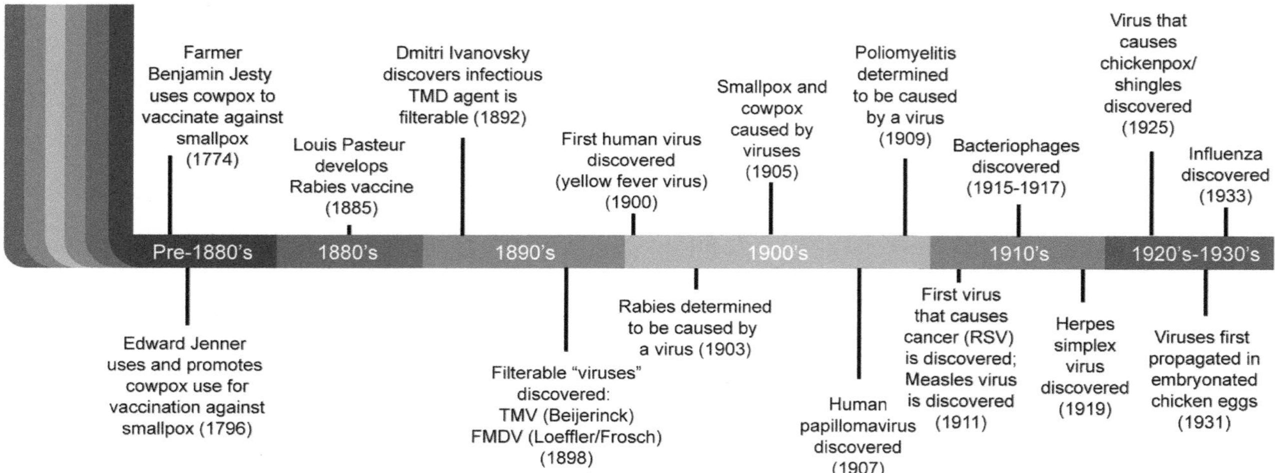

FIGURE 1.10 Timeline of important discoveries in the early history of virology.

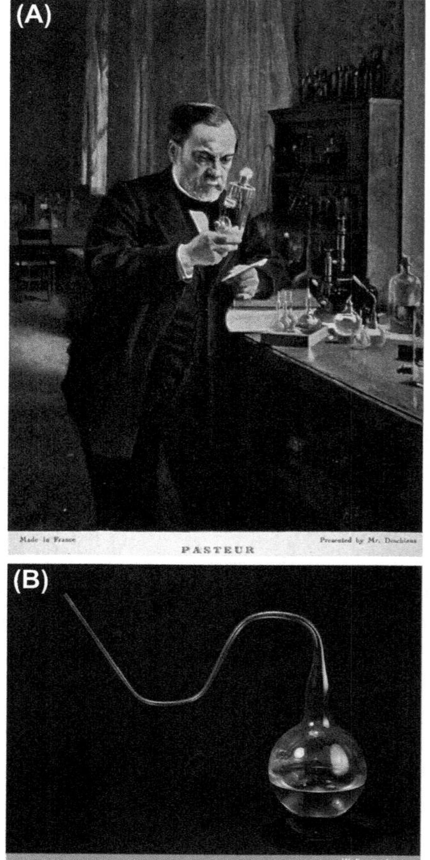

FIGURE 1.11 **Louis Pasteur and the germ theory.** French microbiologist Louis Pasteur (A) conducted experiments with swan-neck flasks (B) in the mid-1800s that convinced his contemporaries that life does not arise by spontaneous generation. *Image of Pasteur courtesy of the U.S. National Library of Medicine; swan-neck flask image courtesy of the Wellcome Library, London.*

FIGURE 1.12 **Robert Koch.** German physician and microbiologist Robert Koch, considered the founder of bacteriology, developed and published "Koch's Postulates" in 1890 to aid in establishing whether a microorganism is the direct cause of a disease. *Image courtesy of the U.S. National Library of Medicine.*

and the bacteria-laden particles within the air were allowed to enter the flask, thereby showing that microorganisms do not spontaneously generate but are able to enter areas by traveling through the air. At this point, the **germ theory**, which states that infectious diseases are caused by microorganisms, gained support and was further substantiated by the discovery of several disease-causing bacteria, such as those that cause anthrax, cholera, and tuberculosis.

The German physician and microbiologist Robert Koch (Fig. 1.12) was the first person to identify that the causative agent of anthrax was a bacterium. Considered the father of bacteriology, he built upon discussions with other scientists and published four criteria, now known as "Koch's Postulates," to scientifically demonstrate that a microbe is the causative agent of a disease:

1. The organism must be present in every case of the disease.
2. The organism must be isolated from the host with the disease and grown in pure culture.
3. The disease must be reproduced when a healthy susceptible host is inoculated with the pure culture.
4. The same organism must again be recovered from the experimentally infected host.

These postulates were regarded as the necessary proof that an infectious agent is responsible for a specific disease. Even though Koch himself admitted that not all bacterial diseases could satisfy the four postulates, the adherence to these would prove to be an obstacle in the discovery of viruses.

Although the discovery of human bacterial diseases set the stage for the discovery of viruses, the history of virology started not with human diseases but with plants. In 1879, the German chemist Adolf Mayer was asked to investigate a disease that was affecting tobacco plants in the Netherlands. He called it the "mosaic disease of tobacco" after the discolored splotches on the leaves that formed a mottled, mosaic-like pattern of light and dark green on the leaves; today, we refer to this as tobacco mosaic disease (Fig. 1.13A). In 1886, Mayer noted that the disease could be transmitted by transferring the sap from an infected plant to a noninfected plant. When he inspected the infectious sap under a microscope, he did not observe any microorganisms, but concluded anyway that the infectious organism must be bacterial in nature. When he was unable to satisfy Koch's third postulate to culture any microorganisms, he did not consider the infectious entity might require cells to reproduce.

Not long after Mayer's work was published, Russian botanist Dmitri Ivanovsky (Fig. 1.13B), a graduate student at the time, was sent to Ukraine and Crimea to investigate the tobacco mosaic disease causing damage to the tobacco plantations located there. After returning to Russia in 1892, he presented a paper to the Academy of Sciences in St. Petersburg, reporting that the infectious agent in tobacco mosaic disease was filterable through a Chamberland "candle," or filter. Charles Chamberland, a colleague of Louis Pasteur (Fig. 1.14A), had created in 1884 a sterilizing filter made of unglazed porcelain (Fig. 1.14B and C). The pores in the porcelain were small enough to let water pass through the filter but not large enough for bacteria, which remained trapped on the filter. In this way, a solution could be made sterile by running it through the filter.

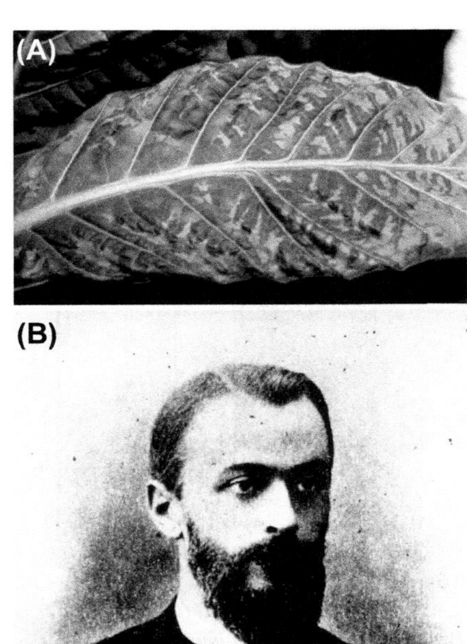

FIGURE 1.13 **The history of virology begins with tobacco mosaic disease.** Tobacco mosaic disease causes a mottled, splotchy appearance on the leaves of tobacco plants (A). *(Image courtesy of R.J. Reynolds Tobacco Company Slide Set, R.J. Reynolds Tobacco Company, Bugwood.org.)* (B) Russian botanist Dmitri Ivanovsky was the first to determine that the infectious agent found in the sap of infected plants retained its infectivity after being filtered through a Chamberland filter. *(Image courtesy of the U.S. National Library of Medicine.)*

Ivanovsky filtered the infectious tobacco plant sap through one of these Chamberland filters and it retained its infectivity, so Ivanovsky knew the infectious agent was smaller than a bacterium.

Like Mayer before him, Ivanovsky failed to satisfy Koch's postulates, then regarded as the standard proof that an infectious agent is responsible for a specific disease, because he was unable to culture the infectious organism. He was hesitant to classify this agent as a new biological entity and instead ascribed the effects to bacterial toxins that would also be small enough to pass through the filter, which he mentioned could also have been defective.

Martinus Beijerinck, a Dutch microbiologist and collaborator of Adolf Mayer, had also been studying tobacco mosaic disease in the Netherlands (Fig. 1.15). Although he was not aware of Ivanovsky's results, he found, like Ivanovsky, that the infectious agent in the tobacco plant sap retained its infectivity after being filtered through a

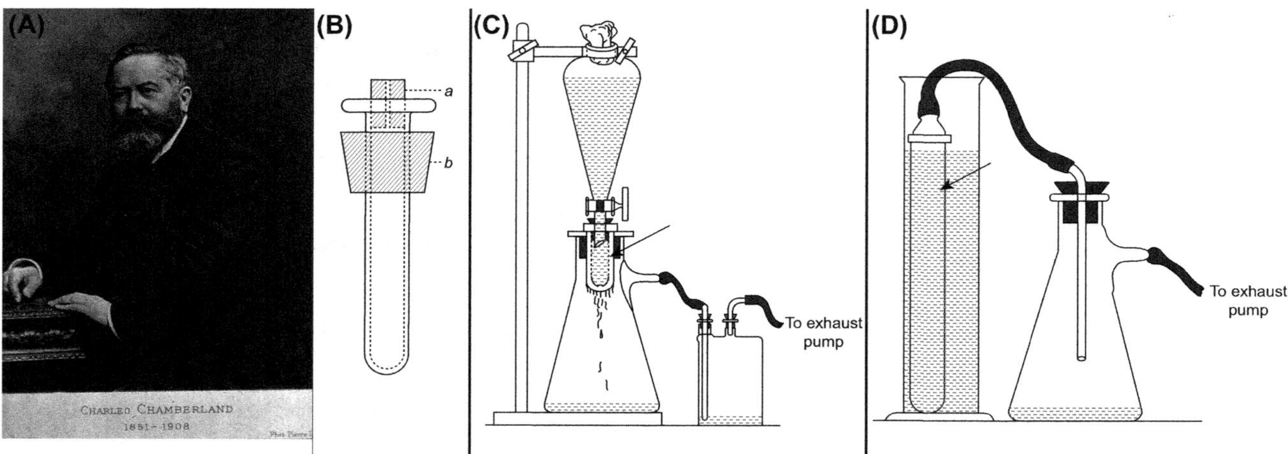

FIGURE 1.14 **Chamberland and his filter.** In 1884, Charles Chamberland (A), a colleague of Louis Pasteur, developed a filter (B) made of unglazed porcelain with pores small enough to allow the passage of liquid but too small for bacteria to pass through. The filter could be fitted into the top of a sterile flask to filter a solution pulled through it (C), or the filter could be placed in a beaker of solution and drawn through the filter in the opposite direction into a sterile flask (D). *Chamberland image courtesy of the U.S. National Library of Medicine; filter diagrams from Eyre, J.W.H., 1913. The Elements of Bacteriological Technique: A Laboratory Guide for Medical, Dental, and Technical Students, second ed. W.B. Saunders and Company, Philadelphia, London.*

FIGURE 1.15 **Martinus Beijerinck.** Martinus Beijerinck in his laboratory.

Chamberland filter. He took this observation one step further, however, by concluding that using living, dividing plant tissue was the only way to replicate the infectious agent. This explained the inability of Mayer and Ivanovsky to culture the infectious organism, as they had tried to culture the filtered plant sap using culture media devoid of cells. In his 1898 publication, Beijerinck declared that "the infection is not caused by microbes, but by a *contagium vivum fluidum*," meaning contagious living fluid. He referred to the agent a *virus* to differentiate to from the word *microbe*, which Pasteur had coined to refer to bacteria. Although the word "virus" dates back to the 14th century and is derived from the Latin *virus*, meaning "poison," Beijerinck was the first to use the word to describe this new, infectious entity.

Together, Mayer, Ivanovsky, and Beijerinck had set the stage for the discovery of viruses. Mayer made the important observations that the infectious sap could cause disease when transferred to a new plant, and Ivanovsky showed that the infectious agent could be filtered through a Chamberland filter, thereby indicating it was distinct from bacteria. Beijerinck was the first to make the jump that a novel biological entity was contained within the infectious sap. This virus is now known as tobacco mosaic virus.

Study Break

What are "Koch's postulates," and why were some of these unable to be satisfied in the initial discovery of viruses?

Meanwhile, also in 1898, Friedrich Loeffler and Paul Frosch, former students of Koch, discovered that the pathogen causing foot-and-mouth disease was also filterable through a Chamberland filter, thereby discovering the first animal virus at the same time that the first plant virus was being discovered. In addition, they determined that the agent was not filterable through a Kitasato filter, which has pores much smaller than the Chamberland filter, and so they were also able to conclude the virus was not a liquid, as described by Beijerinck's "contagious living fluid" explanation, but was actually a very small, solid particle. The virus they discovered, foot-and-mouth disease virus (FMDV), is a highly contagious respiratory virus that infects cloven-hoofed animals such as cattle, pigs, goats, and sheep. It causes blisters on the hooves, mammary glands, tongue, lips, and in the mouth that rupture and produce pain and discomfort. FMDV is usually not lethal to adult animals (although it can cause death in younger animals), but it seriously impacts the growth of the animal and production of milk. Because the virus is so easily transmitted from animal to animal and can spread very rapidly, any outbreak of FMDV leads to massive slaughtering and incineration of infected animals to prevent further transmission.

FMDV remains a formidable challenge to the livestock and dairy industries in the present day. Although the last FMDV outbreak in the United States was in 1929, several outbreaks have occurred throughout Europe and Asia in the last 10 years. Two major outbreaks in the United Kingdom led to the culling of tens of millions of sheep, pigs, and cattle, and similar control measures were instituted during recent outbreaks in Japan and North Korea. Outbreaks are economically devastating because bans are quickly set up that prohibit the export of any animals or animal products from affected regions.

So, by 1898, two separate groups had discovered a new infectious agent that multiplies in living cells and was capable of causing disease. These viruses were filterable through the bacteria-proof Chamberland filter and were too small to be seen with light microscopes. Ivanovsky, Beijerinck, and Loeffler and Frosch are generally mentioned when discussing the founders of the field of virology for the experiments that they performed independently of one another. Ivanovsky discovered the minute nature of viruses by showing the tobacco sap remained infectious following filtration, although he did not consider the agent could be a previously undiscovered entity and was certain it was another kind of bacterium. Beijerinck decisively attributed the infectious nature to a new, nonbacterial source that he termed a "virus," although he incorrectly assumed viruses were a *fluidum* (fluid), rather than a solid. Loeffler and Frosch were the most accurate with their discovery, concluding the filterable infectious entity was a minuscule, solid particle found within the filterable liquid. In any case, all of these scientists made important steps toward the discovery and acceptance of a new infectious entity: the virus.

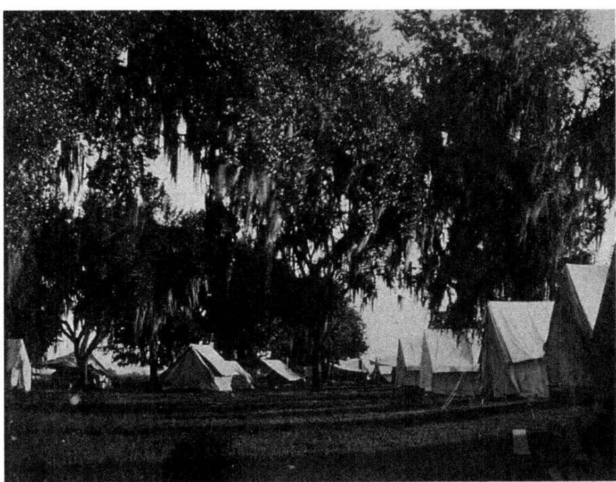

FIGURE 1.16 **Outbreaks of yellow fever.** A temporary tent hospital set up by the U.S. Marine Hospital Service in Franklin, Louisiana, during the yellow fever outbreak of 1898. *Image courtesy of the U.S. National Library of Medicine.*

The first human virus, yellow fever virus, was discovered in 1900 by the U.S. Army Yellow Fever Commission, headed by physician Walter Reed, and found to be transmitted by mosquitoes. The commission was working in Cuba to elucidate the cause of yellow fever, a serious condition that is more prevalent in tropical climates but had caused devastating outbreaks for hundreds of years in U.S. cities, including New York, Philadelphia, Baltimore, and New Orleans (Fig. 1.16).

Within the next 30 years, many diseases and conditions were found to be caused by these "filterable viruses." When American virologist Thomas M. Rivers published the first Virology textbook in 1928, bacteriophages had been discovered, as had the viruses that cause many human diseases, including measles, mumps, rubella, influenza, smallpox, herpes, rabies, and polio. And yet, no one had been able to visualize these pathogens. The electron microscope, invented in 1933, finally allowed the world to see what a virus looks like. Light microscopes use a beam of light to illuminate a specimen and glass lenses to focus and magnify an image, whereas electron microscopes use a magnetic field to focus electrons in order to illuminate a specimen, translating into a more powerful magnifying system with better image resolution. At the time, the new electron microscopes could magnify over than 10 times larger than existing light microscopes, finally allowing for the visualization of virus particles in 1938. Currently, electron microscopes provide excellent resolution and can magnify images over one million times. They have proved to be invaluable in the characterization of viruses.

In-Depth Look: How Light and Electron Microscopes Work

Light microscopes (optical microscopes) were first invented around 1600 and use glass lenses to focus the light that illuminates a specimen. A typical compound light microscope has two sets of lenses: the objective lenses, which can usually be changed to affect magnification in real time, and a fixed lens known as the ocular lens that is present in the eyepiece into which that the observer looks. The light source shines through a specimen that is mounted on a slide, and the lenses focus the light to make the object appear larger (Fig. 1.17A).

Electron microscopes, invented in the 1930s, use magnetic fields to focus a beam of electrons that illuminates a specimen. Electrons have a much smaller wavelength than light does and can therefore create much better resolution and magnification of an object. Instead of glass lenses, electron microscopes use electromagnetic fields to focus the beam of electrons onto a specimen.

For a **transmission electron microscope** (TEM, Fig. 1.17B), the electron beam passes through (is *transmitted* through) an ultra-thin specimen. Areas of dense material in the specimen do not allow electrons to pass through and cast a shadow onto a viewing screen, much in the same way that you can create a shadow of your hand by illuminating it with a flashlight with a wall behind it. Some areas of the specimen easily allow the passage of electrons and appear lighter on the viewing screen. A range of gray colors also occurs, based upon how much of the electron beam is transmitted through the sample.

Scanning electron microscopes (SEM, Fig. 1.17C) work in the same way as TEMs, but the electron beam is used to *scan* the surface of the specimen. Several detectors perceive the electrons that bounce off the specimen, rather than those that are transmitted through the specimen. In this way, SEMs are excellent at revealing the 3D surface of a specimen, in the same way that you can shine a flashlight onto different parts of a complex cave wall to reveal rocky areas, stalactites, and stalagmites.

A major advantage of electron microscopy over light microscopy is greater magnification and resolution. Magnification is the enlargement of an image of an object, and resolution is the ability to see two points as distinct points, rather than one blurry point. Fig. 1.17D shows two identical photographs of a section of liver tissue, and the top image has much better resolution than the bottom image. After the initial invention of the light microscope, scientists were able to create lenses with greater magnification, but poor resolution hampered being able to see the images clearly; there is no advantage in magnifying an object if the resolution is not great enough to discern the different parts of the image. By using electrons instead of light, certain electron microscopes can magnify an object greater than a million times its actual size and achieve resolutions 200 times greater than a light microscope. Needless to say, electron microscopes have been instrumental in the characterization of viruses, which are too small to be observed well with a light microscope.

Light microscopes do have some advantages over electron microscopes, however. Most laboratory tasks do not require the magnification or resolution of electron microscopes, and light microscopes are much less expensive than electron microscopes and do not require specialized training to use. Light microscopes can also observe color, while electron microscopes provide images in gray scale (although they can be artificially colored, or *pseudocolored*, using computer programs). Light microscopes can also observe living things, but specimens must be dead and processed for use in an electron microscope.

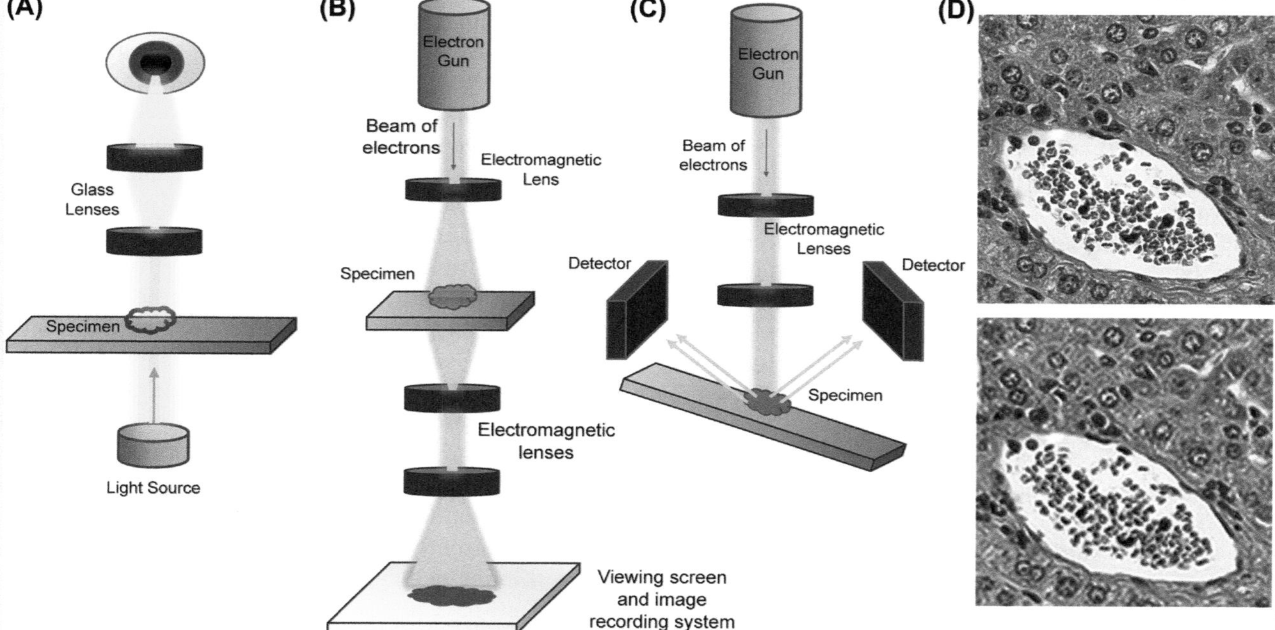

FIGURE 1.17 Light and electron microscopes. Light microscopes use glass lenses to focus the light that illuminates a specimen (A). Electron microscopes use a magnetic field to focus a beam of electrons to illuminate a specimen. In transmission electron microscopy (B), an electron beam passes through an ultra-thin specimen, while scanning electron microscopy (C) uses detectors to measure electrons that are scattered when they hit the surface of the specimen. The term "resolution" refers to the ability to see two points distinctly, rather than as one larger point, in a field of view (D).

1.5 OTHER NONLIVING INFECTIOUS AGENTS

Viruses are not the only nonliving infectious entities, nor are they the most simplistic. **Viroids** are very small circular pieces of RNA, generally only 200–400 nucleotides in size, that have been found only in plants. Unlike viruses, viroids do not have a protective protein coat and their RNA does not encode information to make proteins. Instead, enzymes in the cell copy the viroid RNA, which is then transmitted to the next host plant in the process of infection. The RNA of some viroids has enzymatic activity, so it is thought that viroids may be remnants of the "RNA World" where RNA could function both as genetic material and as enzymes (see Section 1.3). Viroids may also have been the evolutionary precursor of certain types of viruses.

Viroids are transmitted from plant to plant through close contact or by tools and machinery. Viroids have been discovered that infect tomato plants, citrus trees, apple trees, potato plants, coconut trees, and chrysanthemums, among others. Damage to the plant is thought to be caused by one of two mechanisms. In the first scenario, the copying of the viroid RNA ties up the enzymes that the cell needs to copy its RNA, and since the plant RNA is used to make proteins, the result is that the plant ends up with fewer of the proteins that it needs to function. In the second scenario, the viroid RNA is chopped into pieces by the host cell and functions as **small interfering RNA** (siRNA). These small pieces of RNA bind to complementary sequences of plant RNA, creating a double-stranded RNA molecule that the cell degrades. The result is that the plant RNA is no longer available to make proteins, and the plant suffers as a result.

Prions are another type of subviral infectious agent that causes *spongiform encephalopathy*, meaning a disease of the brain (encephalopathy) that looks sponge-like when examined under a microscope. Examples of diseases caused by prions are scrapie in sheep, bovine spongiform encephalopathy (BSE, or "mad cow disease") in cows, and Creutzfeldt–Jakob Disease (CJD) in humans. A prion is not a separate organism or entity and has no nucleic acid genome; it is simply a normal mammalian protein whose shape becomes irreversibly modified. The normal prion protein is referred to as PrP^{C} (for **C**ellular **Pr**ion **P**rotein), and it is transformed into an abnormally folded version designated PrP^{Sc} (for **Sc**rapie-causing **Pr**ion **P**rotein). Once misfolded, PrP^{Sc} is able to continue transforming normal prion proteins into PrP^{Sc} (Fig. 1.18A). The misfolded PrP^{Sc}

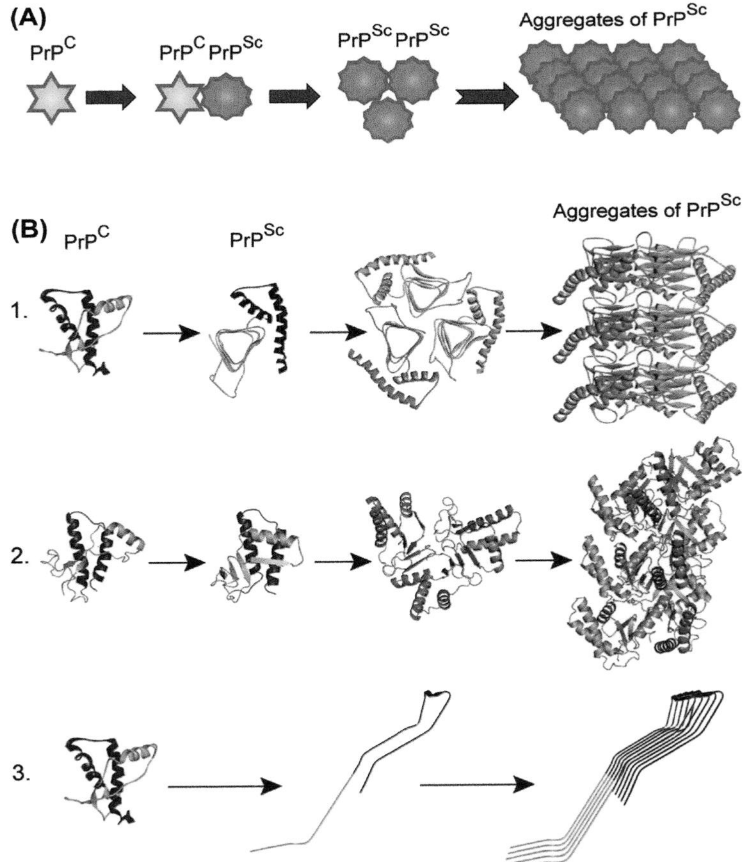

FIGURE 1.18 **Prions.** The misfolded prion protein PrP^{Sc} is able to transform the normal PrP^{C} prion protein into PrP^{Sc}, which accumulates in nervous tissue and causes damage (A). Prion diseases include scrapie in sheep, bovine spongiform encephalopathy (also known as Mad Cow Disease) in cows, and Creutzfeldt–Jakob Disease in humans. (B) Three different models for the structure of the altered PrP protein and how it accumulates into aggregates. *Reprinted with permission from Diaz-Espinoza, R., Soto, C., 2012. High-resolution structure of infectious prion protein: the final frontier. Nat. Struct. Mol. Biol. 19, 370–377, Macmillan Publishers Ltd.*

proteins are unable to be removed from the tissue and build up in the brain (Fig. 1.18B), causing nervous tissue damage that leads to confusion, dementia, and death, usually within months after symptoms appear.

The abnormal PrP^{Sc} protein can be caused by a genetic mutation, or it can be acquired by eating food that contains the PrP^{Sc} protein, such as beef contaminated with the nervous tissue of cattle that had bovine spongiform encephalopathy. Kuru, a disease that was prevalent in the Fore tribe in Papua New Guinea, was found to be caused by the ritualistic cannibalism of deceased family members. Cadaver transplants of corneas and dura mater, the thick covering of the brain and spinal cord, have also been shown to transmit CJD, as has the use of human growth hormone from cadavers. Prions are extremely difficult to destroy, and there have been rare cases of CJD being transmitted with improperly sterilized neurosurgical instruments.

SUMMARY OF KEY CONCEPTS

Section 1.1 The Importance of Studying Viruses

- Viruses are found in all parts of the biosphere and infect all living things. They are the most abundant biological entities on Earth.
- Viruses have been around since the beginning of life on Earth and have shaped the way living organisms have evolved.
- An important reason that viruses are studied is because they cause disease. Epidemics occur when viruses infect more individuals than normal in an area, and a pandemic ensues when a virus spreads throughout a much larger area, such as a country, continent, or the world. Several serious epidemics and pandemics have occurred throughout history.
- Viruses are used in scientific studies to reveal how living systems work. Hershey and Chase used bacteriophages, viruses that infect bacteria, to show that DNA encodes genetic material.
- Viruses can be used therapeutically. Phage therapy uses bacteriophages to kill bacteria, and gene therapy uses viruses to deliver a normal copy of a human gene into a person who is lacking it.

Section 1.2 Viruses Are Not Alive

- In order to be considered alive, an organism must have a genome and metabolism. It must also be able to reproduce and to compensate for changes in the external environment to maintain homeostasis. Populations of living organisms evolve over time.
- Viruses share several characteristics with living things but they are unable to reproduce independently. They also have no metabolism and do not generate their own energy.
- Viruses do not undergo cell division like living things. They completely disassemble after gaining entry into a cell and assemble new virions from scratch.

Section 1.3 The Origin of Viruses

- Viruses appeared at the same time that life began on Earth, around 3.5 billion years ago.
- The precellular hypothesis proposes that viruses existed before cells in an RNA World where RNA, and not DNA, was the genetic material. DNA found in cells may have originated from viruses that evolved DNA genomes and infected RNA cells.
- The escape hypothesis proposes that viruses are pieces of cell genomes that gained the ability to travel from cell to cell.
- The regressive hypothesis proposes that viruses were once living intracellular parasites that lost the ability to reproduce independently. Several giant viruses have been discovered in the last decade that set up complex intracellular virus factories, suggestive of a possible parasitic past.

Section 1.4 The Discovery of Viruses

- In the mid- to late-1800s, scientists like Louis Pasteur and Robert Koch performed experiments that supported the germ theory, that infectious diseases are caused by microorganisms.
- Robert Koch developed four postulates to scientifically demonstrate that a microbe is the cause of a disease. One postulate, that the organism must be isolated from the diseased host and grown in pure culture, would prove to be an impediment in the discovery of viruses, since viruses replicate inside cells.
- In 1886, German chemist Adolf Mayer determined that the sap from tobacco plants with tobacco mosaic disease could transfer the disease to a healthy tobacco plant. Russian scientist Dmitri Ivanovsky soon determined the infectious sap could be filtered through a Chamberland filter, indicating the cause of the disease was smaller than bacteria.
- In 1898, Martinus Beijerinck, a Dutch soil microbiologist, also found the infectious sap was filterable and was convinced it was a new infectious entity, which he used the word "virus" to describe.
- The first animal virus, FMDV, was discovered the same year by Germans Loeffler and Frosch, former students of Robert Koch.
- The first human virus discovered was yellow fever virus, in 1900.
- The invention of the electron microscope finally allowed the visualization of viruses and has been invaluable in the characterization of viruses.

Section 1.5 Other Nonliving Infectious Agents

- Viruses are not the only nonliving infectious entities.
- Viriods are small circular pieces of RNA that are found in plants and cause disease by acting as siRNA or by interfering with normal host processes.
- Prions are normal mammalian proteins found in nervous tissue. A misfolded version of the prion protein, PrP^{Sc}, is able to transform normal prion proteins into the abnormal version, which accumulates in the brain and causes neurological damage and death. The abnormal PrP^{Sc} protein can be caused by a genetic mutation, or it can be acquired by eating food that contains the PrP^{Sc} protein.

FLASH CARD VOCABULARY

Virus	Homeostasis
Virion	Precellular hypothesis
Epidemic	Escape hypothesis
Pandemic	Transposable elements
Deoxyribonucleic acid (DNA)	Horizontal gene transfer

Nucleotide	Retroviruses
Ribonucleic acid (RNA)	Homologous
Protein	Regressive hypothesis
Amino acid	Virus factory
Transforming principle	Germ theory
Bacteriophage (phage)	Light microscope
Phage therapy	Transmission electron microscope
Gene therapy	Scanning electron microscope
Genome	Viroid
Metabolism	Small interfering RNA (siRNA)
Mitosis	Prion

CHAPTER REVIEW QUESTIONS

1. What is a virion? What is the difference between a virus and a virion?
2. Describe three reasons why it is important to study viruses.
3. How was it determined that DNA, and not protein, encodes genetic information?
4. How are viruses used therapeutically?
5. List the characteristics of living things. Describe why viruses do or do not satisfy each criterion.
6. How is virus replication different from cell division?
7. What hypotheses exist for the origin of viruses? Which do you think is most likely, and why?
8. What characteristics of viruses made it difficult for viruses to discover them?
9. How were Chamberland filters useful in the discovery of viruses?
10. How do light and electron microscopes work? What are the advantages and disadvantages of each?

FURTHER READING

Avery, O.T., MacLeon, C.M., Mccarty, M., 1944. Studies in the chemical nature of the substance inducing transformation of pneumococcal types. J. Exp. Med. 79, 137–158.

Bandea, C.I., 2009. The Origin and Evolution of Viruses as Molecular Organisms. pp. 1–16.

Beijerinck, M.W., 1898. Concerning a contagium vivum fluidum as cause of the spot disease of tobacco leaves. In: Phytopathological Classics, Number 7. American Phytopathological Society, pp. 33–53.

Claverie, J.-M., 2006. Viruses take center stage in cellular evolution. Genome Biol. 7, 110.

Diaz-Espinoza, R., Soto, C., 2012. High-resolution structure of infectious prion protein: the final frontier. Nat. Struct. Mol. Biol. 19, 370–377.

Forterre, P., 2010. Giant viruses: conflicts in revisiting the virus concept. Intervirology 53, 362–378.

Forterre, P., 2006. Three RNA cells for ribosomal lineages and three DNA viruses to replicate their genomes: a hypothesis for the origin of cellular domain. Proc. Natl. Acad. Sci. U.S.A. 103, 3669–3674.

Ghigo, E., Kartenbeck, J., Lien, P., et al., 2008. Ameobal pathogen mimivirus infects macrophages through phagocytosis. PLoS Pathog. 4 (6), e1000087. http://dx.doi.org/10.1371/journal.ppat.1000087.

Hershey, A., Chase, M., 1952. Independent functions of viral protein and nucleic acid in growth of bacteriophage. J. Gen. Physiol. 36, 39–56.

Ivanowski, D., 1892. Concerning the mosaic disease of the tobacco plant. In: Phytopathological Classics, No. 7. American Phytopathological Society, pp. 27–30.

Koonin, E.V., Senkevich, T.G., Dolja, V.V., 2006. The ancient virus world and evolution of cells. Biol. Direct 1, 29.

Lustig, A., Levine, A.J., 1992. Minireview: one hundred years of virology. J. Virol. 66, 4629–4631.

Mayer, A., 1886. Concerning the mosaic disease of tobacco. In: Phytopathological Classics, Number 7. American Phytopathological Society, pp. 11–24.

Murphy, F.A., 2014. The Foundations of Virology: Discoverers and Discoveries, Inventors and Inventions, Developers and Technologies, second ed. Infinity Publishing, West Conshohocken, PA.

Prusiner, S.B., 1998. Prions. Proc. Natl. Acad. Sci. U.S.A. 95, 13363–13383.

Rivers, T.M., 1927. Filterable viruses a critical review. J. Bacteriol. 14, 217–258.

Rivers, T.M., 1937. Viruses and Koch's postulates. J. Bacteriol. 33, 1.

Stanley, W.M., 1946. The Isolation and Properties of Crystalline Tobacco Mosaic Virus.

Suzan-Monti, M., La Scola, B., Barrassi, L., Espinosa, L., Raoult, D., 2007. Ultrastructural characterization of the giant volcano-like virus factory of *Acanthamoeba polyphaga Mimivirus*. PLoS One 2 (3), e328. http://dx.doi.org/10.1371/journal.pone.0000328.

Villarreal, L.P., Defilippis, V.R., 2000. A hypothesis for DNA viruses as the origin of eukaryotic replication proteins a hypothesis for DNA viruses as the origin of eukaryotic replication proteins. J. Virol. 74 (15), 7079–7084. http://dx.doi.org/10.1128/JVI.74.15.7079-7084.2000.

Chapter 2

Virus Structure and Classification

2.1 COMMON CHARACTERISTICS OF VIRUSES

As described in Chapter 1, "The World of Viruses" viruses were initially characterized as filterable agents capable of causing disease. Since that time, advances in microscopy and scientific techniques have led to a better classification of viruses and their properties. Electron microscopy has allowed us to visualize viruses in great detail, while molecular and cellular assays have broadened our understanding of how viruses function and are related to one another. Taken together, we have learned that although they can be quite diverse, viruses share several common characteristics:

1. Viruses are Small in Size.
   The smallest of viruses are about 20 nm in diameter, although influenza and the human immunodeficiency virus have a more typical size, about 100 nm in diameter. Average human cells are 10–30 μm (microns) in diameter, which means that they are generally 100 to 1000 times larger than the viruses that are infecting them.
   However, some viruses are significantly larger than 100 nm. Poxviruses, such as the variola virus that causes smallpox, can approach 400 nm in length, and filoviruses, such as the dangerous Ebola virus and Marburg virus, are only 80 nm in diameter but extend into long threads that can reach lengths of over 1000 nm. Several very large viruses that infect amoebas have recently been discovered: megavirus is 400 nm in diameter, and pandoraviruses have an elliptical or ovoid structure approaching 1000 nm in length. It is a common mistake to think that all viruses are smaller than bacteria; most bacteria are typically 2000–3000 nm in size, but certain strains of bacteria called *Mycobacteria* can be 10 times smaller than this, putting them in the range of these large viruses. So although a characteristic of viruses is that they are all small in size, this ranges from only a few nanometers to larger than some bacteria (Fig. 2.1).
2. Viruses are **obligate intracellular parasites**, meaning that they are completely dependent upon the internal environment of the cell to create new infectious virus particles, or **virions**.
   All viruses make contact with and bind the surface of a cell to gain entry into the cell. The virus disassembles and its genetic material (made of nucleic acid) encodes the instructions for the proteins that will spontaneously assemble into the new virions. This is known as de novo replication, from the Latin for "from new." In contrast to cells, which grow in size and divide equally in two to replicate, viruses use the cell's energy and machinery to create and assemble new virions piece by piece, completely from scratch.
3. The genetic material of viruses can be composed of DNA or RNA.
   All living cells, whether human, animal, plant, or bacterial, have double-stranded DNA (dsDNA) as their genetic material. Viruses, on the other hand, have **genomes**, or genetic material, that can be composed of DNA *or* RNA (but not both). Genomes are not necessarily double-stranded, either; different virus types can also have single-stranded DNA (ssDNA) genomes, and viruses with RNA genomes can be single-stranded or double-stranded. Any particular virus will only have one type of nucleic acid genome, however, and so viruses are not encountered that have both ssDNA and ssRNA genomes, for example.
   Similarly to how the size of the virus particle varies significantly, the genome size can also vary greatly from virus to virus. A typical virus genome falls in the range of 7000–20,000 base pairs (bp) (7–20 kilobase pairs (kb)). Smaller-sized virions will naturally be able to hold less nucleic acid than larger virions, but large viruses do not necessarily have large genomes. While most viruses do not contain much nucleic acid, some dsDNA viruses have very large genomes: herpesviruses

Refresher: Orders of Magnitude and Scientific Notation

Virion size: Getting Smaller

1000 millimeters (mm) in a meter (m)	1 mm = 10^{-3} m
1000 micrometers (μm, or microns) in a millimeter	1 μm = 10^{-6} m
1000 nanometers (nm) in a micrometer	1 nm = 10^{-9} m

Virus genome size: Getting Bigger

1000 base pairs (nucleotide pairs, bp) in a kilobase pair (kb)	1 kb = 10^3 bp
1000 kb in a megabase pair (mb)	1 mb = 10^6 bp
1000 mb in a gigabase pair (gb)	1 gb = 10^9 bp

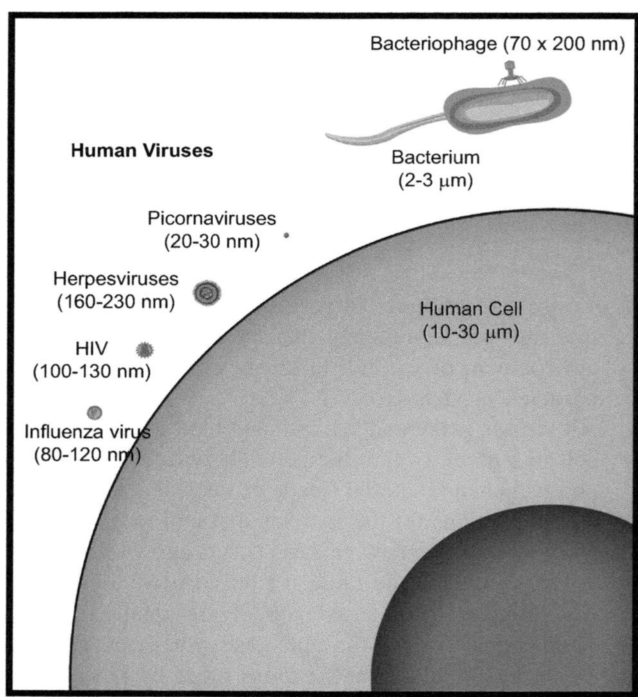

FIGURE 2.1 **Virus and cell size comparison.** Human viruses can vary in size but are generally in the range of 20–200 nm in diameter. In comparison, bacteria are generally 2–3 μM in length, and an average human cell is 10–30 μM.

have genomes that are 120–200 kb in total, and the very large pandoraviruses mentioned previously have the largest genomes: up to 2.5 million bases, rivaling the genome size of many bacteria! In comparison, eukaryotic cells have much larger genomes: a red alga has the smallest known eukaryotic genome, at 8 million base pairs; a human cell contains over 3 billion nucleotides in its hereditary material; the largest genome yet sequenced, at over 22 billion base pairs, is that of the loblolly pine tree.

Study Break
Describe the common characteristics of viruses.

2.2 STRUCTURE OF VIRUSES

The infectious virus particle must be released from the host cell to infect other cells and individuals. Whether dsDNA, ssDNA, dsRNA, or ssRNA, the nucleic acid genome of the virus must be protected in the process. In the extracellular environment, the virus will be exposed to enzymes that could break down or degrade nucleic acid. Physical stresses, such as the flow of air or liquid, could also shear the nucleic acid strands into pieces. In addition, viral genomes are susceptible to damage by ultraviolet radiation or radioactivity, much in the same way that our DNA is. If the nucleic acid genome of the virus is damaged, then it will be unable to produce progeny virions.

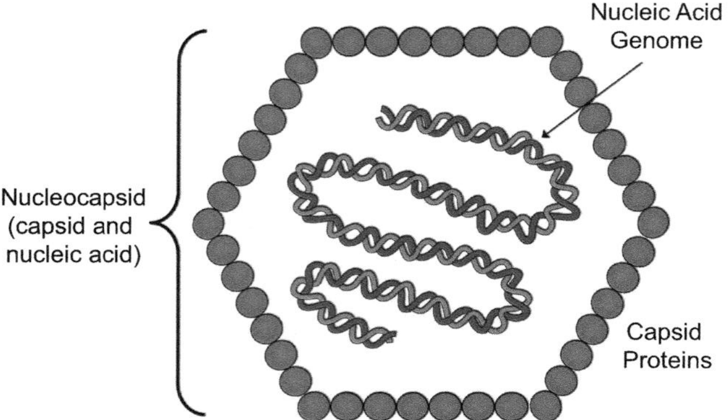

FIGURE 2.2 **Basic virus architecture.** Viral capsid proteins protect the fragile genome, composed of nucleic acid, from the harsh environment. The capsid and nucleic acid together are known as the nucleocapsid.

In order to protect the fragile nucleic acid from this harsh environment, the virus surrounds its nucleic acid with a protein shell, called the **capsid**, from the Latin *capsa*, meaning "box." The capsid is composed of one or more different types of proteins that repeat over and over again to create the entire capsid, in the same way that many bricks fit together to form a wall. This repeating structure forms a strong but slightly flexible capsid. Combined with its small size, the capsid is physically very difficult to break open and sufficiently protects the nucleic acid inside of it. Together, the nucleic acid and the capsid form the **nucleocapsid** of the virion (Fig. 2.2).

Remember that the genomes of most viruses are very small. Genes encode the instructions to make proteins, so small genomes cannot encode many proteins. It is for this reason that the capsid of the virion is composed of one or only a few proteins that repeat over and over again to form the structure. The nucleic acid of the virus would be physically too large to fit inside the capsid if it were composed of more than just a few proteins.

In the same way that a roll of magnets will spontaneously assemble together, capsid proteins also exhibit self-assembly. The first to show this were H. Fraenkel-Conrat and Robley Williams in 1955. They separated the RNA genome from the protein subunits of tobacco mosaic virus, and when they put them back together in a test tube, infectious virions formed automatically. This indicated that no additional information is necessary to assemble a virus: the physical components will assemble spontaneously, primarily held together by electrostatic and hydrophobic forces.

Most viruses also have an **envelope** surrounding the capsid. The envelope is a lipid membrane that is derived from one of the cell's membranes, most often the plasma membrane, although the envelope can also come from the cell's endoplasmic reticulum, Golgi complex, or even the nuclear membrane,

Refresher: Chemical Bonds

Virus capsids are held together by some of the same bonds that are found in living organisms. Rarely are **covalent bonds** found in capsids; these are the strongest of bonds that are formed when atoms share electrons with each other. **Electrostatic** interactions, which are weaker than covalent bonds, occur when two atoms are oppositely charged and therefore attracted to each other, hence the saying *"opposites attract."* **Ionic bonds** are electrostatic interactions because they occur between oppositely charged atoms, or **ions**. **Hydrogen bonds** are also weak electrostatic forces that occur between slightly charged atoms, usually between hydrogen (slightly positively charged) and another atom that is partially negatively charged, such as oxygen. **Van der Waals forces** are weak interactions that occur when an atom becomes slightly charged due to random asymmetry of its electrons.

The properties of water also contribute to virus assembly and attachment to cells. Water is a **polar** molecule, meaning that the molecule has two distinct ends, much like a battery or magnet has a positive and a negative end. Molecules that do not have distinct ends are termed **nonpolar**. Other polar molecules are attracted to water, since water is polar too. These molecules are **hydrophilic**, meaning "water loving." Molecules that are nonpolar are not attracted to water and are **hydrophobic**, or "water avoiding." Hydrophobic molecules will therefore aggregate with each other in an aqueous solution. This explains the phenomenon of oil (nonpolar) not mixing with water (polar).

depending upon the virus. These viruses often have proteins, called **matrix proteins**, that function to connect the envelope to the capsid inside. A virus that lacks an envelope is known as a **nonenveloped** or **naked virus** (Fig. 2.3). Each virus also possesses a **virus attachment protein** embedded in its outermost layer. This will be found in the capsid, in the case of a naked virus, or the envelope, in the case of an enveloped virus. The virus attachment protein is the viral protein that facilitates the docking of the virus to the plasma membrane of the host cell, the first step in gaining entry into a cell.

2.2.1 Helical Capsid Structure

Each virus possesses a protein capsid to protect its nucleic acid genome from the harsh environment. Virus capsids predominantly come in two shapes: **helical** and **icosahedral**. The helix (plural: helices) is a spiral shape that curves cylindrically around an axis. It is also a common biological structure: many proteins have sections that have a helical shape, and DNA is a double-helix of nucleotides. In the case of a helical virus, the viral nucleic acid coils into a helical shape and the capsid proteins wind around the inside or outside of the nucleic acid, forming a long tube or rod-like structure (Fig. 2.4). The nucleic acid and capsid constitute the **nucleocapsid**. In fact, the protein that winds around the nucleic acid is often called the nucleocapsid protein. Once in the cell, the helical nucleocapsid uncoils and the nucleic acid becomes accessible.

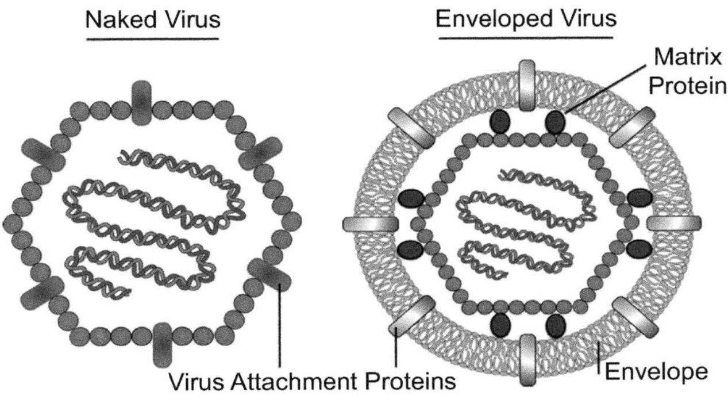

FIGURE 2.3 **Comparison between a Naked and Enveloped Virion.** The capsid of an enveloped virion is wrapped with a lipid membrane derived from the cell. Virus attachment proteins located in the capsid or envelope facilitate binding of the virus to its host cell.

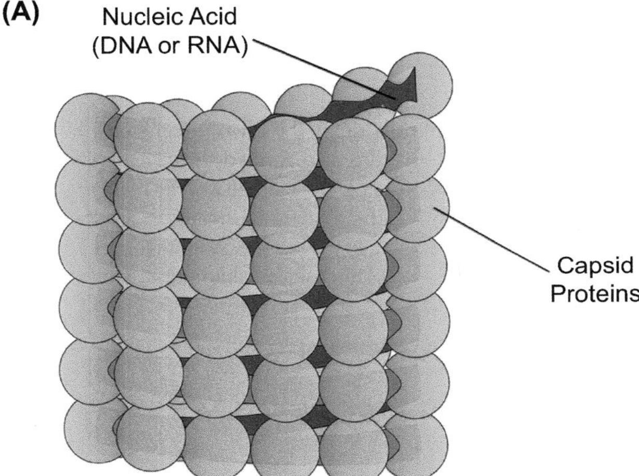

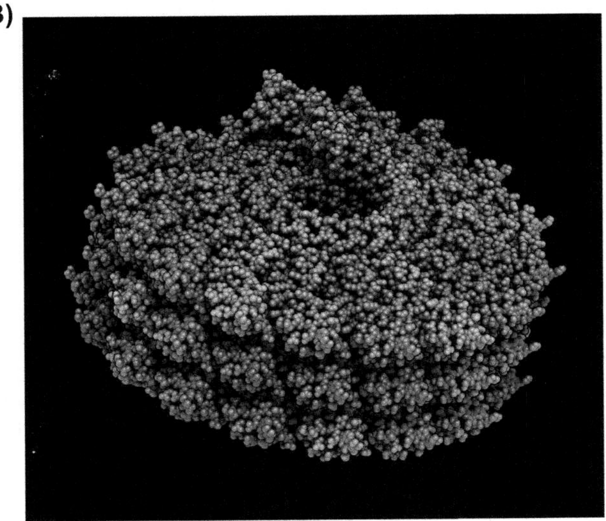

FIGURE 2.4 **Helical capsid structure.** (A) Viral capsid proteins wind around the nucleic acid, forming a helical nucleocapsid. (B) Helical structure of tobacco mosaic virus. Rendering of 49 subunits was performed using QuteMol *(IEEE Trans. Vis. Comput. Graph 2006, 12, 1237–44)* using a 2xea PDB assembly *(J. Struct. Biol. 2010, 171, 303–8)*.

There are several perceived advantages to forming a helical capsid. First, only one type of capsid protein is required. This protein subunit is repeated over and over again to form the capsid. This structure is simple and requires less free energy to assemble than a capsid composed of multiple proteins. In addition, having only one nucleocapsid protein means that only one gene is required instead of several, thereby reducing the length of nucleic acid required. Because the helical structure can continue indefinitely, there are also no constraints on how much nucleic acid can be packaged into the virion: the capsid length will be the size of the coiled nucleic acid.

Helical viruses can be enveloped or naked. The first virus described, tobacco mosaic virus, is a naked helical virus. In fact, most plant viruses are helical, and it is very uncommon that a helical plant virus is enveloped. In contrast, all helical animal viruses are enveloped. These include well-known viruses such as influenza virus, measles virus, mumps virus, rabies virus, and Ebola virus (Fig. 2.5).

2.2.2 Icosahedral Capsid Structure

Of the two major capsid structures, the icosahedron is by far more prevalent than the helical architecture. In comparison to a helical virus where the capsid proteins wind around the nucleic acid, the genomes of icosahedral viruses are packaged completely within an icosahedral capsid that acts as a protein shell. Initially these viruses were thought to be spherical, but advances in electron microscopy and X-ray crystallography revealed these were actually icosahedral in structure.

An icosahedron is a geometric shape with 20 sides (or **faces**), each composed of an equilateral triangle. An icosahedron has what is referred to as **2–3–5 symmetry**, which is used to describe the possible ways that an icosahedron can rotate around an axis. If you hold an icosahedral die in your hand, you will notice there are different ways of rotating it (Fig. 2.6). Let's say you

In-Depth Look: Defining Helical Capsid Structure

A helix is mathematically defined by two parameters, the amplitude and the pitch, that are also applied to helical capsid structures. The **amplitude** is simply the diameter of the helix and tells us the width of the capsid. The **pitch** is the height or distance of one complete turn of the helix. In the same way that we can determine the height of a one-story staircase by adding up the height of the stairs, we can figure out the pitch of the helix by determining the **rise**, or distance gained by each capsid subunit. A staircase with 20 stairs that are each 6 inches tall results in a staircase of 10 feet in height; a virus with 16.3 subunits per turn and a rise of 0.14 nm for each subunit results in a pitch of 2.28 nm. This is the architecture of tobacco mosaic virus.

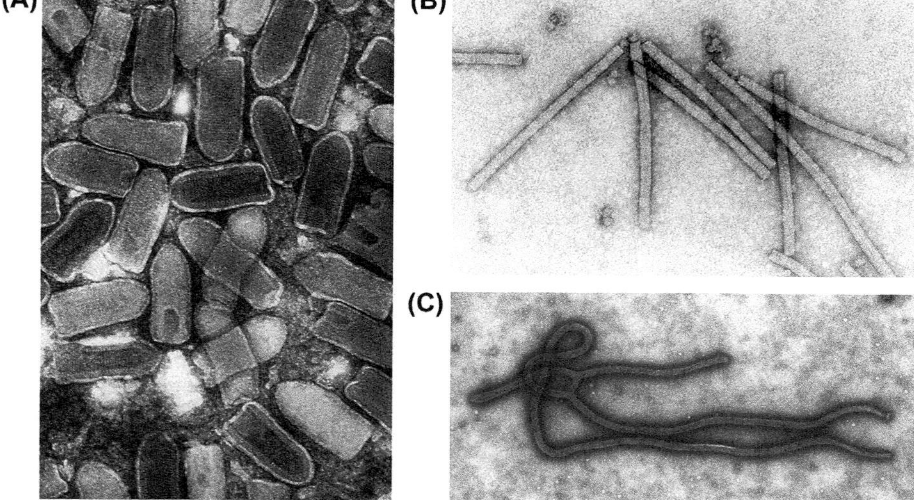

FIGURE 2.5 **Electron micrographs of helical viruses.** (A) Vesicular stomatitis virus forms bullet-shaped helical nucleocapsids. *(Image courtesy of CDC/Dr. Fred A. Murphy.)* (B) Tobacco mosaic virus forms long helical tubes. *(Image courtesy of the USDA Beltsville Electron Microscopy Unit.)* (C) The helical Ebola virus forms long threads that can extend over 1000 nm in length. *(Image courtesy of CDC/Cynthia Goldsmith.)*

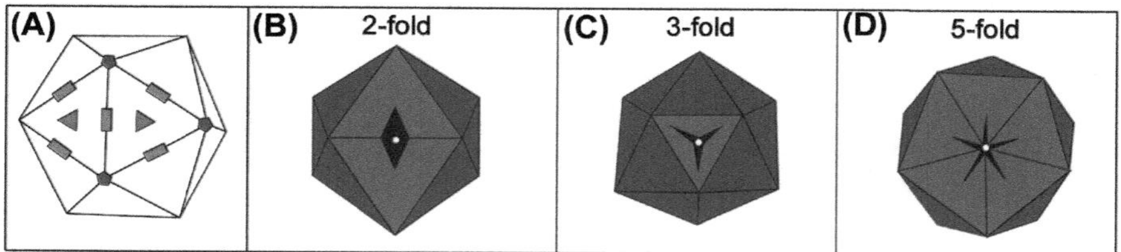

FIGURE 2.6 **Icosahedron terminology and axes of symmetry.** (A) Icosahedron faces (fuchsia triangles), edges (red rectangles), and vertices (violet pentagons) are indicated on the white icosahedron. (B) The twofold axis of symmetry occurs when the axis is placed through the center of an edge. The threefold axis occurs when the axis is placed in the center of a face (C), and the fivefold axis passes through a vertex of the icosahedron (D).

looked straight on at one of the **edges** of the icosahedron and poked an imaginary pencil through the middle of that edge. Your pencil would be right in the middle of a triangle facing up and a triangle facing down. If you rotate the icosahedron clockwise, you will find that in 180 degrees you encounter the same arrangement (symmetry): a triangle facing up and a triangle facing down. Continuing to rotate the icosahedron brings you back to where you began. This is known as the **twofold axis** of symmetry, because as you rotate the shape along this axis (your pencil), you encounter your starting structure twice in one revolution: once when you begin, and again when rotated 180 degrees. On the other hand, if you put your pencil axis directly through the center one of the small triangle **faces** of the icosahedron, you will encounter the initial view two additional times as you rotate the shape, for a total of three times. This is the **threefold axis.** Similarly, if your pencil axis goes through a **vertex** (or tip) of the icosahedron, you will find symmetry five times in one rotation, forming the **fivefold axis**. It is for this reason that an icosahedron is known to have 2–3–5 symmetry, because it has twofold, threefold, and fivefold axes of symmetry. This terminology is useful when dealing with an icosahedral virus because it can be used to indicate specific locations on the virus or where the virion has interactions with the cell surface. For instance, if a virus interacts with a cell surface receptor at the threefold axis, then you know this interaction occurs at one of the faces of the icosahedron. A protein protruding from the capsid at the fivefold axis will be found at one of the vertices (tips) of the icosahedron. All of the illustrations of viruses in Fig. 2.7 are viewed on the twofold axis of symmetry.

Study Break

How many twofold axes of symmetry are found in one icosahedron? How about the number of threefold or fivefold axes? How many faces, edges, and vertices are found in an icosahedron?

Viral proteins form each face (small triangle) of the icosahedral capsid. Viral proteins are not triangular, however, and so one protein subunit alone is not sufficient to form the entire face. Therefore, a face is formed from at least three viral protein subunits fitted together (Fig. 2.8). These can all be the same protein, or they can be three different proteins. The subunits together form what is called

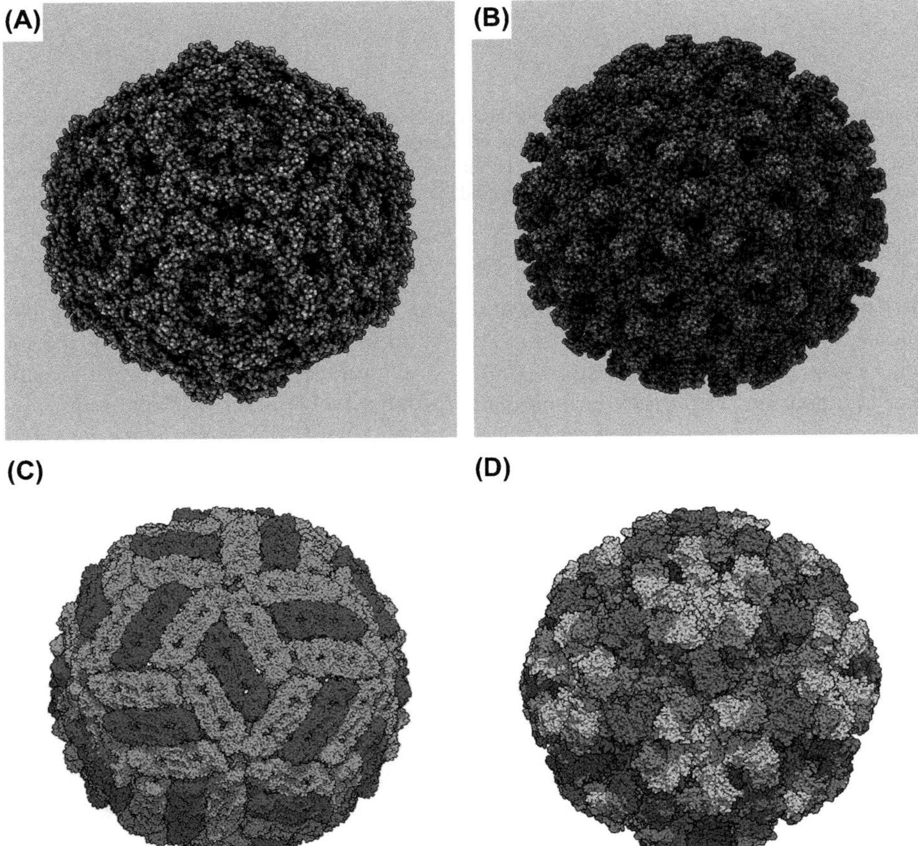

FIGURE 2.7 **Illustrations of viruses, as viewed on the twofold axis of rotation.** 3D surface reconstructions of parvovirus B19 (A), human hepatitis B virus (B), dengue virus (C), and Norwalk virus (D). *Illustrations created with QuteMol (IEEE. Trans. Vis. Comput. Graph 2006, 12, 1237–44) using 2G33 (J. Virol. 2006, 80, 11055–61), 1S58 (Proc. Natl. Acad. Sci. U. S. A. 2004, 101, 11628–33), 1K4R (Cell 2002, 108, 717–25), and 1IHM (Science 1999, 286, 287–90) PDB assemblies.*

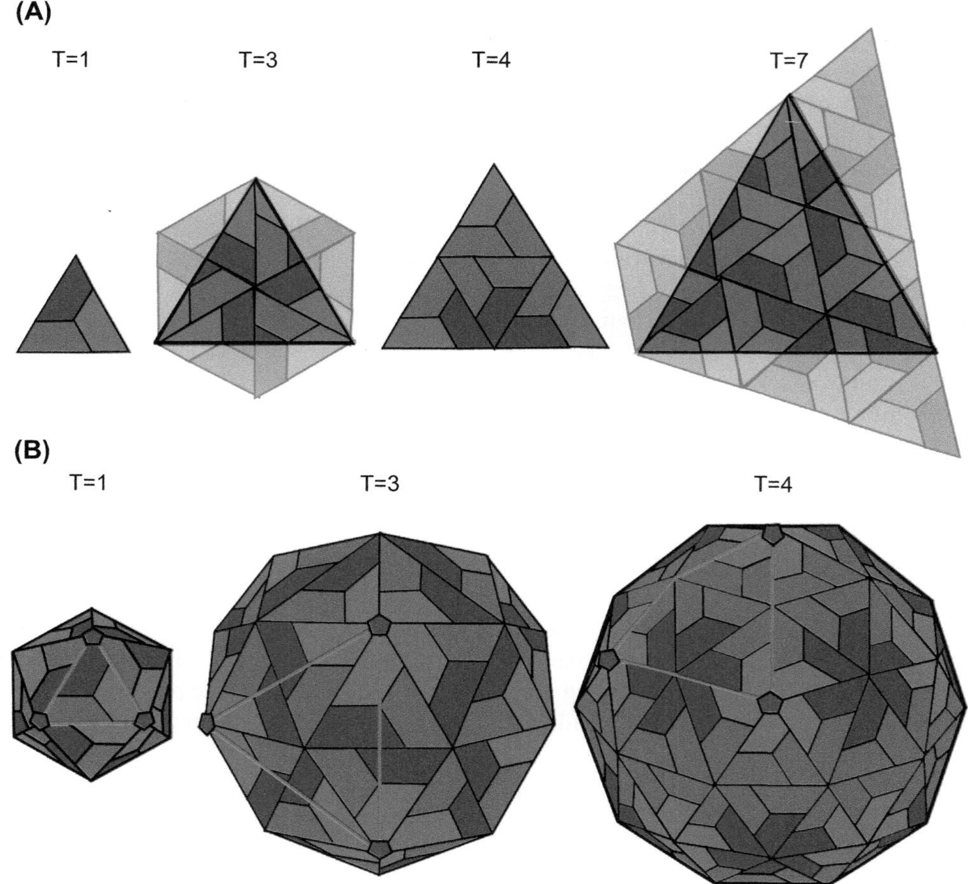

FIGURE 2.8 **Capsid architecture and triangulation number.** (A) Virus capsids are composed of viral protein subunits that form structural units. The triangulation number (T) indicates the number of structural units per face of the icosahedron. In a $T=1$ virus, one structural unit (composed of three different protein subunits: gray, red, and blue) create the icosahedron face. (B) Virion capsids with $T=1$, $T=3$, and $T=4$. The red lines outline a triangular face of the icosahedron, while the purple pentagons indicate the vertices (fivefold axes) of the icosahedron.

the **structural unit**. The structural unit repeats to form the capsid of the virion.

But how can some viruses form very large icosahedral capsids? The answer is repetition. The structural unit can be repeated over and over again to form a larger icosahedron side. The number of structural units that creates each side is called the **triangulation number (*T*)**, because the structural units form the triangle face of the icosahedron. In a $T=1$ virus, only one structural unit forms each icosahedron face (Fig. 2.8). In a $T=4$ virus, four structural units form the face. Sometimes the structural unit overlaps from one face to another: in a $T=3$ virus, three total structural units form the face, although this occurs as six half-units (half of each structural unit forms part of an adjacent face). Similarly, the structural units of a $T=7$ virus are also slightly skewed, compared to the triangle face. The geometry and math involved with icosahedral capsid structure can be complex, and only the very basics are described here. In any case, by increasing the number of identical structural units on each face, the icosahedron can become progressively larger without requiring additional novel proteins to be produced. Some viruses have triangulation numbers over 25, even!

The proteins that compose the structural unit may form three dimensional structures known as **capsomeres** that are visible in an electron micrograph. In icosahedral viruses, capsomeres generally take the form of pentons (containing five units) or hexons (containing six units) that form a visible pattern on the surface of the icosahedron (See Fig. 13.11 for an example). Capsomeres are morphological units that arise from the interaction of the proteins within the repeated structural units.

Why does the icosahedral virus structure appear so often? Research has shown that proteins forming icosahedral symmetry require lesser amounts of energy, compared to other structures, and so this structure is evolutionarily favored.

Many viruses that infect animals are icosahedral, including human papillomavirus, rhinovirus, hepatitis B virus, and herpesviruses (Fig. 2.9). Like their helical counterparts, icosahedral viruses can be naked or enveloped, as well. The type of viral nucleic acid (dsDNA, ssDNA, dsRNA, and ssRNA) does not correlate with the structure of the capsid; icosahedral viral capsids can contain any of the nucleic acid types, depending upon the virus.

2.2.3 Complex Viral Structures

The majority of viruses can be categorized as having helical or icosahedral structure. A few viruses, however, have a **complex** architecture that does not strictly conform to a simple helical or icosahedral shape. Poxviruses, geminiviruses, and many bacteriophages are examples of viruses with complex structure (Fig. 2.10). Poxviruses, including the viruses that cause smallpox or cowpox, are large oval or brick-shaped particles 200–400 nm long. Inside the complex virion, a dumbbell-shaped core encloses the viral DNA and is surrounded by two "lateral bodies," the function of which is currently unknown. The geminiviruses also exhibit complex structure. As their name suggests, these plant-infecting viruses are composed of two icosahedral heads joined together. **Bacteriophages**, also known as **bacterial viruses** or **prokaryotic viruses**, are viruses that infect and replicate within bacteria. Many bacteriophages also have complex structure, such as bacteriophage P2, which has an icosahedral head, containing the nucleic acid, attached to a cylindrical tail sheath that facilitates binding of the bacteriophage to the bacterial cell.

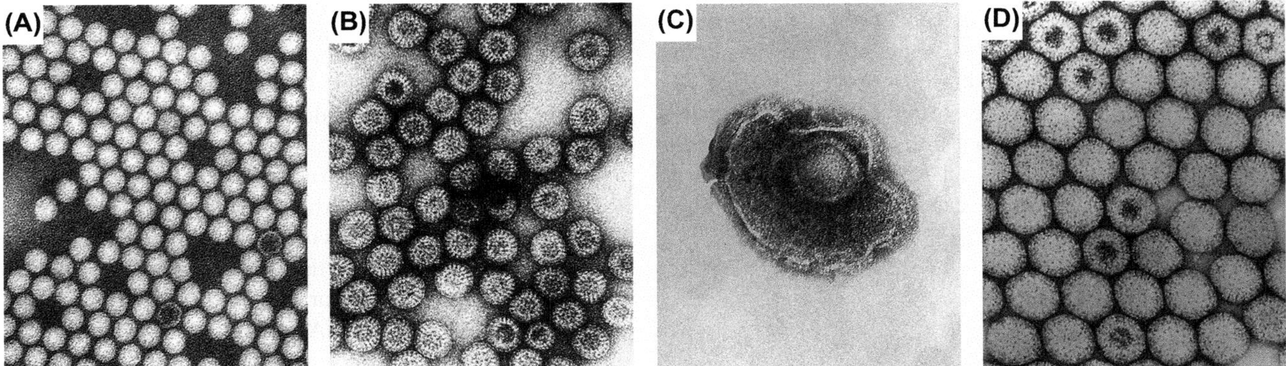

FIGURE 2.9 **Electron micrographs of icosahedral viruses.** Poliovirus (A), rotavirus (B), varicella–zoster virus (C), the virus that causes chickenpox and shingles, and reovirus (D). Note that C is enveloped. *Images courtesy of the CDC: Dr. Fred Murphy and Sylvia Whitfield (A), Dr. Erskine Palmer (B and D), and Dr. Erskine Palmer and B.G. Partin (C).*

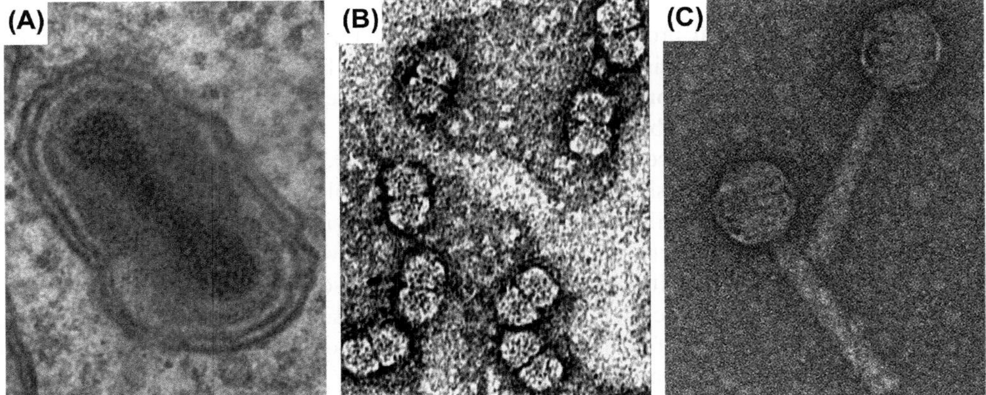

FIGURE 2.10 **Electron micrograph of viruses with complex architecture.** Vaccinia virus (A), a virus belonging to the poxvirus family, has a complex capsid architecture with a dumbbell-shaped core. Geminiviruses (B) have a double-icosahedron capsid. Bacteriophages, such as P2 (C), often have complex capsid structure. *Images courtesy of Ana Caceres et al. (A, PLoS Pathog. 2013, 9(11), e1003719), Kassie Kasdorf (B), and Mostafa Fatehi (C).*

2.3 VIRUS CLASSIFICATION AND TAXONOMY

The classification of viruses is useful for many reasons. It allows scientists to contrast viruses and to reveal information on newly discovered viruses by comparing them to similar viruses. It also allows scientists to study the origin of viruses and how they have evolved over time. The classification of viruses is not simple, however—there are currently over 2800 different viral species with very different properties!

One classification scheme was developed in the 1970s by Nobel laureate David Baltimore. The **Baltimore classification system** categorizes viruses based on the type of nucleic acid genome and replication strategy of the virus. The system also breaks down single-stranded RNA viruses into those that are positive strand (+) and negative strand (−). As will be further discussed in the next chapter, **positive-strand** (also positive-sense or plus-strand) RNA is able to be immediately translated into proteins; as such, messenger RNA (mRNA) in the cell is positive strand. **Negative-strand** (also negative-sense or minus-strand) RNA is not translatable into proteins; it first has to be transcribed into positive-strand RNA. Baltimore also took into account viruses that are able to **reverse transcribe**, or create DNA from an RNA template, which is something that cells are not capable of doing. Together, the seven classes are

- class I: dsDNA viruses
- class II: ssDNA viruses
- class III: dsRNA viruses
- class IV: positive-sense ssRNA viruses
- class V: negative-sense ssRNA viruses
- class VI: RNA viruses that reverse transcribe
- class VII: DNA viruses that reverse transcribe

There are a variety of ways by which viruses could be classified, however, including virion size, capsid structure, type of nucleic acid, physical properties, host species, or disease caused. Because of this formidable challenge, the **International Committee on Taxonomy of Viruses (ICTV)** was formed and has been the sole body charged with classifying viruses since 1966. **Taxonomy** is the science of categorizing and assigning names (**nomenclature**) to organisms based on similar characteristics, and the ICTV utilizes the same taxonomical hierarchy that is used to classify living things. It is important to note that viruses, since they are not alive, belong to a completely separate system that does not fall under the tree of life. Whereas a living organism is classified using domain, kingdom, phylum, class, order, family, genus, and species **taxa** (singular: **taxon**), or categories, viruses are only classified using order, family, genus, and species (Table 2.1).

TABLE 2.1 Taxa Used to Classify Viruses

Taxon	Notes	Example
Order	Ends in -virales suffix; only about half of viruses are currently classified in orders.	*Picornavirales*
Family	Ends in -viridae suffix; subfamilies are indicated with -virinae suffix.	*Picornaviridae*
Genus	Ends in -virus suffix.	*Enterovirus*
Species	Generally the "common name" of the virus. Classifying and cataloging anything below the species classification (such as subtypes, serotypes, strains, isolates, or variants) is the responsibility of the specific field.	*Rhinovirus A* (Serotypes include Human rhinovirus 1, which includes strains human rhinovirus 1A and human rhinovirus 1B)

The ICTV classifies viruses based upon a variety of different characteristics with the intention of categorizing the most similar viruses with each other. The chemical and physical properties of the virus are considered, such as the type of nucleic acid or number of different proteins encoded by the virus. DNA technologies now allow us to sequence viral genomes relatively quickly and easily, allowing scientists to compare the nucleic acid sequences of two viruses to determine how closely related they are. Other virion properties are also taken into account, including virion size, capsid shape, and whether or not an envelope is present. The taxa of viruses that infect vertebrates are shown in Fig. 2.11; notice that some families are not yet classified into orders (refer to Table 2.1 for a refresher on how to distinguish the taxa by their suffixes). Also note the size difference between viruses of different families.

Currently, the ICTV has categorized seven orders of viruses (Table 2.2) that contain a total of 103 families classified within them. Seventy-seven virus families, however, have yet to be assigned to an order, including notable viruses such as the retroviruses, papillomaviruses, and poxviruses. New orders have been proposed, and it is likely that more will be created as the taxonomical process continues.

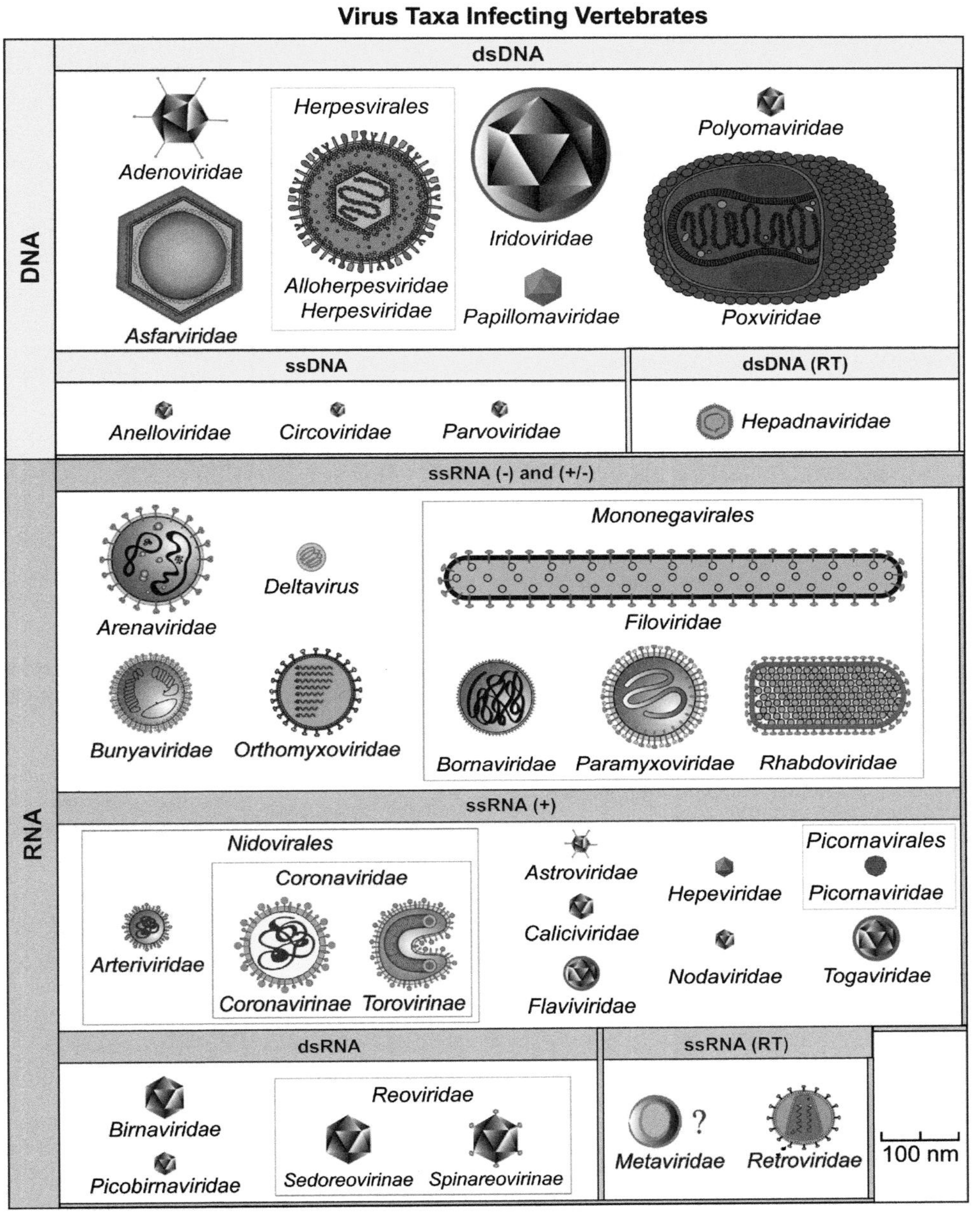

FIGURE 2.11 **Taxa of viruses that infect vertebrates.** Viruses are categorized based upon their type of nucleic acid (DNA viruses in yellow boxes and RNA viruses in blue boxes) and further classified based upon distinguishing characteristics. Note the nucleic acid, size, and architectural differences between viruses of different families. Viruses in color will be discussed in later chapters. *Modified with permission from Virus Taxonomy: Ninth Report of the International Committee on Taxonomy of Viruses, Elsevier, 2012.*

In-Depth Look: How Viruses are Named

The ICTV has established guidelines for naming newly discovered viruses. The Latin binomial names that are used for living organisms, where the genus and species are listed together (such as *Homo sapiens* or *Yersinia pestis*), are not used for naming viruses. Virus names should also not include any person's name (although historically this was how a few viruses were named), and selected names should be easy to use and meaningful.

When directly referring to a viral order, family, genus, or species the virus name should be written in italics with the first letter capitalized. When not referring specifically to viral classification, however, capitalization and italics are not required unless a proper name is encountered. For instance, "Some members of the *Arenaviridae* family can cause severe disease, including lymphocytic choriomeningitis virus and Lassa virus, which is named after the town in Nigeria where the first case occurred." The chart below gives some examples of where virus named originated (Table 2.3).

TABLE 2.2 Current Orders of the Virosphere

Order	Notes
Caudovirales	Tailed dsDNA viruses that infect members of the domains Bacteria and Archaea; name comes from Latin *cauda*, meaning "tail."
Herpesvirales	dsDNA viruses of vertebrates and invertebrates; from Greek *herpes*, meaning "creeping" or "spreading" (describing the rashes of these viruses).
Ligamenvirales	dsDNA viruses that infect the domain Archaea; from Latin *ligamen*, meaning "thread" or "string" (describing the linear structure of the viruses). Newest order, created in 2012.
Mononegavirales	"Negative-strand" ssRNA viruses of vertebrates, invertebrates, and plants; name derives from Latin for "one negative," referring to the single negative-strand RNA genome. Was the first order created, in 1990.
Nidovirales	"Positive-strand" ssRNA viruses of vertebrates and invertebrates; from Latin *nidus* meaning "nest" because they encode several proteins nested within one piece of mRNA.
Picornavirales	"Positive-strand" ssRNA viruses of vertebrates, invertebrates, and plants; from pico (small) + RNA + virales (viruses).
Tymovirales	"Positive-strand" ssRNA viruses of plants and invertebrates; Tymo is an acronym standing for Turnip Yellow Mosaic virus, found within this order.

ds, Double-stranded; *ss*, single-stranded.

TABLE 2.3 Name Origins of Selected Viruses

Name	Origin
Viruses named after the clinical conditions they cause	
Human immunodeficiency virus (HIV)	Causes the decline of the immune system, leading to immunodeficiency
Hepatitis virus	Although they are not in the same family, all hepatitis viruses cause liver inflammation (hepatitis)
Human papillomavirus (HPV)	Causes papillomas, benign epithelial tumors such as warts
Poxviruses	From pockes meaning "sac," referring to the blistery rash observed
Rabies virus	From Latin *rabies*, meaning "madness," describing the symptoms seen with disease progression
Viruses named after their location of discovery	
Coxsackievirus	Named after Coxsackie, New York, the location from where the first specimens were obtained
Ebola virus	Named after the Ebola River in northern Democratic Republic of the Congo (formerly Zaire), where the virus first emerged in 1976
Marburg virus	Named after Marburg, a town in Germany, where an outbreak occurred in 1967
Nipah virus	First identified in the Malaysian village of Kampung Sungai Nipah in 1998
Norwalk virus	Named after a 1968 outbreak in children at an elementary school in Norwalk, Ohio
West Nile virus	First isolated from a woman in the West Nile district of Uganda in 1937
Viruses named after their properties	
Coronavirus	From Latin *corona*, meaning crown, referring to the crown-like appearance of the virions when viewed with an electron microscope
Herpesviruses	From Greek *herpein*, "to creep," referring to the lesions that slowly spread across the skin
Influenza virus	Originated in 15th century Italy, from an epidemic attributed to the "influence of the stars"
Picornaviruses	Pico meaning "small" + RNA viruses
Poliovirus	From Greek *polios*, meaning "gray," referring to the gray matter (cell bodies) in the spinal cord that it infects and damages
Viruses named after people (historically assigned; viruses can no longer be named after individuals)	
Epstein–Barr virus	Named after Michael Anthony Epstein and Yvonne Barr, who discovered the virus
JC Virus	Named after a patient, John Cunningham, from which the virus was isolated
Rous sarcoma virus	Discovered by Peyton Rous in 1911

SUMMARY OF KEY CONCEPTS

Section 2.1 Common Characteristics of Viruses

- Viruses are small. Most viruses are in the range of 20–200 nm, although some viruses can exceed 1000 nm in length. A typical bacterium is 2–3 μM in length; a typical eukaryotic cell is 10–30 μM in diameter.
- Viruses are obligate intracellular parasites and are completely dependent upon the cell for replication. Unlike cells that undergo mitosis and split in two, viruses completely disassemble within the cell and new virions (infectious particles) are assembled de novo from newly made components.
- While living things have dsDNA genomes, the genetic material of viruses can be composed of DNA or RNA, and single- or double-stranded. Most virus genomes fall within the range of 7–20 kb, but they range from 3 kb to over 2 mb.

Section 2.2 Structure of Viruses

- The simplest viruses are composed of a protein capsid that protects the viral nucleic acid from the harsh environment outside the cell.
- Virus capsids are predominantly one of two shapes, helical or icosahedral, although a few viruses have a complex architecture. In addition, some viruses also have a lipid membrane envelope, derived from the cell. All helical animal viruses are enveloped.
- Helical capsid proteins wind around the viral nucleic acid to form the nucleocapsid. A helix is mathematically defined by amplitude and pitch.
- An icosahedron is a geometric shape with 20 sides, each composed of an equilateral triangle. The sides are composed of viral protein subunits that create a structural unit, which is repeated to form a larger side and the other sides of the icosahedron. The triangulation number refers to the number of structural units per side.

Section 2.3 Virus Classification and Taxonomy

- The Baltimore classification system categorizes viruses based upon the type and replication strategy of the nucleic acid genome of the virus. There are seven classes.
- The ICTV was formed to assign viruses to a taxonomical hierarchy. The taxa used for classifying viruses are order, family, genus, and species. Because they are not alive, viruses are not categorized within the same taxonomical tree as living organisms.

FLASH CARD VOCABULARY

Virion	Triangulation number
Genome	Capsomere
Capsid	Bacteriophage
Nucleocapsid	Baltimore classification system
Enveloped virus	Positive-strand (positive-sense)
Naked (unenveloped) virus	Negative-strand (negative-sense)
Matrix proteins	Reverse transcribe
Virus attachment protein	Taxonomy
Helix	Nomenclature
Icosahedron: Face, edge, vertex	International Committee on Taxonomy of Viruses
Structural unit	Taxon

CHAPTER REVIEW QUESTIONS

1. Why are viruses considered obligate intracellular pathogens?
2. How does viral replication differ from cell replication?
3. What is the function of the capsid? Why must viruses repeat the same capsid protein subunits over and over again, rather than having hundreds of different capsid proteins?
4. Explain what 2–3–5 symmetry is, pertaining to an icosahedron.
5. What is a structural unit? In a $T=3$ virus that has three subunits per structural unit, how many total subunits form the capsid?
6. List the seven groups of the Baltimore classification system.
7. What taxa are used to classify viruses? How does this differ from the classification of a living organism?
8. What viral properties are used to classify viruses?

FURTHER READING

Bourne, C.R., Finn, M.G., Zlotnick, A., 2006. Global structural changes in hepatitis B virus capsids induced by the assembly effector HAP1. J. Virol. 80, 11055–11061.

Cáceres, A., Perdiguero, B., Gómez, C.E., et al., 2013. Involvement of the cellular phosphatase DUSP1 in vaccinia virus infection. PLoS Pathog. 9. http://dx.doi.org/10.1371/journal.ppat.1003719.

Clare, D.K., Orlova, E.V., 2010. 4.6 A Cryo-EM reconstruction of tobacco mosaic virus from images recorded at 300 keV on a 4k×4k CCD camera. J. Struct. Biol. 171, 303–308.

Dixon, L.K., Alonso, C., Escribano, J.M., et al., 2012. Virus taxonomy. In: Virus Taxonomy: Ninth Report of the International Committee on Taxonomy of Viruses, pp. 153–162.

Kaufmann, B., Simpson, A.A., Rossmann, M.G., 2004. The structure of human parvovirus B19. Proc. Natl. Acad. Sci. U. S. A. 101, 11628–11633.

Kuhn, R.J., Zhang, W., Rossmann, M.G., et al., 2002. Structure of dengue virus: implications for flavivirus organization, maturation, and fusion. Cell 108, 717–725.

Prasad, B.V., Hardy, M.E., Dokland, T., Bella, J., Rossmann, M.G., Estes, M.K., 1999. X-ray crystallographic structure of the Norwalk virus capsid. Science 286, 287–290.

Prasad, B.V.V., Schmid, M.F., 2012. Principles of virus structural organization. Adv. Exp. Med. Biol. 726, 17–47.

Tarini, M., Cignoni, P., Montani, C., 2006. Ambient occlusion and edge cueing to enhance real time molecular visualization. IEEE Trans. Vis. Comput. Graph 12, 1237–1244.

Van Regenmortel, M.H., 1990. Virus species, a much overlooked but essential concept in virus classification. Intervirology 31, 241–254.

Chapter 3

Features of Host Cells: Cellular and Molecular Biology Review

As obligate intracellular parasites, viruses are completely dependent upon a host cell for their replication. They use energy generated by the host cell, and they exploit the host's machinery to manufacture viral proteins. This chapter takes us for a tour inside a eukaryotic cell, highlighting the processes that viruses take advantage of during infection.

3.1 THE BASIC ORGANIZATION OF THE CELL

There are three domains of life—*Bacteria*, *Archaea*, and *Eukarya*. The organisms within these groups are divided depending on the presence or absence of a **nucleus** within the cell(s) of the organism. **Prokaryotes** are organisms without a nucleus to wall off their genetic material from the rest of the cell, while **eukaryotes** are organisms that contain a nucleus within their cells. All organisms within *Bacteria* and *Archaea* are prokaryotes, whereas *Eukarya*—as the name suggests—contains eukaryotes. Viruses exist that infect cells of all three domains. Most of the viruses that are discussed in this book infect humans and other animals, which are eukaryotes.

The defining characteristic of a eukaryotic cell is the nucleus, which is generally located in the center of a cell. Many structures, called **organelles**, are distributed in the liquid cytosol between the nucleus and the plasma membrane of the cell (Fig. 3.1). In the same way that each organ of our body performs a specialized function, the organelles within a cell each play a specific role in maintaining an operational cell.

Most organelles are composed of the same lipid membrane that creates the plasma membrane of the cell. This membrane is only two molecules thick and made of **phospholipids**. Phospholipids are a class of lipids; fats, oils, and waxes are other lipids. They are so named because the molecule has two parts: a polar head that contains a phosphate group, and a nonpolar portion that is composed of two fatty acid lipid tails (Fig. 3.2A). As described in the Chapter 2, "Virus Structure and Classification," Refresher on Chemical Bonds, water is a polar molecule and readily associates with other polar molecules. The phospholipid head of the phospholipid, being polar, is **hydrophilic**, while the nonpolar tails do not associate with water and are **hydrophobic**. Because of this **amphipathic** nature of a phospholipid, a group of phospholipids placed in an aqueous solution (such as the environment of a cell) will spontaneously assemble into a double layer of phospholipid molecules with the hydrophilic polar heads of the molecules facing the aqueous solution and the hydrophobic nonpolar tails associating with each other (Fig. 3.2B). This forms an effective barrier to prevent large molecules from escaping or

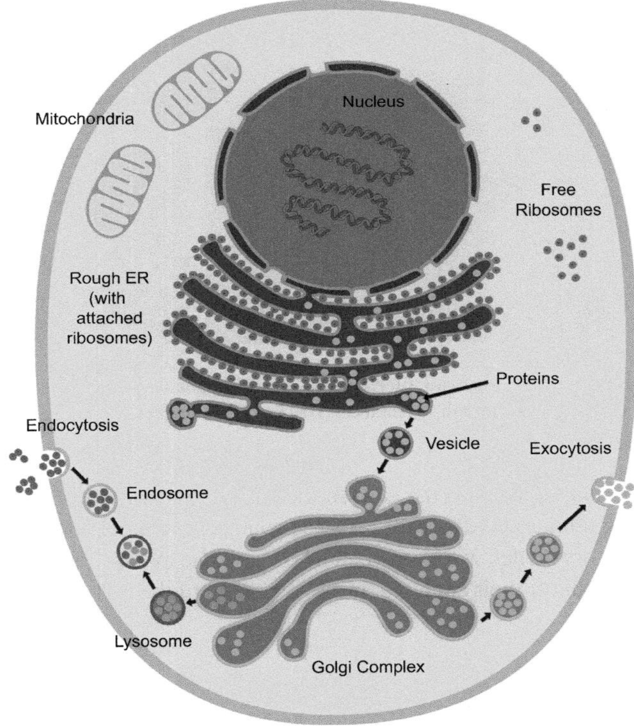

FIGURE 3.1 **A typical eukaryotic cell.** A eukaryotic cell contains a centrally located nucleus and several important organelles within the cytosol of the cell that viruses take advantage of during infection. Ribosomes in the cytosol manufacture proteins, and the ribosomes attached to the rough endoplasmic reticulum (rER) make proteins that are folded and modified within the rER. These proteins are then packaged in vesicles and travel to the Golgi complex, where they are finished and shipped to other locations within the cell or outside the cell, through the process of exocytosis. Proteins and other molecules enter the cell via endocytosis. The endocytic vesicles become endosomes, which may fuse with lysosomes that contain enzymes to degrade biological molecules. Viruses will also use the ATP generated by the mitochondria within a cell. *Note that not all cellular organelles are described here.*

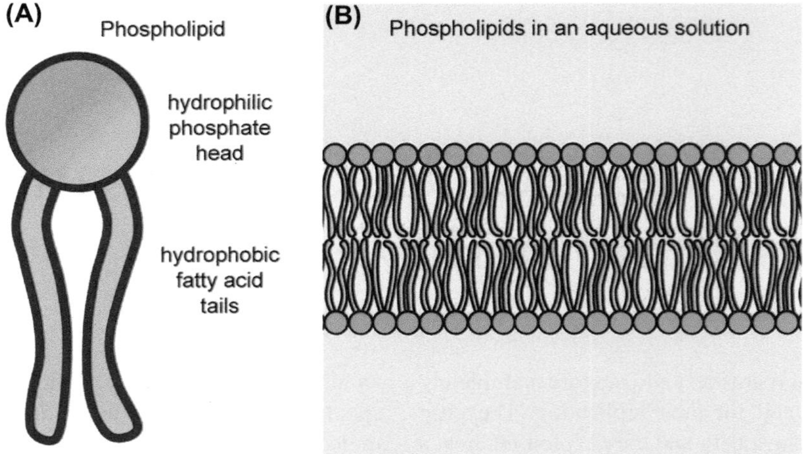

FIGURE 3.2 **Phospholipids and membrane structure.** The plasma membrane of the cell is composed of many phospholipid molecules, which are amphipathic: the phosphate head is hydrophilic, while the fatty acid tails of the molecule are hydrophobic (A). As such, when they are placed in an aqueous solution, like that of the cell environment (B), they will spontaneously form a double layer, or bilayer, with the polar heads toward the aqueous solution and the fatty acid tails facing each other, forming a barrier between the extracellular and intracellular environment.

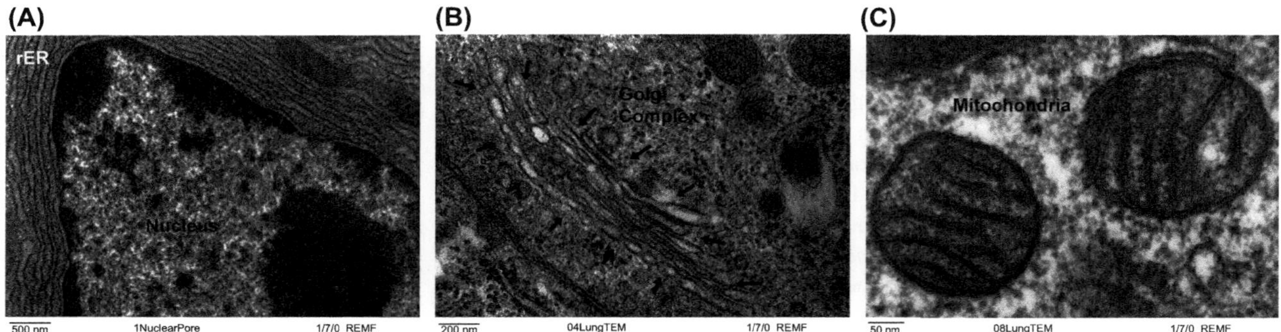

FIGURE 3.3 **Electron micrographs of cell organelles.** (A) The nucleus of a lung cell, surrounded by rER. Note the ribosomes on the rER that give it its characteristic rough appearance. (B) The Golgi complex of a lung cell, with its characteristic flattened sacs of membrane. (C) Two mitochondria within a lung cell. *Images courtesy of Louisa Howard.*

gaining entry into the cell. In the same way that the plasma membrane acts as a barrier for the contents of the cell, most of the organelles within the cell also use a phospholipid bilayer to wall off their contents from the cytosol of the cell.

Word Origin: Endoplasmic Reticulum

From the Greek *endo*, meaning "within," and Latin *reticulum*, meaning "little net"—the little network within the cytoplasm.

The **rough endoplasmic reticulum** (rER) is the first organelle encountered outside of the nucleus (Fig. 3.1). It is composed of connected sacs of membrane and is studded with ribosomes, giving it its characteristic "rough" appearance (Fig. 3.3A). **Ribosomes** make proteins after binding to messenger RNA (coming from the nucleus). They can be found attached to the rough ER or "free" (not attached) within the cytosol. Ribosomes attached to the rER will make protein that are subsequently transported into the **lumen**, or hollow inside, of the rER. Here, proteins are folded and modified; those that are modified with carbohydrates (including sugars) are known as **glycoproteins**, and proteins modified with lipids are termed **lipoproteins**. At the end of the rER, the proteins bud off in a pod of the rER membrane, known as a **vesicle**, that is transported to the Golgi complex (Fig. 3.1).

Word Origin: Golgi Complex

Also known as the Golgi apparatus or Golgi body, the Golgi complex was discovered in 1898 by Italian physician Camillo Golgi and named after him.

The **Golgi complex** is created from flattened sacs of membrane (Fig. 3.3B). The membrane vesicle inbound from the rER fuses with the Golgi to deliver the protein contents to the interior of the Golgi. The Golgi complex functions as a finishing and shipping company: the enzymes contained within it complete the protein modifications that began in the rER, and the proteins are then packaged into vesicles that travel to various locations within or outside the cell.

Some proteins are transported to the plasma membrane and released from the cell, while other proteins become permanently embedded into the plasma membrane.

At the Golgi complex, certain enzymes are packaged into specific vesicles called **lysosomes**. Lysosome enzymes are able to digest complex biological molecules that are delivered to the lysosome. These molecules can come from outside the cell in endosomes, which will be discussed in Section 3.2, or even from vesicles containing malfunctioning organelles. The 30+ enzymes found in the lysosome function best at a pH of ~5, which is more acidic than the neutral pH of the cell (~7.2), reducing the risk to the cell if the lysosome enzymes were to enter the cytosol.

Word Origin: Lysosome
From the Greek words *lysis*, meaning to break down or destroy, and *soma*, meaning body.

The organelles described above facilitate the creation, modification, packaging, and transport of proteins. Viruses do not have their own organelles, so after gaining entry into the cell, viruses will take advantage of these organelles to manufacture the viral proteins necessary to create more infectious virions.

There are other important parts of the cell that are not directly involved in protein synthesis, and viruses will utilize these components, as well. The majority of cellular respiration, which generates cellular energy in the form of ATP, takes place within the **mitochondria** (singular: mitochondrion) of the cell (Fig. 3.3C). Viruses do not have their own mitochondria and so will use the ATP generated by the cell. Cells also have a **cytoskeleton** made of different-sized protein components: microtubules, intermediate filaments, and microfilaments, from largest to smallest diameter. In the same way that a human skeleton shapes the form of the body and provides support for its organs, these cytoskeletal components provide structure for the cell and its organelles (Fig. 3.4). They are also involved in the movement of vesicles within the cell, and the movement of the cells themselves. Some viruses use the cytoskeleton system for transport to different parts of the cell.

Word Origin: Cytoskeleton
Cyto refers to "cell." The cytoskeleton is the skeleton (structural support) of the cell.

3.2 THE PLASMA MEMBRANE, EXOCYTOSIS, AND ENDOCYTOSIS

The plasma membrane is the primary zone of contact between the cell and the extracellular world. As such, this is the first place a virus interacts with a cell.

As mentioned above, the plasma membrane is made of a phospholipid bilayer. The current view of how the membrane is assembled is known as the **fluid mosaic model**, proposed by Singer and Nicolson. The "mosaic" part of the model refers to the presence of proteins suspended in the membrane bilayer. Many proteins, including glycoproteins, are embedded into the lipid bilayer (Fig. 3.5). Known as **integral proteins**, these proteins have a variety of functions, including being receptors for extracellular substances or facilitating the adhesion of one cell to another. **Peripheral** membrane proteins associate closely with the surface of the membrane but are not integrated within it. The "fluid" part of this model refers to the proteins and phospholipid molecules that are noncovalently associated with each other and are therefore not static within the membrane but move around freely. Cholesterol is a lipid that is found in the phospholipid bilayer that helps to maintain the fluidity and movement in the membrane. Cholesterol is also enriched in lipid rafts, portions of the membrane that contain integral proteins involved in transmitting signals to the interior of the cell.

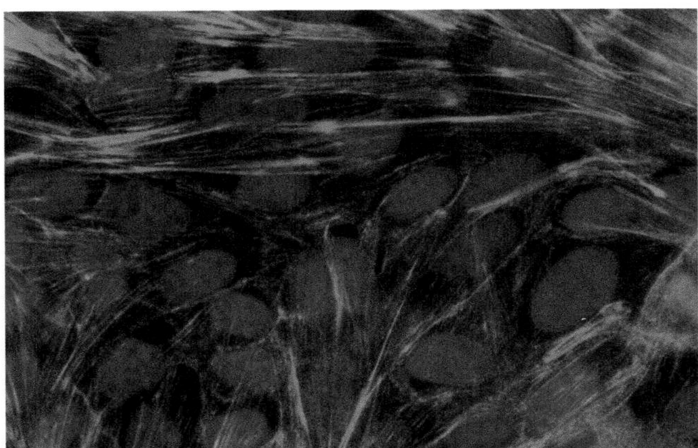

FIGURE 3.4 **Fluorescent image of actin, a component of the cytoskeleton.** The cytoskeleton functions to provide support and structure to the cell. There are three types of cytoskeletal elements: microtubules, intermediate filaments, and microfilaments (from largest to smallest). Microfilaments are composed of a protein called actin, which is dyed green in these ~30 MDCK cells using a fluorescently labeled actin-binding protein. The nuclei of the cells are labeled with a blue fluorescent dye. *Image courtesy of Michael A. Frailey.*

The plasma membrane forms an effective barrier, but certain substances must be transported from one side of the membrane to the other. Certain integral proteins transport ions and small molecules into or out of the cell, but many molecules are too large for these channel or carrier proteins. To address this problem, eukaryotic cells export and import larger molecules by **exocytosis** and **endocytosis**. In the process of exocytosis, proteins packaged into secretory vesicles by the Golgi complex travel to the plasma membrane. The secretory vesicles fuse with the plasma membrane, releasing the vesicle contents to the cell exterior (Figs. 3.1 and 3.6A). The vesicle membrane, also composed of a phospholipid bilayer, becomes part of the plasma membrane.

Word Origin: Endocytosis and Exocytosis
Endo is Greek for "within" and *cyto* means "cell," so endocytosis is the process of bringing substances within the cell's plasma membrane. *Exo* is Greek for "outside," so exocytosis is the opposite process of exporting substances to outside the cell.

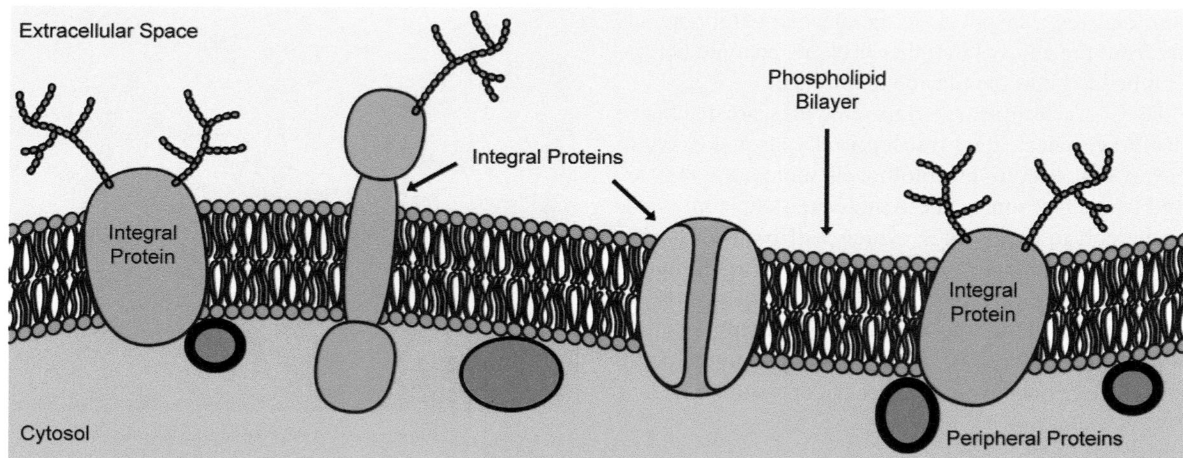

FIGURE 3.5 **The fluid mosaic model.** The current model of the plasma membrane architecture, called the fluid mosaic model, proposes that integral proteins can move freely within the phospholipid bilayer. Peripheral proteins associate with the intracellular portion of the membrane.

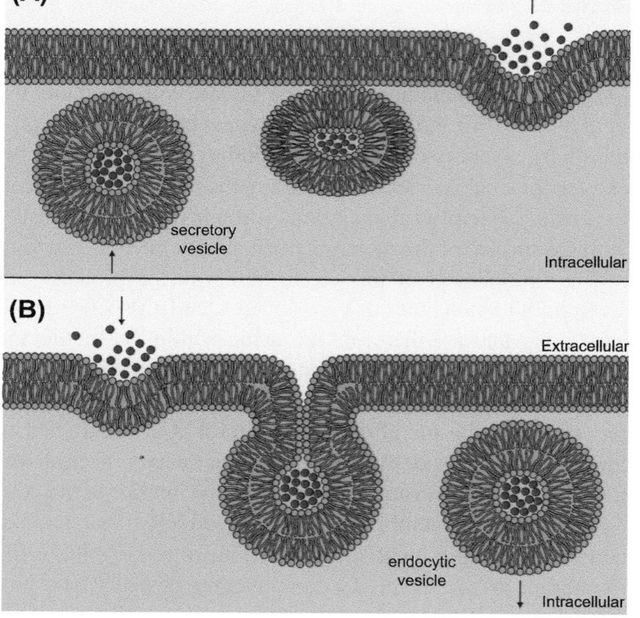

FIGURE 3.6 **Exocytosis and endocytosis.** Eukaryotic cells export and import larger biological molecules through exocytosis and endocytosis. In the process of exocytosis (A), proteins packaged into secretory vesicles by the Golgi complex travel to the plasma membrane, where the vesicle fuses with the plasma membrane. This releases the vesicle contents to the cell exterior. In endocytosis (B), material from the cell exterior is enclosed in a cavity formed by the plasma membrane, which pinches off to form an endocytic vesicle.

In endocytosis, material from the cell exterior is enclosed in a cavity formed by the plasma membrane, which pinches off to form an endocytic vesicle (Fig. 3.6B). There are two broad categories of endocytosis: bulk-phase endocytosis and receptor-mediated endocytosis. In bulk-phase endocytosis, the cell forms a vesicle that engulfs whatever molecules are present in the extracellular fluid, and so the process is nonspecific. On the other hand, receptor-mediated endocytosis is initiated when specific ligands bind to receptors that are present on the cell surface. The cell has receptors on its surface for many biological factors, including growth factors and hormones. The cell imports the ligands by forming a vesicle that includes the membrane area with the receptors (Fig. 3.7A). These endocytic vesicles form in a specific area of the membrane called **clathrin-coated pits**. Clathrin is a protein that forms a honeycomb-shaped lattice on the intracellular membrane of the endocytic vesicle (Fig. 3.7B). A similar functioning protein, caveolin, forms membrane pits known as **caveolae** (singular: caveola).

Once inside the cell, the endocytic vesicle soon loses its clathrin or caveolin coating and fuses with a membrane vesicle known as an **endosome**. The early endosome becomes increasing acidic to form a late endosome, which then fuses with an enzyme-packed lysosome to degrade the contents of the endosome (Fig. 3.1).

Phagocytosis is a form of receptor-mediated endocytosis that is used by specialized cells to engulf entire cells. Amoebae use phagocytosis to ingest their prey via phagocytosis. In defense against pathogens, several immune system cells are able to phagocytose whole bacteria and dead cells.

To replicate, viruses must gain entry into a cell. Many viruses enter the cell through receptor-mediated or bulk-phase endocytosis and have mechanisms to escape from endosomes before they fuse with lysosomes. A few viruses are also able to gain entry into the cell via phagocytosis. These viral processes will be explained in detail in Chapter 4, "Virus Replication."

3.3 THE CELL CYCLE

Viruses take advantage of the cell's transcription and/or translation machinery in the process of virus replication. After gaining entry into a cell, a virus will need to replicate its nucleic acid genome and manufacture viral proteins in order to assemble new infectious virions. Different types of viruses use different aspects of the host cell; the basic cell processes will be discussed here, and the specifics of each virus type will be discussed in the following chapter.

The human genome is composed of over 3 billion nucleotides of DNA, arranged in a double-stranded format where

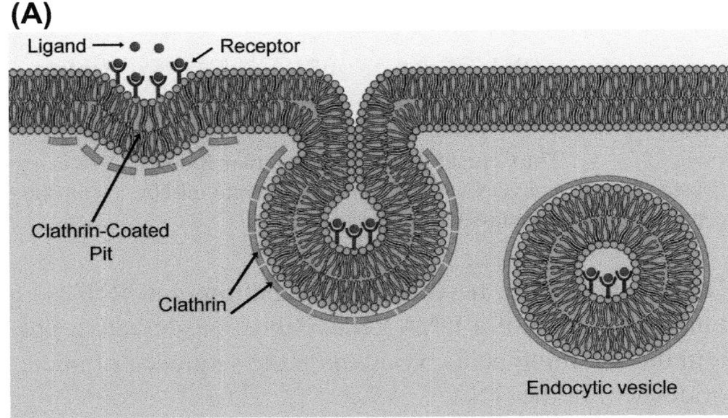

FIGURE 3.7 **Receptor-mediated endocytosis.** (A) In receptor-mediated endocytosis, receptors gather in clathrin-coated pits of plasma membrane. Appropriate ligand binding triggers the endocytosis of the receptors into clathrin-coated endocytic vesicles. The clathrin proteins are involved in the formation of the endocytic vesicle and break down soon after endocytosis. (B) Clathrin proteins form a polyhedral lattice around the vesicle. Illustrated here is a reconstruction of the structure using QuteMol (*Tarini* et al., *IEEE Trans. Vis. Comput. Graph 2006; 12: 1237–44*) to render data from PDB 1xi4 assembly 1, deposited from *Fotin et al., Nature 2004; 432(7017): 573–579.*

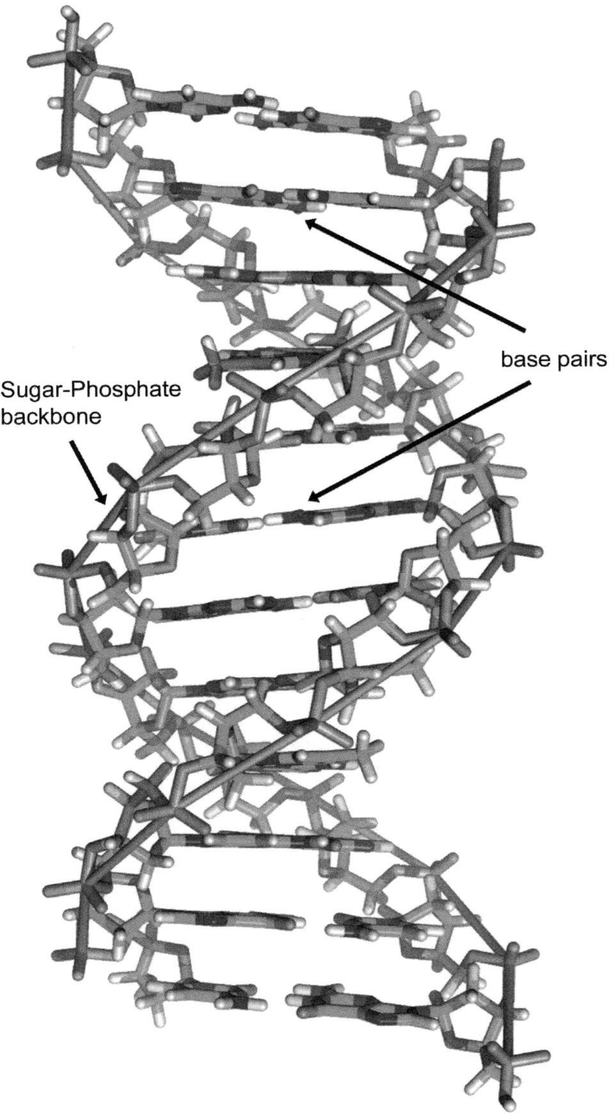

FIGURE 3.8 **The 3D structure of DNA.** DNA is a double-helix composed of nucleotides. The phosphate and sugar groups in the nucleotides form the backbone of the DNA strand (outlined in orange and green), while the bases of the nucleotides of one strand form base pairs with the bases on the other strand of DNA. *Image by Richard Wheeler.*

the phosphate and sugar portions of the nucleotides form the backbone of the strands and the nucleotide bases of one strand bind to the nucleotide bases of the other strand, forming a **base pair** (Fig. 3.8). Instead of having one long piece of nucleic acid, however, the DNA is broken up into pieces called **chromosomes**. Human cells are **diploid**, meaning that each cell has two copies of each chromosome, one passed along in the mother's egg and the other from the father's sperm (Fig. 3.9).

The first cell of a human being is the fertilized egg, or **zygote**, and all the cells that exist within an organism arise from the growth and division of previously existing cells. The **cell cycle** is the sequential stages through which a cell grows, replicates its DNA, and divides into two **daughter** cells. The cell cycle is divided into four phases (Fig. 3.10):

1. Gap 1, or G_1: Normal cellular growth occurs. Certain cells, such as neurons, will never continue the cell cycle and enter a stage known as Gap Zero (G_0). Cells that will divide continue to the next phase.
2. Synthesis, or S: The cell creates an additional copy of its chromosomes through **DNA replication.**
3. Gap 2, or G_2: The cell further enlarges and prepares for cell division.
4. Mitosis, or M: The two sets of chromosomes are separated as the one cell divides into two cells.

The cell cycle stage at which a virus infects a cell can be a crucial determinate of whether infection proceeds within the cell. Certain viruses require cells to be undergoing cell division because the viruses require the enzymes that are present during cell replication in order to replicate

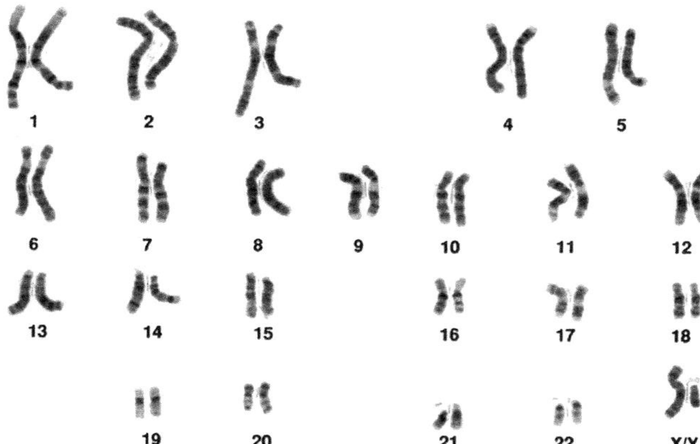

FIGURE 3.9 **A karyotype of the 46 human chromosomes.** Humans have two sets of 23 chromosomes, one inherited from each parent, for a total of 46 chromosomes of DNA. The 23rd set is known as the "sex chromosomes" and determine the sex of the individual. *Karyotype courtesy of the National Cancer Institute.*

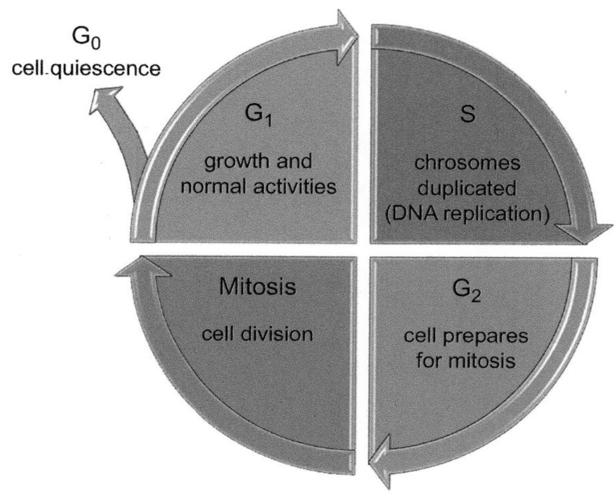

FIGURE 3.10 **The cell cycle.** The sequential stages through which a cell grows and divides are known as the cell cycle. It is divided into four main phases: Gap 1, Synthesis, Gap 2, and Mitosis. In G_1, cells undergo growth and normal activities. Cells that will divide enter S phase, where the 46 chromosomes are replicated. In G_2, the cell prepares for mitosis, which separates the replicated chromosomes and divides the cell into two cells.

their own genomes. A number of viruses also interfere with the stages of the cell cycle to increase the efficiency of virus replication.

3.4 THE CENTRAL DOGMA OF MOLECULAR BIOLOGY: DNA REPLICATION

DNA replication, which occurs during S phase of the cell cycle, is the first tenet of the **Central Dogma of Molecular Biology**: DNA is *replicated* in the nucleus to create a copy of the DNA, DNA is *transcribed* into messenger RNA in the nucleus, and messenger RNA is *translated* by ribosomes in the cytosol to create a protein (Fig. 3.11).

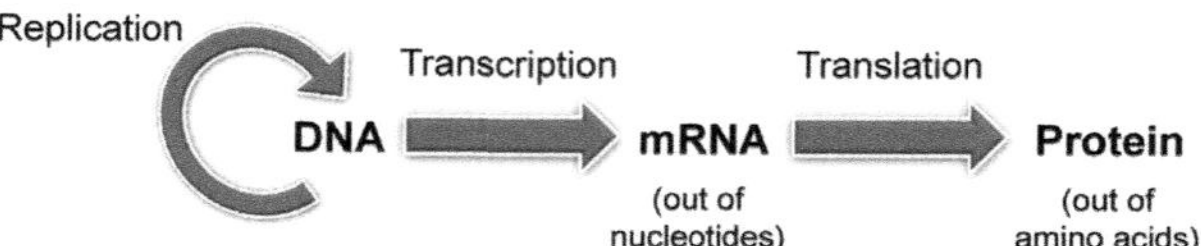

FIGURE 3.11 **The Central Dogma of Molecular Biology.** DNA is replicated to create more DNA; DNA is transcribed into mRNA, a temporary copy; and mRNA is translated into proteins.

DNA contains the hereditary information, and RNA is a temporary copy of a DNA gene. Ribosomes create a protein out of amino acids based upon the sequence of nucleotides within the RNA.

Consider the following analogy: you have a desktop computer at home in your bedroom with thousands of files on the hard drive. One of those files is a document that explains how to complete your final class project. The machine and supplies you need to complete your final project are located at school, however. Instead of taking your whole desktop computer with you, you copy the single file onto a USB drive and leave the house. Once you arrive at school, you read the instructions and use the machine and supplies to complete your final project. In this analogy, your hard drive is your DNA that contains thousands of genes, and the temporary copy that left the house (nucleus) is the mRNA. That temporary copy provided the instructions that were used to create your final class project (the protein) with the machine (ribosome) and its supplies (amino acids) found at school (in the cytoplasm).

Word Origin: The Central Dogma: Replication, Transcription, and Translation

The word *dogma* means a set of accepted principles. The Central Dogma of Molecular Biology is the main set of scientific principles that underlies the field of molecular biology, which deals with DNA, RNA, and proteins.

A *replicate* is an exact copy, and *DNA replication* is the process of copying DNA. To *transcribe* something is to rewrite it, and *transcription* is the process of creating a temporary RNA copy of an original DNA sequence. To *translate* something, on the other hand, is to change it from one language to another. Protein *translation* is the process of translating the language of RNA, made of nucleotides, into the language of proteins, made of amino acids.

The first part of the Central Dogma is DNA replication. The two strands of DNA are **antiparallel**, meaning that they are parallel to each other but going in opposite directions, much like the lanes of a two-way road. The directionality of the strand is determined by the position of the sugar deoxyribose in the nucleotide. The carbons within the nucleotide base are numbered, and the carbons within the sugar are also numbered but each number is followed by a prime symbol (similar to an apostrophe) to distinguish the sugar carbons from the carbon in the base (Fig. 3.12A). In a growing strand of nucleic acid, the phosphate group of the nucleotide attaches to the sugar at the 5′ (pronounced "five prime") carbon atom, and a new nucleotide is added to the 3′ (pronounced "three prime") carbon of the sugar. This "forward" direction is referred to as 5′→3′ ("five prime to three prime"). All replication of DNA occurs in this forward direction.

Since the two strands of a DNA molecule are antiparallel, if one strand is going forward (5′→3′) left-to-right, then the other strand is going forward (5′→3′) from right-to-left. As such, the 5′ end of one strand is matched with the 3′ end of the other strand (Fig. 3.12B).

DNA replication occurs during the S phase of the cell cycle, when the chromosomes are replicated. During the process of DNA replication, cellular enzymes unwind the DNA molecule and separate the two DNA strands from each other. DNA replication is **semiconservative**: each current strand of DNA functions as a template for a new strand, and so a copied piece of DNA will be composed of one old and one new strand (Fig. 3.13A). After the two strands of the DNA separate, **DNA polymerase** is the enzyme that lays down the complementary nucleotides of the new strand of DNA, *always in the 5′→3′ direction* (Fig. 3.13B). Note that since the two strands of DNA are antiparallel, the old strand is read 3′→5′ while the new strand grows 5′→3′. DNA polymerase adds new nucleotides based upon the complementary base pair rules discussed in Chapter 1, "The World of Viruses," and shown in Fig. 3.12A: adenine bonds with thymine, and cytosine bonds with guanine. Cellular DNA polymerases are *DNA-dependent DNA polymerases* because they synthesize DNA using a DNA template.

DNA polymerases have **high fidelity**, meaning that they do not often place an incorrect base in the growing strand of replicating DNA. They also have proofreading ability: in the same way you may type an incorrect letter on a keyboard and hit the "Backspace" key to replace it with the correct letter, DNA polymerase can reverse and replace an incorrectly placed nucleotide. DNA polymerase and repair enzymes can also cut out a section around an incorrect nucleotide and replace the section of DNA with the correct nucleotides. Taken together, DNA polymerase makes one mistake for every 1 million nucleotides copied, on average.

Several other proteins and enzymes are involved in DNA replication. For instance, DNA polymerase cannot bind to a single-stranded portion of DNA, so when the two existing DNA strands are separated, an enzyme known as **primase** lays down a short complementary RNA fragment onto the DNA strand, creating a double-stranded portion to which DNA polymerase can bind. Other enzymes are also required for the process of replication: since DNA polymerase can only add to a nucleotide chain in the 5′→3′ direction, it can only create short fragments of DNA on one strand of the replicating DNA (known as the lagging strand), until the double-stranded DNA opens farther down the strand. The enzyme

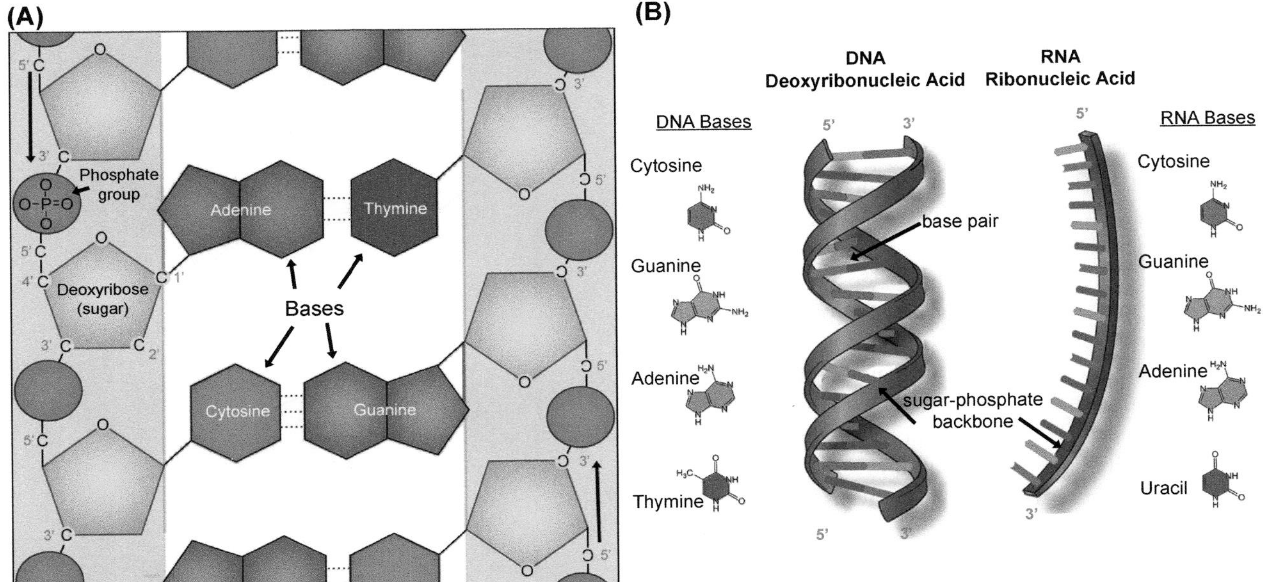

FIGURE 3.12 **The structure and organization of DNA.** (A) DNA is composed of nucleotides, which each have a phosphate group, a sugar, and a base. The "forward" direction of DNA is from 5′ to 3′, referring to the numbered carbons within the nucleotide sugars. The two DNA strands are antiparallel and are thus going in opposite directions. (B) In a molecule of DNA, the bases of one DNA strand form base pairs with the bases on the other strand of DNA. The four bases found in DNA are cytosine, guanine, adenine, and thymine; in RNA, which is single-stranded, uracil substitutes for thymine. *Part B modified from an image by Darryl Leja, National Human Genome Research Institute.*

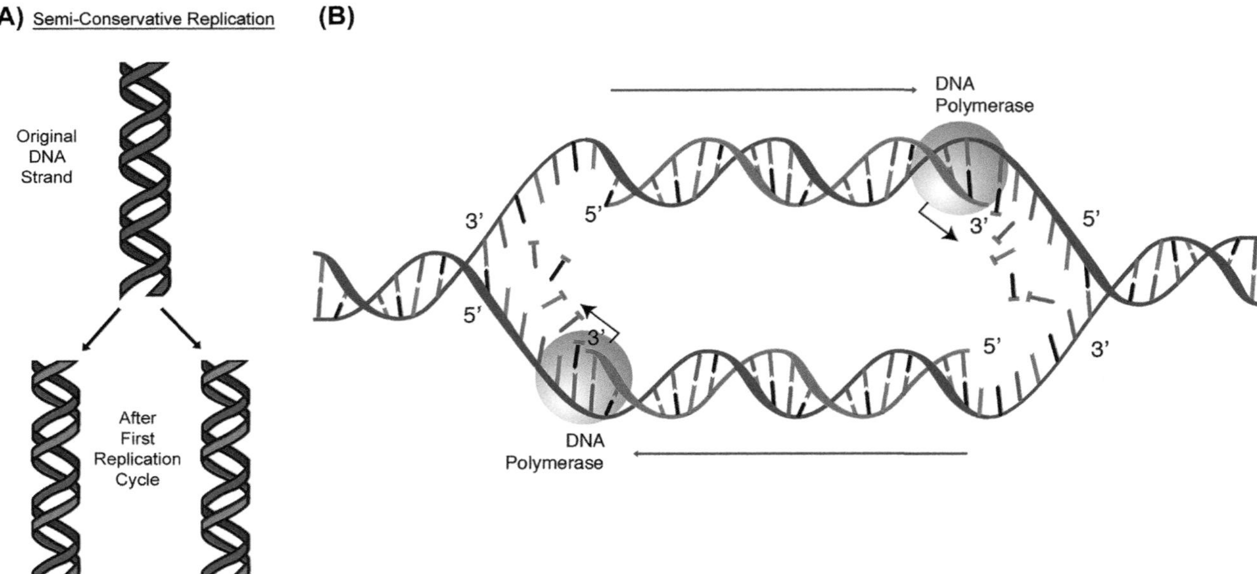

FIGURE 3.13 **DNA replication.** (A) DNA replication is semiconservative, meaning that each old strand (in blue) is used as the template to create a new strand (in red). The replicated DNA is therefore composed of one old and one new strand. (B) In the process of DNA replication, DNA polymerase moves along the existing strand of DNA and lays down the nucleotides of the new strand in the 5′ to 3′ direction. The old strand is therefore read from 3′ to 5′. Both DNA strands are copied simultaneously. *Illustration in (B) by Darryl Leja, National Human Genome Research Institute.*

ligase joins together these short fragments (known as **Okazaki fragments**) to create a contiguous DNA strand.

Several DNA viruses take advantage of cellular DNA polymerase and replication enzymes to replicate their genomes. Because DNA replication takes place within the nucleus, these viruses must gain entry into the nucleus to replicate their genomic DNA.

3.5 THE CENTRAL DOGMA OF MOLECULAR BIOLOGY: RNA TRANSCRIPTION AND PROCESSING

Sections of DNA called **genes** encode the information needed to create proteins. There are over 20,000 protein-encoding genes within the 46 chromosomes that constitute the human genome. There are three steps in the process of generating a protein from the information stored within DNA: transcription, RNA processing, and translation.

DNA replication occurs in the nucleus because DNA is located in the nucleus, and **transcription**, the process of creating a temporary RNA copy of the DNA, also occurs in the nucleus for the same reason (Fig. 3.14). A complex of **transcription factor** proteins binds the DNA immediately upstream of the gene start site at a location called a **promoter**. **RNA polymerase II** then associates with the transcription factors and the DNA (Fig. 3.15A). Transcription factors bind to specific sequences of DNA within the promoter region, ensuring that transcription of the DNA begins at the correct location. Other transcription factors can bind to **enhancer** regions that, as their

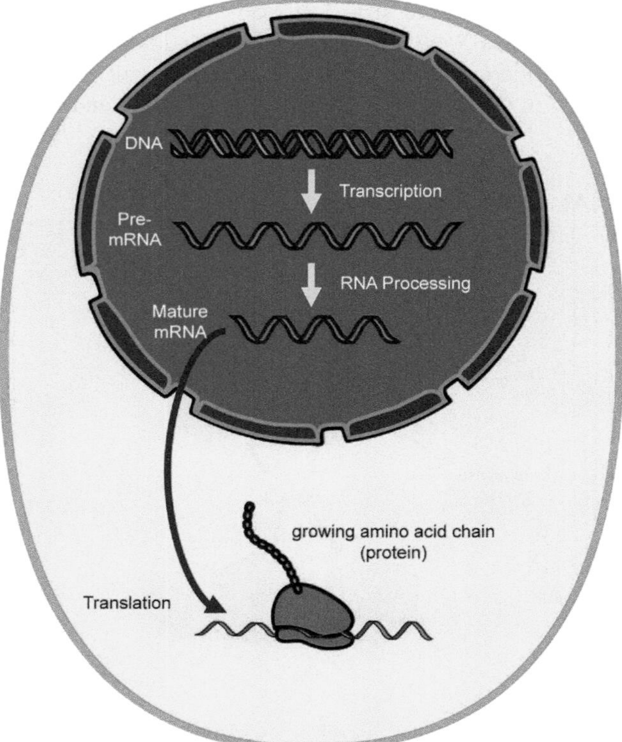

FIGURE 3.14 **The cellular location of transcription, RNA processing, and translation.** DNA is located in the nucleus, and therefore transcription occurs in the nucleus. The precursor mRNA also undergoes RNA processing in the nucleus before departing the nucleus through a nuclear pore. In the cytosol, a ribosome binds to the mRNA and translates it into a protein.

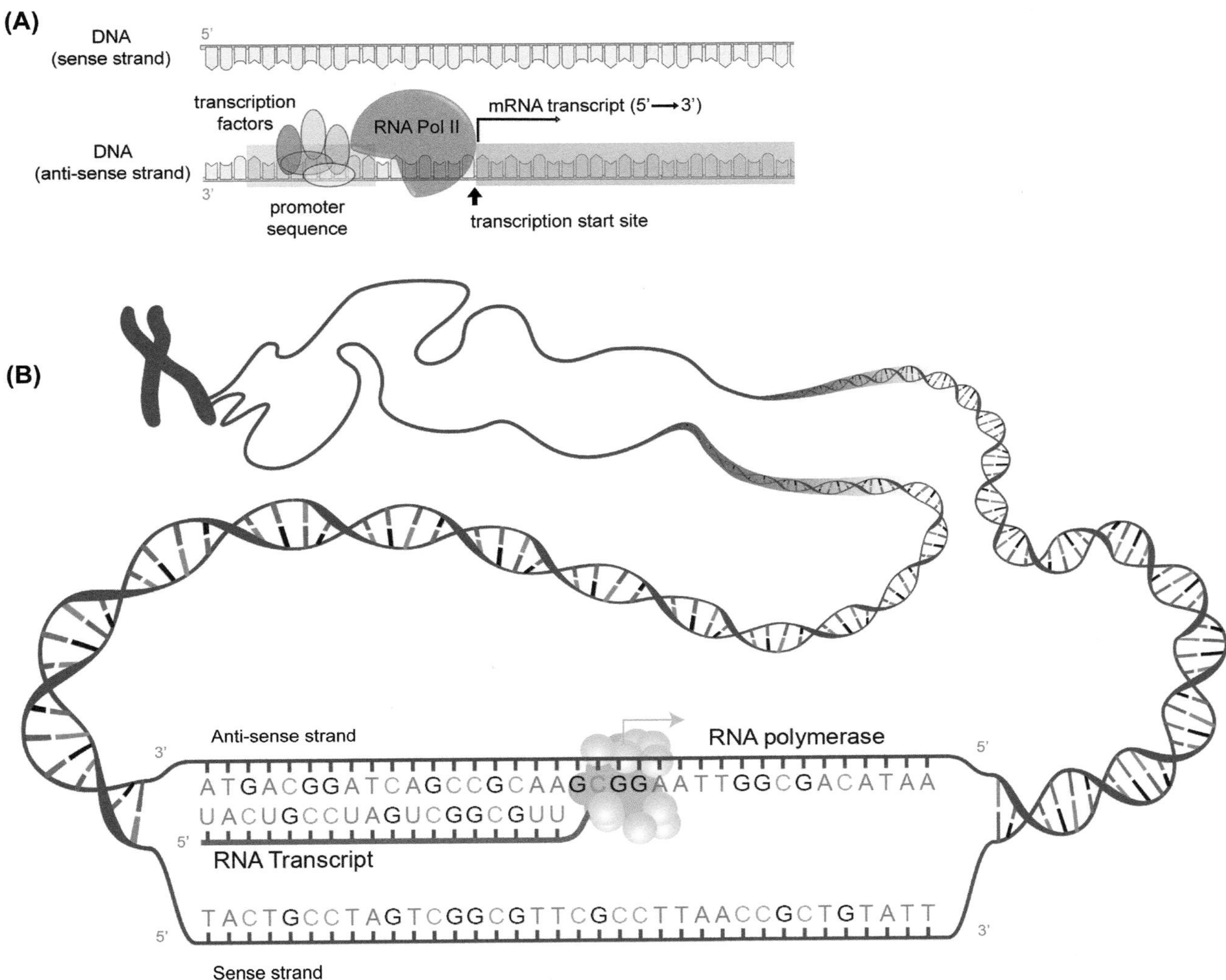

FIGURE 3.15 **Transcription.** Transcription is the process of copying a segment of DNA into mRNA. Only one of the two DNA strands acts as the template for transcription (known as the antisense or template strand). Transcription factors bind to the promoter sequence upstream of the transcription start site and recruit RNA Polymerase II to the complex (A). RNA Polymerase II reads the antisense strand of DNA in the 3′ to 5′ direction to create a single-stranded RNA transcript in the 5′ to 3′ direction using complementary base pairing rules (B). *Illustration in (B) by Darryl Leja, National Human Genome Research Institute.*

name suggests, can increase the rate of transcription. Unlike the promoter, enhancer regions can be thousands of nucleotides away, either upstream or downstream from the gene start site.

Cellular RNA polymerases are *DNA-dependent RNA polymerases* because they synthesize RNA based on a DNA template. Much in the same way that DNA polymerase uses a strand of DNA to create the complimentary strand, RNA polymerase uses the DNA template to create a strand of RNA, adding nucleotides in the 5′→3′ direction using the same complementary base pair rules as DNA replication, except that the base uracil substitutes for thymine (Figs. 3.12B and 3.15B). Because only one of the two DNA strands, known as the **template strand** or **antisense strand**, acts as the template for RNA polymerase, the resulting RNA is single-stranded (Figs. 3.12B and 3.15B). RNA polymerase terminates transcription when it reaches a consensus sequence at the end of the gene. At this point, the RNA **transcript** is known as precursor messenger RNA (mRNA). It is termed "messenger RNA" because it is the message, encoded within the DNA, of how to create a specific protein.

RNA polymerases do not have as high fidelity as DNA polymerases and place an incorrect base on average once per 100,000 nucleotides transcribed, 10 times more often than DNA polymerase. These RNA polymerases are *DNA-dependent RNA polymerases.* Eukaryotic cells do not contain *RNA-dependent* RNA polymerases for the creation of mRNA, and so several types of RNA viruses encode their own RNA polymerases, with error rates of 1 in 100 to 1 in

100,000 nucleotides. A high mutation rate is the result of the low fidelity of several RNA viruses that encode their own RNA polymerase.

Following transcription, the precursor mRNA undergoes **RNA processing**, also known as *posttranscriptional modification*, to convert the precursor mRNA into mature mRNA. The first modification, which occurs while RNA polymerase is still transcribing the mRNA transcript, is the addition of a "cap" to the 5′-end of the transcript (Fig. 3.16). The **5′-cap** consists of a methylated guanine nucleotide (known as 7-methylguanosine, m^7G) that protects the 5′-end of the RNA transcript. Ribosomes will also bind to the 5′-cap to begin translation. The second modification is the addition of a **3′ poly(A) tail**. The "tail" consists of 50–250 adenine nucleotides added to the 3′-end of the mRNA to protect the mRNA transcript. The final modification is the removal of **introns** by a process known as **RNA splicing**. Within most eukaryotic mRNA transcripts, there are sequences of mRNA that will not be translated into proteins. These sequences, known as **introns**, are removed during posttranscriptional modification, leaving behind the **exons** or coding sequences (Fig. 3.16). More than one protein can be created from a single mRNA through RNA splicing because different introns can be removed from an mRNA transcript, resulting in different RNA sequences and subsequently, different proteins. This process is known as **alternative splicing**. The mRNAs of some viruses, including HIV, also undergo alternative RNA splicing.

Study Break

Pertaining to DNA and RNA architecture, explain what "five prime" and "three prime" mean and what these phrases have to do with DNA replication, transcription, and RNA processing.

3.6 THE GENETIC CODE

Now processed, the mature mRNA transcript leaves the nucleus and is delivered to the ribosome, which is located in the cytosol. The ribosome acts as a protein factory, and the mature mRNA functions as the instructions for manufacturing. Proteins are made of amino acids, and most human proteins are 50–1000 amino acids in size. There are 20 different amino acids, and the sequence of mRNA determines the order in which the ribosome will assemble the amino acids into a protein.

The ribosome initially moves down the transcript one base at a time, reading the sequence in three-base words known as **codons** (Fig. 3.17). The ribosome starts **translation**, the assembly of a protein out of amino acids, when it encounters the **start** codon in the mRNA, which is the sequence *AUG*. The AUG codon is usually within the context of a slightly larger sequence, called the **Kozak consensus sequence**, which generally has the sequence GCC<u>A</u>CC**AUG**G (the underlined adenine can also be a guanine). AUG codes for the amino acid methionine, and so all protein translation begins with methionine.

The start codon sets the **reading frame**: instead of continuing to move down the mRNA transcript one base at a

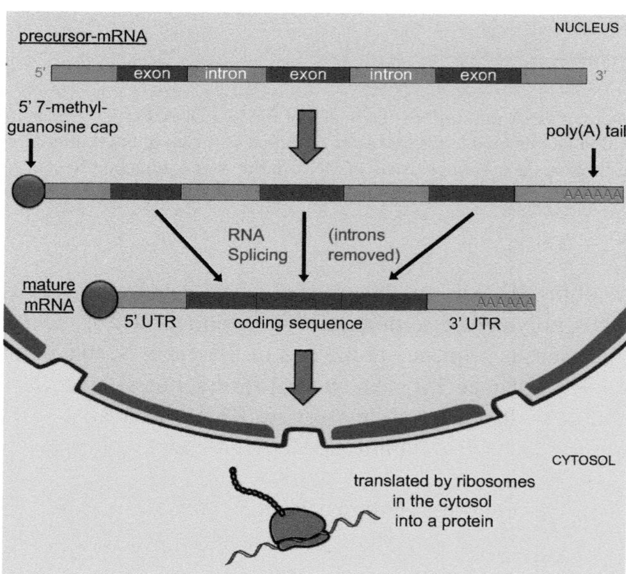

FIGURE 3.16 **RNA processing.** The precursor mRNA, produced by transcription, undergoes RNA processing while still within the nucleus. A 7-methylguanosine cap is added to the 5′ end of the transcript, while 50–250 adenine nucleotides are added to the 3′ end of the transcript. Introns are removed to produce a mature mRNA molecule that leaves the nucleus and is translated by ribosomes in the cytosol. The removal of different introns results in an mRNA transcript that is translated into a different protein.

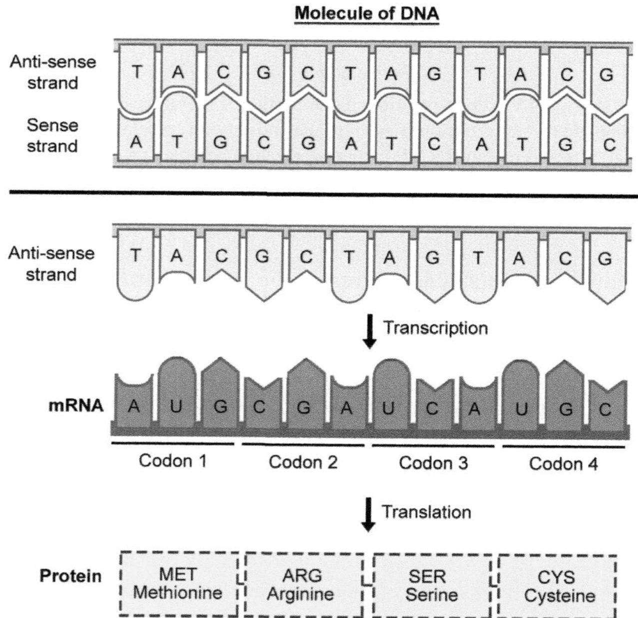

FIGURE 3.17 **The flow of data, from DNA to protein.** The antisense or template strand of DNA acts as a template to transcribe mRNA. The ribosome reads the mRNA in three nucleotide codons, beginning with the start codon, AUG, which codes for the amino acid methionine. The order of the bases within the codons determines which amino acid will be added to the growing protein by the ribosome.

(A) 5'- UACCAUGGCACUGGUCGAUCAUAAAGGGGGUCGAUGACC -3'

1) 5'- UAC CAU GGC ACU GGU CGA UCA UAA AGG GGG UCG **AUG** ACC -3'
2) 5'- U ACC **AUG** GCA CUG GUC GAU CAU AAA GGG GGU CGA **UGA** CC -3'
3) 5'- UA CCA UGG CAC UGG UCG AUC AUA AAG GGG GUC GAU GAC C -3'

(B) Human S100 calcium binding protein A1 (S100A1) mRNA

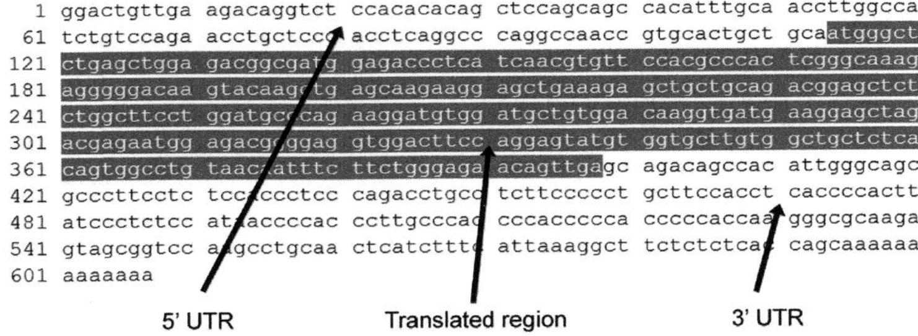

FIGURE 3.18 **mRNA organization and reading frames.** The ribosome reads the mRNA in three nucleotide chunks known as codons. Within a piece of mRNA, however, there are three possible reading frames, shown in (A). Because the ribosome starts at the 5′ end of the mRNA transcript, it will first encounter the start codon highlighted in reading frame two and translate in this reading frame, thereby missing other start codons because they will be out of frame. (B) The mRNA sequence of a small human protein, called human S100 binding protein A1. (Note that the database uses "t" instead of "u" for simplicity, but remember that uracil replace thymine in RNA.) The translated region is highlighted in brown; observe that the sequence starts with AUG (atg) and ends with UGA (tga). Not all of the mRNA is translated, leaving 5′ and 3′ UTRs. Also note the poly(A) tail at the end of the transcript. *From NCBI GenBank Reference Sequence NM_006,271.1.*

time, the ribosome now reads the mRNA codons consecutively, three bases at a time (Fig. 3.18). The sequence of the triplet codon determines which amino acid is added next to the growing protein. When the ribosome reaches a **stop codon**, it falls off the mRNA, and the protein is complete. There are three variations of the stop codon: UGA, UAA, and UAG. The segment of mRNA before this starting point is not translated and is known as the **5′ untranslated region** (5′ UTR) (Fig. 3.18B). Any mRNA past the stop codon will not be translated; this region is known as the **3′ UTR**. The sequence from the start codon to the stop codon is known as an **open reading frame** because it is translatable.

Genetic Code Analogy

After binding to the mRNA, the ribosome begins translation at the start codon, AUG, and then moves down the mRNA transcript one codon (three nucleotides) at a time until it reaches a stop codon. Try finding the translated codons in the following sentence. The start codon—THE—will set the reading frame. The three stop codon possibilities are OKK, OOK, and OKO.

5′- GPTHOAEGUTHEDOGANDFATCATATETHERED HAMOKONZOMIOLVGN -3′

The letters before and after the sentence found in the letters above are not part of the sentence, in the same way that the nucleotides before and after the translated region do not encode any of the amino acid sequence. These are the 5′ and 3′ UTRs.

Based on the four nucleotides in RNA—adenine, guanine, cytosine, and uracil—there are 64 possible different 3-letter permutations (Fig. 3.19). There are only 20 amino

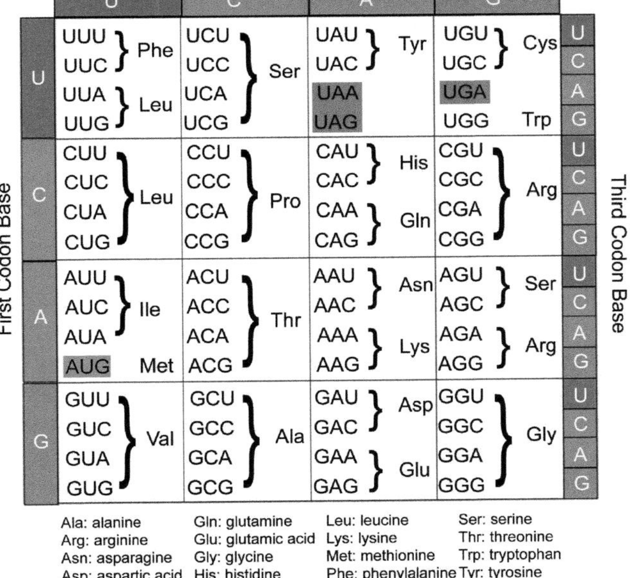

First Codon Base	Second: U		Second: C		Second: A		Second: G		Third Codon Base
U	UUU	Phe	UCU	Ser	UAU	Tyr	UGU	Cys	U
U	UUC	Phe	UCC	Ser	UAC	Tyr	UGC	Cys	C
U	UUA	Leu	UCA	Ser	UAA		UGA		A
U	UUG	Leu	UCG	Ser	UAG		UGG	Trp	G
C	CUU	Leu	CCU	Pro	CAU	His	CGU	Arg	U
C	CUC	Leu	CCC	Pro	CAC	His	CGC	Arg	C
C	CUA	Leu	CCA	Pro	CAA	Gln	CGA	Arg	A
C	CUG	Leu	CCG	Pro	CAG	Gln	CGG	Arg	G
A	AUU	Ile	ACU	Thr	AAU	Asn	AGU	Ser	U
A	AUC	Ile	ACC	Thr	AAC	Asn	AGC	Ser	C
A	AUA	Ile	ACA	Thr	AAA	Lys	AGA	Arg	A
A	AUG	Met	ACG	Thr	AAG	Lys	AGG	Arg	G
G	GUU	Val	GCU	Ala	GAU	Asp	GGU	Gly	U
G	GUC	Val	GCC	Ala	GAC	Asp	GGC	Gly	C
G	GUA	Val	GCA	Ala	GAA	Glu	GGA	Gly	A
G	GUG	Val	GCG	Ala	GAG	Glu	GGG	Gly	G

Ala: alanine, Arg: arginine, Asn: asparagine, Asp: aspartic acid, Cys: cysteine, Gln: glutamine, Glu: glutamic acid, Gly: glycine, His: histidine, Ile: isoleucine, Leu: leucine, Lys: lysine, Met: methionine, Phe: phenylalanine, Pro: proline, Ser: serine, Thr: threonine, Trp: tryptophan, Tyr: tyrosine, Val: valine

FIGURE 3.19 **The genetic code.** The sequence of amino acids within a protein is determined by the nucleotide sequence of the mRNA. To use the table, find the first base of the codon in the leftmost column. Next, find the second base of the codon on the top row. The intersection of the column and row will be the target box in which the codon is located. Next, find the third base of the codon in the rightmost column to identify on which line of the target box the codon is located. Next to the codon sequence in the target box is the amino acid that corresponds to the codon. The list of amino acid abbreviations is located below the table. AUG, as the start codon, is in green and codes for methionine. The three stop codons are UAA, UAG, and UGA. Stop codons encode a release factor, rather than an amino acid, that causes translation to cease.

acids, however, and so some of the codons are **redundant**, meaning that two or more codons encode the same amino acid. There are three stop codons, which end translation and do not encode any amino acid.

Many scientists worked to decipher the **genetic code**. Robert W. Holley, Har Gobind Khorana, and Marshall W. Nirenberg shared a Nobel Prize in physiology or medicine in 1968 for their work in determining the "key" to deciphering the genetic code. The Table in Fig. 3.19 reveals which amino acids are encoded by each codon. The code is *universal*: all living things have the same 20 amino acids that are encoded by these codons, indicating that this system originated very early in the development of life and has been evolutionarily conserved over time. Being that viruses take advantage of the host translational machinery and ribosomes, viral mRNAs use these same codons to encode the same amino acids in their proteins as do living things.

Study Break
Translate the two mRNA sequences found in Fig. 3.18 A and B.

3.7 THE CENTRAL DOGMA OF MOLECULAR BIOLOGY: PROTEIN TRANSLATION

Three major components are required for translation to occur: mRNA, the ribosome, and **transfer RNAs** (tRNAs). The ribosome is an organelle with two subunits—a small and large subunit (Fig. 3.20A)—that are made of another type of RNA, termed ribosomal RNA (rRNA), and over 50 proteins. Transfer RNAs are also made of RNA (Fig. 3.20B) and act as adaptors between the mRNA and the ribosome. It is the tRNA that brings amino acids to the ribosome so they may be joined together into a protein. Within a tRNA is an **anticodon** sequence that is complementary to the mRNA codon, and at the 3′ end of the tRNA is attached an amino acid. The mRNA codon sequence binds to the anticodon sequence within the tRNA, which has attached a specific amino acid (Fig. 3.20A). It is this adaptor molecule that actually determines which codon encodes which amino acid.

Refresher: RNA
mRNA: messenger RNA. The temporary copy of DNA that will be translated by the ribosome.

rRNA: ribosomal RNA. The ribosome is made of rRNA and proteins.

tRNA: transfer RNA. Acting as an adaptor, it transports the amino acid to the ribosome and has an anticodon region that binds the mRNA codon.

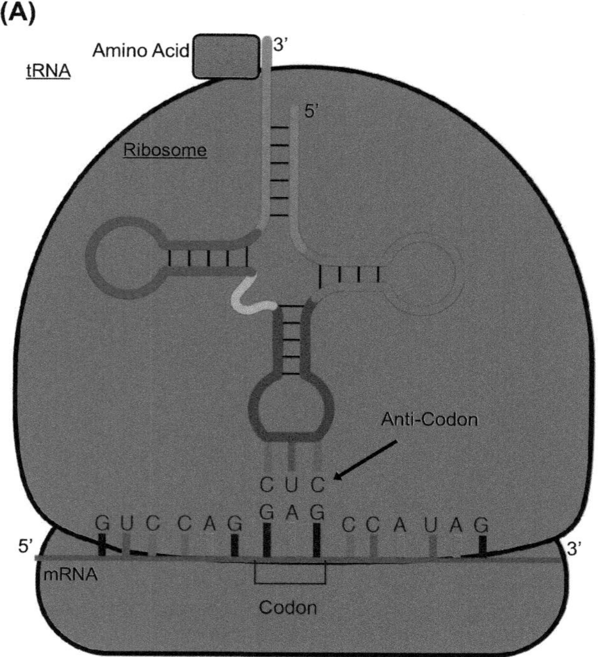

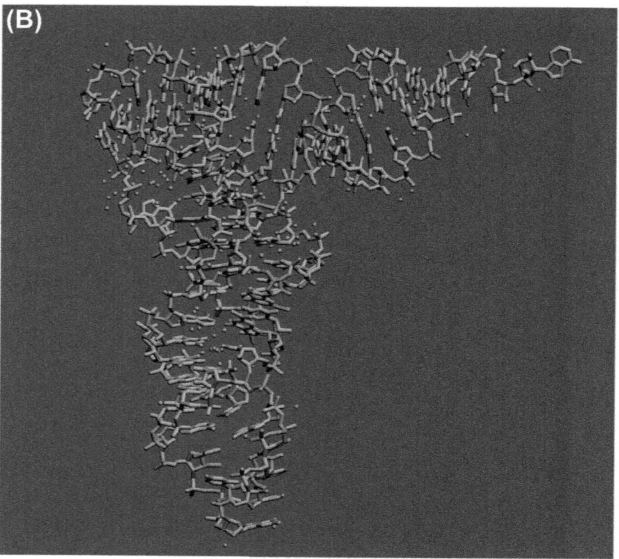

FIGURE 3.20 **The ribosome and tRNA.** (A) The ribosome is composed of a small and large subunit. The mRNA transcript fits between the two subunits and contains three nucleotide codons. Transfer RNAs (tRNAs) enter the ribosome and interact with the mRNA using an anticodon sequence found within the tRNA. The corresponding amino acid is attached to the 3′-end of the tRNA. (B) The 3D structure of a tRNA. *Image in Part B created using PDB 1EHZ (deposited by Shi et al., RNA 2000; 6: 1091–1105) using QuteMol (Tarini et al., IEEE Trans Vis Comput Graph 2006; 12: 1237–44).*

There are three stages of translation (Fig. 3.21):

1. **Initiation**: the start of translation. The ribosome small unit, containing the tRNA holding methionine, binds at the 5′-cap of the processed mRNA molecule. It scans the mRNA until the start codon, AUG, is encountered.

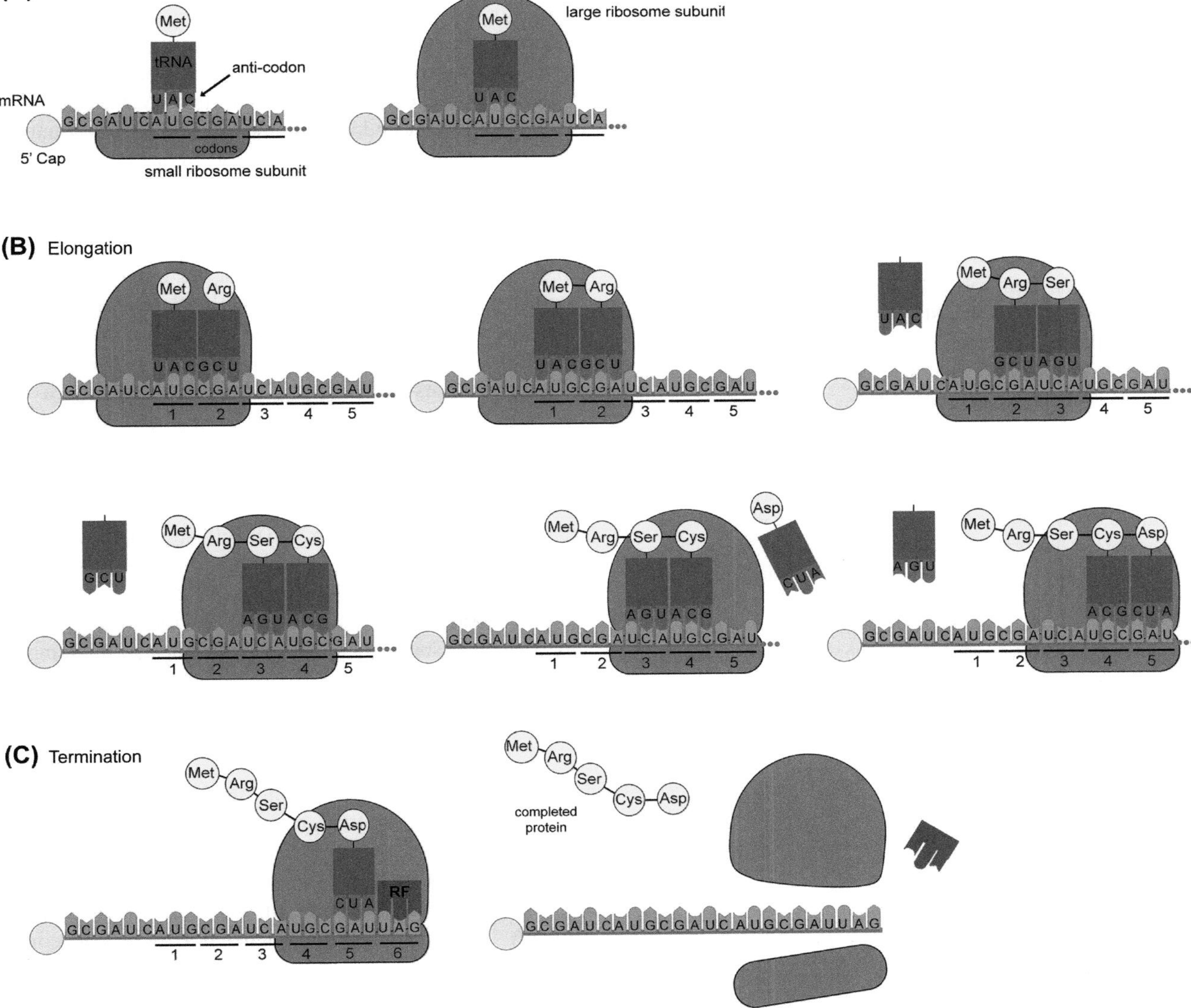

FIGURE 3.21 Translation. Translation, the assembly of a protein out of amino acids by the ribosome, is divided into three parts. In *Initiation* (A), the ribosome attaches to the 5′-cap of the mRNA transcript and scans until the start codon, AUG, is encountered. The corresponding tRNA (containing the anticodon UAC and carrying the amino acid methionine) and large ribosomal subunit join the complex. During *Elongation* (B), the ribosome begins by moving one codon down the mRNA from the start codon. The corresponding tRNA delivers the subsequent amino acid, which is joined to methionine, the first amino acid of the protein. The ribosome then moves to the next codon, and the next corresponding tRNA delivers the subsequent amino acid, which is joined to the first two amino acids. The empty tRNA once carrying methionine is released from the ribosome and will be recharged by other enzymes within the cell. The ribosome continues moving down the mRNA one codon at a time, and an amino acid is added to the growing chain for each new codon. (C). The final stage of translation is *Termination*. When the ribosome encounters one of the three stop codons (UGA, UAA, or UAG), a release factor enters the ribosome. The protein is released and the ribosome leaves the mRNA.

In the same way that transcription factors were necessary for RNA polymerase II to bind to the DNA gene to be transcribed, an assortment of translation **initiation factors** assists in recruiting the ribosome and the first tRNA to the mRNA transcript. The large ribosome subunit joints the small subunit.

2. **Elongation**: the synthesis of the protein out of amino acids. The ribosome moves along the mRNA strand. For each codon, a tRNA with a complementary anticodon enters the ribosome, delivering the corresponding amino acid. The ribosome joins the growing amino acid strand to the new amino acid. It continues moving along the mRNA, one codon at a time, and a tRNA containing the corresponding amino acid enters the ribosome for each codon. The new amino acid is joined to the previous amino acids, elongating the amino acid chain.

3. **Termination**: the end of translation. When a stop codon in the mRNA is encountered by the ribosome, a release factor enters the ribosome and translation ceases. The now completed protein is released, and the ribosome falls off the mRNA.

As described above, many of these proteins undergo **posttranslational modification** in the rER and Golgi complex to add lipid or sugar residues to the molecule.

Eukaryotic mRNA is **monocistronic**, meaning that any mRNA transcript codes for only one protein. *All* viruses are dependent upon their host cells for the translation of their proteins, so viral mRNAs must conform to the biological constraints of the host cell machinery. Viruses have several tactics, however, to ensure the preferential transcription and translation of their viral mRNAs and proteins over those of the host.

In-Depth Look: Peptides, Polypeptides, and Proteins

A peptide is a short chain of amino acids. A long peptide chain is known as a polypeptide. When a ribosome translates mRNA, it creates a polypeptide of amino acids.

So what is the difference between a polypeptide and a protein? A protein is a complete, functional entity. Some proteins are made of only one polypeptide chain, but other proteins are made of more than one polypeptide chain. Hemoglobin, the protein that transports oxygen in our red blood cells, is a protein composed of four polypeptide chains in total. While discussing protein translation in this chapter, we assume that the polypeptide created by the ribosome will be a functional protein, but keep in mind that many proteins are composed of more than one polypeptide chain.

3.8 PROMOTION OF VIRAL TRANSCRIPTION AND TRANSLATION PROCESSES

Viruses have evolved several mechanisms to ensure the successful transcription and translation of their gene products, necessary to create more infectious virions. Some viruses take advantage of the host splicing machinery to produce several mRNA transcripts from one precursor mRNA. HIV-1, for instance, produces most of its mRNAs from alternative splicing. Some viruses, like influenza, can snatch the 5′-caps from host mRNAs to gain the necessary cap for the viral mRNA, leaving the host mRNA untranslatable without a 5′-cap. Some viruses also create mRNAs that are translated into one long **polyprotein** that is then cleaved into several viral proteins after translation.

Other viruses have evolved tactics for protein translation that are not customarily used by the host. Ribosomes recognize and bind to the 5′-cap of mRNAs, but some viral mRNAs contain **internal ribosome entry sites** (IRES) that allow ribosomes to bind within the mRNA sequence, without a 5′-cap (Fig. 3.22A and B). IRES are used to initiate translation in internal sections of viral mRNA that are in a different reading frame, or when the virus has interfered with the normal translation process to inhibit host protein synthesis. **Ribosomal frameshifting** occurs when a ribosome pauses or meets a "slippery sequence" within a piece of mRNA. Instead of continuing to read the codons in frame, the ribosome encounters a problem and moves backward or forward one nucleotide. The result is that the ribosome begins translating a different reading frame than it was previously, producing a different mRNA and protein. Viruses like HIV utilize ribosomal frameshifting to encode several proteins within just one portion of DNA. A similar mechanism occurs with **termination suppression**, in which a stop codon is suppressed and the ribosome continues translating the mRNA, creating a polyprotein. **Ribosomal skipping** occurs while the ribosome is translating a viral mRNA. Viral proteins prevent the ribosome from joining a new amino acid to the growing protein chain, which releases the protein from the ribosome. Having not encountered a stop codon, however, the ribosome continues translating the remainder of the viral mRNA. The effect is that several viral proteins can be synthesized with only one piece of mRNA. **Leaky scanning** happens when a host ribosome encounters a start codon (AUG) within a piece of viral mRNA, but the Kozak consensus sequence (the nucleotides surrounding the start codon) is not in a favourable configuration. The ribosome may begin translation at this site, but the next ribosome that binds to the viral mRNA may continue past this weak start codon to begin translation at the next AUG encountered (Fig. 3.22C). The result is that one viral mRNA can encode two viral proteins. All of these transcription and translation processes promote the synthesis of viral proteins, often faster and with less energy than normal cellular mechanisms.

In addition to evolving mechanisms to ensure their mRNA is processed and recognized by host ribosomes, viruses have also evolved ways to interfere with the transcription and translation of host proteins:

1. Some viruses can interfere with the host's RNA polymerase II. They can do so by interfering with transcription factors, by preventing the activation of the enzyme, or by causing the breakdown of the enzyme. Several herpesviruses have strategies to interfere with RNA polymerase II.
2. Viruses can interfere with the processing of precursor RNA. HIV-1 protein Vpr inhibits the splicing of host mRNA, and influenza inhibits the addition of a poly(A) tail to the host's mRNA transcripts.
3. Many viruses interfere with the export of the mRNA transcript from the nucleus. They do so by interfering with or causing the breakdown of the proteins that export the processed mRNA from the nucleus.
4. Certain viral proteins can cause the degradation of host mRNA. For instance, severe acute respiratory syndrome (SARS) is caused by a coronavirus named SARS-CoV,

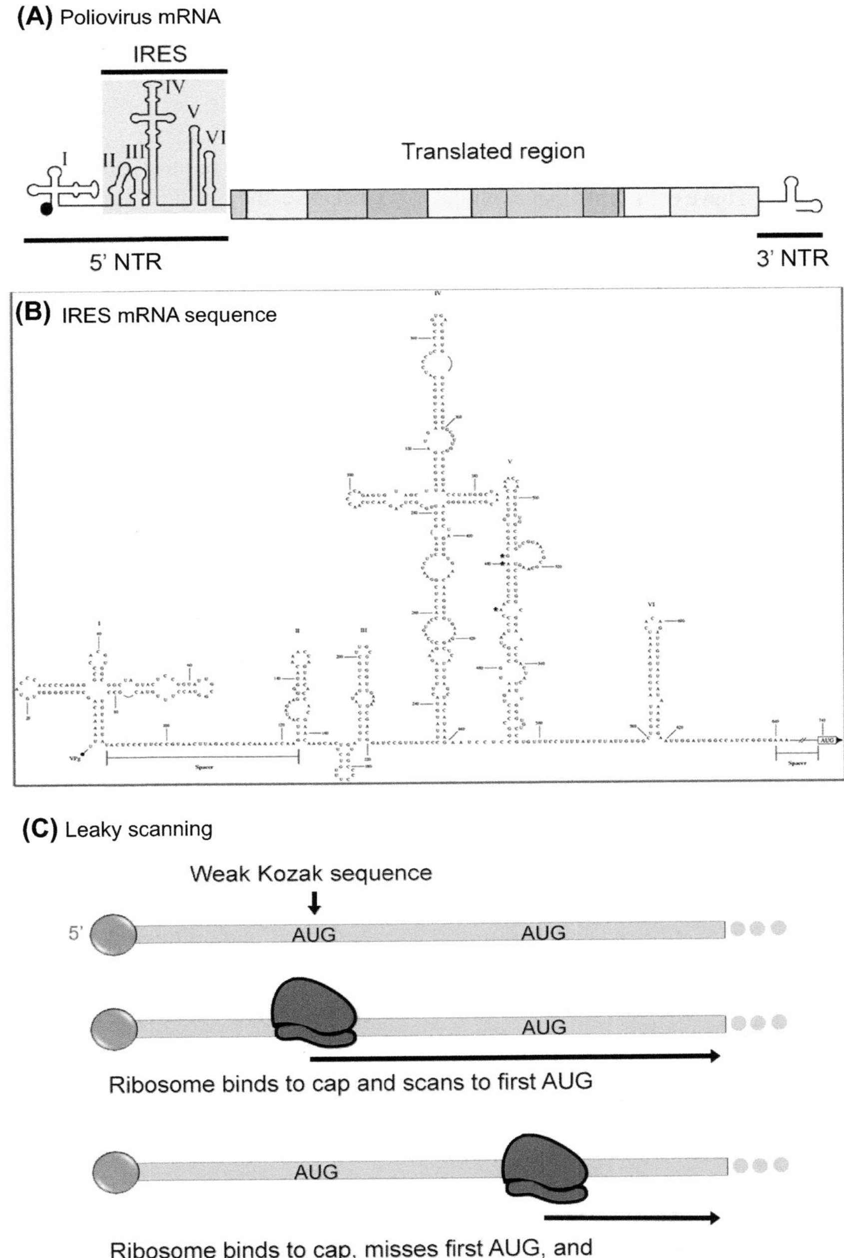

FIGURE 3.22 **Viral translation strategies.** Viruses employ several strategies to ensure the translation of their proteins. (A) Some viruses, such as poliovirus, have internal ribosome entry sites (IRES) within their mRNA that allows ribosomes to begin translation within the mRNA sequence, without a 5′-cap. (B) shows how the poliovirus mRNA forms the IRES. *(A) and (B) reprinted with permission from De Jesus* et al., *Virology J. 2007; 4: 70.* (C) In leaky scanning, the ribosome begins scanning at the 5′-end of the mRNA and encounters a start codon (AUG) within a piece of viral mRNA, but the Kozak consensus sequence is not in a favorable configuration. The ribosome may begin translation at this site, but the next ribosome that binds to the viral mRNA may continue past this weak start codon to begin translation at the next AUG encountered. The result is that one viral mRNA can encode two viral proteins.

which has a protein named nsp1 that induces the breakdown of host mRNAs. Interestingly, the SARS-CoV mRNA transcripts are protected from degradation.

5. Several viral proteins prevent translation of host mRNA. This can happen by interfering with host translation initiation factors or by removing the 5′-caps from host mRNAs.

The host replication, transcription, and translation machinery is complex and involves a multitude of enzymes and molecules. Viruses must conform to the limitations of the host cell in order to replicate, but they have evolved many strategies to ensure the preferential transcription of their mRNA transcripts and efficient translation of viral proteins.

SUMMARY OF KEY CONCEPTS

Section 3.1 The Basic Organization of the Cell

- Each cell of a eukaryote has a central nucleus that separates its DNA from the rest of the cell. A prokaryote does not have a nucleus. All organisms within the domain *Eukarya*, including humans, have cells with nuclei, while those within *Bacteria* and *Archaea* are prokaryotes.
- Many organelles ensure the efficient functioning of the cell. Ribosomes make proteins and can be free within the cytosol or attached to the rER, which folds and modifies proteins into glycoproteins or lipoproteins after they have been made by the ribosome. These are shipped in vesicles to the Golgi complex, which packages them in vesicles to be delivered to the various parts of the cell or the extracellular space.
- Lysosomes are vesicles that are filled with enzymes that can digest complex biological molecules. These organelles break down nonfunctional organelles or material coming from outside the cell.
- Mitochondria are the powerhouses of the cell: they generate ATP. The cytoskeleton of the cell is made of protein components that lend support to the cell and its organelles. They are also involved in cell and organelle movement.
- Viruses use cell-generated ATP, take advantage of cellular organelles to manufacture viral proteins, and use the cytoskeleton for transport to different parts of the cell.

Section 3.2 The Plasma Membrane, Exocytosis, and Endocytosis

- The plasma membrane is composed of phospholipids, which have a polar head and nonpolar tails. As such, they are amphipathic: the phosphate head is hydrophilic and the fatty acid tails are hydrophobic.
- The fluid mosaic model describes the current thinking on the plasma membrane. This model states that the phospholipids of the membrane move freely within the membrane, as do the many integral proteins embedded in the membrane.
- The process of secreting large molecules, like proteins, from the cell is known as exocytosis. Endocytosis imports large molecules into the cell. Bulk-phase endocytosis forms a vesicle that encapsulates the extracellular fluid, while receptor-mediated endocytosis is initiated when ligands bind to receptors on the cell surface. The ligands are imported in vesicles that form in clathrin-coated pits.
- Endocytic vesicles soon fuse with endosomes that increase their acidity as they travel into the cell.
- In order to infect a cell, viruses must get through the plasma membrane. Some viruses that enter the cell through endocytosis must also escape from endosomes.

Section 3.3 The Cell Cycle

- Human DNA is diploid and divided into chromosomes.
- The cell cycle is the sequential stages through which a cell grows, replicates its DNA, and divides in two through the process of mitosis.
- The four phases of the cell cycle are Gap 1, Synthesis, Gap 2, and Mitosis.
- Some viruses require cells to be undergoing cell division because they require enzymes present during mitosis. Many viruses interfere with the stages of the cell cycle to increase the efficiency of viral replication.

Section 3.4 The Central Dogma of Molecular Biology: DNA Replication

- The Central Dogma of Molecular Biology states that DNA is replicated to create more DNA, DNA is transcribed into mRNA, and mRNA is translated by ribosomes to create proteins.
- DNA is composed of four different nucleotides (with bases adenine, cytosine, guanine, and thymine) bonded together. It is double stranded; each strand has directionality and the "forward" direction is termed $5'\rightarrow3'$. Because DNA is antiparallel, the 5′ end of one strand is matched with the 3′ end of the other strand.
- DNA replication is semiconservative, so each old strand acts as a template for the new DNA strand. DNA polymerase is the enzyme that reads the old strand in the $3'\rightarrow5'$ direction and creates the new strand out of nucleotides in the $5'\rightarrow3'$ direction. Nucleotides are added according to complementary base pair rules.
- DNA polymerases have high fidelity and make one error in every 1 million nucleotides added, on average.
- A few viral families take advantage of cellular DNA polymerases to replicate their DNA genomes.

Section 3.5 The Central Dogma of Molecular Biology: RNA Transcription and Processing

- The process of creating an mRNA copy of a portion of DNA is known as transcription. RNA polymerase II binds to transcription factors that assemble on the promoter of the gene, and the enzyme joins RNA nucleotides (adenine, guanine, cytosine, and uracil) in the $5'\rightarrow3'$ direction to create the precursor mRNA transcript.
- RNA polymerases have lower fidelity than DNA polymerases.
- Precursor mRNA is processed before leaving the nucleus. The transcript receives a 5′ 7-methylguanosine cap and a 3′ poly(A) tail, and introns are removed via RNA splicing.

Section 3.6 The Genetic Code

- The "genetic code" refers to which amino acids correspond to a sequence of processed mRNA.
- The eukaryotic ribosome translates an mRNA transcript in the $5'\rightarrow3'$ direction and begins at the start codon, AUG. This codon is found within a larger sequence

known as the Kozak consensus sequence. Translation will continue until a stop codon (UGA, UAA, or UAG) is reached.

- The region of mRNA that is translated is known as the open reading frame. The untranslated parts of the sequence are known as the 5′ UTR and 3′ UTR.
- There are 20 amino acids but 64 possible codons. Some codons are redundant and code for the same amino acid.

Section 3.7 The Central Dogma of Molecular Biology: Protein Translation

- Within the ribosome, tRNAs act as adaptor molecules because their anticodon sequence recognizes the mRNA transcript. The corresponding amino acid is attached to the 3′ end of the tRNA molecule.
- There are three stages of translation: initiation, elongation, and termination.
- During initiation, translation initiation factors recruit the ribosome and tRNA to the mRNA transcript. The ribosome scans the mRNA until the start codon is encountered.
- During elongation, the ribosome moves along the mRNA. For each new codon, a tRNA carrying an amino acid enters the ribosome. The growing amino acid chain is bonded to the new amino acid.
- During termination, the ribosome encounters a stop codon and translation ends. The protein is complete.

Section 3.8 Promotion of Viral Transcription and Translation Processes

- Viruses have evolved many tactics to take advantage of the cellular transcription and translation machinery. Viruses use cellular transcription, RNA processing, and translation mechanisms to ensure the translation of their proteins. They also interfere with the transcription, RNA processing, and translation of host gene products to ensure the preferential translation of viral products.

FLASH CARD VOCABULARY

Nucleus	Amphipathic
Prokaryote	Rough endoplasmic reticulum
Eukaryote	Ribosome
Organelle	Glycoprotein
Phospholipid	Lipoprotein
Hydrophilic	Vesicle
Hydrophobic	Golgi complex
Lysosome	Semiconservative replication
Mitochondrion	DNA polymerase
Cytoskeleton	Fidelity
Fluid mosaic model	Primase
Integral protein	Ligase
Peripheral protein	Okazaki fragments
Exocytosis	Transcription factors
Endocytosis	Promoter
Clathrin-coated pit	Enhancer
Endosome	RNA polymerase II
Base pair	Transcript
Chromosome	RNA processing
Diploid	5′-methylguanosine cap
Zygote	3′-Poly(A) tail
Cell cycle	RNA splicing
Daughter cells	Intron
Mitosis	Exon
Central Dogma of Molecular Biology	Alternative splicing
DNA replication	Codon
Transcription	Start codon
Translation	Stop codon
Antiparallel	Kozak consensus sequence
Reading frame	Termination
Open reading frame	Translation initiation factors
5′-Untranslated region	Posttranslational modification
3′-Untranslated region	Monocistronic
Redundant codons	Polyprotein
Genetic code	Internal ribosome entry site
Transfer RNA	Ribosomal frameshifting
Anticodon	Ribosomal skipping
Initiation	Leaky scanning
Elongation	

CHAPTER REVIEW QUESTIONS

1. Describe the general process of expressing a gene (in a chromosome) into a protein. Where does each step take place within the cell?
2. A secreted protein has been synthesized by a ribosome. Describe the pathway it will take to leave the cell and what happens at each step.
3. What molecules do you know that are hydrophobic or hydrophilic?

4. Explain the fluid mosaic model of plasma membrane assembly.
5. Which cellular organelles or processes are utilized by viruses?
6. Describe what happens during each of the four stages of the cell cycle.
7. What is the Central Dogma of Molecular Biology?
8. Draw a double-stranded piece of DNA. Make sure to label the 5′- and 3′-ends. Now draw out the process of DNA replication, paying attention to the 5′- and 3′-ends and the direction that DNA Polymerase lays down the new strand.
9. Describe the three steps involved in RNA processing.
10. Use the genetic code in Fig. 3.19 to translate the following piece of mRNA: 5′- GCCGCCAUGGCCAU AGCCGAUUGACCCGGA -3′
11. Determine the 5′-UTR and 3′-UTR in the sequence above.
12. Describe what happens during the three stages of translation.
13. Explain at least three translational processes involving the ribosome that occur with viral translation but do not normally occur with cellular translation of a protein.
14. How do viruses ensure the preferential translation of their gene products over cellular gene products?

FURTHER READING

De Jesus, N., 2007. Epidemics to eradication: the modern history of poliomyelitis. Virol. J. 4, 70.

Fotin, A., Cheng, Y., Sliz, P., et al., 2004. Molecular model for a complete clathrin lattice from electron cryomicroscopy. Nature 432, 573–579.

Nirenberg, M., 2004. Historical review: deciphering the genetic code - a personal account. Trends Biochem. Sci. 29, 46–54.

Tarini, M., Cignoni, P., Montani, C., 2006. Ambient occlusion and edge cueing to enhance real time molecular visualization. IEEE Trans. Vis. Comput. Graph 12, 1237–1244.

Vereb, G., Szöllosi, J., Matkó, J., et al., 2003. Dynamic, yet structured: the cell membrane three decades after the Singer-Nicolson model. Proc. Natl. Acad. Sci. U.S.A. 100, 8053–8058.

Watson, J.D., Crick, F.H.C., 1953. Molecular structure of nucleic acids. A structure for deoxyribose nucleic acid. Nature 171, 737–738.

Chapter 4

Virus Replication

A virus must undergo the process of **replication** to create new, infectious virions that are able to infect other cells of the body or subsequent hosts. After gaining entry into the body, a virus makes physical contact with and crosses the plasma membrane of a target cell. Inside, it releases and replicates its genome while facilitating the manufacture of its proteins by host ribosomes. Virus particles are assembled from these newly synthesized biological molecules and become infectious virions. Finally, the virions are released from the cell to continue the process of infection.

The seven stages of virus replication are categorized as follows:

1. Attachment
2. Penetration
3. Uncoating
4. Replication
5. Assembly
6. Maturation
7. Release

A mnemonic to remember the stages of virus replication is the sentence "**A PUR**ple **A**pple **M**ight **R**edden." The letters in bold are the first letters of the names of the seven stages in order.

All viruses must perform the seven stages in order to create new virions. Some stages may take place simultaneously with other stages, or some stages may take place out of order, depending upon the virus. This chapter describes the details of what occurs during each stage of viral replication.

4.1 ATTACHMENT

A cell interacts with the extracellular world at the plasma membrane, and it is at this location that a virus first makes contact with a target cell. As described in Chapter 3, "Features of Host Cells: Cellular and Molecular Biology Review," the plasma membrane of the cell is composed of a phospholipid bilayer that has numerous proteins protruding from the membrane. These surface proteins have a variety of functions that include transporting ions and molecules, facilitating the binding of one cell to another, or acting as receptors for incoming proteins. The majority of plasma membrane proteins are **glycosylated**, meaning that they have been modified with sugars and carbohydrates. To infect a cell, it is critical that a virus initiates **attachment**—the binding of the virus to the host cell. This interaction is specific: the virus contains a **virus attachment protein** that adsorbs to a **cell surface receptor** on the cell (Table 4.1). The target receptor molecules on the cell surface are normal molecules required for cellular functions that viruses have evolved to exploit, usually glycoproteins or the sugar/carbohydrate residues present on glycoproteins or the plasma membrane. For instance, rhinovirus binds a protein known as intercellular adhesion molecule 1 (ICAM-1), involved in the attachment of one cell to another. Influenza A virus strains bind to the sialic acid sugars found at the ends of cellular carbohydrate chains, and herpes simplex viruses (HSV) reversibly bind to glycosaminoglycans (GAGs), such as heparan sulfate, in order to bind to the herpesvirus entry mediator protein or nectins on the cell surface (Fig. 4.1).

Some viruses also require **coreceptors** to infect cells. HIV initially binds to a protein known as CD4 on the surface of T lymphocytes ("T cells") but requires one of two coreceptor proteins to continue the process of infection. As will be described later in Chapter 11, "Human Immunodeficiency Virus," humans that have a modified version of CCR5, one of these coreceptors, are largely

TABLE 4.1 Cell Surface Receptors for Attachment of Human Viruses

Virus	Cell surface receptor(s)
Rhinoviruses	Intercellular adhesion molecule 1 (ICAM-1) (90%), low-density lipoprotein receptor (10%)
Poliovirus	Poliovirus receptor (PVR) CD155
Human immunodeficiency virus	CD4 (receptor); CCR5 or CXCR4 (coreceptors)
Influenza A virus	Sialic acid
Measles virus	CD46, CD150
Herpes simplex virus-1	Heparan sulfate, HVEM, Nectin-1
Dengue virus	DC-SIGN
Hepatitis B virus	Sodium taurocholate–cotransporting polypeptide
Human papillomavirus	Heparan sulfate, integrins

Essential Human Virology.

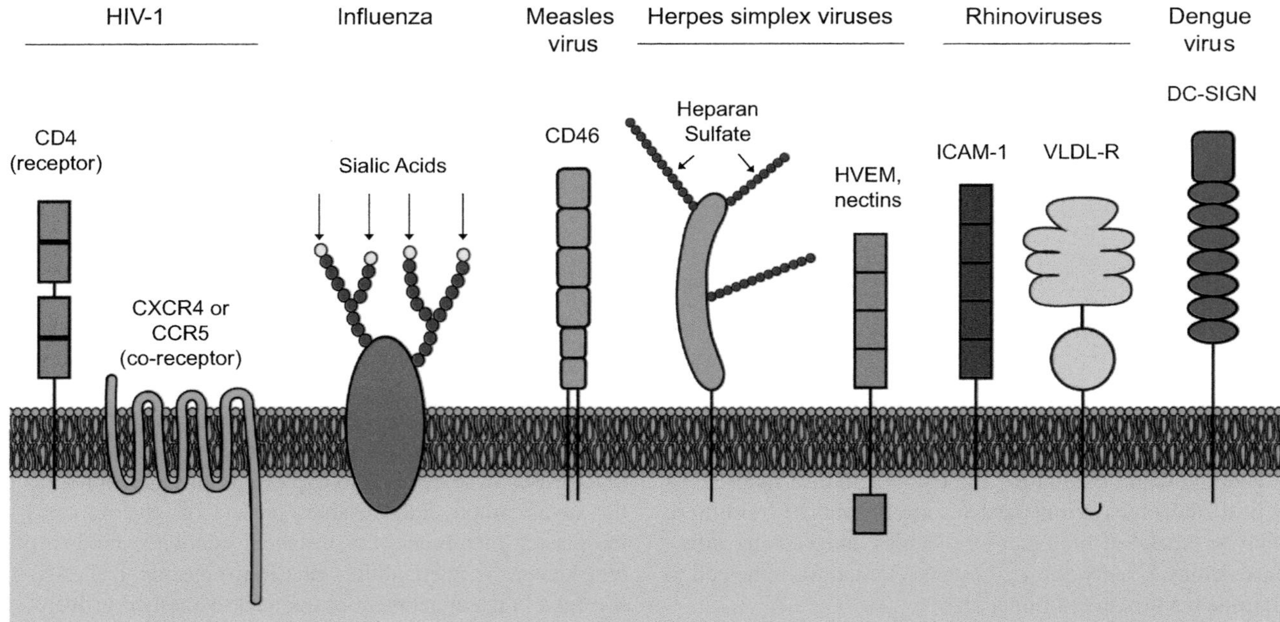

FIGURE 4.1 **Cell surface receptors.** Different viruses use specific cell surface receptors for attachment. HIV-1 requires CD4 as a receptor and chemokine receptors CCR5 or CXCR4 as coreceptors. Influenza viruses bind to terminal sialic acid residues found on cell surface glycoproteins. Laboratory strains of measles virus bind CD46 (although CD150 is also a receptor for the virus). Herpes simplex virus-1 initially binds to heparan sulfate on GAGs in order to specifically bind entry receptors, such as HVEM or nectins. Ninety percent of rhinoviruses use ICAM-1 as a receptor, while 10% use the VLDL receptor. Dengue virus attaches using DC-SIGN. Note the different structures and types of receptors that viruses use for entry. The tropism of the virus is determined by which cells in the body express the cell surface receptor.

resistant to infection with HIV because the virus cannot use the modified CCR5 as a coreceptor and so infection is blocked. Infection of a cell can be prevented if attachment of the virus can be inhibited, and virus attachment proteins are the target of many antiviral drugs in use and in development.

Attachment involves opposing electrostatic forces on the virus attachment protein and the cell surface receptor. The virus attachment protein is located in the outermost portion of the virus, since this is where contact with the cell occurs. The attachment protein protrudes from the envelope of an enveloped virus, whereas nonenveloped viruses have one or more capsid proteins that interact with the cell surface receptor. The viral attachment proteins can extend from the surface of the virion or can be within "canyons" formed by capsid proteins. For example, 90% of human rhinovirus serotypes bind to ICAM-1 on the surface of cells. Instead of binding to the outside of the rhinovirus capsid, the molecule docks into a deep canyon formed by the rhinovirus VP1, VP2, and VP3 proteins (Fig. 4.2A). In contrast, 10% of human rhinoviruses attach to the very low-density lipoprotein (VLDL) receptor. This interaction does not occur in canyons formed by the viral proteins, however. Instead, several VP1 proteins at the vertices of the icosahedral capsid bind to the receptor (Fig. 4.2B). Even if the binding affinity between the VP1 protein and the VLDL receptor is low, the multiple VP1 proteins increase the total binding strength of the interaction. This example also illustrates that different strains of the same virus can take advantage of different cell surface receptors for attachment.

4.2 PENETRATION

Following attachment, successful viruses quickly gain entry into the cell to avoid extracellular stresses that could remove the virion, such as the flow of mucus. **Penetration** refers to the crossing of the plasma membrane by the virus. In contrast to virus attachment, penetration requires energy, although this is contributed by the host cell, not the virus.

Several different mechanisms are utilized by viruses to gain entry into a cell (Fig. 4.4, Table 4.2). One of these takes advantage of a normal host process: endocytosis. As described in Chapter 3, "Features of Host Cells: Cellular and Molecular Biology Review," cells are able to import molecules through the process of endocytosis. Receptor-mediated endocytosis occurs when receptors on the cell surface are bound by their ligands and internalized in clathrin-coated pits or caveolae that become endocytic vesicles. Eventually, these vesicles lose their clathrin or caveolin coating and fuse with "early endosomes," slightly acidic vesicles (pH of 6.0–6.5) that become "late endosomes" as their acidity increases (pH of 5.0–6.0). Late endosomes deliver materials to lysosomes, larger vesicles full of digestive enzymes.

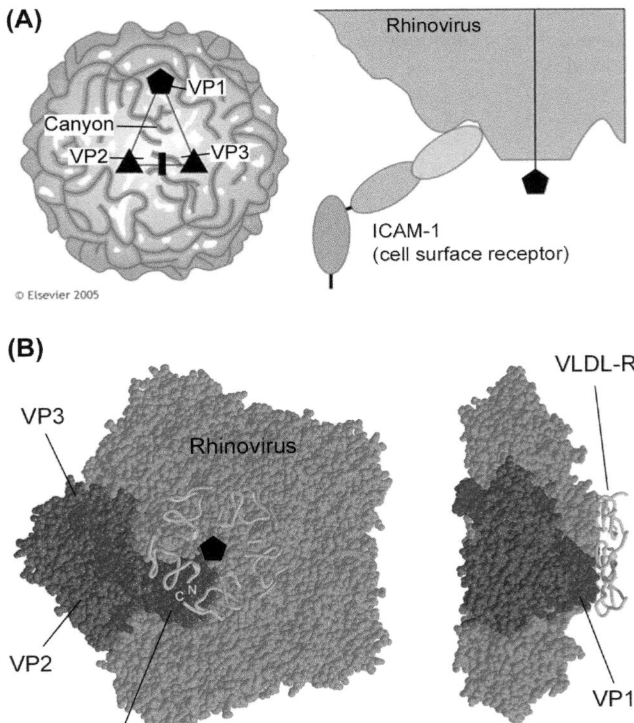

FIGURE 4.2 **Rhinovirus attachment.** (A) 90% of rhinovirus serotypes use ICAM-1 as a cell surface receptor. The cellular protein binds into a canyon formed by capsid proteins VP1, VP2, and VP3. *(Reproduced with permission from Elsevier Academic Press: Alan J. Cann, 2005. Principles of Molecular Virology, fourth ed.)* (B) 10% of rhinovirus serotypes bind the very low-density lipoprotein receptor (VLDL-R). In contrast to ICAM-1 binding, the binding of these rhinovirus serotypes occurs on the fivefold axis at the vertex of the capsid icosahedron, formed by repeating VP1 proteins. This space-filling model shows the surface of the rhinovirus capsid (gray) with one structural unit highlighted, formed by VP1 (blue), VP2 (green), and VP3 (red). The gold molecules represent the VLDL receptors, showing where they bind to rhinovirus protein VP1. *(Reprinted with permission from: Verdaguer et al., 2004. Nature Structural and Molecular Biology. Macmillan Publishers Ltd, 11(5), 429–434.)*

In-Depth Look: Tropism

Different cells perform different functions within a multicellular organism. As such, not all cells within the body display the same types of cell surface proteins. The **tropism** of a virus refers to the specificity of a virus for a particular host cell or tissue. Viruses will only be able to infect the cells that display the molecules to which their virus attachment proteins bind. Similarly, one reason that certain viruses have a narrow **host range** is because different host species may lack the cell surface proteins that a particular virus uses for attachment. For instance, humans are the only known natural hosts of poliovirus. Because of this, poliovirus has historically been a difficult virus to study because the cell surface receptor it uses for attachment, called CD155 or the poliovirus receptor, is not present in small animal models, such as mice. In 1990, a transgenic mouse strain was engineered to express the human CD155 molecule. These mice were susceptible to infection, whereas the normal nontransgenic mice were not (Fig. 4.3).

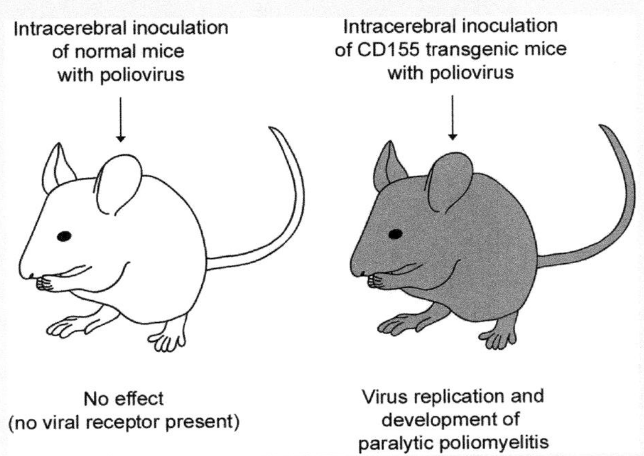

FIGURE 4.3 **Transgenic mice for the human poliovirus receptor CD155.** In 1990, the Racaniello research group generated mice that expressed human CD155, the poliovirus receptor. Poliovirus does not replicate in normal mice because they lack this receptor. When the CD155-transgenic mice were infected intracerebrally with poliovirus, the virus replicated in the brain and spinal cord, causing paralysis. These transgenic mice have proved useful in studying poliovirus infection in a nonprimate animal model. *(Study documented in Ren et al., 2012. Cell 63(2), 353–362.)*

Receptor-mediated endocytosis is commonly used by viruses to penetrate the plasma membrane. As the pH of the endosome drops, the viral proteins change configuration, which allows them to escape from the endosome. Depending upon the virus, this can happen in early endosomes, late endosomes, or lysosomes. Both enveloped and nonenveloped viruses take advantage of receptor-mediated endocytosis to gain entry into the cytoplasm of the cell (Fig. 4.4). Most types of viruses use clathrin-mediated endocytosis to enter the cell, including dengue virus, hepatitis C virus, and reoviruses. A few well-known viruses that infect humans, such as SV40 and papillomaviruses (that cause warts or cervical cancer), use caveolae-mediated endocytosis; this was discovered by using a drug that inhibited the formation of caveolae. Blocking clathrin-mediated endocytosis did not prevent the entry of these viruses into cells. Still other viruses undergo receptor-mediated endocytosis that is independent of both clathrin and caveolin.

Other forms of endocytosis, such as bulk-phase endocytosis and phagocytosis, are also exploited by viruses to enter the cell. In bulk-phase endocytosis, the cell forms a vesicle that engulfs whatever molecules are present in the extracellular fluid, including viruses. **Phagocytosis** is a form of receptor-mediated endocytosis that is used by specialized cells to engulf entire cells. Recently, two large DNA viruses, HSV-1 and mimivirus, were shown to enter cells through phagocytosis-like pathways.

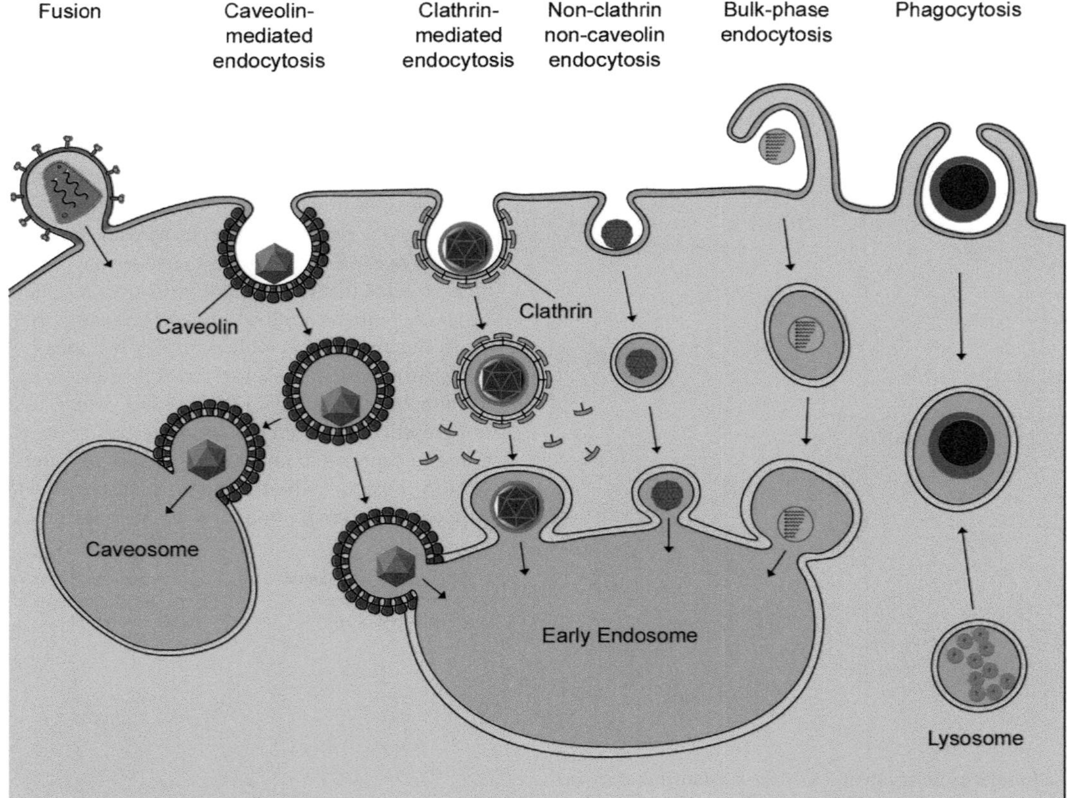

FIGURE 4.4 **Viral penetration into the cell.** Different viruses take advantage of various cellular mechanisms to gain entry into the cell after binding their specific cell surface receptors. Some enveloped viruses undergo fusion, which fuses the viral envelope with the plasma membrane. Both enveloped and nonenveloped viruses take advantage of receptor-mediated endocytosis in caveolin- or clathrin-coated pits to gain entry into the cytoplasm of the cell. Still other viruses undergo receptor-mediated endocytosis that is independent of both clathrin and caveolin. Bulk-phase endocytosis and phagocytosis are also utilized by viruses to gain entry into the cell.

TABLE 4.2 Methods of Penetration for Select Human Viruses

Type of penetration (entry)	Virus examples
Clathrin-mediated endocytosis	Dengue virus, hepatitis C virus, reovirus, adenovirus, parvovirus B19, West Nile virus
Caveolin-mediated endocytosis	Human papillomavirus, SV40, hepatitis B virus
Fusion	HIV, influenza, respiratory syncytial virus, herpes simplex viruses, dengue virus, Ebola virus

A method of penetration that is used exclusively by enveloped viruses is **fusion**. Fusion of the viral envelope can occur at the cell membrane or within endocytosed vesicles, such as the endosome, and is mediated by the same viral protein that is used by the virus for attachment or by a different viral protein, depending upon the virus. For instance, HIV has a protein known as gp120 that binds to CD4 and one of the two coreceptors for entry, CCR5 or CXCR4. Once this occurs, a different viral protein, gp41, fuses the virus envelope with the cell membrane, releasing the nucleocapsid into the cytoplasm.

Study Break
Describe the different ways that viruses can gain entry into the cytosol.

4.3 UNCOATING

Uncoating refers to the breakdown or removal of the capsid, causing the release of the virus genome into the cell to wherever genome replication and transcription will take place. Uncoating can be separated from or tightly linked with penetration, and viruses achieve uncoating in a variety of different ways (Fig. 4.5). For example, rhinoviruses are taken into the cell by receptor-mediated endocytosis in clathrin-coated vesicles. Within the acidic endosome, the virus expands in size about 4%, and one of the capsid proteins, VP1 (viral

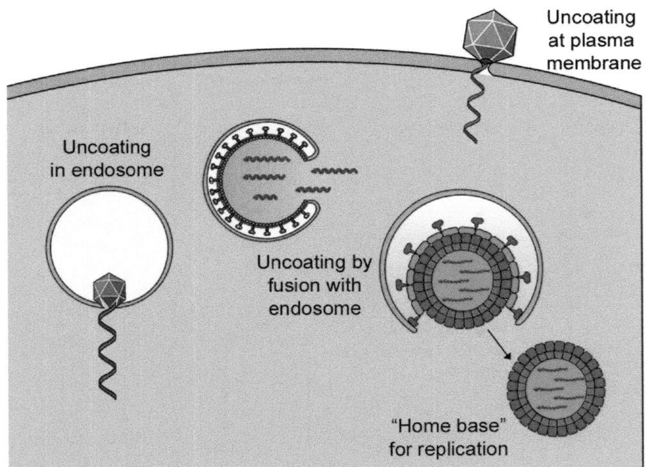

FIGURE 4.5 **Uncoating of virion capsids.** Certain viruses, including rhinoviruses, expand to form pores in the endosome through which the viral genome can escape. Like influenza virus, other viruses induce fusion of the virion envelope with the endosomal membrane, releasing the viral genome. Historically, it has been though that poliovirus capsids do not enter the cell at all: binding of the capsid to the cell surface receptor induces a conformation change that creates a pore in the membrane through which the genome is transported. Many viruses maintain a partially intact capsid in the cytosol that acts as a "home base" for replication, like reoviruses do.

protein 1), forms pores in the endosome that allow the release of the rhinovirus RNA genome. On the other hand, influenza virus has a viral protein known as hemagglutinin (HA) embedded into the virus envelope. HA binds to sialic acid residues found on the surface of respiratory epithelial cells, and penetration occurs through receptor-mediated endocytosis. The low pH of the endosome causes a conformational change in the viral HA protein, revealing a fusion peptide that brings the two membranes close together and fuses the viral envelope with the endosomal membrane. In this case, the HA protein facilitates both attachment and uncoating of the virus. The released viral RNA genome segments are transported to the nucleus and enter through nuclear pores. Other viral capsids, such as those of poliovirus, have been thought to not enter the cell at all: the binding of the poliovirus capsid to the cell surface receptor causes a conformational change in the virion that creates a pore in the cell membrane through which the viral RNA is released into the cytoplasm. In contrast, many viruses remain largely intact after penetration. Reoviruses do not completely uncoat within the cytoplasm, providing a "home base" for genome replication.

Many herpesviruses infect neurons but must replicate in the nucleus, which can be quite a distance from their site of entry at the plasma membrane. After fusion of the viral envelope with the plasma membrane, the intact nucleocapsids of HSV-1 are transported along microtubules to the nucleus. HSV proteins bind to dynein, a host cell protein that "walks" vesicles of cargo along microtubules. At the nucleus, the HSV capsid docks at a nuclear pore and its viral DNA is transported into the nucleus (Fig. 4.6).

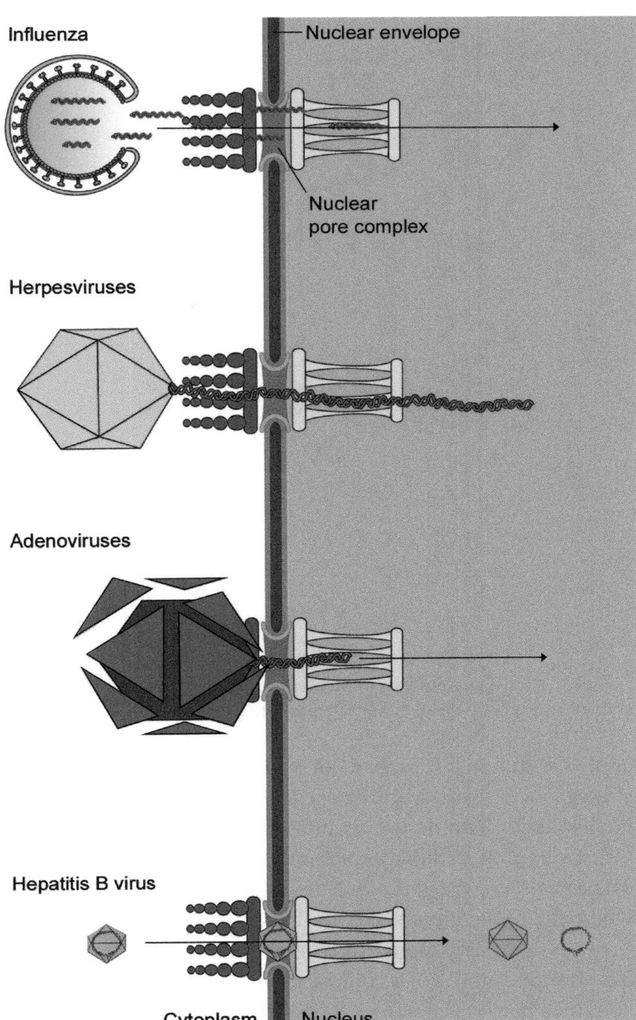

FIGURE 4.6 **Transport of viral genomes into the nucleus.** Several viruses must transport their genomes into the nucleus for viral transcription and/or replication to occur. Influenza genome segments are transported through the nuclear pore into the nucleus. Herpesvirus capsids are transported along microtubules to the nuclear pore, where uncoating occurs. Adenovirus capsids disassemble at the nuclear pore and the viral DNA is transported into the nucleus. Other viruses, including hepatitis B virus, are small enough that the entire capsid might be able to pass through the nuclear pore.

Still other capsids are small enough to pass through nuclear pores: the hepatitis B capsid, with a diameter around 30 nm, may be imported intact through a nuclear pore to uncoat within the nucleus.

4.4 REPLICATION

In the same way that our DNA encodes the information to manufacture our proteins, a virus's genome acts as the instructions for the synthesis of virus proteins. To create new virions, the proteins that will be incorporated into the virion are made through expression of viral genes, and the virus genome is

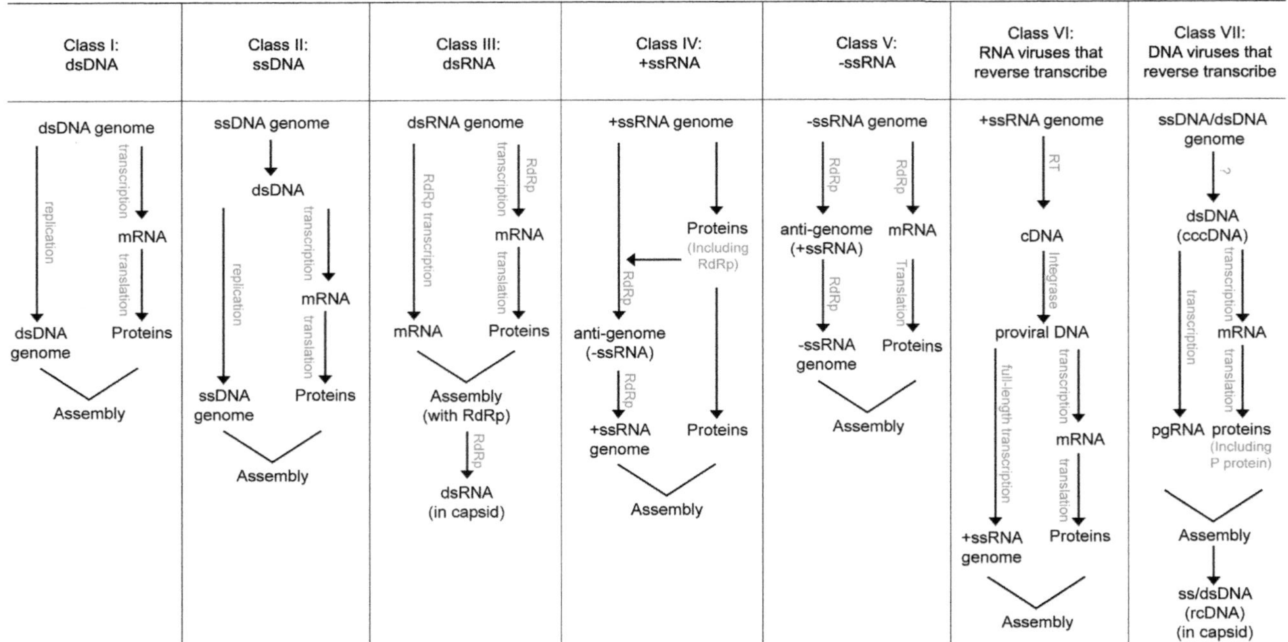

FIGURE 4.7 **The replication strategies of the viruses in the Baltimore replication classes.** All viruses, regardless of their type of nucleic acid genome, must copy the genome and translate proteins in order to assemble nascent virions.

copied through the process of **replication**. The replication strategy of a virus is generally dependent upon the type of nucleic acid genome it contains (Fig. 4.7). Many classes of viruses exploit cellular proteins to replicate their genomes. The location of these proteins will therefore generally dictate where the replication of the viral nucleic acid will take place.

All living organisms have genomes composed of double-stranded (ds) DNA, but viruses are unique in that their genomes can be made of single-stranded (ss) DNA, as well as dsRNA or ssRNA, which can be positive sense or negative sense. The Baltimore classification system, as discussed in Chapter 2, "Virus Structure and Classification," is useful when discussing the general replication strategies of viruses because it categorizes the viruses into seven classes based upon their type of genome:

1. Double-stranded DNA viruses
2. Single-stranded DNA viruses
3. Double-stranded RNA viruses
4. Positive-sense RNA viruses
5. Negative-sense RNA viruses
6. RNA viruses that reverse transcribe
7. DNA viruses that reverse transcribe

As the replication strategies of these classes are discussed, it may be helpful to refer to Table 4.3 and Fig. 2.11 to keep track of which vertebrate-infecting viral families are found in each class.

Viral nucleic acids are found in a variety of configurations. They can be linear or circular, and they can be **segmented** into several smaller pieces within the virion, as occurs with influenza viruses, or **nonsegmented** like rabies virus, containing one molecule of nucleic acid that encodes all necessary genes. Longer molecules are more subject to breaking, but segmented viruses must package all genome segments into a virion for it to be infectious. Regardless of the structure of their nucleic acid, all viruses need to express their viral proteins and replicate their genome within the cell in order to create new virions.

4.4.1 Class I: dsDNA Viruses

All living organisms have double-stranded DNA genomes. Viruses with dsDNA genomes therefore have the most similar nucleic acid to living organisms and often use the enzymes and proteins that the cell normally uses for DNA replication and transcription, including its DNA polymerases and RNA polymerases. These are located in the nucleus of a eukaryotic cell, and so all dsDNA viruses that infect humans (with the exception of poxviruses) enter the nucleus of the cell, using the various mechanisms of entry and uncoating mentioned above. Many recognizable human viruses have dsDNA genomes, including herpesviruses, poxviruses, adenoviruses, and polyomaviruses.

Transcription of viral mRNA (vmRNA) must occur before genome replication if viral proteins are involved in replicating the virus genome. In addition, certain translated viral proteins act as **transcription factors** to direct the transcription of other genes. As discussed in Chapter 3, "Features of Host Cells: Cellular and Molecular Biology Review, transcription factors bind to specific sequences within the **promoters** of

TABLE 4.3 Families of Human Viruses Within Each Replication Class

Family	Virus examples
Class I: dsDNA viruses	
Adenoviridae	Adenovirus
Herpesviridae	Herpes simplex virus, Epstein–Barr virus, varicella zoster virus
Papillomaviridae	Human papillomavirus
Polyomaviridae	JC polyomavirus, BK polyomavirus, SV40
Poxviridae	Variola, vaccinia
Class II: ssDNA viruses	
Parvoviridae	Parvovirus B19
Anelloviridae	Torque teno virus
Class III: dsRNA viruses	
Picobirnaviridae	Human picobirnavirus
Reoviridae	Rotavirus
Class IV: +ssRNA viruses	
Astroviridae	Human astrovirus
Caliciviridae	Norwalk virus
Coronaviridae	Human coronavirus
Flaviviridae	Dengue virus, yellow fever virus, West Nile virus, hepatitis C virus
Hepeviridae	Hepatitis E virus
Picornaviridae	Poliovirus, rhinovirus, enterovirus, hepatitis A virus
Togaviridae	Eastern equine encephalitis, Chikungunya virus, rubella virus
Class V: −ssRNA viruses	
Arenaviridae	Lymphocytic choriomeningitis virus, Lassa virus, Machupo virus
Bunyaviridae	Hantavirus, Crimean–Congo hemorrhagic fever virus
Filoviridae	Ebola virus, Marburg virus
Orthomyxoviridae	Influenza A virus, influenza B virus
Paramyxoviridae	Nipah virus, Hendra virus, measles virus, mumps virus
Rhabdoviridae	Rabies virus
Class VI: RNA viruses that reverse transcribe	
Retroviridae	Human immunodeficiency virus-1 and -2
Class VII: DNA viruses that reverse transcribe	
Hepadnaviridae	Hepatitis B virus

cellular genes immediately upstream of the transcription start site to initiate transcription of those genes. **Enhancers**, regulatory sequences also involved in transcription, are located farther away from the transcription start site and can be upstream or downstream. dsDNA viruses also have promoter and enhancer regions within their genomes that are recognized not only by viral transcription factors but by host transcription factors, as well. These proteins initiate transcription of the viral genes by the host RNA polymerase II.

Processing of viral precursor mRNA (also known as post-transcriptional modification) occurs through the same mechanisms as for cellular mRNA. Viral transcripts receive a 5′-cap and 3′-poly(A) tail, and some viruses' transcripts are spliced to form different vmRNAs. For example, the genes of herpesviruses are each encoded by their own promoter and are generally not spliced, but the human adenovirus E genome has 17 genes that encode 38 different proteins, derived by alternative splicing of vmRNA during RNA processing.

The dsDNA viruses transcribe their viral gene products in waves, and the **immediate early** and/or **early** genes are the first viral genes to be transcribed and translated into viral proteins. These gene products have a variety of functions, many of which help to direct the efficient replication of the genome and further transcription of the **late** genes that encode the major virion structural proteins and other proteins involved in assembly, maturation, and release from the cell. The replication of the viral genome requires many cellular proteins; having the late genes transcribed and translated after the virus genome has been replicated ensures that the host enzymes needed for replication are not negatively affected by the translation of massive amount of virion structural proteins.

To create new virions, viral proteins must be translated and the genome must also be copied. With the exception of poxviruses, the genome replication of all dsDNA viruses takes place within the nucleus of the infected cell. Eukaryotic DNA replication, also reviewed in more detail in Chapter 3, "Features of Host Cells: Cellular and Molecular Biology Review," is also carried out by DNA polymerases and other proteins within the nucleus. DNA polymerases, whether they are cell derived or virus derived, cannot carry out de novo synthesis, however. They must bind to a short primer of nucleic acid that has bound to the single-stranded piece of DNA, forming a short double-stranded portion that is then extended by DNA polymerase (Fig. 4.8A). Primase is the enzyme that creates primers during cellular DNA replication, and some viruses, such as polyomaviruses and some herpesviruses, take advantage of the cellular primase enzyme to create primers on their dsDNA genomes during replication. Other herpesviruses, such as HSV-1, provide their own primase molecule, although this process occurs less commonly. Still other viruses, such as the adenoviruses, encode a viral protein primer that primes its own viral DNA polymerase (Fig. 4.8B). Cellular DNA polymerases are used by polyomaviruses and papillomaviruses, while all other dsDNA viruses encode their own DNA polymerases to replicate the viral genome. Many other cellular enzymes and proteins are required for DNA synthesis, and viruses are dependent on these to varying degrees, depending upon the specific virus. The poxviruses are a notable exception to this: they encode all the proteins necessary for DNA replication. In fact, they also encode the proteins needed for transcription of RNA, and so, unlike all other dsDNA viruses, they do not need to gain entry into the nucleus of a host cell for either genome replication or transcription and processing of viral genes, allowing their replication to take place entirely in the cytoplasm.

Abbreviations	
cccDNA	Covalently closed circular DNA
cDNA	Complementary DNA
DR	Direct repeat
dsDNA	Double-stranded DNA
dsRNA	Double-stranded RNA
IN	Integrase
LTR	Long terminal repeat
PBS	Primer-binding site
pgRNA	Pre-genomic RNA
PPT	Polypurine tract
rcDNA	Relaxed circular DNA
RdRp	RNA-dependent RNA polymerase
RT	Reverse transcriptase
ssDNA	Single-stranded DNA
+ssRNA	Positive-sense RNA
−ssRNA	Negative-sense RNA
vmRNA	Viral mRNA

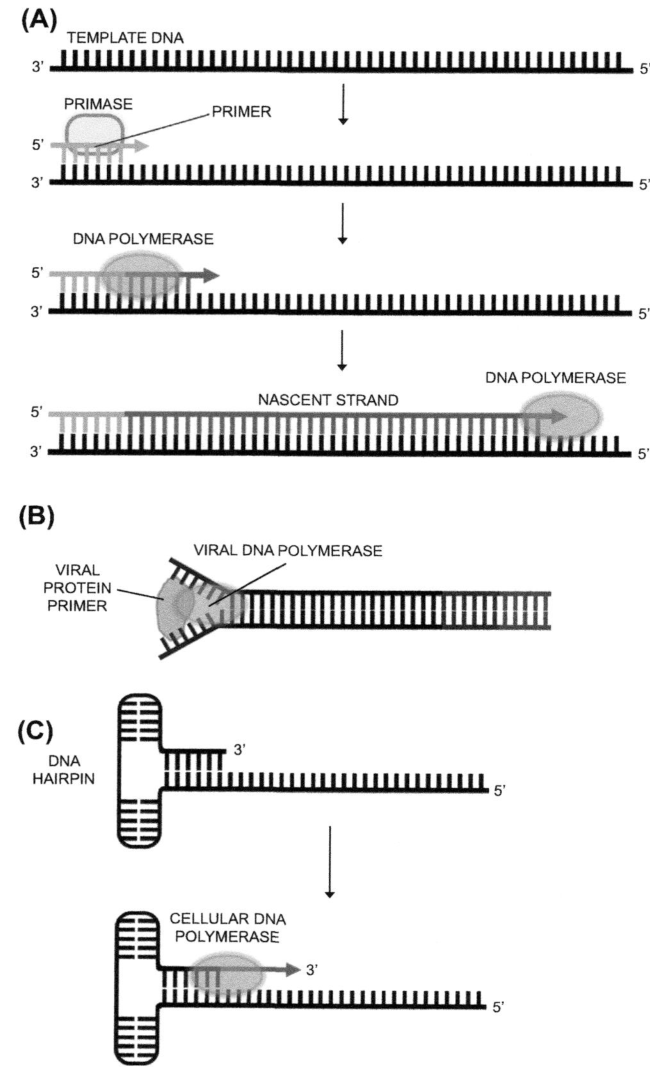

FIGURE 4.8 **DNA priming.** DNA polymerases cannot carry out de novo synthesis and so need a primer in order to replicate DNA. Some viruses take advantage of the cellular primase in order to create primers (A), while other viruses, such as adenoviruses, encode a protein primer that primes its own DNA polymerase (B). In the process of self-priming, the ssDNA genomes of parvoviruses fold back upon themselves to form hairpin ends that act as a primer for host DNA polymerase (C).

4.4.2 Class II: ssDNA Viruses

Viruses with ssDNA genomes infect primarily bacteria and plants, although two families, Anelloviridae and Parvoviridae, infect humans. These viruses are some of the smallest known viruses, with nonenveloped icosahedral capsids of 18–30 nm in diameter, and correspondingly small genomes of 4000–6000 nucleotides. Because they encode only a few genes, they are completely dependent on host cell enzymes for genome replication and transcription.

During replication, the ssDNA genome enters the nucleus of the host cell, where the ssDNA is converted to dsDNA by DNA polymerase during S phase of the cell cycle. This occurs because the ssDNA genome of parvoviruses has "hairpin" ends that fold back and complementarily bind to the ssDNA (Fig. 4.8C). This process, known as **self-priming,** creates a primer for DNA polymerase to extend. After the ssDNA genome becomes double-stranded, RNA polymerase II is able to transcribe the viral genes, which are then translated into viral proteins, and DNA polymerase replicates the genome so assembly of nascent virions can occur.

4.4.3 Class III: dsRNA Viruses

dsRNA viruses are all nonenveloped and possess icosahedral capsids. They have segmented genomes, and two families of dsRNA viruses infect humans. Viruses in the *Reoviridae* family include rotavirus, so named because

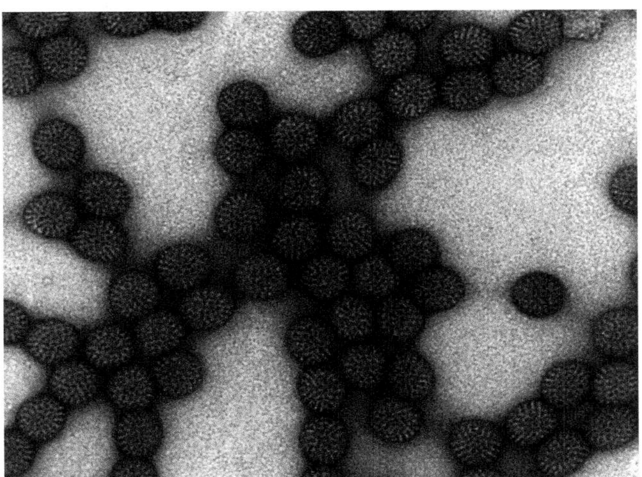

FIGURE 4.9 **Rotavirus virions.** Negatively stained transmission electron micrograph of rotavirus particles at 446,428× magnification. *Micrograph courtesy of the CDC/Dr. Erskine Palmer.*

the virion looks like a wheel (*rota* means "wheel" in Latin; Fig. 4.9). Rotavirus has 11 genome segments and is the major cause of childhood diarrhea. Picobirnaviruses are another family of dsRNA viruses that infect humans, but they are bisegmented, only having two genome segments that together are around 4.2 kb in length (the name of the viral family means "small two-RNA viruses," referring to the two dsRNA genome segments). Human picobirnaviruses have been isolated from diarrhea, although the association of the virus as a cause of a specific disease is currently unclear.

Unlike DNA viruses, viruses with RNA genomes do not usually enter the nucleus of an infected cell. Because they do not have a DNA intermediate, none of the host enzymes involved in DNA replication are required for the replication of the RNA genome. However, RNA viruses must still transcribe vmRNA so that viral proteins can be translated by host ribosomes and new virions can be formed. Cellular mRNA is transcribed by a **DNA-dependent RNA polymerase** called RNA polymerase II. As the name suggests, a DNA-dependent RNA polymerase requires a DNA template to make RNA, so it cannot transcribe mRNA from an RNA genome. Therefore, cells do not contain the enzymes necessary to transcribe mRNA from an RNA template, and so all RNA viruses must carry *or encode* their own **RNA-dependent RNA polymerase** (RdRp) to transcribe viral mRNA (Fig. 4.10A). dsRNA viruses contain an RdRp that is carried into the cell within the virion.

As mentioned in Section 4.3, reoviruses do not completely uncoat within the cytoplasm of the cell, providing a "home base" for transcription. In fact, free viral dsRNA or mRNA is not observed within the cytoplasm of the cell. The RdRp is closely associated with the partially uncoated capsid, which contains pores through which the transcribed

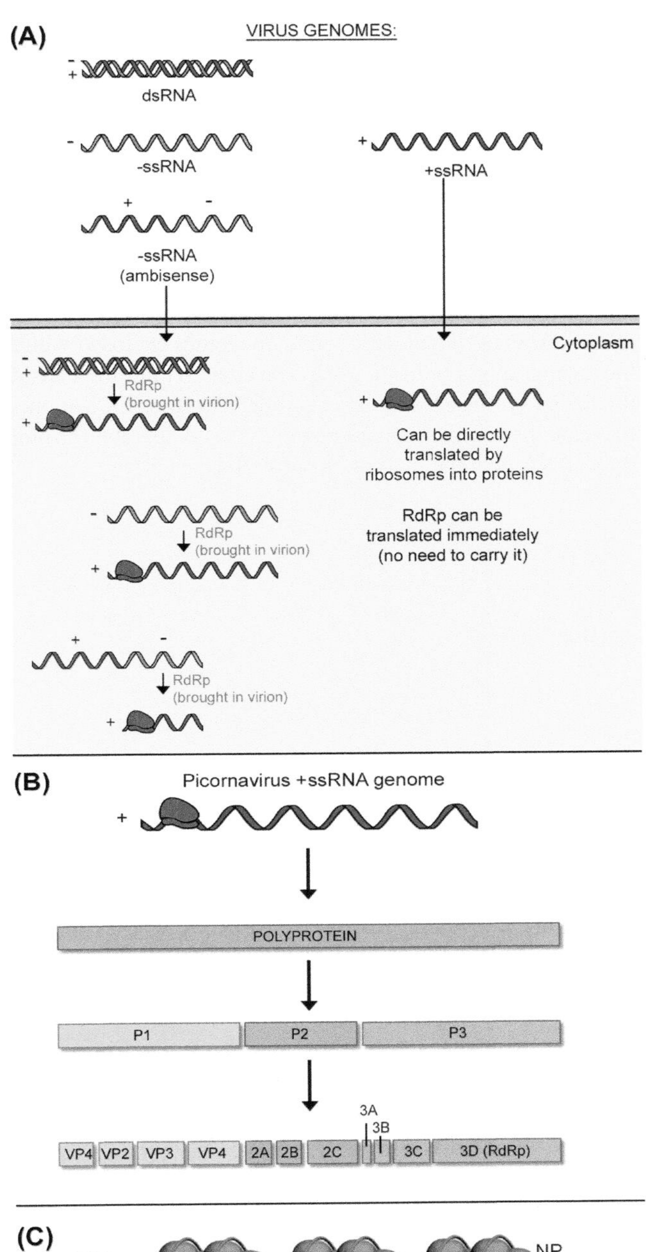

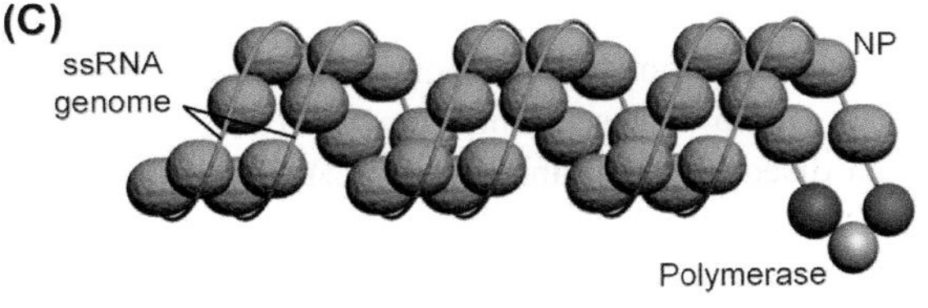

FIGURE 4.10 **Details of RNA virus replication.** (A) Like mRNA, +ssRNA viruses have infectious genomes that can immediately be translated by ribosomes. Consequently, they do not need to bring an RdRp into the cell. RNA viruses with dsRNA, −ssRNA, or ambisense genomes must carry their own RdRp protein into the cell in order for transcription to occur. (B) The +ssRNA picornaviruses encode a single polyprotein that is cleaved several times to create all the proteins necessary for replication. (C) The ribonucleoprotein complex of helical viruses, such as this one from influenza A virus, is composed of the −ssRNA genome, the protecting nucleocapsid protein (NP), and associated proteins, including the RdRp complex. *Graphic from Fig. 1 of Tao et al., 2010. PLoS Pathogens 6(7), e1000943.*

mRNA passes to enter the cytoplasm of the cell, where the mRNA associates with host ribosomes.

Only one of the two RNA strands within the rotavirus genome, the negative strand, is used as a template by the virus's RdRp to transcribe mRNA. The viral mRNA is translated by host ribosomes to produce structural and nonstructural viral proteins. Each genome segment is transcribed into mRNA that is **monocistronic**, meaning that each mRNA transcript encodes one protein only, as is the case with eukaryotic mRNA transcripts. As new capsids are forming, a viral mRNA from each genome segment becomes enclosed within the capsid, along with the RdRp protein. Within the capsid, the RdRp synthesizes along each mRNA transcript just once to create the complementary negative strand, thereby forming the dsRNA genome in the newly formed capsid.

Study Break

Describe the difference between a DNA-dependent RNA polymerase, an RdRp, and an RNA-dependent DNA polymerase.

4.4.4 Class IV: +ssRNA Viruses

RNA viruses are unique in that their genetic information is encoded using RNA, not DNA. As we have seen, this can occur as dsRNA, but many ssRNA viruses also exist. Viruses with ssRNA genomes that can act directly as mRNA are known as **positive-sense** RNA viruses (abbreviated +ssRNA). Similarly, ssRNA viruses with genomes that are not able to be immediately translated by ribosomes are known as **negative-sense** RNA viruses (abbreviated -ssRNA). Negative-sense RNA must be copied into positive-sense RNA by a viral RdRp before it can be translated by ribosomes (Fig. 4.10A). The terms *positive strand* and *negative strand* are also used interchangeably with these two terms.

+ssRNA viruses are more abundant than any other class of viruses and infect a wide range of host species. They include seven different human viral families, including the coronaviruses, flaviviruses, and picornaviruses, that cause significant disease in humans (Table 4.3). Their abundance indicates that +ssRNA viruses have been very successful evolutionarily.

Because the genome of +ssRNA viruses acts as mRNA, these viruses have genetic information that is **infectious**. Their genomes are translatable by host ribosomes and have 5′-caps (or proteins that act similarly to a 5′-cap) and often contain poly(A) tail sequences at the 3′-end. Experiments that delivered only the genome of poliovirus into the cytoplasm of a cell resulted in new virions being formed, because translation of the genome is the first activity that takes place with +ssRNA genomes. This produces all the viral proteins necessary for orchestrating the remainder of the replication cycle. Where dsRNA viruses must carry an RdRp within the virion, +ssRNA viruses *encode* an RdRp within their +ssRNA genome. The RdRp protein is produced immediately upon entry into the cell by translation of the viral genome. It is important to note, however, that even though the virus encodes its own RdRp protein, cellular proteins are often also required for replication to take place. For example, despite that the poliovirus genome is infectious, it is not replicated when injected into a *Xenopus* frog oocyte (ovum/egg) unless the cytoplasm from a human cell is injected alongside the genome, indicating that at least one human cellular component is required for poliovirus genome replication.

A common characteristic of +ssRNA viruses is that their infectious genome encodes a **polyprotein**, meaning that the genome is translated by ribosomes into a long chain of amino acids that is then cleaved into several smaller proteins. This provides an economical method of deriving several proteins from the translation of only one piece of mRNA. In the case of picornaviruses, the positive-strand genome is translated in its entirety, and then proteases cleave the polyprotein in different locations to create several different proteins (Fig. 4.10B). Alternative cleavage of certain sections results in additional proteins.

Other +ssRNA viruses, such as the togaviruses that include rubella virus, begin by translating only a portion of the +ssRNA genome to create an initial set of proteins that direct the later replication of the genome and translation of other viral proteins. This allows for the creation of "stages" of virus replication, similar to what is observed with immediate early, early, and late gene transcription of certain DNA viruses. Creation of polyproteins also commonly accompanies this method of translation, and termination suppression results in the production of different polyprotein chains. This happens at a low rate (approximately 10% of proteins initially synthesized from the togavirus +ssRNA genome) but results in the generation of important viral proteins, including the viral RdRp.

As the replication proteins of +ssRNA viruses are synthesized, they tend to gather at or within certain membranes in the cell, creating **replication complexes**. For example, the viral proteins of poliovirus remain bound to rough ER (rER) membranes or secretory vesicles. Poliovirus is a nonenveloped virus, so the function of this appears to be to concentrate viral proteins in one location of the cell to better facilitate replication processes. Another reason this may have evolved is to shield the viral ssRNA from intracellular immune responses, discussed further in Chapter 6, "The Immune Response to Viruses."

The viral genome of +ssRNA viruses is used to create a complementary negative strand, the **antigenomic RNA**, that is used as a template to create many copies of the +ssRNA genome. Along with viral protein production, copying of the viral genome is a necessary step in generating the required elements for creating new virions.

4.4.5 Class V: –ssRNA Viruses

In contrast to +ssRNA viruses, negative-sense RNA viruses (–ssRNA viruses) have genomes that do not act as mRNA. Therefore, like their dsRNA counterparts, they must carry an RdRp within the virion into the cell. There exist six –ssRNA virus families that include some of the most well-known disease-causing viruses, including Ebola virus, Marburg virus, measles virus, mumps virus, rabies virus, and influenza virus. These viruses have enveloped, helical capsids and can have segmented genomes (like influenza) or nonsegmented genomes (like rabies virus).

Viruses with –ssRNA genomes generally do not enter the nucleus (although the –ssRNA influenza viruses are a notable exception to this rule). The –ssRNA genomes are not capped and do not have poly(A) tails, because the –ssRNA does not function as mRNA. Instead, the genome must first be transcribed by the viral RdRp into mRNA, which is then translated. –ssRNA viruses have helical nucleocapsids, where the viral RNA is coated with a repeating nucleocapsid protein, termed NP or N. In addition, the viral RdRp and other proteins necessary for transcription also associate with the nucleocapsid protein and viral RNA. Together, the complex of viral RNA and proteins is termed the viral **ribonucleoprotein** complex, because it contains RNA and viral proteins (Fig. 4.10C). Within the cell, the –ssRNA is immediately transcribed into viral mRNAs by the viral RdRp and any other required helper proteins to produce the virus's proteins.

As with +ssRNA viruses, antigenomic RNA is created to act as a template for replication of the genome. In the case of –ssRNA viruses, the complementary antigenomic RNA is +ssRNA. This antigenome is not identical to the positive-sense viral mRNAs produced during infection, however, since viral mRNAs are capped and polyadenylated. At some point during viral replication, a switch occurs so the RdRp drives genome replication over mRNA transcription. This can occur because certain translated viral proteins bind to the –ssRNA genome at sites that would normally stop the polymerase, allowing it to continue copying the entire genome into the antigenome. In some cases, newly translated viral proteins join the RdRp complex to promote genome replication over viral mRNA transcription.

Certain RNA viruses, termed **ambisense** viruses, have genomes that are partially negative sense and partially positive sense. They are still considered within the class of –ssRNA viruses, however, because the positive-sense portion of their genome is not directly infectious: it must first be copied into an antigenome segment that is used to create the viral mRNA. The arenaviruses are the only ambisense viruses that infect humans (although plant viruses in the *Tospovirus* genus within the *Bunyaviridae* family also have an ambisense genome). The arenaviruses, which include Lassa virus and lymphocytic choriomeningitis virus, have two genome segments, a long (L) segment and short (S) segment, which are each ambisense. Positive-sense viral mRNA is transcribed by the RdRp from the negative-sense portions of these segments. A complementary antigenome is also transcribed for each segment, and this is used to create the viral mRNA from the positive-sense portions of the genome segments (Fig. 4.11).

RNA viruses are more prone to mutation than DNA viruses. All polymerases, whether they use DNA or RNA as a template, introduce errors as they incorporate an incorrect nucleotide, but DNA-dependent DNA polymerases have **proofreading** ability: they can remove an incorrectly placed nucleotide and replace it with the correct one. RdRps, on the other hand, do not have proofreading ability. This raises the overall error rate of the enzyme, from 1 error per 10^9 bases for a DNA polymerase to greater than 1 error per 10^5 bases for an RdRp, which results in lower enzyme **fidelity**, or accuracy. RNA viruses have some of the highest mutation rates of all biological entities. Mutations generated by RdRps may result in mutated viral proteins and, subsequently, slightly different strains of the virus that may survive better under environmental pressures.

When more than one strain of virus enters a cell, **recombination** can occur. Recombination is the process by which a virus exchanges pieces of its genetic material with another strain of the same virus. This process, which can occur in dsRNA, +ssRNA, or –ssRNA viruses, occurs during genome replication when the RdRp, while copying one RNA genome template, switches to the template of another strain of the virus and continues replicating, thereby creating a hybrid genome that is different from either parent strain. The switching of

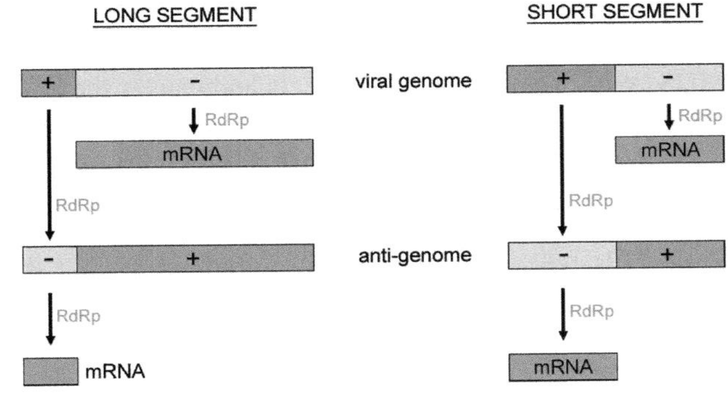

FIGURE 4.11 Replication of ambisense genomes. Ambisense genomes are composed of both –ssRNA and +ssRNA. The viral RdRp transcribes the mRNA from the –ssRNA portion. The +ssRNA portion is not directly translatable by ribosomes and must first be transcribed into the antigenome, which has the opposite sense as the ambisense genome. The –ssRNA portion of the antigenome is then transcribed into mRNA by the viral RdRp.

templates has been shown to occur at random sequences or at complementary sequences on the two genome templates that base pair with each other.

Segmented viruses can also undergo **reassortment** when two strains of virus with segmented genomes enter and replicate within the same cell. When the genome segments are copied, segments from one virus may mix with segments from another virus when they are being packaged into new virions, creating a new strain of virus. This can be potentially dangerous when two strains of viruses from different subtypes reassort to create a viral strain that has not previously circulated within the human population. This occurs with influenza A virus and has been the cause of several influenza epidemics, discussed in more detail in Chapter 10, "Influenza Viruses."

Study Break
What is the difference between positive-sense RNA and negative-sense RNA? What does this have to do with the replication of different types of viruses?

4.4.6 Class VI: RNA Viruses That Reverse Transcribe

The first event that occurs after a +ssRNA or −ssRNA virus enters a host cell is translation or transcription, respectively. **Retroviruses** also have RNA genomes, but must **reverse transcribe** their genome before using host enzymes to transcribe it. In human cells, DNA is used as a template to create mRNA. Retroviruses, on the other hand, encode and carry within their virions an enzyme called **reverse transcriptase** (RT) that is an RNA-dependent DNA polymerase. Reverse transcriptase is able to reverse transcribe the ssRNA genome into a linear strand of double-stranded complementary DNA (cDNA), which is then integrated into a host chromosome.

There exist only a handful of retroviruses, and even fewer that infect humans (Table 4.3). The most well-studied human retrovirus is HIV, the virus that causes AIDS. Within the body, HIV infects and causes the slow decline of T lymphocytes, unarguably one of the most important immune system cells in the defense against pathogens of all kinds. People with HIV are diagnosed with AIDS when the number of T lymphocytes in the blood falls below a certain number, indicating that the person's immune system is severely compromised (hence the origin of "immune deficiency syndrome" in the name of the disease). Without a functioning immune system, people with AIDS often succumb to **opportunistic infections** that healthy people would manage. It is often these opportunistic infections that ultimately cause death in people infected with HIV.

Retroviruses are unlike any other viruses because their genome is diploid, meaning that two copies of the genome are present within the virion. Other viruses are segmented and have their genomes in several segments, but the segments all encode different viral genes. The retrovirus genome is +ssRNA, although it does not serve as mRNA, like the genomes of +ssRNA viruses do. They are also unique among the RNA viruses because their genome will be copied by cellular enzymes, rather than an RdRp.

The retrovirus genome has several different domains that are of importance during viral replication. The two ends of the genome are flanked by redundant sequences, termed R. Inside of the R domain is the U5 (unique to the 5′) and U3 (unique to the 3′) domains on the 5′- and 3′-ends of the RNA, respectively. These domains will end up forming **long terminal repeats** (LTRs) on each side of the cDNA that will be important for **integration** of the HIV cDNA into the host's DNA.

Because the HIV reverse transcriptase enzyme is a target for drugs against HIV, it has been extensively studied and is the paradigm among retrovirus reverse transcriptases. RT is a heterodimer composed of two different-sized polypeptide chains, although one chain is just a slightly shorter version of the other. It is a unique enzyme because it can perform several enzymatic functions. It acts as an *RNA-dependent* DNA polymerase, a *DNA-dependent* DNA polymerase, and it also has RNase H activity, meaning that it is able to degrade RNA when bound to DNA. Although it is a DNA polymerase, it does not have proofreading ability, and is as error-prone as the RdRps, introducing an incorrect nucleotide once every 10^5 bases. It is also slow, about a 10th of the speed of DNA polymerase.

Only one of the two viral ssRNA strands is reverse transcribed by RT, although recombination can occur if the reverse transcriptase jumps to the other strand of viral ssRNA during reverse transcription. Initiation of reverse transcription requires a primer, which is provided by one of around 100 tRNAs (most of which are specific for lysine or proline) that the virus obtained from the previous host cell. The tRNA partially unwinds and complementary base pairs to 18 nucleotides within the **primer-binding site** (PBS), located toward the 5′-end of the viral RNA (Fig. 4.12). RT binds to the primer and extends its sequence, adding DNA nucleotides complementary to the viral RNA, until it reaches the end. This forms what is referred to as the **negative-strand strong-stop DNA**, because the RT has extended the negative strand and it stops when it reaches the end of the RNA template. The U5 and R sequences are copied in this segment. The RNase H activity of RT degrades the RNA from the RNA–DNA hybrid, leaving the ssDNA U5 and R sequences. Because the R sequence in the DNA is complementary to the R sequence found on the 3′-end of the viral RNA, the two R sequences bind and the RT is able to continue extending the negative strand until the end, where the PBS sequence

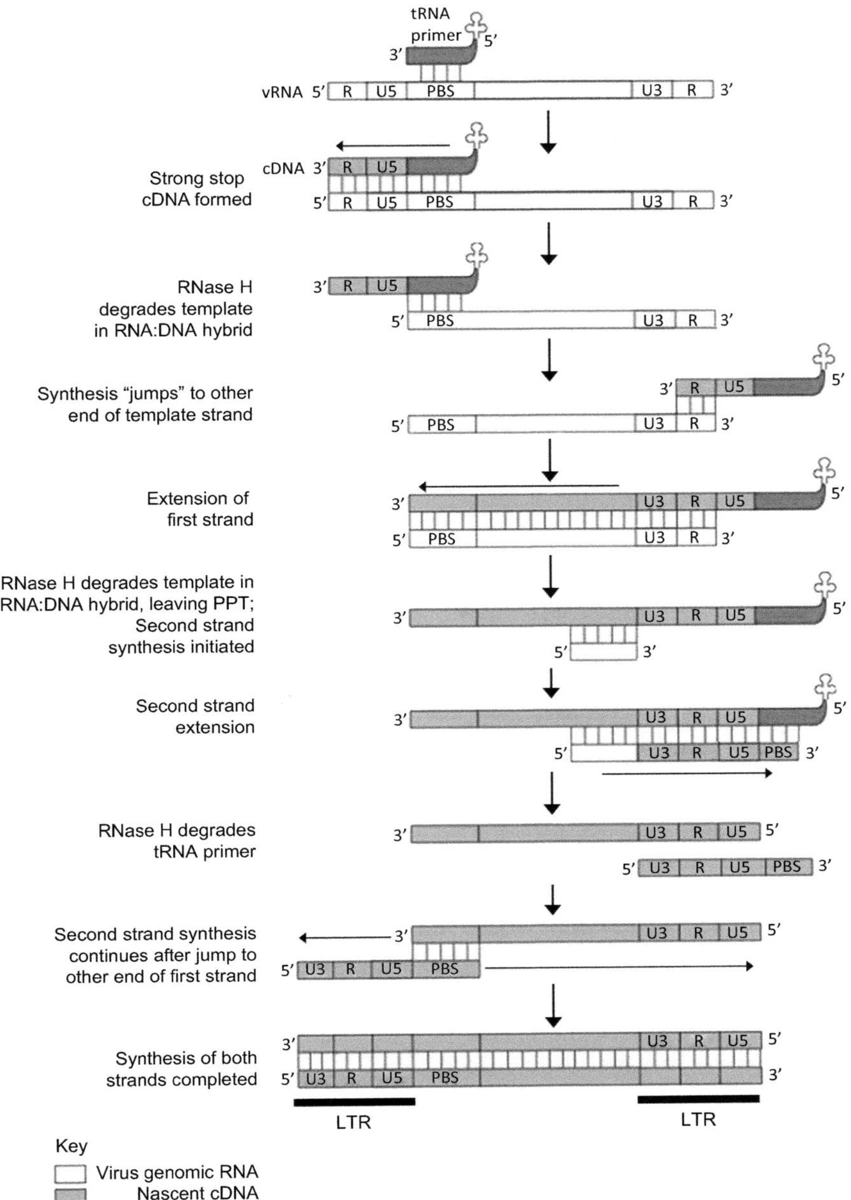

FIGURE 4.12 **Retrovirus reverse transcription.** Reverse transcription begins with a cellular tRNA binding to the primer-binding site (PBS) in the genomic ssRNA. RT binds to and adds DNA nucleotides to the primer in the 5′→3′ direction, copying the U5 and R sequences and forming the "negative-strand strong-stop cDNA." RT RNase H activity degrades the RNA of the RNA:DNA section. The strong-stop cDNA is transferred to the 3′-end of the RNA and binds it because of the complementary R sites. Synthesis of DNA continues in the 5′→3′ direction, completing the entire negative-sense DNA strand. RNase H activity of RT degrades the RNA template, with the exception of the polypurine tract (PPT), which acts as a primer for the synthesis of the positive-sense DNA strand. RT binds to the PPT and extends it through the U3, R, and U5 domains, completing 18 nucleotides into the tRNA primer to create the new PBS. The tRNA is digested. This "positive-strand strong-stop cDNA" is transferred to the other end of the DNA template; RT completes replication of both strands, creating long terminal repeats (LTRs) on both ends of the double-stranded cDNA. (It is likely that the RNA template circularizes during reverse transcription, which facilitates the "transfer" of the strong-stop cDNAs to the other end of the strand.) *(Image used with permission from Alan J. Cann, 2005. Principles of Molecular Virology, fourth ed. Elsevier/Academic Press, Figure 3.18, Copyright 2005.)*

is found. After being reverse transcribed, the viral RNA is degraded by the RNase H activity of the enzyme, although a small portion resists being digested. This **polypurine tract (PPT)** is composed of purine nucleotides, namely adenine and guanine, and it acts as a primer for RT to begin reverse transcribing the DNA positive strand. This PPT is extended by RT in the 5′→3′ direction, through the U3, R, and U5 domains, until it reaches the tRNA primer, where it finishes the strand by copying 18 nucleotides into the tRNA primer. (You will recall that these 18 nucleotides

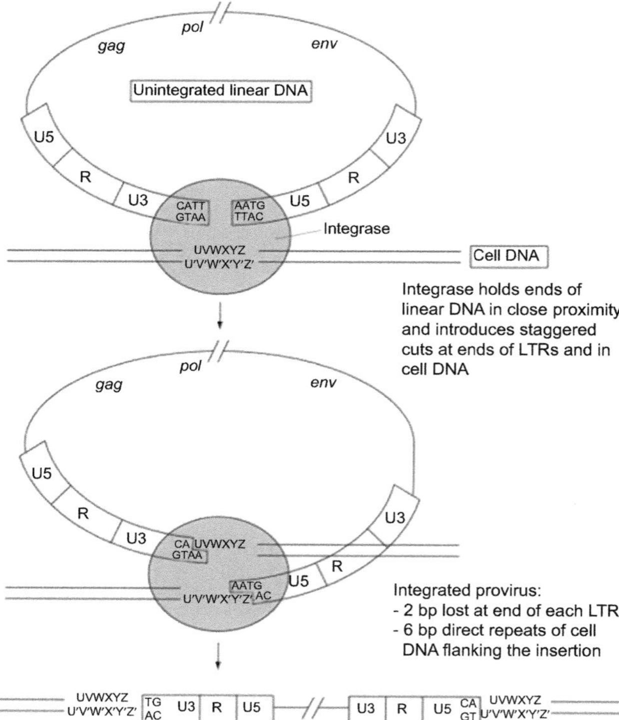

FIGURE 4.13 Integration of the retrovirus genome into the host chromosome. Viral integrase removes two base pairs from each end of the proviral DNA, creates a nick in the host chromatin, and joins the proviral DNA to the host DNA. Cellular DNA repair enzymes seal the nick in the sugar–phosphate backbone of the DNA. *Image used with permission from Elsevier/Academic Press: Alan J. Cann, 2005. Principles of Molecular Virology, fourth ed., Fig. 3.20, Copyright 2005.*

of the tRNA initially unwound and bound to the PBS. By copying this segment of the tRNA, another PBS sequence has been created.) This segment is termed the **positive-strand strong-stop DNA**, because it is the positive DNA strand and came to a stop because it reached the end of the viral DNA template.

The tRNA is digested by the RT RNase H, which leaves available the PBS sequence. This anneals to its complementary sequence found at the end of the newly copied negative strand. This acts as the primer for RT to complete the positive strand. Similarly, the negative-strand is extended to the end of the positive-strand strong-stop DNA.

The resulting DNA is double stranded with repeated ends, termed LTRs, that are composed of the U3, R, and U5 domains (Fig. 4.12). These LTRs are important during the integration of this **proviral DNA** into the genome of the host cell. **Integrase** (IN), another necessary retroviral enzyme found within the virion, carries out the process of integration (Fig. 4.13). Having removed two base pairs from each end of the proviral DNA, IN creates a nick in the host chromatin and joins the proviral DNA to the host DNA. DNA repair enzymes within the cell seal the break.

At this point, the integrated viral DNA is like any other cellular gene and will be transcribed by the host RNA polymerase II. Promoter sequences within the U3 region are bound by host transcription factors that are recruited within activated cells, causing the production of several viral mRNAs through splicing and ribosomal frameshifting. The functions of HIV genes will be discussed in more detail in Chapter 11, "Human Immunodeficiency Virus." Full-length mRNA is produced, complete with a 5′-cap and 3′-poly(A) tail, and two copies are packaged into **nascent** (newly formed) virions as the diploid viral genome.

In-Depth Look: Reverse Transcriptase

Reverse transcriptase was independently discovered in 1970 by two separate groups led by Howard Temin and David Baltimore, who both received Nobel Prize in physiology or medicine in 1975 for their discovery. Reverse transcriptase possesses the activity of three different enzymes in one: an RNA-dependent DNA polymerase, a DNA-dependent DNA polymerase, and RNase H, which degrades the RNA from an RNA:DNA duplex.

The enzyme has been of much utility in molecular biology since its discovery. It was first isolated from viruses but now is produced by bacteria that have been engineered to express the gene. In the laboratory, RT allows for the production of cDNAs from mRNAs. Because cDNA is more stable and easier to manipulate than mRNA, this discovery has been vital in the characterization of gene expression and the study of mRNAs (Fig. 4.14).

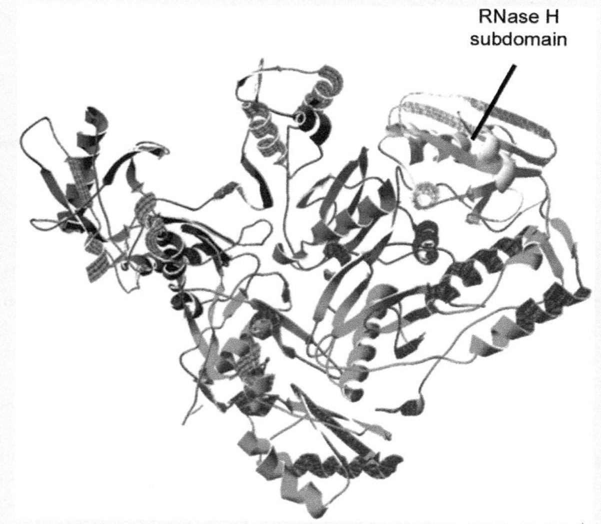

FIGURE 4.14 HIV reverse transcriptase. Crystal structure of HIV-1 RT (PDB 1RTD). The p51 subunit is shown in orange; the slightly larger p66 subunit is divided into the fingers (cyan), connection (blue), and RNase H (gray) subdomains. *Image courtesy of Beilhartz et al., 2010. Viruses. 2, 900–926.*

4.4.7 Class VII: DNA Viruses That Reverse Transcribe

Retroviruses are RNA viruses that reverse transcribe, but two families of DNA viruses also undergo reverse transcription during their replication within the cell. However, in contrast to retroviruses that undergo reverse transcription as one of the first events in the cell, DNA viruses that reverse transcribe do so at the end of their replication cycle in order to generate their DNA genomes.

Whether RNA or DNA, any virus that reverse transcribes is termed a **retroid virus**. Two DNA virus families fall into this category, *Caulimoviridae*, which infect plants, and the *Hepadnaviridae*, which infect animals. The only human virus within this family is hepatitis B virus (HBV). The virus infects the liver and can cause **hepatitis**, inflammation of the liver. Approximately 5% of infections with HBV become a long-term, **chronic** infection that can lead to liver scarring (cirrhosis) and even liver cancer.

Like other DNA viruses, the genome of hepadnaviruses must also be transported into the nucleus. The DNA genome is partially double stranded and partially single stranded. The complete strand is the negative-sense DNA, and the incomplete strand is positive-sense DNA (Fig. 4.15A). This **relaxed circular DNA** (rcDNA) is transported into the nucleus, where the gapped segment is repaired into double-stranded DNA by a yet unidentified enzyme, completing the **covalently closed circular DNA** (cccDNA). The cccDNA is ligated and maintained as an **episome**, meaning that the circular dsDNA does not integrate into the host DNA but remains as a separate entity within the nucleus.

From the negative strand of episomal cccDNA, host RNA polymerase II transcribes viral mRNAs that leave the nucleus and are translated by host ribosomes to create viral proteins. RNA polymerase II also creates an RNA **pregenome** that leaves the nucleus. Interestingly, it functions as mRNA but also acts as the template for reverse transcription. Within the cccDNA are two identical sequences of 12 nucleotides, termed DR1 (direct repeat 1) and DR2 (direct repeat 2) (Fig. 4.15A). The pregenomic RNA (pgRNA) begins being translated on the negative strand of the DNA at a site upstream of DR1, and RNA polymerase II transcribes the entire negative strand, including the DR2 site. The dsDNA episome is circular, however, and RNA polymerase II continues past its initial starting point, terminating downstream of DR1 at a polyadenylation signal. The result is a long pgRNA that has repeating sequences at both ends that each include a copy of DR1. Part of the repeated sequence folds into a physical structure, known as epsilon (ε), that serves as the initial start of reverse transcription.

The pgRNA and the recently translated P protein are packaged into forming capsids. Similarly to the RT protein of retroviruses, the **P protein** functions as an RNA-dependent DNA polymerase, a DNA-dependent DNA polymerase, and an RNase H, removing RNA from RNA–DNA hybrids. The P protein binds to the ε structure and reverse transcribes a few base pairs from it (Fig. 4.15B). This functions as a primer that binds to the positive-sense pgRNA and reverse transcribes it, creating the complete negative strand of the DNA genome. The RNase H activity of P protein degrades the pgRNA from this strand, leaving a short RNA segment at the end that includes the repeated DR1 sequence. This RNA primer relocates to the other end of the DNA and binds to the identical DR2 sequence there (Fig. 4.15C). P protein extends this positive-sense DNA strand to the end of the negative-sense template. This creates the shorter positive-sense DNA strand within the HBV genome, which bridges the gap between the 5′- and 3′-ends of the negative-sense strand using complementary sequences. This completes genome replication and the creation of the rcDNA.

4.5 ASSEMBLY

Viruses are created from newly synthesized components, and to be released from the cell, those components must be collected at a particular site of the cell and undergo **assembly** to form an immature virus particle. In the same way that penetration and uncoating are difficult to separate in the cycle of some viruses, assembly can often occur alongside maturation and release.

The location of virion assembly will depend upon the particular virus. It can take place within the nucleus of the cell, at the plasma membrane, or at a variety of intracellular membranes, such as the Golgi complex. Most nonenveloped DNA viruses assemble their nucleocapsid in the nucleus, since that is the site of genome replication. Viral proteins are imported through nuclear pores to reach the site of assembly. When assembled, most DNA viruses are too large to fit through nuclear pores, however. At this point, some viruses are able to traverse the double-membraned nuclear envelope, while others induce cell lysis or apoptosis to escape the nucleus. On the other hand, viruses with envelopes derived from the plasma membrane usually assemble there.

The nucleic acid genome of a helical virus is protected by repeating capsid proteins. Because of this, capsid proteins can begin wrapping the genome as soon as it is copied (or vice versa, depending upon the virus: the genome can be wrapped around capsid proteins; Fig. 4.16). In contrast, some icosahedral viruses nearly complete the assembly of their capsids before the nucleic acid genome is inserted.

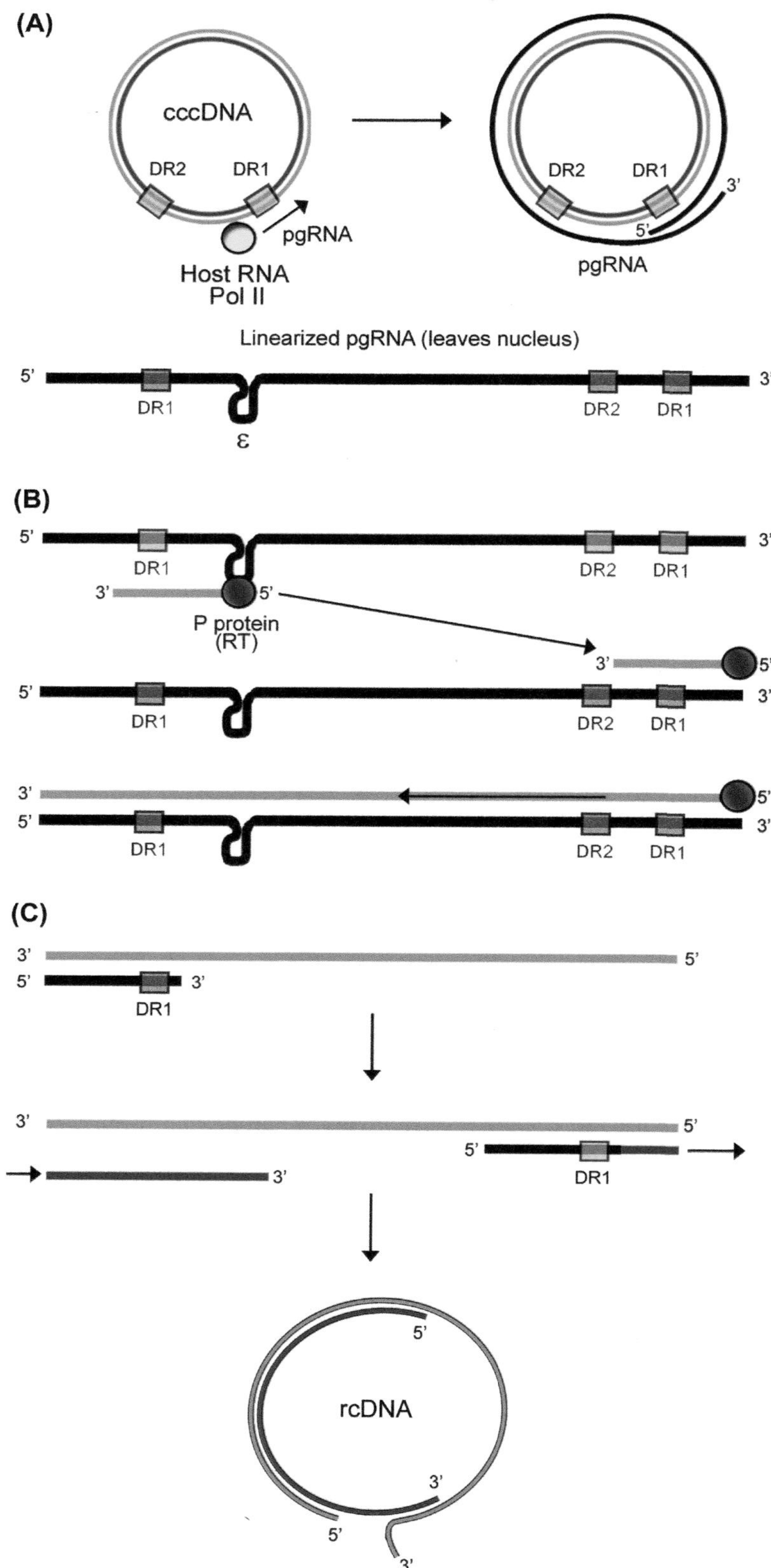

FIGURE 4.15 **Reverse transcription of the HBV genome.** The HBV genome has identical repeated sequences termed DR1 and DR2. Once the rcDNA is repaired into closed circular DNA (cccDNA), cellular RNA polymerase II transcribes an RNA pregenome (A). It starts upstream of the DR1 site and completes a full circle, ending past the DR1 site where it began. This results in an RNA pregenome that has a 5′-DR1 site and 3′-DR2 and DR1 sites. In the cytoplasm, the pregenomic RNA (pgRNA) is encapsidated. The HBV reverse transcriptase, known as P protein, binds to the ε site and reverse transcribes DNA in the 5′→3′ direction (B). This small piece acts as a DNA primer, which jumps to the other end of the pgRNA, binding to the complementary sequence found there. P protein complex the reverse transcription of the negative DNA strand and degrades most of the RNA:DNA duplex. (C) The remaining RNA segment is transferred to the other end of the negative-sense DNA strand, binding to the complementary DR2 sequence found there. It continues synthesizing DNA in the 5′→3′ direction; because of complementary sequences, the genome circularizes and the P protein completes the synthesis of the shorter positive-sense DNA strand.

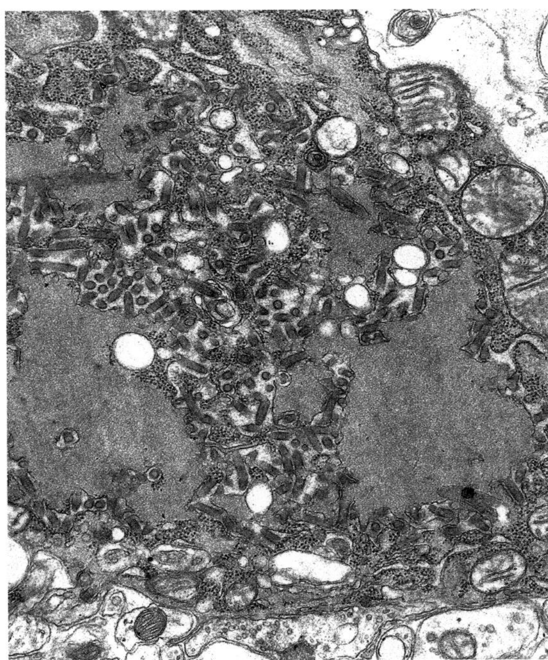

FIGURE 4.16 **Assembly of rabies virions.** This electron micrograph of a cell infected with the rabies virus shows the bullet-shaped capsids assembling within the cytoplasm of the cell. The rabies virus subsequently buds from the plasma membrane, obtaining its envelope. *(Micrograph courtesy of CDC/Dr. Fred Murphy.)*

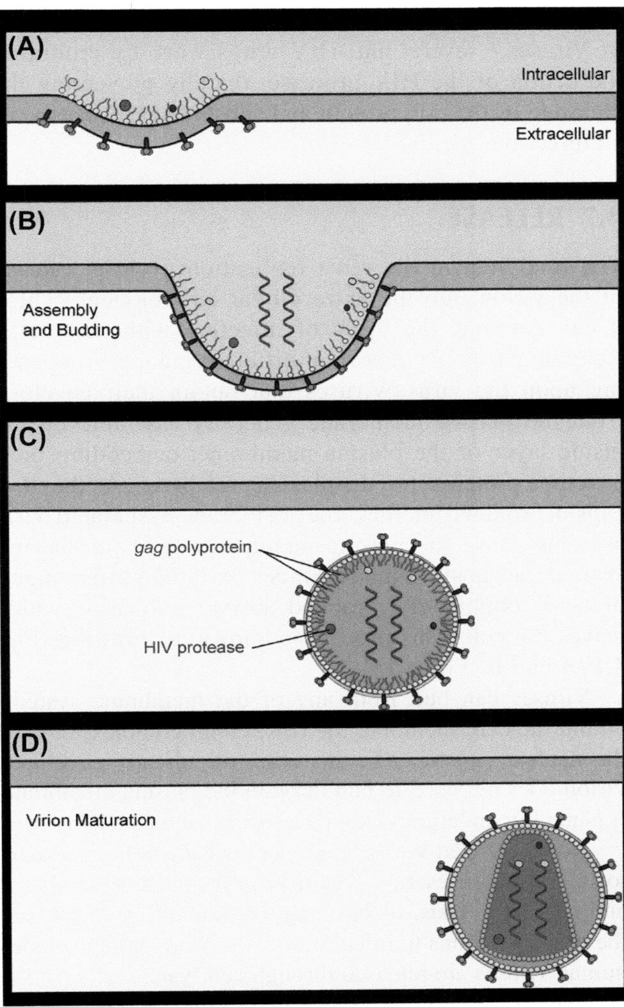

FIGURE 4.17 **Assembly, release, and maturation of HIV virions.** (A) HIV proteins congregate at the plasma membrane of the cell, causing a bud to form in the membrane. The diploid RNA genome is packaged into the assembling capsid (B). The virus is released from the cell membrane, but the Gag polyprotein has not yet been cleaved to separate the capsid and matrix proteins of the virion (C). The HIV protease cleaves the polyprotein, allowing the proteins to complete the infectious virion architecture (D).

Spontaneous assembly of the capsid, termed "self-assembly," occurs with the capsid proteins of simple icosahedral viruses, such as the picornaviruses and parvoviruses. The assembly of viruses with more complex architecture is orchestrated by a variety of viral chaperone proteins called scaffolding proteins. Herpesviruses and adenoviruses are examples of large icosahedral viruses that assemble with scaffolding protein assistance.

4.6 MATURATION

After the nucleic acid genome and other essential proteins are packaged within the capsid, which was assembled from one or several translated viral proteins, the final steps of virus replication occur: maturation and release. Up to this point, the virion had been in the process of forming, and if the cell were broken open at this point, the virions would not be able to initiate infection of new cells. **Maturation** refers to the final changes within an immature virion that result in an infectious virus particle. Structural capsid changes are often involved, and these can be mediated by host enzymes or virus-encoded enzymes. A good example involves the influenza HA protein. It is involved in attachment to the cell's sialic acid, as described above, and the HA protein is able to bind sialic acid after being glycosylated (via posttranslational modification). However, the HA protein must be cleaved into two portions, HA_1 and HA_2, to become infectious, because although the HA_1 portion binds the cell surface receptor, the HA_2 portion is what fuses the viral envelope to the endosomal membrane to release the virus into the cytoplasm. This cleavage of HA into HA_1 and HA_2 is carried out by cell proteases (enzymes that cleave proteins). In contrast, the HIV core particle is composed of proteins encoded by the *gag* gene. The gene is translated into a polyprotein that is cleaved by the viral protease to form the capsid, matrix, and nucleocapsid proteins of the virion. In this case, maturation occurs after the virion has been released from the cell surface (Fig. 4.17) and is required to form an infectious virion. Discussed in

Chapter 8, "Vaccines, Antivirals, and the Beneficial Uses of Viruses," several anti-HIV drugs work by inhibiting the action of the HIV protease, thereby preventing the cleavage of the polyprotein and subsequent formation of an infectious virion.

4.7 RELEASE

The final step in the virus replication cycle is **release** of the virion into the extracellular environment, where it can continue the cycle of infection with new cells. Release can occur in several different manners, depending upon the virus. Viruses that obtain their envelope from the plasma membrane generally assemble on the inside layer of the plasma membrane, embedding their envelope proteins into the plasma membrane. As the viral capsid proteins interact, the membrane-associated viral proteins cause the plasma membrane to begin curving around the capsid. This continues until the plasma membrane is completely wrapped around the virus, which leaves the cell. This process is known as **budding** (Fig. 4.17A and B; Fig. 4.18).

Viruses can bud from any of the membrane systems within the cell, including the rER, Golgi complex, or even the nuclear envelope. In this case, the already enveloped virion does not need to bud through the plasma membrane. It generally undergoes exocytosis to leave the cell.

Nonenveloped viruses can also exit the cell via exocytosis. **Lytic** viruses, however, disrupt the plasma membrane and cause the **lysis**, or bursting, of the cell. This releases the nascent virions to infect new cells. Many nonenveloped human viruses are released through cell lysis.

The processes of assembly, maturation, and release are closely linked, but all are required to create progeny infectious virions able to continue the cycle of infection.

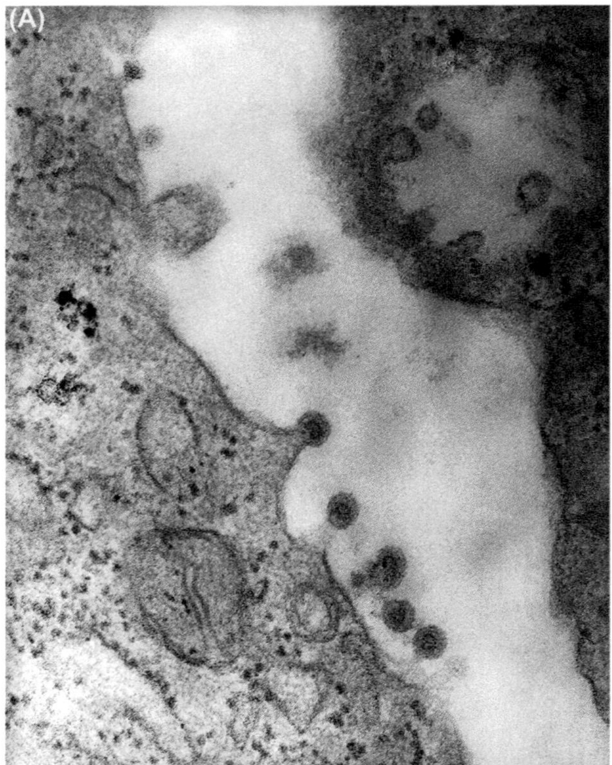

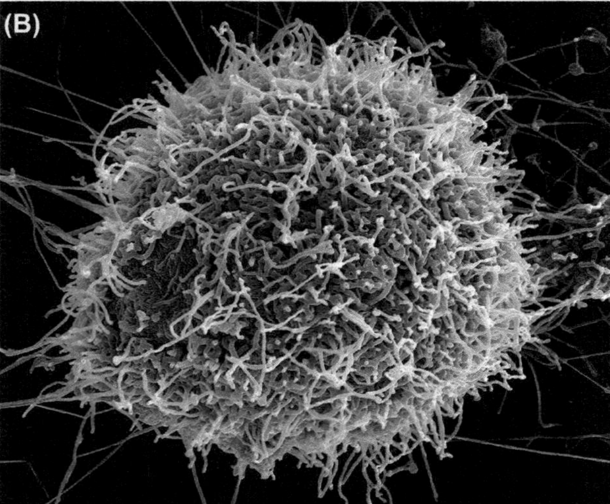

FIGURE 4.18 **Virion budding.** (A) Rubella virus virions are observed budding from the host plasma membrane in this transmission electron micrograph. *(Image courtesy of CDC/Dr. Fred Murphy and Sylvia Whitfield.)* (B) In this digitally enhanced pseudocolored scanning electron micrograph, helical Ebolavirus virions (blue) are budding from an infected cell (yellow). *(Courtesy of the National Institute of Allergy and Infection Diseases (NIAID).)*

4.8 VIRUS GROWTH CURVES

In the laboratory, scientists can infect cells with virus to observe the kinetics of the viral replication cycle. **One-step growth curves** are used to study the replication cycle of a virus infection. While at the California Institute of Technology, Emory Ellis and Nobel laureate Max Delbrück devised the "one-step growth curve" using bacteriophages, so named because the viruses replicated simultaneously, all together in one step. To synchronize the infection of many cells at once, a high ratio of virus to cells is used. The **multiplicity of infection** (MOI) refers to this ratio: an MOI of 1 means that 1 virus particle is used per cell for infection, while an MOI of 10 means that 10 virus particles are used per cell. For one-step growth curves, a high MOI is generally used to ensure infection of all cells.

When infecting bacteria, the bacteriophages infect the cells immediately. After a very short period of time, generally 1 min, the culture is diluted with media to prevent continued infection, or alternatively, the cells can

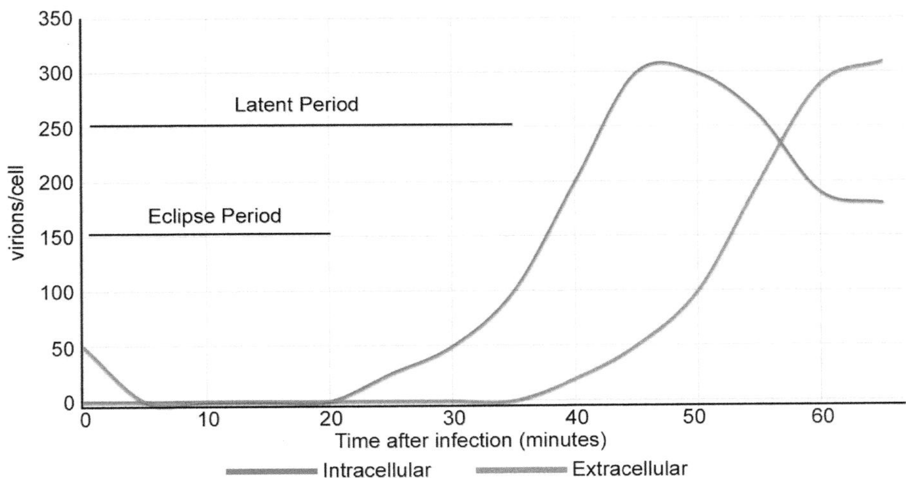

FIGURE 4.19 **The one-step growth curve experiment.** In this experiment, a high MOI is used to synchronize infection of cells with bacteriophages. Once infection occurs, no extracellular virus is observed because all the viruses are replicating internally. No infectious virions are observed intracellularly until assembly occurs within the cell. The amount of time this takes to occur is known as the eclipse period. The infectious virions are released from the cell, at which time they can be observed extracellularly. The amount of time it takes to produce extracellular infectious virions is known as the latent period. A plaque assay, discussed in Chapter 7, "Detection and Diagnosis of Viral Infections," is used to determine the number of infectious virions present.

be centrifuged so the liquid media with any nonadsorbed bacteriophages can be removed and replaced with fresh media. A sample of the culture is taken every minute, and the number of infectious virions is determined using a **plaque assay**, which will be described in detail in Chapter 7, "Detection and Diagnosis of Viral Infections." Initially, there are no additional infectious virions in the culture, because all of the bacteriophages have infected the bacterial cells and are in the process of intracellular replication. This period is known as the **latent period** (Fig. 4.19). Then, the first **burst** occurs as the bacteria are lysed by the bacteriophages, which escape from the cells and are now present in the media. The **burst size** is the number of infectious virions that are released per cell. A typical bacteriophage infection releases 50–200 virions per infected cell, and the latent period is typically 20–30 min.

Because the virions assemble from newly synthesized proteins, scientists can also artificially lyse the bacterial cells at various timepoints to see when the virions are assembled internally and infectious, but not yet released. This period is known as the **eclipse period** and represents the amount of time it takes to form infectious virions within the cell (Fig. 4.19).

One-step growth curves have also been performed with viruses that infect eukaryotes. These viruses take significantly longer to replicate because eukaryotic cells do not replicate as quickly as bacteria; as a result, the enzymes required by some viruses are not necessarily immediately available. Eukaryotic cells are also more complex and involve processes and organelles, such as the rER or Golgi complex, that are not found in bacteria. Consequently, the viral replication cycle takes longer in eukaryotic cells, and the latent period of viruses that infect eukaryotes is significantly longer, about 18–24 h. The eclipse period, where intracellular infectious virions are observed, will vary depending upon whether or not the virus is enveloped. Nonenveloped viruses will be infectious upon assembly and maturation, whereas viruses that obtain their envelope from the plasma membrane will not be infectious until they have budded from the cell.

SUMMARY OF KEY CONCEPTS

Section 4.1 Attachment

- Viruses make initial contact with cells at the plasma membrane. The binding of a virion is a specific process that involves the virus attachment protein binding to a cell surface receptor, which will vary depending upon the virus. This determines the tropism of the virus. Some viruses require coreceptors for entry.
- The binding of a virus attachment protein to a cell surface receptor involves electrostatic forces. Virus attachment proteins will be located on the outermost surface of the virion, whether that is the envelope or capsid (for nonenveloped viruses).

Section 4.2 Penetration

- Penetration is the method by which viruses cross the plasma membrane of the cell.
- Viruses take advantage of normal cellular processes to gain entry into the cytoplasm. One common method involves endocytosis. This can occur through endocytosis of caveolin- or clathrin-coated pits, bulk-phase endocytosis, or phagocytosis.
- Some enveloped viruses use fusion to enter the cell. This process is mediated by viral fusion proteins and merges the viral envelope with the cell membrane.

Section 4.3 Uncoating

- Uncoating refers to the breakdown of the viral capsid, releasing the genome into the cell. Unlike a cell, which divides into two, nascent virions are assembled de novo by packaging a copied genome into a newly created capsid.
- Many viruses achieve uncoating by escaping from the endosome that they used to enter the cell. Other viruses do not completely uncoat and use the remaining capsid as a home base for replication processes.
- Some viruses uncoat at the nuclear envelope immediately before transporting the genome into the nucleus. A few viruses are small enough to pass through the nuclear pores and uncoat in the nucleus.

Section 4.4 Replication

- There are seven classes of viruses in the Baltimore classification system, which is based upon the type of nucleic acid and replication strategy of viruses: dsDNA, ssDNA, dsRNA, +ssRNA, −ssRNA, RNA viruses that reverse transcribe, and DNA viruses that reverse transcribe.
- All human dsDNA viruses, with the exception of poxviruses, must gain entry into the nucleus for replication because of the DNA or RNA polymerases that are present there. dsDNA viruses often transcribe their gene products in waves in order to ensure an ordered process that does not put too much overall stress upon the cell. Herpesviruses, poxviruses, and adenoviruses are examples of dsDNA viruses that infect humans.
- The genome of ssDNA viruses is converted into dsDNA by the host cell DNA polymerase after self-priming. At this point, the replication of DNA and transcription of mRNA are carried out in the same way as dsDNA viruses. Anelloviruses and parvoviruses are examples of human ssDNA viruses.
- Human dsRNA viruses, namely reoviruses and picobirnaviruses, are nonenveloped and icosahedral with segmented genomes. To transcribe their genome into vmRNA, they must carry their own RdRp, since the cell does not contain an enzyme that will transcribe mRNA from an RNA template.
- The genomes of +ssRNA viruses are infectious, since positive-sense RNA is able to be directly translated by ribosomes. Consequently, +ssRNA viruses do not carry an RdRp protein but instead encode it so it is immediately translated by ribosomes into the RdRp protein. +ssRNA viruses encode a single polyprotein that is cleaved into several different individual proteins. Seven different families of +ssRNA viruses infect humans and include poliovirus, rhinovirus, Norwalk virus, hepatitis C virus, and rubella virus, among many others.
- −ssRNA viruses are not infectious and must be transcribed into vmRNA before translation can occur. Therefore, −ssRNA viruses must also carry an RdRp into the cell. −ssRNA viruses have helical ribonucleoprotein complexes that are composed of the viral RNA, the protecting capsid protein, and any associated enzymes. As with +ssRNA viruses, an antigenome is transcribed and acts as a template to create the genome. Ambisense viruses are partially positive sense and partially negative sense. Several well-known human viruses have −ssRNA genomes, including influenza viruses, Ebola virus, rabies virus, measles virus, and mumps virus.
- Due to the lack of proofreading ability, RNA polymerases have lower fidelity than DNA polymerases.
- New strains of virus can occur when two different strains infect one cell. Recombination occurs when the genome of an RNA virus is being replicated and the RdRp jumps from the template of one strain to the template of the other strain, creating a hybrid genome. Reassortment occurs when the genome segments of segmented viruses are mixed while being packaged into new capsids.
- Retroviruses are viruses that reverse transcribe an RNA genome into cDNA. Reverse transcriptase is the enzyme that carries this out and has the activity of an RNA-dependent DNA polymerase, DNA-dependent DNA polymerase, and RNase H. HIV, a human retrovirus, becomes a provirus when its IN protein merges the viral cDNA into the host's chromosomal DNA. Transcription and genome replication is carried out by host enzymes.

- Hepatitis B virus is also a retroid virus but instead reverse transcribes in order to create its DNA genome, which is partially single-stranded and partially double-stranded and known as rcDNA. This is repaired to a completely double-stranded episome (cccDNA) in the nucleus of the cell. RNA polymerase II transcribes an RNA pregenome that is reverse transcribed, after being packaged into the capsid, into the rcDNA genome.

Section 4.5 Assembly

- Assembly refers to the packaging of the copied viral genome with newly manufactured viral proteins to create a virion. This can occur in the nucleus, within an organelle like the rER or Golgi complex, or in the cytosol, depending upon the virus.

Section 4.6 Maturation

- Maturation refers to the final changes that must occur within the virion to create an infectious virion rather than an inert particle. This can involve the modification of cell surface receptors, the cleavage of viral polyproteins, or changes to the viral capsid. Maturation is often tightly linked with assembly and/or release.

Section 4.7 Release

- Release refers to the exit of the virion from the cell. This most often occurs through the budding of enveloped viruses or via cell lysis.

Section 4.8 Virus Growth Curves

- One-step growth curves rely on synchronous infection of cells so that the replication process and virion release occur simultaneously in all cells. The eclipse period refers to the amount of time required to assemble infectious virions intracellularly, and the latent period is the amount of time before infectious virions are observed outside the cell. Viruses that infect eukaryotic cells take much longer to replicate than bacteriophages. The eclipse and latent periods will be the same for eukaryotic viruses that obtain their envelope from the plasma membrane, unless maturation occurs after release.

FLASH CARD VOCABULARY

Attachment	Release
Penetration	Glycosylation
Uncoating	Virus attachment protein
Replication	Cell surface receptor
Assembly	Coreceptor
Maturation	Tropism
Host range	Fidelity
Phagocytosis	Recombination
Fusion	Reassortment
Segmented genome	Retroviruses
Nonsegmented genome	Reverse transcriptase
Self-priming	Opportunistic infections
Monocistronic	Long terminal repeats
Polycistronic	Integration/Integrase
Positive-sense RNA	Proviral DNA
Negative-sense RNA	Nascent
RNA-dependent RNA polymerase	Retroid virus
Polyprotein	Hepatitis
Replication complexes	One-step growth curves
Antigenomic RNA	Multiplicity of infection
Ribonucleoprotein complex	Latent period
Ambisense	Eclipse period
Proofreading	Burst size

CHAPTER REVIEW QUESTIONS

1. List what takes place at each of the seven steps of viral replication.
2. Considering that each virus must bind to a specific cell surface receptor for attachment, explain how you would create a drug that prevents viral attachment.
3. Focusing on the nucleic acids and enzymes involved, draw out the replication strategies of the seven classes of viruses.
4. Regardless of the type of nucleic acid, what are the general requirements for a virus to create functional nascent virions?
5. Make a chart that lists the location of transcription for each of the seven classes of viruses.
6. Explain why +ssRNA viruses do not have to carry their own RdRp within their virions.
7. What is the difference between recombination and reassortment?
8. List the steps involved in the reverse transcription and integration of a retrovirus genome.
9. Describe the steps involved in replicating the genome of HBV.
10. Both HIV and HBV use reverse transcription. Explain how reverse transcription is used differently in the replication of these two viruses.
11. What generally determines whether or not a virus needs to gain entry into the nucleus to replicate?
12. Make a table of the seven classes of viruses and list what the *first* event is that occurs after the virus gains

entry into the cell. Transcription? Reverse transcription? Translation?

13. Which of the cellular processes described in this chapter are limited only to enveloped viruses compared to nonenveloped viruses?
14. Which classes of viruses are more prone to introducing mutations during genome replication?
15. What would be the result of interfering with the maturation of virions?
16. Looking at the one-step growth curves, extracellular virus disappears because the virus enters the cell. Why does the virus initially disappear from the intracellular samples, too?

FURTHER READING

Baltimore, D., 1970. RNA-dependent DNA polymerase in virions of RNA tumour viruses. Nature 226, 1209–1211.

Beilhartz, G.L., Götte, M., 2010. HIV-1 ribonuclease H: structure, catalytic mechanism and inhibitors. Viruses 2, 900–926.

Brenkman, A.B., Breure, E.C., van der Vliet, P.C., 2002. Molecular architecture of adenovirus DNA polymerase and location of the protein primer. J. Virol. 76, 8200–8207.

Clement, C., Tiwari, V., Scanlan, P.M., Valyi-Nagy, T., Yue, B.Y.J.T., Shukla, D., 2006. A novel role for phagocytosis-like uptake in herpes simplex virus entry. J. Cell Biol. 174, 1009–1021.

Ellis, E.L., Delbruck, M., 1939. The growth of bacteriophage. J. Gen. Physiol. 22, 365–384.

Ghigo, E., Kartenbeck, J., Lien, P., et al., 2008. Ameobal pathogen mimivirus infects macrophages through phagocytosis. PLoS Pathog. 4. http://dx.doi.org/10.1371/journal.ppat.1000087.

Goff, S.P., 2013. Chapter 47: Retroviridae. In: Knipe, D.M., Howley, P.M. (Eds.), Fields Virology, sixth ed. Wolters Kluwer Lippincott Williams and Wilkins, pp. 1424–1473.

Haywood, A.M., 2010. Membrane uncoating of intact enveloped viruses. J. Virol. 84, 10946–10955.

Kielian, M., 2014. Mechanisms of virus membrane fusion proteins. Annu. Rev. Virol. 1, 171–189.

Lyman, M.G., Enquist, L.W., 2009. Herpesvirus interactions with the host cytoskeleton. J. Virol. 83, 2058–2066.

Rabe, B., Vlachou, A., Panté, N., Helenius, A., Kann, M., 2003. Nuclear import of hepatitis B virus capsids and release of the viral genome. Proc. Natl. Acad. Sci. U.S.A. 100, 9849–9854.

Ren, R.B., Costantini, F., Gorgacz, E.J., Lee, J.J., Racaniello, V.R., 1990. Transgenic mice expressing a human poliovirus receptor: a new model for poliomyelitis. Cell 63, 353–362.

Risco, C., Fernández de Castro, I., Sanz-Sánchez, L., Narayan, K., Grandinetti, G., Subramaniam, S., 2013. Three-dimensional imaging of viral infections. Annu. Rev. Virol. 1, 140718093033004.

Sanjuán, R., 2012. From molecular genetics to phylodynamics: evolutionary relevance of mutation rates across viruses. PLoS Pathog. 8, 1–5.

Stöh, L.J., Stehle, T., 2013. Glycan engagement by viruses: receptor switches and specificity. Annu. Rev. Virol. 1, 140707224641009.

Verdaguer, N., Fita, I., Reithmayer, M., Moser, R., Blaas, D., 2004. X-ray structure of a minor group human rhinovirus bound to a fragment of its cellular receptor protein. Nat. Struct. Mol. Biol. 11, 429–434.

Temin, H.M., Mizutani, S., 1970. RNA-dependent DNA polymerase in virions of Rous sarcoma virus. Nature 226, 1211–1213.

Tao, Y.J., Ye, Q., 2010. RNA virus replication complexes. PLoS Pathog. 6, 1–3.

Suomalainen, M., Greber, U.F., 2013. Uncoating of non-enveloped viruses. Curr. Opin. Virol. 3, 27–33.

Will, H., Reiser, W., Weimer, T., et al., 1987. Replication strategy of human hepatitis B virus. J. Virol. 61, 904–911.

Chapter 5

Virus Transmission and Epidemiology

The previous chapter described how viruses enter, replicate within, and exit host cells. This chapter takes a broader look at how viruses enter a host, spread throughout the body, and exit a host to infect other individuals within a population.

Viral pathogenesis is how viruses cause disease within a host. Several factors must be overcome, however, for a virus to initiate a successful infection. First, sufficient numbers of virions must enter the host. A single virion is theoretically enough to initiate infection, but there are many other factors that make it unlikely that a single virion will be successful in establishing an infection. The host cells must be **accessible** to the virus, and those cells must be **susceptible** to infection, meaning that the cells express the receptors to which the virus can bind. This affinity for susceptible tissues is known as **tropism**. The cells must also be **permissive** to infection, meaning that they contain the proteins and molecules within the cell that are necessary for replication to occur. There are also mechanical, chemical, and microbiological barriers to infection at every site within the body, and the host's immune system is quickly activated to eradicate the virus. The viruses that exist today have evolutionarily been selected for their traits that allow them to circumvent host factors and initiate infection, although the most successful viruses are not the most virulent: an extremely pathogenic virus will kill its host, thereby eliminating its reservoir and interrupting the chain of infection to another susceptible host.

5.1 PORTALS OF VIRUS ENTRY

To establish infection, a virus must come in contact with host cells that are susceptible and permissive to infection. There are several different **portals of entry** that are used by different viruses (Fig. 5.1 and Table 5.1). Most viruses interact with the cells of the host **epithelium**, the layers of cells that line the outside surface and inner cavities of the body. The epithelium acts as the main barrier between the outside world and the internal environment of the body. **Mucosal epithelium**, so named because the epithelium is covered in a protective layer of mucus, lines all the internal surfaces of the body, including the respiratory tract, gastrointestinal tract, and genital tract. The epithelium can be bypassed, however, when viruses are delivered to internal sites through penetration of the skin, as happens with an insect or animal bite, or through transplantation of a virally infected organ. Several viruses are also able to cross the placenta to a fetus or be transmitted to a child during or after birth.

5.1.1 Respiratory Tract

The **respiratory tract** is the most common portal of entry for viruses into the human body. It is a system of tubes that allows for gas exchange between the body and the external environment. The mucosal surfaces of the respiratory tract translate to a very large surface area with which viruses can interact. A resting human inhales around 2 gallons of air every minute, and within each breath are aerosolized droplets and particles that could contain viruses, such as from a cough or sneeze of an infected individual.

The respiratory tract is subdivided into the upper respiratory tract, which consists of the nose, nasal passages, sinuses, pharynx, and larynx (voice box), and the lower respiratory tract, which consists of the trachea (windpipe), bronchi (singular: bronchus), and lungs (Fig. 5.2A). Within the lungs, the two bronchi branch into smaller-diameter bronchioles that lead to an estimated 300 million alveoli (singular: alveolus), where gas exchange occurs (Fig. 5.2B). Viruses contained in larger droplets are deposited in the upper respiratory tract, while smaller aerosolized particles or liquids are able to travel into the lower respiratory tract. The upper respiratory tract epithelium contains abundant **goblet cells** that produce mucus, a thick fluid that traps inhaled particulate matter. The majority of the upper respiratory epithelium is lined with **cilia**, small hairlike structures that move together like oars to push the mucus and its trapped contents to the throat, where it is swallowed (Fig. 5.2C). Each cell has about 300 cilia. In the lower respiratory tract, mucus-secreting goblet cells become less abundant, and ciliated cells are present at the beginning of the lower respiratory tract but are absent in the alveoli of the lungs. The flow of mucus in the upper and lower respiratory tract traps many viral particles, and antibodies (particularly of the IgA isotype) produced by immune system cells bind to virus particles, preventing them from interacting with the cells of the respiratory epithelium. Within the alveoli of the lung are found many **alveolar macrophages**, another kind of immune system cell that is specialized in **phagocytosis**, a type of receptor-mediated endocytosis that is used by macrophages to endocytose whole cells or pathogens that are then digested by lysosomes within the macrophage.

Essential Human Virology.

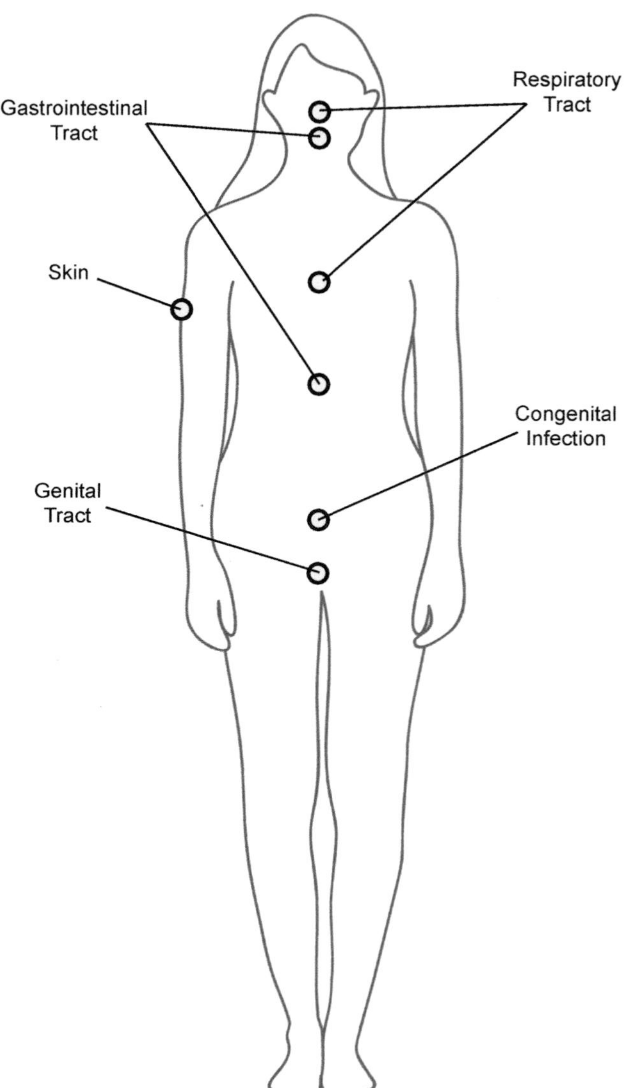

FIGURE 5.1 **Common portals of virus entry.** Viruses are able to gain entry into the body through a variety of different portals. These include the respiratory tract, gastrointestinal tract, or genital tract, as well as infection of the skin or underlying subcutaneous tissue. Congenital infections are those that are passed from mother to fetus and present at birth. Other less common pathways to infection, including transplants or infection of the eye, can also occur.

The ciliated epithelial cells of the respiratory tract display receptors for respiratory viruses, such as influenza or rhinovirus (Table 5.1). To initiate infection, however, a virus must avoid being trapped within the mucus lining the epithelium or eliminated by antibodies and macrophages, further emphasizing the importance of having a sufficient number of virions present.

5.1.2 Gastrointestinal Tract

The human digestive or **gastrointestinal tract** is a hollow tube that stretches from the oral cavity (mouth) to the anus (Fig. 5.3A). Food enters the mouth, where it is chewed and begins being digested by the enzymes found in saliva, provided by the salivary glands. The pharynx moves food via the swallowing reflex into the esophagus, which has several mucus-secreting glands. Muscle contractions by the esophagus move the food into the stomach, which contains numerous pits of cells, all of which produce mucus to protect the cells from the acidic gastric juices secreted by the stomach. From there, food moves to the small intestine, where the food finishes being digested and is absorbed. The small intestine is roughly 6 ft long and is composed of fingerlike projections called **villi** (singular: villus) that increase the surface area of the epithelium (Fig. 5.3B and C). Moreover, the epithelial cells that form the villi have multiple hairlike **microvilli** at their apical (outermost) end, forming what looks like the teeth of a comb. Each cell of a villus has an estimated 3000 microvilli. Together, the total surface area of the small intestine is similar to that of a tennis court. This increases the area of contact for absorption of food, but some viruses also take advantage of this area of exposure for infection (Table 5.1).

Being a mucosal epithelium, the small intestine also contains numerous goblet cells and glands that secrete mucus, which lines the epithelium. Under the epithelium of the small intestine, lymph node–like masses called Peyer's patches contain millions of antibody-secreting lymphocytes (of the IgA antibody variety, as in the lungs), macrophages, and other immune system cells (Fig. 5.3D). Interspersed within the epithelial layer are M (microfold) cells, specialized epithelial cells that constantly survey the contents of the small intestine lumen. These cells transfer the molecules from the lumen to the immune system cells found in the lymphoid tissue below (Fig. 5.3E). However, poliovirus, reovirus, and HIV are thought to exploit M cells to gain entry past the epithelium.

At this point of the gastrointestinal tract, the food has been broken down completely and all nutrients have been absorbed by the small intestine. The food remnants pass into the colon (large intestine), which lacks the villi of the small intestine but has abundant goblet cells for secreting mucus. Although most of the water in the food has been absorbed by the small intestine, absorptive cells in the colon reclaim any remaining water from the digested food, which becomes compacted into feces. At the end of the colon, the rectum stores feces, which are expelled via the anus.

Viruses that enter via the gastrointestinal tract must be able to survive its hostile environment. The flow of food, water, and saliva provides a mechanical barrier to infection, and mucus produced by the stomach, small intestine, and large intestine provides a physical barrier to infection. Macrophages phagocytose virions, and antibodies neutralize virions to prevent their interaction with host cell receptors. Successful viruses must also be resistant to the low pH of stomach acid and the detergent qualities of bile, which is

TABLE 5.1 Common Portals of Entry and the Viruses That Use Them

Portal	Selected human viruses that use this portal for local or systemic infection
Respiratory tract	Adenovirus, measles, mumps, rubella, enterovirus D68, influenza A virus, influenza B virus, rhinovirus, respiratory syncytial virus, varicella zoster virus, variola
Gastrointestinal tract	Norwalk virus, rotavirus, poliovirus, enteric enteroviruses, hepatitis A virus, hepatitis E virus, sapovirus
Genital tract	Human papillomaviruses, HIV, hepatitis B virus, hepatitis C virus, herpes simplex virus-2 (HSV-2)
Skin	
• Direct contact	Human papillomaviruses, HSV-1, Molluscum contagiosum virus
• Penetration into dermis or subcutaneous tissue	*Injection/cuts*: hepatitis B virus, hepatitis C virus, HIV, Ebola virus
	Mosquito: Dengue virus, West Nile virus, eastern equine encephalitis virus, Chikungunya virus, yellow fever virus
	Ticks: Heartland virus, Powassan virus, Colorado tick fever virus
Through placenta (trans-placental)	Cytomegalovirus, variola virus, HSV-1 and -2, measles virus, Zika virus, rubella virus
Eye	Adenoviruses, HSV-1, cytomegalovirus, enterovirus 70, Coxsackievirus A24, rubella virus, measles virus, vaccinia virus
Transplants	
• Solid organs	Hepatitis B virus, hepatitis C virus, HIV, cytomegalovirus, West Nile virus, rabies virus, lymphocytic choriomeningitis virus, HSV, varicella zoster virus
• Blood	HIV, hepatitis B virus, hepatitis C virus, human T-lymphotrophic virus-I and -II, dengue virus, Ebola virus

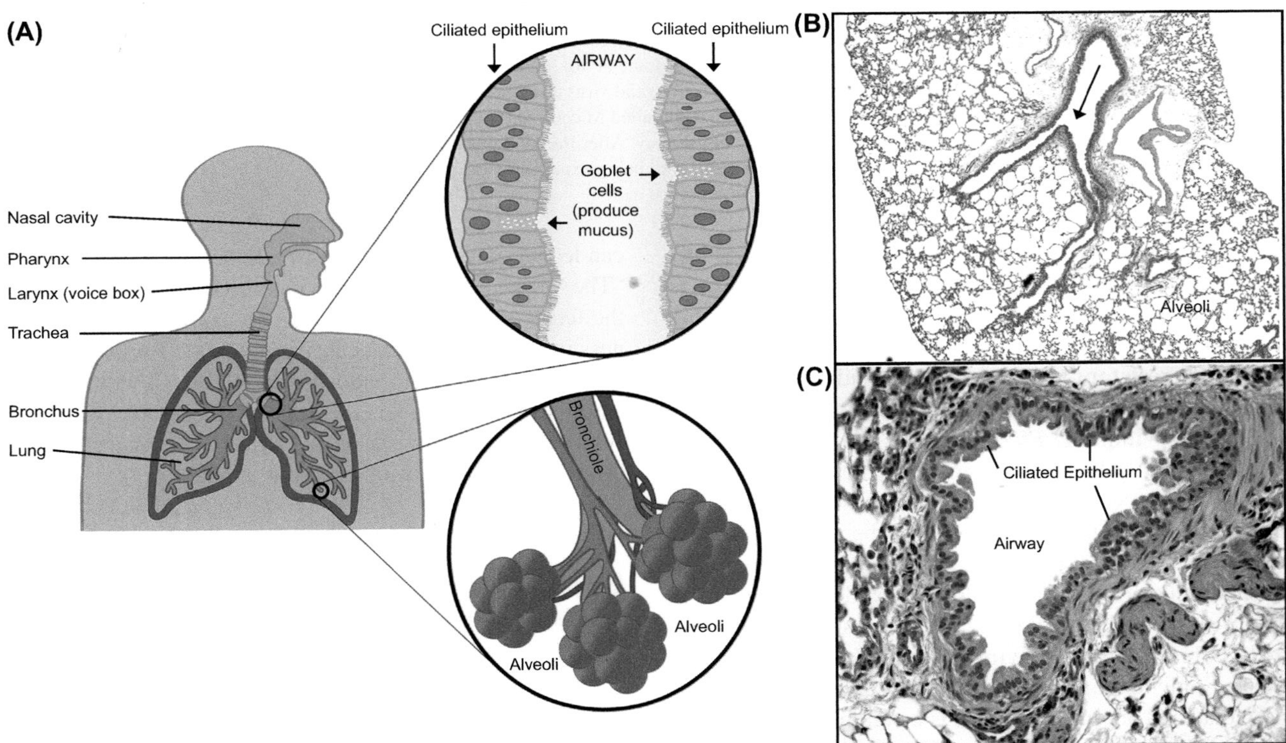

FIGURE 5.2 **The respiratory tract.** (A) The respiratory tract is subdivided into the upper respiratory tract, which consists of the nose, nasal passages, sinuses, pharynx, and larynx (voice box), and the lower respiratory tract, which consists of the trachea, bronchi, and lungs. The airways of the trachea and bronchi are lined with goblet cells that produce mucus to trap particles and pathogens, and ciliated epithelial cells push the mucus out of the lungs. The two bronchi branch into smaller-diameter bronchioles that lead to an estimated 300 million alveoli, where gas exchange occurs. (B) Cross section of a normal mouse lung at 40× magnification, showing a large bronchus branching into two smaller bronchi, which eventually branch into hundreds of alveoli. (C) Cross section of a normal mouse bronchus at 400× magnification showing the ciliated mucosal epithelium that lines the airways.

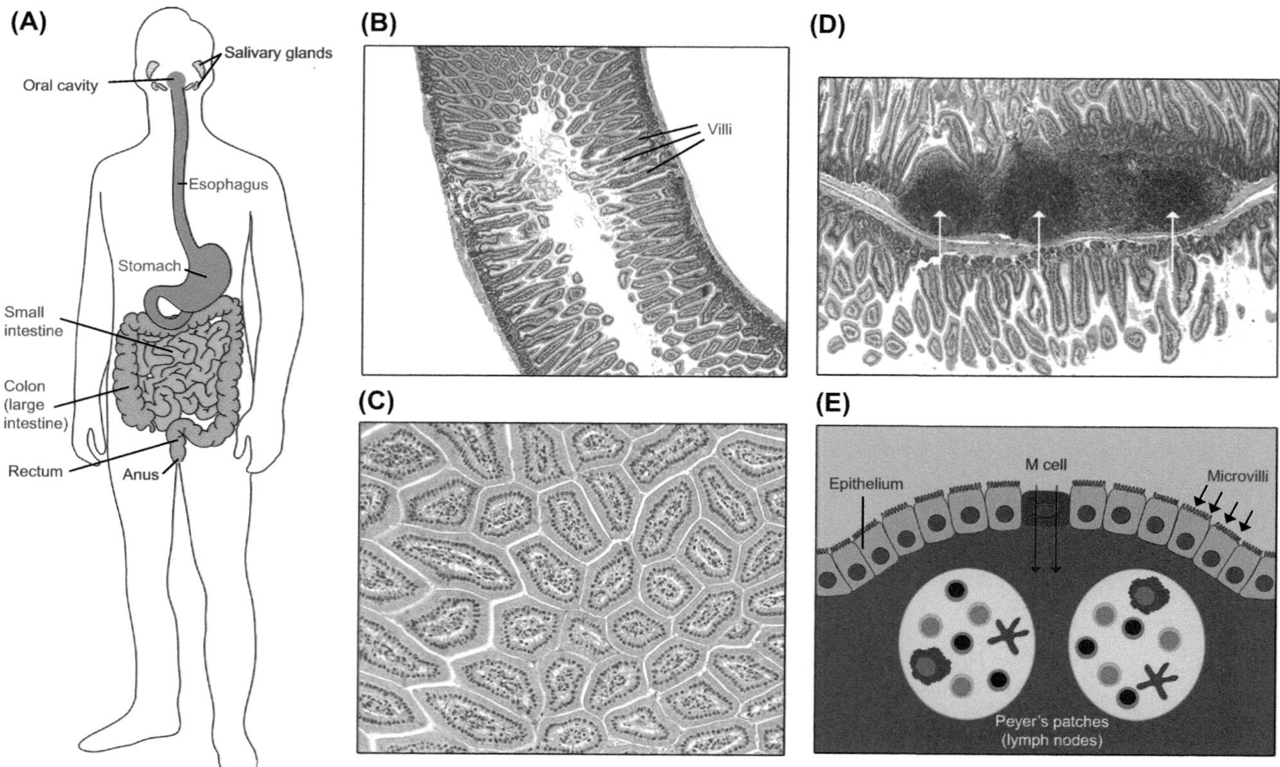

FIGURE 5.3 **The gastrointestinal tract.** (A) The human gastrointestinal tract, which is used for entry of several viruses into the body, stretches from the oral cavity to the anus. Food is chewed in the mouth and begins being digested by salivary enzymes. The pharynx moves food into the esophagus; muscle contractions here move the food into the stomach, where acidic gastric juices digest the food molecules. It passes into the small intestine, where bile from the gall bladder acts as a detergent. Numerous fingerlike villi (B, mouse tissue at 40× magnification), lined by a layer of epithelium (C, cross section at 200×), increase the surface area of the small intestine for greater absorption (and virus attachment). Lymph nodes called Peyer's patches are found beneath the epithelium in the small intestine (D), Here, specialized epithelial cells called M cells pass molecules from the lumen of the small intestine to immune cells located in the Peyer's patches (E). Some viruses gain entry in this way. After the food has been broken down completely in the small intestine, it passes into the colon and is compacted into feces, stored in the rectum until being expelled by the anus (A).

produced by the liver, stored by the gallbladder, and secreted into the small intestine. The membrane envelopes of most enveloped viruses are disintegrated by bile. **Acid-labile** viruses are unable to withstand the low pH of the stomach, while **acid-resistant** viruses contain capsid proteins that are not denatured by low pH (or their protein denaturation is reversible). Within the *Picornaviridae* family, rhinoviruses are acid labile, whereas poliovirus is acid resistant.

Viruses can be transmitted via the gastrointestinal tract in several different ways. Viruses can be transmitted from mother to child in breast milk, either as free virions or within infected cells; HIV, cytomegalovirus, and West Nile virus are three such viruses. Other viruses enter via the **fecal–oral route**, meaning that virions present in the feces of an infected individual gain entry into the oral cavity of another individual and are ingested. This is a common occurrence in nations without water treatment plants and has been one of the major roadblocks in eradicating polio from undeveloped nations. In developed countries, touching contaminated surfaces or ingesting food or water that has been contaminated can lead to infection. Norwalk virus, described in Chapter 1, "The World of Viruses," is another virus that is transmitted by the fecal–oral route. It is the leading cause of foodborne illness in the United States, causing roughly 20 million cases of acute gastroenteritis (a sudden stomach illness with vomiting and diarrhea) each year, according to the CDC. Worldwide, acute gastroenteritis caused by viruses is a major cause of death, leading to 1.5 million deaths annually.

Study Break

What characteristics must a gastrointestinal virus possess in order to effectively infect through this route?

5.1.3 Genital Tract

The genital tract refers to the organs that are involved in reproduction. In males, this includes the penis, testicles, and several associated glands and connecting tubes. The female genital tract includes the vagina, cervix, uterus, fallopian tubes, and ovaries. Viruses that are transmitted via the genital tract as a result of sexual activity are **sexually transmitted diseases**. Cells can be infected, exhibited by the tropism of human papillomavirus (HPV) for the epithelium of the cervix or penis, or viruses can gain entry into the body through breaks in the genital epithelium or by binding local cell receptors, as occurs with hepatitis B virus or HIV (Table 5.1). Viruses infecting via the genital tract have to overcome local barriers to infection, such as mucus and the low pH of the vagina.

5.1.4 Skin

The skin is a unique organ, a covering to the body that creates 1.5–2 m^2 of surface area exposed to the external environment. It is composed of two layers of tissue: the outermost **epidermis** and the underlying **dermis** (Fig. 5.4A and B). **Subcutaneous tissue** is found beneath the skin and contains primarily fat and loose connective tissues.

The epidermis consists of five layers, or strata, of keratin-producing cells (Fig. 5.4C). The innermost layer, the stratum basale, consists of living cells that undergo mitosis. The cells become filled with thick keratin filaments and die as they continue their differentiation through the layers, finally becoming the outermost layer of the skin, the stratum corneum. The cells continuously slough off from the outermost stratum and are reconstituted from the cells progressing through the lower strata.

The epidermis has barrier mechanisms to prevent infection. The flow of fluid or perspiration over the skin makes viral attachment difficult, and the sebum (oil) produced by sebaceous glands creates an acidic environment. In addition, the cells in the outermost strata of skin are not alive and thus cannot support viral replication. Viruses that replicate in the epidermis, such as HPV, gain access through small cuts or abrasions in the skin that allow access to the lower, dividing layers of skin where viral replication can occur.

The epidermis lacks blood and lymph vessels, but the underlying dermis and subcutaneous tissue are highly vascularized and contain lymphatic vessels that drain lymph to regional lymph nodes. Viruses can be introduced to these areas through penetration of the epidermis (Table 5.1). Bites of insect vectors (mosquitoes, ticks, mites) can introduce viruses into the dermis, and the subcutaneous tissue can be accessed by viruses through animal bites, needle punctures, or improperly sterilized tattooing or piercing equipment. In the dermis and subcutaneous tissue, viruses can easily gain entry into the bloodstream, either directly or through the draining lymph that eventually empties into the bloodstream. The epidermis primarily supports localized viral infections, but introduction of the virus to the underlying dermis and subcutaneous tissue can result in dissemination of the virus to other locations within the body.

(A)

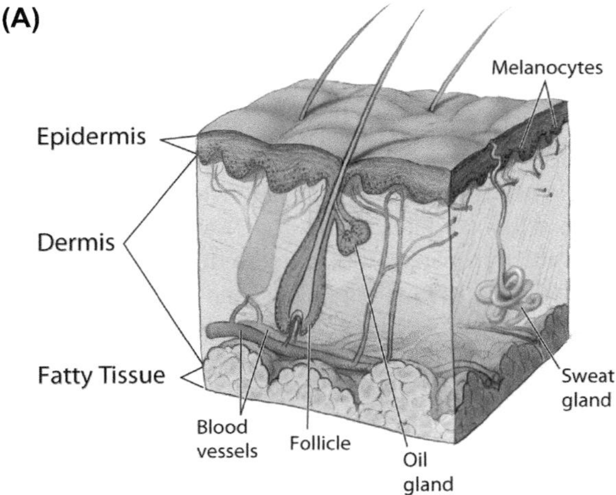

(B)

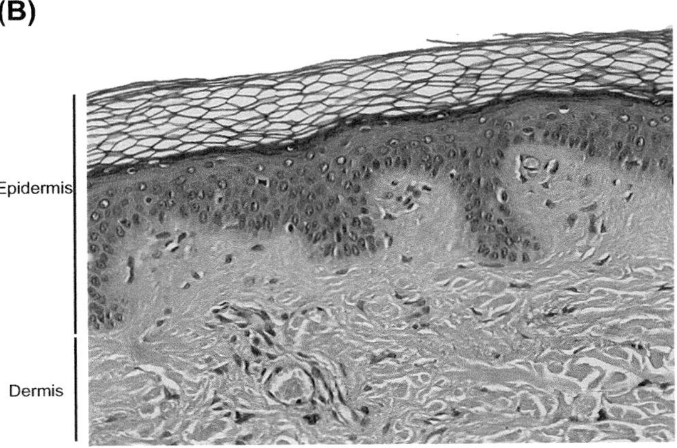

(C)

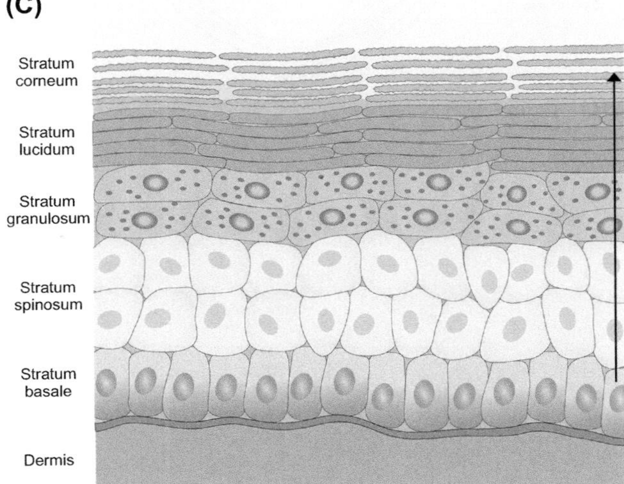

FIGURE 5.4 **The skin.** The skin is composed of two layers, the epidermis and dermis (A and B). The epidermis is composed of five layers (C); the innermost layer consists of living cells that undergo mitosis. The cells differentiate as they progress through the different layers, eventually dying as they are filled with thick keratin filaments. The cells continuously slough off from the outermost layer, being replenished with cells from the next lower layer. Some viruses, like human papillomaviruses, are able to replicate within the living part of the epidermis, but other viruses must gain entry to the dermis or underlying subcutaneous tissue in order to replicate or disseminate. *Image in Part A by Don Bliss, National Cancer Institute. Magnification of Part B is 100×.*

5.1.5 Eyes

The eye is a complex organ used for perceiving shapes, light, and color. As an interface with the outside world, the eyes can also be a portal of entry for viruses (Table 5.1).

The external layer of the eye is composed of the **sclera** and the **cornea** (Fig. 5.5A). The tough white covering of the eye is the sclera ("whites of the eyes"), which becomes the colorless and transparent cornea in the front and center area of the eye that covers the pupil, lens, and iris. The **conjunctiva** is a thin layer of epithelium that covers the sclera and the part of the eyelid that abuts the eye. An eye blinks every 5 seconds, on average, and tears function to keep the outer surface of the eye moist while washing away any potential pathogens. It is rare to have a viral infection of the eye itself without a traumatic event (a puncture wound, for example) to provide entry into the eye. However, infection of the cornea can occur with herpesvirus exposure, and herpes simplex virus (HSV) infection of the cornea is the most common infectious cause of corneal blindness in the United States. More often, viruses infect the conjunctiva of the eye or eyelid, causing **conjunctivitis** (Fig. 5.5B). Viral conjunctivitis, also known as "pink eye," is usually caused by adenoviruses.

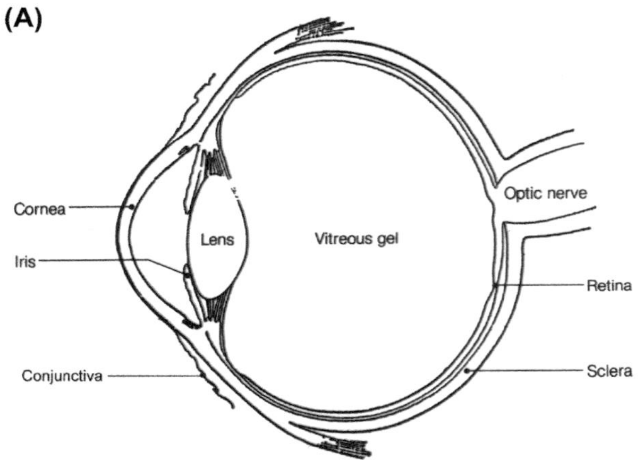

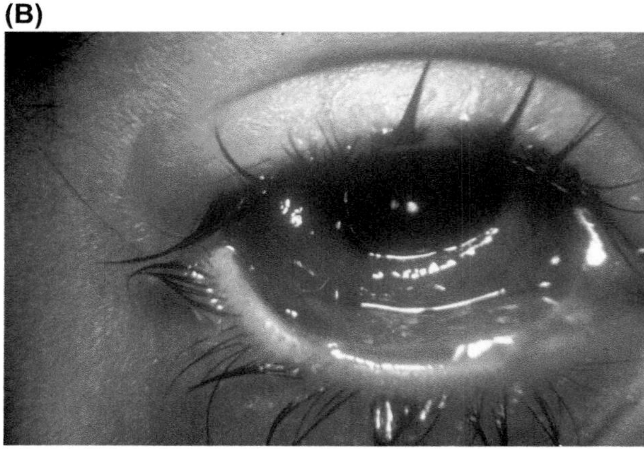

FIGURE 5.5 **The eye.** (A) The eye is composed of several layers. The white covering of the eye is the sclera, which becomes the transparent cornea in the front of the eye. A thin layer of epithelium, called the conjunctiva, covers part of the sclera and connects to the eyelid. (B) This photograph shows conjunctivitis, the inflammation of the conjunctiva, caused by accidental infection with vaccinia, the virus in the smallpox vaccine. *Courtesy of CDC/Arthur E. Kaye.*

5.1.6 Placenta

Congenital infections occur when a mother infects a fetus before its birth. Congenital infections occur via **vertical transmission**, meaning that the virus is spread from one generation to the next generation (Fig. 5.6). In contrast, most viral infections exhibit **horizontal transmission**, meaning that direct host-to-host transmission occurs. Viruses with horizontal transmission rely upon a high rate of infection to sustain the virus population, while vertical transmission often leads to long-term persistence of the virus within the child.

Congenital infection can occur when a virus crosses the placenta during pregnancy. The blood of the mother is not mixed with the blood of the fetus; instead, the placenta is the interface between the mother and developing fetus (Fig. 5.7), allowing oxygen, waste products, and nutrients to pass between mother and fetus. Similarly, some viruses are able to pass through the placenta (Table 5.1). Cytomegalovirus, a herpesvirus, is the most common cause of congenital infections, occurring in about 2.5% of live births. Several other viruses can be transmitted transplacentally, including variola (smallpox), rubella, measles, Zika, and parvovirus B19. The effects

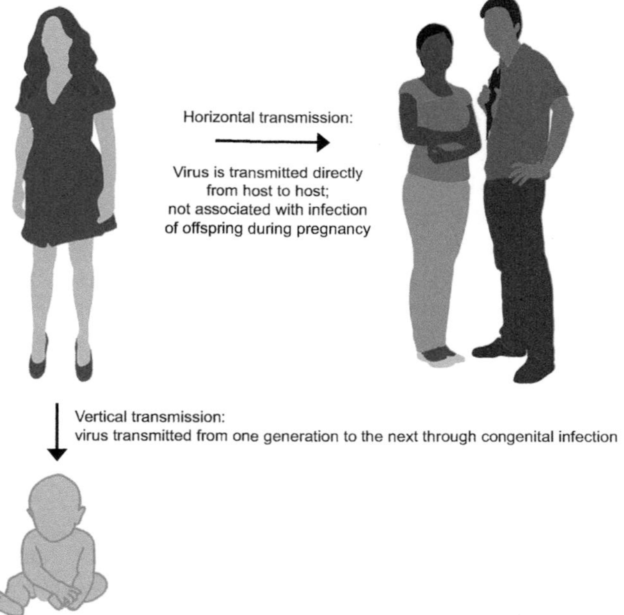

FIGURE 5.6 **Horizontal transmission versus vertical transmission.** Horizontal transmission refers to the transmission of an infectious agent among individuals within a population. In vertical transmission, however, an infectious agent is transmitted from one generation to another through congenital infection.

can be severe, including miscarriage, low birth weight, intellectual deficiencies, hearing loss, and death of the infant.

Intrapartum transmission occurs when the child is infected during the birthing process due to contact with the mother's infected blood, secretions, or biological fluids. Vertical transmission of HIV most often occurs by intrapartum transmission, although breastfeeding can also transmit the virus via the gastrointestinal tract. HSV-1 and HSV-2 cause genital warts but are most often asymptomatic in adults. Intrapartum transmission of HSV to the child is highest when the mother has active lesions or contracted a new HSV infection in her third trimester of pregnancy. Antiviral drugs are used to reduce viral load, but a pregnant mother may be encouraged to deliver the child by C-section if there are active signs of maternal infection.

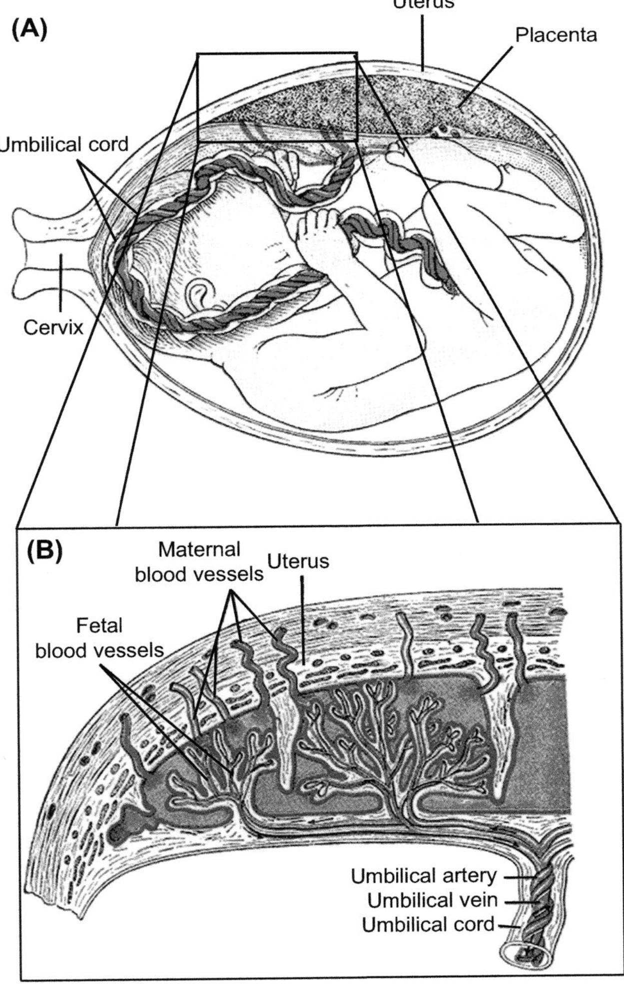

FIGURE 5.7 **The placenta.** The placenta is a fetal tissue that allows the transfer of oxygen, carbon dioxide, waste products, and nutrients between the mother and fetus (A). Although maternal and fetal blood does not mix, the fetal arteries and veins come into close contact with material blood vessels in the placenta, allowing diffusion to occur (B). Several viruses are able to travel from mother to fetus through the placenta. *Illustrations from Henry Gray, Anatomy of the Human Body. Philadelphia and New York, Lea and Febinger, 1918.*

5.1.7 Transplants

Although the rate is low compared to other portals of entry, transplanted organs and tissues can also harbor viruses that can be transmitted to a new host (Table 5.1). Blood is the most commonly transplanted tissue, and before the advent of sensitive screening tests, **transfusion-transmitted infections** (TTIs) were a low probability but possible result of receiving blood and blood products. Several viruses can be transmitted through blood, including hepatitis A virus, hepatitis B virus, hepatitis C virus, HIV, West Nile virus, and dengue virus. The blood supply is currently screened for many of these, and the current risk of receiving contaminated blood is very low (Table 5.2). However, before tests were developed and screening was instituted, hepatitis B and C viruses contaminated the blood supply. In the early 1980s, when AIDS was reported, blood industry supporters underestimated the risk of HIV having entered the blood supply and were slow to take action. At the time, more than half of the 16,000 people in the United States with hemophilia, a genetic bleeding disease that requires blood plasma transfusions to transfer clotting factors, contracted HIV from contaminated plasma products. Similarly, it is estimated that over 20,000 people contracted HIV from contaminated blood before March 1985, when the U.S. Food and Drug Administration (FDA) approved the first test to detect HIV. Incidents like this one have proven valuable in emphasizing the importance of detecting new viruses in the blood supply. When transfusion-transmitted West Nile virus infections were reported in 2002, a test was rapidly developed and the blood industry began regular testing of donated blood for the virus.

Although infrequent, virus transmission through transplanted organs has also been documented. To prevent the rejection of the transplant, organ recipients are given potent

TABLE 5.2 Pathogens for Which the Blood Supply Is Currently Tested

Pathogen type	Specific pathogens tested for
Virus	Hepatitis B virus
	Hepatitis C virus
	Human immunodeficiency virus-1 and -2 (HIV-1/HIV-2)
	Human T-lymphotropic virus-I and -II (HTLV-I/HTLV-II)
	West Nile virus
Bacterium	*Treponema pallidum* (syphilis)

Case Study: Lymphocytic Choriomeningitis Virus Infection in Organ Transplant Recipients—Massachusetts and Rhode Island, 2005

Excerpted from the Morbidity and Mortality Weekly Report, May 26, 2005 54 (Dispatch):1–2.

On May 3, 2005, CDC received a report of severe illness in four patients who had received solid organ transplants from a common donor. All four organ recipients subsequently were found to have evidence of infection with lymphocytic choriomeningitis virus (LCMV), a rodent-borne Old World arenavirus. Preliminary findings from the ensuing investigation indicate the source of infection likely was an infected hamster in the donor's home.

In early April, in Rhode Island, a woman with a medical history remarkable only for hypertension (*high blood pressure*) and 1 week of headache had sudden onset of hemiplegia (*paralysis of one half of the body*) caused by a stroke, followed by brain death within 3 days. A thorough evaluation was not suggestive of infection.

Family members of the woman consented to donation; organs and tissues were recovered, including the liver, the lungs, both the kidneys, both the corneas, and the skin. Within 3 weeks after transplantation, the four persons who received the liver, lungs, and two kidneys had abnormalities of liver function and blood coagulation and dysfunction of the transplanted organ. Signs, symptoms, and clinical laboratory test results varied in these patients and included fever, localized rash, diarrhea, hyponatremia (*low blood sodium*), thrombocytopenia (*low blood platelets*), hypoxia (*low oxygen levels*), and kidney failure. Three of the four organ recipients died, 23–27 days after transplantation. The fourth patient, a kidney recipient, survived.

When the cause of illness among the recipients was not identified through extensive diagnostic testing and suspicion of transplant-transmitted infection arose, tissue and blood samples from the donor and recipients were sent from the Rhode Island Department of Health and the Massachusetts Department of Public Health to CDC. LCMV was identified as the cause of illness in all four organ recipients. Sequencing of the virus genome confirmed its identity as LCMV.

Epidemiologic Investigation

To determine the source of LCMV infection, investigations were conducted at the hospitals involved in organ recovery and transplantation and at the coordinating organ procurement organization. Interviews also were conducted at locations where the donor had spent substantial time in the month preceding her death.

Interviews with hospital and organ bank staff members revealed no likely sources of LCMV infection in the hospital or organ-recovery settings. Environmental assessment at locations the donor frequented (eg, home and work) revealed limited opportunities for exposure to wild rodents; the sole location noted with rodent infestation was a garden shed at her home. Interviews with family members of the donor determined that a pet hamster had been acquired recently. The hamster was cared for primarily by another family member. No illnesses compatible with LCMV had been reported in the donor or family members during the month preceding the donor's death.

Laboratory Investigation

Family members of the donor were tested for LCMV antibodies. The family member who cared for the hamster had specific IgM and IgG antibodies to LCMV. No other family member had detectable IgG or IgM antibodies to LCMV. All available donor tissues were tested, and no evidence of LCMV was determined. However, the pet hamster was determined positive for LCMV.

Reported by Rhode Island Hospital, Providence; Rhode Island Dept of Health. New England Organ Bank, Newton; Massachusetts General Hospital, Brigham and Women's Hospital, Boston; Massachusetts Dept of Public Health. Infectious Disease Pathology Activity, Special Pathogens Br, Div of Viral and Rickettsial Diseases, Div of Healthcare Quality Promotion, National Center for Infectious Diseases; EIS officers, CDC.

drugs to suppress the immune system, which causes the host to become **immunocompromised**. When transferred through organ transplantation, these viruses can reemerge and infect the immunocompromised host. Herpesviruses, which remain in tissues or cells in a dormant state after infecting a healthy host, are common viral pathogens in transplants, although a variety of other viruses have also been transmitted through transplantation, including rabies, West Nile virus, HIV, hepatitis viruses, lymphocytic choriomeningitis virus, and several respiratory viruses. Risk questionnaires and screening of donor tissues are recommended to reduce the transmission of certain of these viruses.

5.2 DISSEMINATION WITHIN A HOST

As we have seen, viruses can gain entry into a host through a variety of ways. Viruses that infect and replicate only within cells at the site of infection cause **localized infections**. Rhinovirus is an example of a virus that causes a localized infection: it infects the epithelial cells of the upper respiratory tract and replicates there. Similarly, papillomavirus strains that infect the skin replicate locally in the epidermis. On the other hand, viruses that initiate infection through one organ but then spread to other sites within the body cause **systemic infections**. These viruses infect cells within the initial organ, where they either replicate locally before spread or use infected local cells to travel to other locations within the body.

Virions spread to other organs through one of two ways. In **hematogenous spread**, viruses spread to target organs using the bloodstream. This can occur through direct injection into the blood, as would happen with animal or insect bites, or it can occur from virions entering the interstitial fluid that bathes all our cells within the tissues. This interstitial fluid, or lymph, is collected in lymphatic vessels that lead back to lymph nodes (Fig. 5.8). Immune system cells filter the lymph within the lymph nodes, but high virion concentrations can mean that some escape these cells and continue within the lymph, which is eventually returned to the bloodstream. **Viremia** is the term used to describe the presence of virus within the bloodstream. Since blood circulates throughout the entire body, viruses can use the bloodstream to gain access to their cellular targets in organs other than the ones through which they entered the body. **Primary viremia** indicates the first time that the virus is found in the bloodstream. After replicating within the target organ, additional virions may enter the bloodstream. This is known as a **secondary viremia**.

In **neurotropic spread**, viruses spread through the body using neurons. Viruses rarely infect neurons directly because it is difficult for viruses to directly access these cells. Most often, viruses replicate in cells at the local site of infection and then infect neurons located nearby. Several herpesviruses spread in this manner, infecting and replicating within the local epithelium until sufficient virions are present to infect nerves associated with the tissue.

Herpesviruses are dsDNA viruses and must gain entry into the cell's nucleus for replication. However, the nucleus of a neuron, can be quite a distance from the axon terminal where the virus initiated attachment to the cell. As mentioned in Chapter 4, "Virus Replication," herpesviruses have evolved mechanisms to overcome this challenge. Viral proteins bind to dynein, a host cell protein that "walks" vesicles of cargo along microtubules to the nucleus. At the nucleus, the capsid docks at a nuclear pore and its viral DNA is transported into the nucleus for replication or latency.

Viruses first infect neurons of the peripheral nervous system, which can be used to access the central nervous system. Certain viruses can cause devastating results if they reach the central nervous system. For instance, death occurs within days of the rabies virus reaching the brain.

5.3 PORTALS OF VIRUS EXIT

In order to persist within a population, a virus must spread from an infected host to a susceptible host. The **shedding** of virus refers to the release of infectious virions from the host. During localized infections, the virus is shed from the primary site of infection. Viruses that infect the skin are spread through skin-to-skin contact, and respiratory viruses are shed within respiratory secretions, passed along through a cough or sneeze to a new, susceptible host. Gastrointestinal viruses are shed within aerosolized vomit or diarrhea, potentially contaminating food or water that could infect a subsequent individual. Viruses that replicate in the lungs, nasal cavity, or salivary glands can be shed in saliva, and viruses such as HIV and herpesviruses can replicate within genital compartments and be shed in semen or vaginal secretions.

Viremia is a common occurrence of infection with several viruses, including HIV and hepatitis. Consequently, these viruses can be transmitted through blood. **Viruria**,

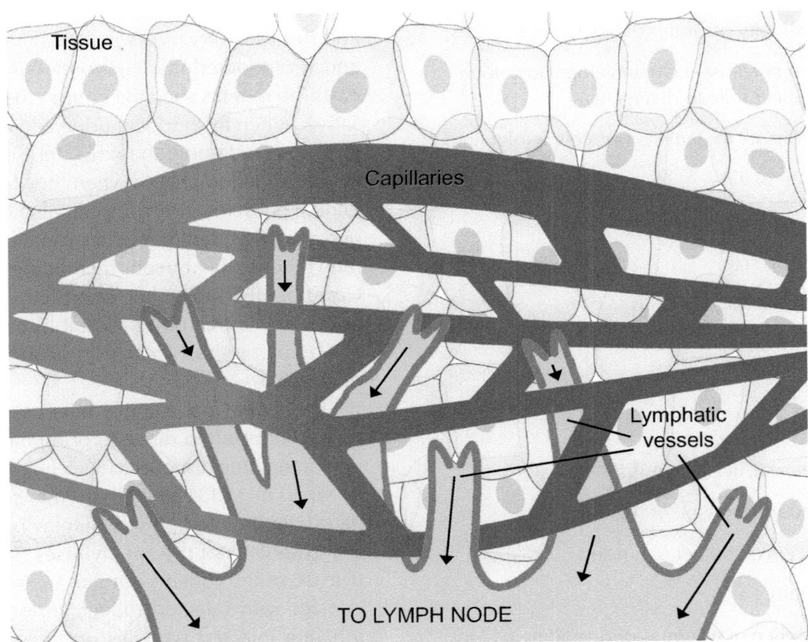

FIGURE 5.8 **The lymphatic system.** Fluid with nutrients and oxygen diffuses through small capillaries within a tissue. This interstitial fluid, or lymph, is collected by one-way lymphatic vessels that drain to lymph nodes, where immune cells filter the lymph. It is eventually returned to the bloodstream. Viruses within tissues can travel within the lymph and enter the blood if not successfully filtered by the lymph node immune cells.

the presence of virus within the urine, occurs with several systemic viral infections, including measles and mumps. Other viruses replicate within and are shed by cells of the urogenital tract, such as JC polyomavirus (JCPyV) and BK polyomavirus. Some viruses that cause severe disease are transmitted to people through aerosolized virions found in rodent urine or droppings (see *In-Depth Look*).

Although viruses cannot replicate independently outside a host, virions can remain infectious outside the body. The stability of virions within the environment, however, depends upon several factors, both of the environment and of the virion itself (Table 5.3). The biochemical characteristics of the virion and its contents, including the type of nucleic acid, the sensitivity of viral proteins to pH changes, and the presence or absence of a lipid envelope, play a role in the sensitivity of the virion to inactivation by environmental factors. Temperature, humidity, moisture content, sunlight, pH, and the presence of organic matter all affect the inactivation of virions within the environment.

Viruses within contaminated feces or urine can be transmitted via the fecal–oral route or shed within feces and urine and then aerosolized and inhaled. The neutral pH of human waste generally protects virions, and organic matter within feces also buffers the chemical makeup and temperature of the environment. Temperature plays a large part in the persistence of viruses within feces or waste water. Viruses are inactivated within minutes or hours at high temperatures (above 121°F or 50°C), but certain viruses, particularly those that are nonenveloped, can remain infectious for days or months at ambient temperatures.

Airborne viruses can be transmitted within liquid droplets or aerosolized particles that are released when a person sneezes, coughs, speaks, or breathes. Droplets

TABLE 5.3 Characteristics That Affect Virion Stability

Factor	Characteristic
Virion	Type of nucleic acid (DNA is more stable than RNA)
	Sensitivity of viral proteins to pH changes
	Presence/absence of envelope (envelope is susceptible to detergents)
	Sensitivity to damage from ultraviolet light
	Strain of virus
	Adsorption to other materials
	Particle size
	Aggregation of virions
Environment	Temperature
	pH
	Humidity
	Seawater/freshwater/distilled water
	Amount and type of organic matter
	Presence of other organisms
	Salinity
	Presence of enzymes or degrading factors
	Type of surface/medium

Study Break

Describe the difference between respiratory droplets and aerosolized particles. How can these affect the dissemination of virions in the environment?

In-Depth Look: Transmission of Serious Diseases Through Rodent Urine and Droppings

Several potentially dangerous viruses replicate primarily in rodent populations and only become problematic when the virus is inadvertently transmitted to humans. One such illness is hantavirus pulmonary syndrome (HPS), caused by a hantavirus known as Sin Nombre virus. This virus causes a fever ≥101°F (38.3°C), chills, muscle aches, headache, and severe respiratory problems. In the United States, an average of 29 people have been infected each year since recording began in 1993. Of these infections, 36% have been fatal.

Hantaviruses are transmitted by infected deer mice (*Peromyscus maniculatus*) present throughout the western and central United States, cotton rats (*Sigmodon hispidus*) and rice rats (*Oryzomys palustris*) in the southeastern states, and white-footed mice (*Peromyscus leucopus*) in the northeast. The rodents shed the virus in their urine, droppings, and saliva. When fresh rodent urine and droppings are disturbed, particles containing the virus get into the air, and the virus is transmitted to people when they breathe in the virus-containing particles. The virus can also be transmitted when the person inadvertently touches rodent urine or droppings and then touches the mucous membranes of their nose or mouth. Cases of HPS most often occur in rural areas where forests, fields, and farms offer a suitable habitat for these rodents. No human-to-human transmission of hantaviruses has yet occurred.

The *Arenaviridae* family of viruses is also associated with rodent-transmitted diseases of humans (see case study). The first arenavirus discovered was lymphocytic choriomeningitis virus, in 1933. Since that time, seven other arenaviruses have been isolated. As with hantaviruses, rodents are the natural reservoirs of the arenaviruses. The viruses do not cause disease in the rodents but can be transmitted to humans by contact with infected rodent urine or droppings, either by inhaling infected particles or ingesting contaminated food. Lassa virus, Machupo virus, and Lujo viruses have been associated with person-to-person spread.

are larger in size, about 20 μm, and therefore tend to only be spread short distances before they succumb to gravity and fall out of the air (Fig. 5.9). At 5 μm or less, aerosolized particles remain airborne for much longer periods of time, and the evaporation of liquid from aerosols creates smaller particles that can persist in the air for an extended period of time. Some viruses are better protected from inactivation within droplets. Aerosols, being smaller, generally have greater success in reaching the lower respiratory tract.

For airborne viruses, humidity and temperature often play a role in the persistence of the virion in the environment. It has been shown for several enveloped respiratory viruses, including influenza A virus, measles virus, severe acute respiratory syndrome coronavirus (SARS-CoV), and Middle East respiratory syndrome (MERS)-CoV, that lower temperature and humidity are more conducive to maintaining airborne virions. Virions are inactivated faster at higher temperatures and humidity, and droplets tend to fall out of the air more readily with higher humidity. On the other hand, nonenveloped viruses like rhinovirus and adenovirus remain infectious longer in higher-humidity environments. Organic matter, such as proteins and carbohydrates derived from mucus or aerosolized fecal matter, can also slow the inactivation of viruses within droplets or aerosols. Other viruses are sensitive to light, primarily the ultraviolet component.

FIGURE 5.9 **Respiratory droplets.** A sneeze transmits liquid droplets, seen here, that can directly transmit viruses from person to person. *Photo courtesy of CDC/James Gathany.*

5.4 PATTERNS OF INFECTION

A host typically goes through four stages of disease development when it is infected with a virus (Fig. 5.10). The **incubation period** is the time between when the virus initially infects the host and when symptoms appear. For example, the incubation period of rhinovirus, a cause of the common cold, tends to be about 1–3 days. The incubation periods of well-known viruses are listed in Table 5.4.

The **prodromal period** occurs after the incubation period and is when symptoms first appear. These tend to be nonspecific, mild symptoms that are not clinically indicative of the type of virus, such as malaise, muscle aches, or a low-grade fever. During this period, however, the virus is replicating quickly within the host. The **illness period** occurs when specific symptoms of the disease occur. At this point, the virus is multiplying to high levels and the immune system has been activated, but the response takes time. In **immunocompetent** hosts with functioning immune systems, infected cells will eventually be eliminated and the amount of virus within the host will decline. At this point, the symptoms of the disease subside as the host begins feeling better, having entered into the **convalescent period**. This period may last for days or months, depending upon the severity of the infection.

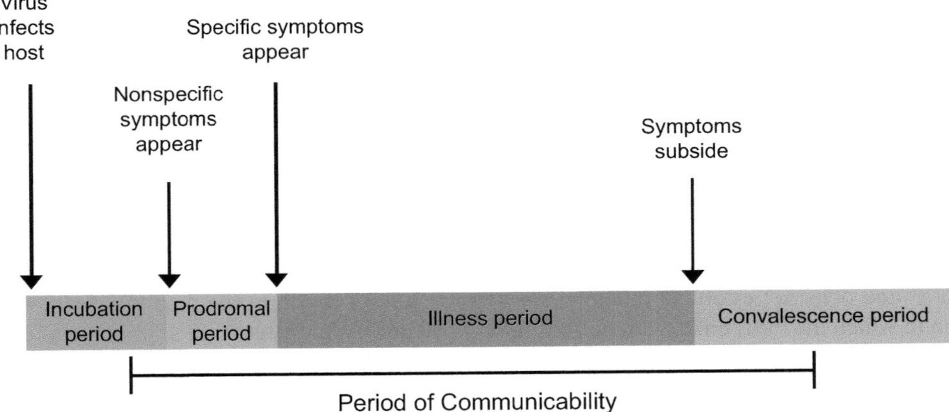

FIGURE 5.10 **Stages of disease development.** It takes time for symptoms to appear after a virus infects a host. This period is known as the incubation period. The prodromal period begins when nonspecific signs of illness appear. The appearance of infection-specific symptoms marks the beginning of the illness period. The convalescence period begins when specific symptoms subside. The period of communicability is the range in which a person is infectious and can transmit the virus to a new host. This may include before symptoms appear or after they have subsided.

TABLE 5.4 Incubation Period and Period of Communicability for Selected Human Viruses

Virus	Average incubation period (range)	Period of communicability
Rhinovirus	~24 h (1–3 days)	24 h before to 5 days after symptoms begin
Influenza A virus	2 days (1–4 days)	24 h before to 5–10 days after symptoms begin
Variola virus (smallpox)	7–17 days	24 h before fever begins until disappearance of all scabs
Ebola virus	8–10 days (2–21 days)	Infectious as long as blood or secretions contain the virus
Measles virus	10–12 days	5 days before to four days after onset of rash
Rubella virus	14 days (12–23 days)	1 week before until at least 4 days after rash appears
HIV	2–4 weeks	Early during infection and continues indefinitely
Mumps virus	16–18 days (12–25 days)	1–2 days before until 5 days after salivary gland swelling
Hepatitis A virus	28 days (15–50 days)	Last half of the incubation period to a week into jaundice (skin yellowing)
Hepatitis C virus	6–9 weeks (2 weeks–6 months)	1+ week before symptoms and continues indefinitely
Hepatitis B virus	~4 months (1.5–6 months)	Weeks before onset of symptoms and continues indefinitely

The duration of each period of disease can vary depending upon the virulence of the pathogen, the site of infection, and the strength of the host immune system. The period of **communicability**, when a person is contagious and able to spread the virus to new hosts through virion shedding, will also vary depending upon the virus and can even include part of the incubation period, when symptoms are not yet present, all the way into convalescence (Table 5.4). For example, influenza is contagious from 1 day *before* symptoms arise until about a week after becoming sick. Ebola virus is communicable in breast milk and semen for weeks to months during the convalescent period.

The replication and persistence of a virus within a host generally follow one of two different patterns of disease. In an **acute infection**, the virus replicates rapidly within the host and is spread to other individuals, but the immune system clears the virus, generally within 7–10 days. Epidemics are most often caused by viruses that cause acute infection. Some acute infections are **inapparent** or **subclinical**, meaning that they produce no symptoms of disease, although the virus still replicates and activates the immune system. Spread to other hosts can still occur with inapparent acute infections. This is different from an unsuccessful infection in which the virus is not able to successfully replicate within a host and does not establish infection, or an abortive infection in which the virus enters susceptible cells but cannot complete replication, usually due to the absence of a protein in the infected cell required for replication.

On the other hand, **persistent infections** occur when the host immune system is unable to effectively clear the virus, but the virus does not replicate to levels that kill the host. Persistent infections often last for the lifetime of an individual and occur for a variety of reasons. Viral proteins can modulate the immune response, and certain viruses, including HIV, infect immune cells and interfere with their proper functioning. Persistent infections can also result from the production of **defective interfering (DI) particles**. DI particles are virions that are created during the replication process but contain incomplete or deleted genomes. Although these are defective, they are still released from the cell and act as a sponge to sequester antibodies. Although the mechanism is unknown, they also interfere with the apoptosis of infected cells, which can lead to a persistent state. The production of DI particles is more common in RNA viruses. Finally, certain organs of the body, including the brain, have highly controlled mechanisms to safeguard against cellular damage caused by inflammation. Viral infection of these tissues may lead to persistent infections due to the protection of these cells against apoptosis.

Persistent infections can also result from viral **latency**, a state in which the virus becomes dormant within host cells. A hallmark of all herpesviruses is that they establish latency. For example, varicella zoster virus causes chickenpox upon primary infection of an individual, but the virus is never completely cleared by the immune system. Instead, it becomes latent in the sensory neurons of the individual. Although the genome persists in the neurons, new virions are not made, and a separate latency gene program is expressed. The virus is largely undetected by the immune system during latency. Later in life, **reactivation** from latency may occur, at which point the virus switches on a productive infectious cycle. Varicella zoster virus reactivation causes the painful skin rash known as shingles. Latency and reactivation will be discussed in more detail in Chapter 13, "Herpesviruses."

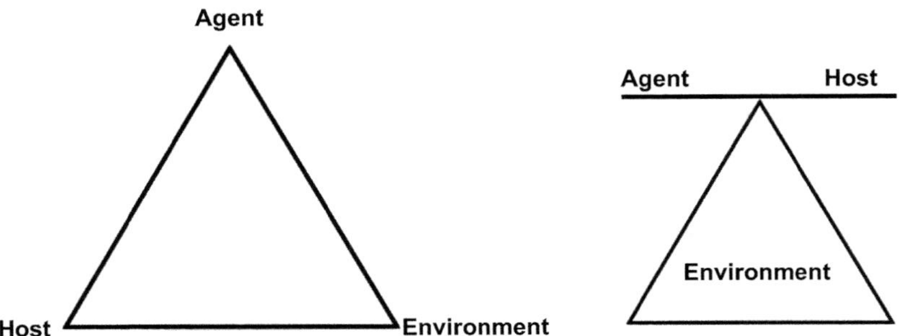

FIGURE 5.11 **Epidemiologic triad model.** The epidemiologic triad model is used to represent how infectious diseases are caused and spread. It illustrates that for infection to occur, an external agent and susceptible host must be brought together by the environment. This can also be thought of as a balance between agent and host factors within an environment.

An unusual variation of persistent infection occurs with **slow infections**. As the name suggests, these viruses can take years to reach a symptomatic phase (if one ever occurs). HIV establishes a slow infection: without antiretroviral drugs, it takes around 8–10 years for an individual to progress to a stage of disease where symptoms are apparent. In this case, symptoms arise as a result of opportunistic infections due to immunosuppression. JCPyV is a very common virus, infecting around half of the general population. JCPyV establishes a slow (and possibly latent) infection in kidney cells. It can reactivate in immunosuppressed individuals to cause progressive multifocal leukoencephalopathy, a brain disease with a high mortality rate that damages the white matter of the brain.

5.5 EPIDEMIOLOGY

Epidemiology is the study of how diseases, including those caused by viruses, spread throughout a population. In fact, the word "epidemiology" means "the study of what falls upon a people." Epidemiology is the field that studies the spread of noninfectious and infectious diseases, with a goal of identifying how diseases occur and can be controlled. Epidemiologists are the detectives of the public health world: during a disease outbreak, they determine how a disease is being transmitted from person to person and establish **control measures** that interrupt the continued transmission of the pathogen.

Within a population, infections can be classified based upon their frequency of occurrence. A **sporadic** disease occurs infrequently and without a consistent pattern. A single case of hantavirus in Nevada would be considered sporadic in nature. **Endemic** refers to the usual presence of a disease in a population at any given time. It is not necessarily a desired level, but it is the norm for a particular area. For example, during a normal September, 3.5 out of every 1000 people in the United States present with symptoms of rhinovirus. An **epidemic** occurs when there are clearly more cases of disease in a particular area than are expected during endemic periods. This is also referred to as an **outbreak**. From August 2014 through January 2015, 1153 people in 49 US states and the District of Columbia contracted enterovirus D68; only small numbers had been previously reported each year. When an epidemic spreads throughout several countries or the world, it is referred to as a **pandemic**. A major pandemic caused by viruses was the 1918 "Spanish Flu," which is estimated to have killed 50 million people worldwide in the course of months.

5.5.1 Causation of Disease

Assuming that diseases do not occur completely randomly, models are helpful in understanding how viral diseases are caused and spread. The **epidemiologic triad** model consists of three factors: an external **agent**, a susceptible **host**, and an **environment** that brings the host and agent together. This can be represented as a triangle or as a balance (Fig. 5.11).

In this model, *agent* refers to the pathogen and its characteristics that could affect its ability to be spread throughout a population. For instance, how is the pathogen transmitted? How stable is it in the environment? Does it have enhanced virulence factors? Is it susceptible to current antiviral drugs? On the other hand, *host* refers to the human that may come into contact with the agent. The presence of the agent is required for disease to occur, but coming into contact with the agent does not necessarily mean that disease will occur in the host. There are often intrinsic host factors that impact whether disease will occur, including the age and sex of an individual; the immune status of an individual (immunocompetent or immunocompromised); whether or not the person is malnourished; and if the person's behaviors are more likely to expose them to the agent. The *environment* refers to the extrinsic factors that affect whether or not the host will come into contact with the agent. These include socioeconomic factors (proper sanitation, crowding, availability of health services), biological factors (presence of vectors that transmit the agent, other animals that spread the virus), and physical factors (climate, physical environment). For disease to occur, the environment brings together the agent and a susceptible host.

5.5.2 Chain of Infection

The specific details of the epidemiologic triad model can be elaborated upon by examining the factors found within the **chain of infection**: transmission of the *agent* to the host occurs when the agent leaves its *reservoir* through a *portal of exit*, is conveyed by a *mode of transmission*, and enters a *susceptible host* through a *portal of entry* (Fig. 5.12).

Since viruses cannot reproduce outside of a host, humans are most often the reservoirs of human viruses. Some viruses replicate exclusively in humans, including smallpox or polio, while other viruses are able to infect humans and other animals. An example is influenza A viruses, which can infect humans, waterfowl, pigs, and other animals. A **zoonosis** is an infectious disease that can be transmitted from an animal to a human. HIV, Ebola virus, MERS-CoV, and SARS-CoV are examples of viruses that are thought to have been initially transmitted into the human population from animal hosts.

A **carrier** is a reservoir that can transmit the pathogen but shows no symptoms of the infection. This can occur with healthy hosts that are asymptomatic, or because a person has been infected but is still within the incubation period of the illness, before symptoms appear. Similarly, people in the convalescent period may also be capable of transmitting the disease, even though their symptoms have subsided.

Within the chain of infection, the agent leaves its reservoir through a portal of exit. As described in detail in Section 5.3, viruses commonly leave their reservoir hosts in respiratory secretions, urine, feces, or blood. The virus is conveyed through a **mode of transmission**, which can be through direct or indirect means. **Direct transmission** refers to the transfer of the virus by direct contact or droplet spread. Skin-to-skin contact, sexual intercourse, or kissing would be considered direct contact, while droplet spread includes the transmission of virions in respiratory droplets that are sneezed or coughed out of one person and immediately enter the respiratory tract of another person. **Indirect transmission**, on the other hand, requires the presence of an intermediary between hosts. In comparison to droplet spread, which passes directly from one person to another, airborne transmission of viruses carried by dust or aerosolized particles that remain suspended in the air for long periods of time are considered indirect modes of transmission (see Section 5.3 for a review of droplets vs aerosolized particles). **Vehicles** refer to nonliving physical substances—such as food, water, blood, or inanimate objects (fomites)—that can indirectly transmit virions. **Vectors** are living intermediaries that can also transmit viruses. Mosquitos, fleas, and ticks are examples of vectors. Since these are arthropods, the terms **arbovirus** is used to denote **ar**thropod-**bo**rne viruses. "Arbovirus" is a general categorization term, however, and is not a taxonomical term (eg, Flaviviridae).

The agent (virus) is transmitted to and enters a subsequent host through a portal of entry. As described in Section 5.1, examples of portals of entry include the respiratory system,

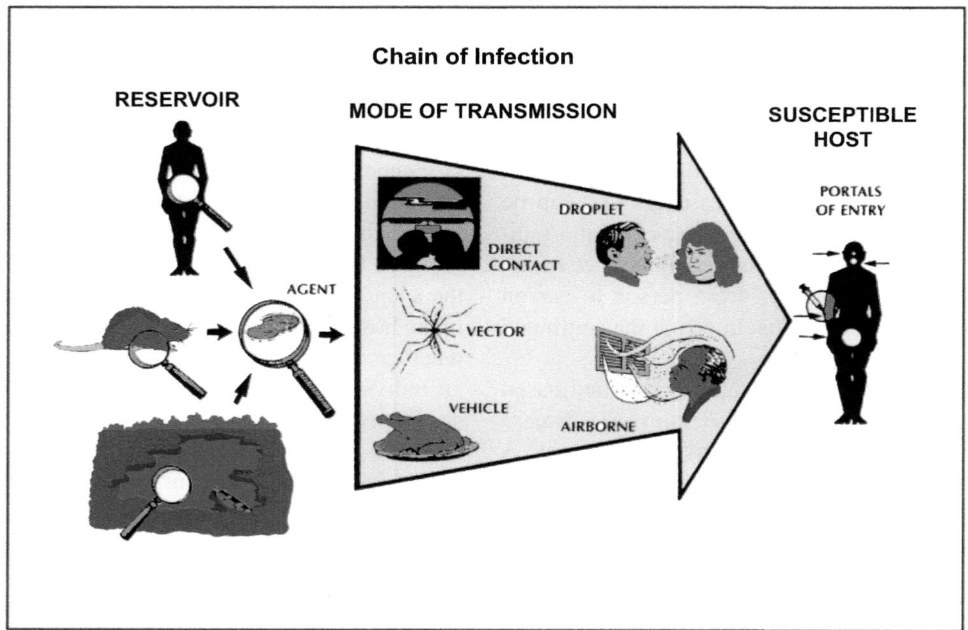

FIGURE 5.12 **Chain of infection.** The chain of infection represents that an infectious agent leaves its reservoir through a portal of exit, is conveyed by a mode of transmission, and enters a susceptible host through a portal of entry. Epidemiologists identify and examine each step in order to institute control measures that prevent further transmission of the virus. *Reproduced from Centers for Disease Control and Prevention, Updated May, 2012. Principles of Epidemiology in Public Health Practice, third ed. Department of Health and Human Services, Atlanta, GA, US.*

gastrointestinal system, or skin. The susceptibility of the host will depend upon those host factors described above, including age, sex, nutritional status, and immune status.

Identifying the factors involved at each step within the chain of infection allows epidemiologists to devise **control measures** that interfere with the transmission of the virus from the reservoir to a susceptible host (Table 5.5). These measures can be instituted at any point in the chain but are most often directed at controlling/eliminating the virus at the source, preventing the transmission of the virus, protecting portals of entry, and increasing host defenses.

The spread of a virus can be interrupted immediately by preventing the virus from leaving the infected individual. Antiviral drugs are available against HIV, HSV, and certain strains of influenza, and the use of antiviral drugs can be effective in reducing viral load and transmission of the virus, even if the host is not cured of the disease. Those coming into direct contact with infected individuals, such as hospital employees, must also be diligent about handwashing and sterilization measures. For some zoonotic infections, infected animal hosts can be removed or relocated. In other cases, infected patients can be isolated from other individuals to control the direct transmission of the virus. During the 2014 Ebola outbreak, infected individuals were quarantined in an attempt to prevent further transmission of the virus.

Other control measures attempt to interrupt the indirect transmission of the virus. For example, ventilation systems can be modified to prevent airborne transmission, waste water can be treated to kill viruses, clean drinking water can be provided, or insect spraying programs can be instituted to reduce vector populations. For viruses that are transmitted through the fecal–oral route, the environment can be rearranged to prevent transmission.

If the reservoir and environment cannot easily be modified, portals of entry can be protected to prevent infection of the host. Appropriate precautions and proper personal protective equipment (PPE), such as gloves or safety glasses, can be used to protect portals of virus entry. Other physical barriers can be instituted: bed nets can protect infection by mosquitos, and wearing long pants and using insect repellent can prevent interaction with mosquitos, fleas, and ticks.

The epidemiologic triad model requires the interaction of the agent and host in an environment that brings the two into contact, but this does not mean that infection will necessarily occur. A susceptible host is required, and control measures to increase host immune defenses can interrupt the chain of infection. Decreasing malnutrition can positively affect the immune system, but the majority of viruses that cause disease have evolved mechanisms to infect individuals with perfectly functional immune systems. Vaccination is by far the most effective means of preventing susceptibility to these pathogens. As will be discussed in more detail in Chapter 6, "The Immune Response to Viruses," vaccination works by exposing an individual to noninfectious parts of a pathogen to prepare the immune system for infection with the actual pathogen. Vaccination can protect the individual, but it can also protect a population through **herd immunity** (Fig. 5.13). If a large enough

TABLE 5.5 **Examples of Some Recommended Control Measures for Preventing MERS-Coronavirus Transmission in U.S. Hospitals**

Component	Recommendation
Patient placement	Room for isolating airborne infections
Personal protective equipment (PPE)	• Gloves • Gowns • Eye protection (goggles or face shield) • Fitted respiratory face mask (respirator)
Hand hygiene	• Healthcare personnel should perform hand hygiene frequently, including before and after all patient contact, contact with potentially infectious material, and before putting on and upon removal of PPE, including gloves. • Healthcare facilities should ensure that supplies for performing hand hygiene are available.
Environmental infection control	Follow standard procedures, per hospital policy and manufacturers' instructions, for cleaning and/or disinfection of environmental surfaces and equipment, textiles and laundry, and food utensils and dishware.
Monitoring and management of potentially exposed personnel	Healthcare personnel who care for patients with MERS-CoV should be advised to monitor and immediately report any signs or symptoms of acute illness to their supervisor or a facility designated person (eg, occupational health services) for a period of 14 days after the last known contact with the sick patient.

From Interim Infection Prevention and Control Recommendations for Hospitalized Patients with Middle East Respiratory Syndrome Coronavirus (MERS-CoV). 2014, Centers for Disease Control and Prevention. http://www.cdc.gov/coronavirus/mers/infection-prevention-control.html.

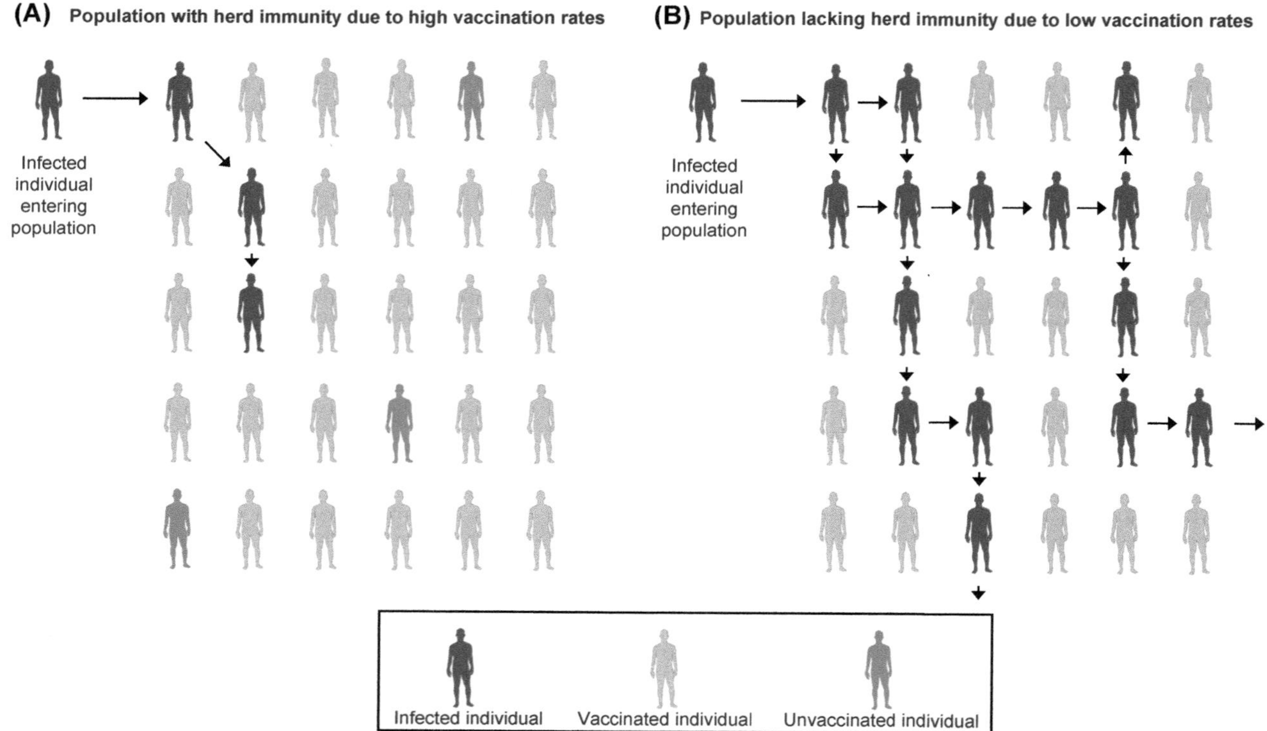

FIGURE 5.13 **Herd immunity.** (A) If a high proportion of the population is vaccinated against a pathogen, then transmission of the pathogen will soon cease due to lack of susceptible hosts. Note that herd immunity is able to protect those individuals who are unable to be vaccinated because they are too young or immunocompromised (in blue). (B) If a population does not have high vaccination rates, however, then the pathogen continues to easily spread throughout the population.

proportion of the population is vaccinated, then an infected individual may not come into contact with any susceptible hosts and the chain of infection will be terminated. Smallpox, caused by the variola virus, was eradicated from the human population by vaccinating any individuals that came into contact with an infected person. Eventually there were no new susceptible hosts for the virus, and it died out. Herd immunity is effective in preventing epidemics, but a virus is still able to cause an outbreak if a specific population within the herd chooses to not vaccinate its individuals. This also compromises the individuals within the herd that were unable to receive the vaccine due to medical reasons, such as allergy or immunosuppression.

5.6 EPIDEMIOLOGICAL STUDIES

An orderly examination of all the facts surrounding an outbreak is required for epidemiologists to accurately investigate the variables within the chain of infection. Epidemiologists also determine the **morbidity** (rate of illness) and **mortality** (rate of death) associated with an illness. The **incidence** of the disease refers to the number of *new* cases within a population during a specified time, while the **prevalence** of a disease refers to the total number of individuals with the disease at that time. For example, the US CDC reports that in 2012, the incidence of HIV in the United States was 18.4 cases per 100,000 people, while the prevalence of the disease was 342.1 cases per 100,000. In other words, 18.4 new cases were diagnosed and 342.1 people were living with HIV in 2012, per 100,000 people.

Before counting cases, however, an epidemiologist must determine what qualifies as a case of the illness. Epidemiological studies rely upon a **case definition** to determine whether or not a person has a particular disease. The case definition is a set of clinical and laboratory criteria that rely upon the symptoms the person presents with and the results of virus-specific blood tests. For a specific outbreak, limitations on the time and location may also be included within the case definition. Nationally and internationally, use of a standard case definition allows for proper diagnosis and also ensures comparability among different hospitals and locations. Classification of results can categorize the case as suspected, probable, or confirmed (Fig. 5.14).

Case definitions can change over time and often do so as more information about the illness becomes available. Case definitions can also possess "loose" or "strict" requirements. A sensitive ("loose") case definition is used for containment of viruses with potentially serious effects upon public health; this type of case definition is not

Measles (Rubeola): 2013 Case Definition

Clinical Description
An acute illness characterized by
- Generalized, maculopapular rash lasting ≥3 days;
- Temperature ≥101°F or 38.3°C; **and**
- Cough, coryza, or conjunctivitis.

Case Classification
Probable
In the absence of a more likely diagnosis, an illness that meets the clinical description with:
- No epidemiologic linkage to a laboratory-confirmed measles case; **and**
- Noncontributory or no measles laboratory testing.

Confirmed
An acute febrile rash illness with:
- Isolation of measles virus from a clinical specimen; or
- Detection of measles virus-specific nucleic acid from a clinical specimen using polymerase chain reaction; or
- IgG seroconversion or a significant rise in measles IgG antibody; or
- A positive serologic test for measles IgM antibody not explained by MMR vaccination during the previous 6-45 days; or
- Direct epidemiologic linkage to a case confirmed by a method above.

FIGURE 5.14 **Measles (Rubeola) case definition.** A case definition is a set of clinical and laboratory criteria to classify an potentially infected person. In this CDC case definition, a probable case of measles virus is an illness that meets the clinical criteria but has not been confirmed with laboratory testing. A confirmed case occurs when a person has a rash and one of several confirmatory laboratory tests showing the presence of the measles virus or a recent immune response against it.

very specific, in an attempt to include all possible cases of the virus even if other viral infections may fall within the case definition. When trying to determine the specific cause of an outbreak, however, epidemiologists employ a strict case definition that will only confirm those infected with the particular pathogen. For instance, an epidemiologist studying the cause of a Norwalk virus outbreak, which is a fecal–oral disease that causes diarrhea, will not want to include diarrhea as the only requirement—many other infectious and noninfectious diseases also cause diarrhea as a symptom. Loose and strict case definitions may overestimate and underestimate the total number of cases, respectively, but are necessary for containing serious viral outbreaks or determining the definitive source of an infection.

In the United States and in many countries around the world, state and national public health departments must be notified when a patient is diagnosed with certain infectious and noninfectious conditions. This surveillance system is used to monitor disease trends, identify populations at high risk, formulate control measures, and create public health policies. The list of notifiable viral diseases is listed in Table 5.6.

Just like reporters, epidemiological studies seek to identify the 5 W's of the epidemiological story: What (the agent), Who (the person infected), Where (the location), When (the time), and Why/How (the causes and modes of transmission). These are also known as **epidemiological variables.** Epidemiological studies are divided into **descriptive** and **analytic studies**. Both types of study start with the agent,

TABLE 5.6 Notifiable Infectious Viral Diseases

Viruses
Arboviral diseases
Chikungunya virus
Eastern equine encephalitis virus
Powassan virus
St. Louten encephalitis virus
West Nile virus
Western equine encephalitis virus
Dengue virus
Hantavirus
Hepatitis A, acute
Hepatitis B, acute
Hepatitis B, chronic
Hepatitis C, acute
Hepatitis C, past or present
HIV
Influenza-associated pediatric mortality
Measles (Rubeola)
Mumps
Novel influenza A virus infections
Poliovirus infection
Rabies (animal)
Rabies (human)
SARS coronavirus
Variola (smallpox)
Varicella (chickenpox)
Varicella deaths
Viral hemorrhagic fevers
Crimean–Congo hemorrhagic fever virus
Ebolavirus
Lassa virus
Lujo virus
Marburg virus
Guanarito virus
Junin virus
Machupo virus
Sabia virus
Yellow fever virus

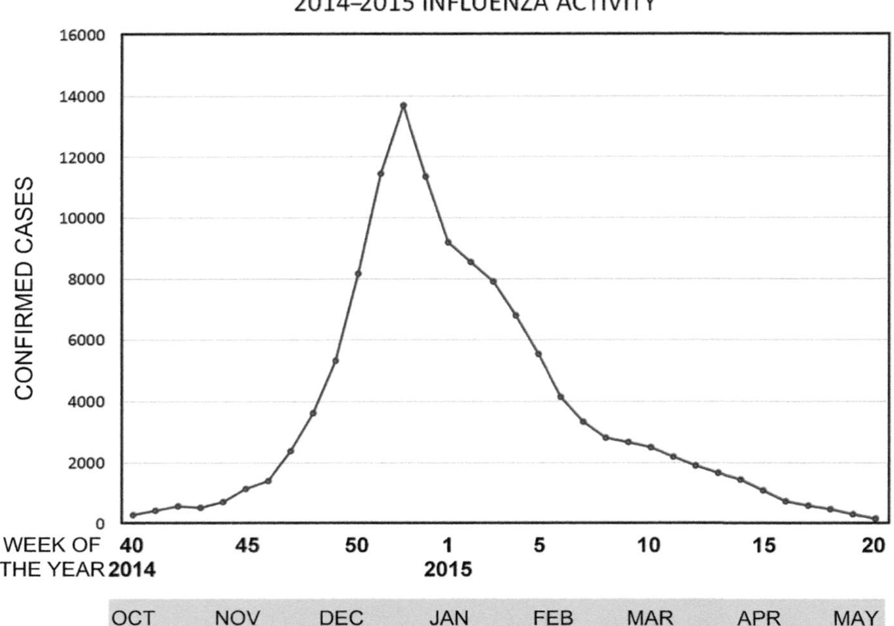

FIGURE 5.15 **2014–15 Confirmed cases of seasonal influenza.** Many epidemiological variables affect the transmission of viral infection and are worthwhile to consider while analyzing disease transmission. This graph shows that in 2014–15, laboratory-confirmed cases of influenza were at their highest in the United States from November to March, peaking in December of 2015. *Data obtained from FluView (internet), Centers for Disease Control and Prevention, 2015. Available at: http://www.cdc.gov/flu/weekly.*

defined by the case definition, and identify the who, where, and when (host, location, and time). In addition, analytic studies try to determine the cause and transmission.

Time refers not only to the hour and minute of the day, but also to the season and time of year. Some diseases occur more frequently during a particular time of the year; for example, influenza viruses peak during the winter months (Fig. 5.15), when drier weather supports the increased dissemination of aerosolized particles in the environment and people congregate inside more often. In contrast, the incidence of diseases transmitted by mosquitoes, such as West Nile virus or eastern equine encephalitis virus, is higher at the end of the summer when mosquito populations are highest. Other viral infections show no association with season or time of year, such as hepatitis B virus, HIV, or HPV. Determining the typical yearly or seasonal pattern of viral infections is important in creating a baseline that can be used to compare future occurrence of the disease or monitor the effectiveness of control measures. Graphing the time of the incident versus the occurrence of symptoms can also be useful in assessing the incubation period of the virus, which can be helpful in identifying the specific virus.

Place refers to the local location of the case as well as the larger geographic location. A gastrointestinal virus outbreak may occur at a restaurant (the place), suggesting that the food may have been involved in the transmission of the virus. A person presenting with neurological manifestations in a rural location may suggest different viruses than in a big city—for example, rabies transmission is much more likely in areas where wildlife is prevalent. It would also be important to note that a patient presenting with hepatitis recently traveled to a remote location in South America. This also highlights that, because of viral incubation periods, the location and timing of symptoms does not always correlate with when infection occurred (Fig. 5.16).

Details concerning the *person* (host) can affect the chain of infection and are important to consider in epidemiological studies. Almost every health-related event varies with age because it is a factor in exposure, immune status, and physiological response. For example, older individuals are much more likely to reactivate varicella zoster virus, leading to shingles, whereas children are most likely to show symptoms of the "childhood diseases" to which adults have already become immune. Influenza A virus causes higher morbidity and mortality rates in young children and the elderly.

Many other personal attributes can contribute to infection. The sex of the individual can sometimes be a consideration—cervical cancer caused by HPV will not occur in men—and being part of different genetic, cultural, or social groups can contribute to the exposure of an individual to a particular virus. The socioeconomic status of an individual (income, education, and occupation) can also play a role in exposure to the virus and access to medical care.

(A) **Reported cases of SARS in the United States through 11/3/2004: by case definition and state of residence**

Location	Total Cases Reported	Total Suspect Cases Reported	Total Probable Cases Reported
Alaska	1	1	0
California	29	22	5
Colorado	2	2	0
Florida	8	6	2
Georgia	3	3	0
Hawaii	1	1	0
Illinois	8	7	1
Kansas	1	1	0
Kentucky	6	4	2
Maryland	2	2	0
Massachusetts	8	8	0
Minnesota	1	1	0
Mississippi	1	0	1
Missouri	3	3	0
Nevada	3	3	0
New Jersey	2	1	0
New Mexico	1	0	0
New York	29	23	6
North Carolina	4	3	0
Ohio	2	2	0
Pennsylvania	6	5	0
Rhode Island	1	1	0
South Carolina	3	3	0
Tennessee	1	1	0
Texas	5	5	0
Utah	7	6	0
Vermont	1	1	0
Virginia	3	2	0
Washington	12	11	1
West Virginia	1	1	0
Wisconsin	2	1	1
Puerto Rico	1	1	0
Total	**158**	**131**	**19**

(B) **Reported cases of SARS in the United States through 11/3/2004: by high-risk area visited**

Area	Count*
Hong Kong City, China	45
Toronto, Canada	35
Guangdong Province, China	34
Beijing City, China	25
Shanghai City, China	23
Singapore	15
China, mainland	15
Taiwan	10
Anhui Province, China	4
Hanoi, Vietnam	4
Chongqing City, China	3
Guizhou Province, China	2
Macoa City, China	2
Tianjin City, China	2
Jilin Province, China	2
Xinjiang Province	1

FIGURE 5.16 **2004 United States cases of severe acute respiratory syndrome (SARS) by location of diagnosis and travel area.** In the 2004 outbreak of SARS, caused by the SARS coronavirus, case patients were distributed throughout the United States (A). A clearer correlation with *place* could be determined when looking at the location to which each individual recently traveled: the majority of infected individuals traveled to areas of China, Singapore, or Taiwan (B). *Modified from Tables 1.3 and 1.4 of Centers for Disease Control and Prevention, Updated May, 2012. Principles of Epidemiology in Public Health Practice, third ed. Department of Health and Human Services, Atlanta, GA, US.*

Descriptive studies are effective in chronicling patterns and developing hypotheses as to the cause of an illness or outbreak. In addition to the *what, when, where*, and *who* epidemiological variables, analytic studies are also concerned with the *why/how* of the illness. A hallmark of analytic studies is the presence of a comparison (control) group that can be used to generate baseline data to which the outbreak or illness can be compared. With a comparison group, statistical analyses can be performed to determine a cause with good certainty.

Epidemiological studies fall into two general categories: **experimental** and **observational**. Experimental studies are planned, controlled studies; a clinical trial to test new vaccines that enrolls participants into the study, randomly assigns them into one of three groups (vaccine A, vaccine B, or placebo), and then gathers data is an experimental study. As the name implies, observational studies involve the observation of subjects and subsequent recording of data. In comparison to experimental studies, the epidemiologist does not have any influence over what exposure the participant receives. Observational studies are more common in epidemiology than experimental studies.

Observational studies fall into three categories: **cohort studies**, **case-control studies**, and **cross-sectional studies**. Cohort studies are similar to experimental studies in that two groups are compared in real time, except that, being an observational study, the epidemiologists do not assign participants to **cohorts**, or groups. Instead, they allow the natural course of things to proceed, tracking whether or not the two cohorts have different results. An example of a cohort study would be to observe consistent users of hookah pipes, a water pipe used to smoke flavored tobacco. One cohort of participants uses individual disposable plastic mouthpieces on their pipes, while the other group uses the attached metal mouthpiece. This study might observe whether HSV-1 transmission is more common in those using the metal, shared mouthpieces compared to those who each have their own disposable plastic mouthpiece. If this were an experimental study, the epidemiologists would have assigned each participant to a specific group, either the group that uses the shared mouthpiece or the disposable plastic mouthpiece.

A second type of observational study, the case-control study, is always retrospective, meaning that it analyzes past events. A case-control study happens after an event (for example, a viral outbreak) has occurred. Being that the outbreak has already happened, a control group is assembled retroactively with a group of similar people in a similar place as the outbreak to see if, in fact, the "outbreak" was different from the norm. An example of a case-control study would involve the infection of several people with hepatitis C virus at a tattoo parlor. After noting that all the **case patients** received their tattoos from one tattoo artist, epidemiologists retroactively enrolled a group of people that received tattoos from the other tattoo artist at the parlor. The control group allows them to have a baseline group to determine the typical infection rate. In this case, it was determined that one of the tattoo artists, but not the other, was improperly sterilizing tattoo equipment.

The final type of observational study is the cross-sectional study. In this type of study, data are gathered from a random sample of individuals at one time (a "cross section" of the population) and correlations are made. For example, a cross-sectional study might find that a high proportion of those individuals that have liver scarring have hepatitis C infection. Although the obvious conclusion seems to be that the liver scarring must be caused by the virus, it is also possible that the scarring makes the liver more susceptible to hepatitis C infection, and that is why these individuals have high rates of both liver scarring and hepatitis C. Because the study compares individuals at only one point in time, cause and effect (causation—the how/why) are difficult to determine. *Correlation does not equal causation*! Therefore, a cross-sectional study is not effective as an analytic study, but they are used routinely for descriptive studies.

Since both the cohort and case-control study have comparison groups, they would be considered analytic studies. Cross-sectional studies often do not have a comparison group as a control and are therefore most often carried out as descriptive studies.

SUMMARY OF KEY CONCEPTS

Section 5.1 Portals of Virus Entry

- For successful infection to occur, a host must come in contact with the virus (accessibility). The cells must express the cell surface receptors used by the virus (susceptibility), and they must also contain all the internal requirements for viral replication to proceed (permissivity).
- Portals of entry are the locations through which viruses gain entry into the body. Unless delivered directly into the tissues through a mosquito bite or needle, most viruses first interact with the epithelium at the site of entry.
- The majority of viruses enter humans through inhalation into the respiratory tract. The gastrointestinal tract, genital tract, skin, blood, and eyes are also points of entry.
- Acid-labile viruses break down within the low pH of the stomach, while acid-resistant viruses are resistant.
- Some viruses are able to be transmitted through organ transplants or through the placenta to a developing fetus.

Section 5.2 Dissemination Within a Host

- Localized viral infections replicate at the initial site of infection. Systemic infections spread to additional areas throughout the body, generally through hematogenous or neurotropic spread.

Section 5.3 Portals of Virus Exit

- Viruses are shed through portals of exit to infect new hosts. For localized infections, this is generally at the same location through which the virus entered: viruses that infect the skin are spread through skin-to-skin contact, respiratory viruses are shed within respiratory secretions, and gastrointestinal viruses are shed within aerosolized vomit or diarrhea.
- The stability of virions within the environment is dependent upon many factors, including the type of nucleic acid; the presence of a viral envelope; and the sensitivity of virions to pH, humidity, moisture content, and sunlight.
- Respiratory droplets are larger in size (~20 μm) and are spread only short distances before falling to the ground. Smaller aerosolized particles (<5 μm) can travel extended distances within the air.

Section 5.4 Patterns of Infection

- Upon infection, a host generally passes through four stages of disease: the incubation period, prodromal period, illness period, and convalescent period.
- Acute infections, during which viruses multiply and spread quickly, are often the cause of epidemics.
- Persistent infections result when the immune system is unable to clear the virus from the body. Persistent infections can also result from viral latency or slow infections.

Section 5.5 Epidemiology

- Epidemiology is the study of how disease occurs in populations. Sporadic diseases occur irregularly, while endemic refers to the normal rate of infection. An epidemic occurs when there are more cases than the normal endemic rate. A pandemic is an epidemic that spreads throughout several countries or the world.
- The epidemiologic triad model represents the factors that bring an agent and susceptible host together in a particular environment. There are many factors that affect each component of the model.
- The chain of infection represents how an agent leaves its reservoir through a portal of exit, is conveyed via a mode of transmission, and enters a susceptible host through a portal of entry. Epidemiologists identify the factors involved at these stages in an effort to institute control measures.

Section 5.6 Epidemiological Studies

- Epidemiological studies are divided into descriptive and analytic studies. Both types use a case definition to confirm persons with a case of the disease. Both also analyze the time, place, and persons involved in the chain of infection, but analytic studies use a comparison group to determine the cause of the outbreak.
- Experimental studies are planned ahead of time and participants are randomly assigned to predetermined groups. Observational studies—which fall into cohort, case-control, and cross-sectional studies—acquire data and track cohorts as they occur within normal situations.

FLASH CARD VOCABULARY

Accessibility	Mode of transmission
Susceptibility	Portal of exit
Permissivity	Epithelium
Tropism	Mucosal epithelium
Portal of entry	Respiratory tract
Goblet cells	Neurotropic spread
Alveolar macrophages	Viremia
Phagocytosis	Primary/secondary viremia
Gastrointestinal tract	Shedding
Villi	Viruria
Microvilli	Incubation period
Acid labile	Prodromal period
Acid resistant	Illness period
Fecal–oral route	Convalescent period
Sexually transmitted diseases	Immunocompetent

Continued

FLASH CARD VOCABULARY—Cont'd

Epidermis	Immunocompromised
Dermis	Period of communicability
Subcutaneous tissue	Acute infection
Conjunctiva/conjunctivitis	Persistent infection
Congenital infection	Defective interfering particles
Vertical transmission	Latency/reactivation
Horizontal transmission	Slow infection
Intrapartum transmission	Epidemiology
Transfusion-transmitted infections	Sporadic
Localized infection	Endemic
Systemic infection	Epidemic/outbreak
Hematogenous spread	Pandemic
Epidemiologic triad model	Mortality
Agent	Incidence
Host	Prevalence
Environment	Case definition
Chain of infection	Epidemiological variables
Carrier	Descriptive studies
Zoonosis	Analytic studies
Vehicles	Experimental studies
Vectors	Observational studies
Direct transmission	Cohort studies
Indirect transmission	Case-control studies
Control measures	Cross-sectional studies
Herd immunity	Case patient
Morbidity	

CHAPTER REVIEW QUESTIONS

1. Make a table listing each portal of entry. What defenses does the host have at each location and how are viruses able to successfully bypass them?
2. Norwalk virus causes significant morbidity in developed nations, despite that these countries have clean water supplies. How do you think the virus is transmitted, and why it is so successful?
3. Describe the architecture of the skin and how viruses gain access to each layer and the subcutaneous tissue.
4. How are vertical and horizontal transmission of viruses different from each other?
5. Which viruses are capable of initiating transplacental or intrapartum infections?
6. How do localized infections become systemic infections?
7. Describe how different factors could affect the inactivation of virions within the environment.
8. Draw out the stages of infection after a person is infected with a virus.
9. Your friend walks into class, still sniffling occasionally from a respiratory viral illness. She reassures you that she's "not infectious anymore." You have your concerns. Why?
10. How can persistent infections arise?
11. Describe the difference between endemic and epidemic.
12. Make a list of the three major aspects of the epidemiologic triad model and what factors could affect each aspect in promoting infection.
13. Create a list of control measures that interfere with each variable within the chain of infection.
14. Design a case-control study that attempts to determine the precise food product that was the cause of a viral gastrointestinal illness.

FURTHER READING

Bausch, D.G., Towner, J.S., Dowell, S.F., et al., 2007. Assessment of the risk of Ebola virus transmission from bodily fluids and fomites. J. Infect. Dis. 196 (Suppl. 2), S142–S147.

Casanova, L.M., Jeon, S., Rutala, W.A., Weber, D.J., Sobsey, M.D., 2010. Effects of air temperature and relative humidity on coronavirus survival on surfaces. Appl. Environ. Microbiol. 76, 2712–2717.

Centers for Disease Control and Prevention, 2012. Principles of Epidemiology in Public Health Practice, third ed. Department of Health and Human Services, Atlanta, GA, US.

Centers for Disease Control and Prevention, National Center for Emerging and Zoonotic Infectious Diseases. Hantavirus. http://www.cdc.gov/hantavirus/ (accessed 31.05.15.).

Galel, S.A., Lifson, J.D., Engleman, E.G., 1995. Prevention of AIDS transmission through screening of the blood supply. Annu. Rev. Immunol. 13, 201–227.

Gensberger, E.T., Kostić, T., 2013. Novel tools for environmental virology. Curr. Opin. Virol. 3, 61–68.

Koopmans, M., 2013. Viral abundance and its public health implications. Curr. Opin. Virol. 3, 58–60.

Pirtle, E.C., Beran, G.W., 1991. Virus survival in the environment. Rev. Sci. Tech. 10, 733–748.

Stramer, S.L., 2007. Current risks of transfusion-transmitted agents: a review. Arch. Pathol. Lab. Med. 131, 702–707.

Susser, E., Bresnahan, M., December 2001. Origins of epidemiology. Ann. N. Y. Acad. Sci. 6–18.

Villarreal, L.P., Defilippis, V.R., Gottlieb, K.A., 2000. Acute and persistent viral life strategies and their relationship to emerging diseases. Virology 272, 1–6.

Chapter 6

The Immune Response to Viruses

It is a constant evolutionary battle between a virus and the host immune system. An **immune response** is activated immediately by the immune system upon the entry of a virus into the body. On one hand, viruses that elicit a strong immune response are cleared immediately from the body, unable to spread to new hosts. On the other end of the spectrum, aggressive viruses that are very **virulent** may end up killing the host before spread can occur. The most successful viruses are those that are able to escape from the immune system long enough to replicate and spread. Viruses have evolved a variety of mechanisms to do just this.

As discussed in Chapter 5, "Virus Transmission and Epidemiology," there are a variety of intrinsic defense mechanisms at epithelial surfaces of the body that interfere with infection, such as mucus, low pH, or the flow of air or fluid. If the virus is able to bypass these barriers and initiate infection of a cell, then the host immune system will become activated. The immune system is a coordinated system of white blood cells, proteins, and receptors that sense the presence of the virus, control infection, and provide long-term **immunity**, or resistance, against the virus. There are two arms of the immune system, the **innate immune system** and **adaptive immune system**, that work together to accomplish this goal (Fig. 6.1). The innate immune system is activated immediately upon infection with a virus, but the response is nonspecific, meaning that the mechanisms employed are general and not customized for any specific virus. On the other hand, the adaptive immune response is tailored for the specific virus. The adaptive immune response against any one virus will be different than the adaptive immune response against another virus. If the immune system is successful, the specific response will ultimately result in the control of infection. The downside of this customized adaptive response is that it takes time to happen, in the order of days to weeks. Although the innate immune response is nonspecific, it is critical to properly activate the adaptive immune system and control infection until the adaptive response is generated. A host without an adaptive immune system would not clear an infection, but a host lacking an innate immune system would not have the means to control virus replication immediately after infection or activate the adaptive immune system.

6.1 THE INNATE IMMUNE SYSTEM

6.1.1 Pattern Recognition Receptors

The first step in combatting infection involves recognition of the virus by the host. Immune cells within host tissues have a variety of protein receptors known as **pattern recognition receptors** (PRRs) that recognize **pathogen-associated molecular patterns** (PAMPs) from a variety of pathogens, including viruses. Some PRRs have evolved to interact with viral nucleic acid structures that are not normally found within cells, such as dsRNA, or to recognize them in cellular locations in which the cell's nucleic acid would not be found, such as within endosomes. Other PRRs

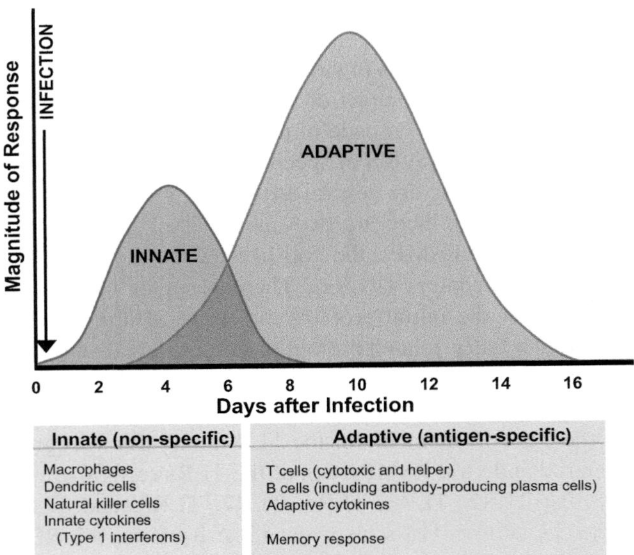

FIGURE 6.1 **The innate and adaptive arms of the immune system.** When a host is infected with a pathogen, the innate immune system is activated immediately. It is composed of cells that produce cytokines, stimulate adaptive cells, and function in concert to control the pathogen until the adaptive immune system is fully activated. The adaptive immune system is composed of two major types of cells, T lymphocytes and B lymphocytes. Although it takes days to become fully activated, the adaptive immune response is tailored to the specific pathogen. Cytotoxic T cells kill virally infected cells and helper T cells activate B cells to differentiate into plasma cells, which produce antibody that neutralizes the virus.

Essential Human Virology.

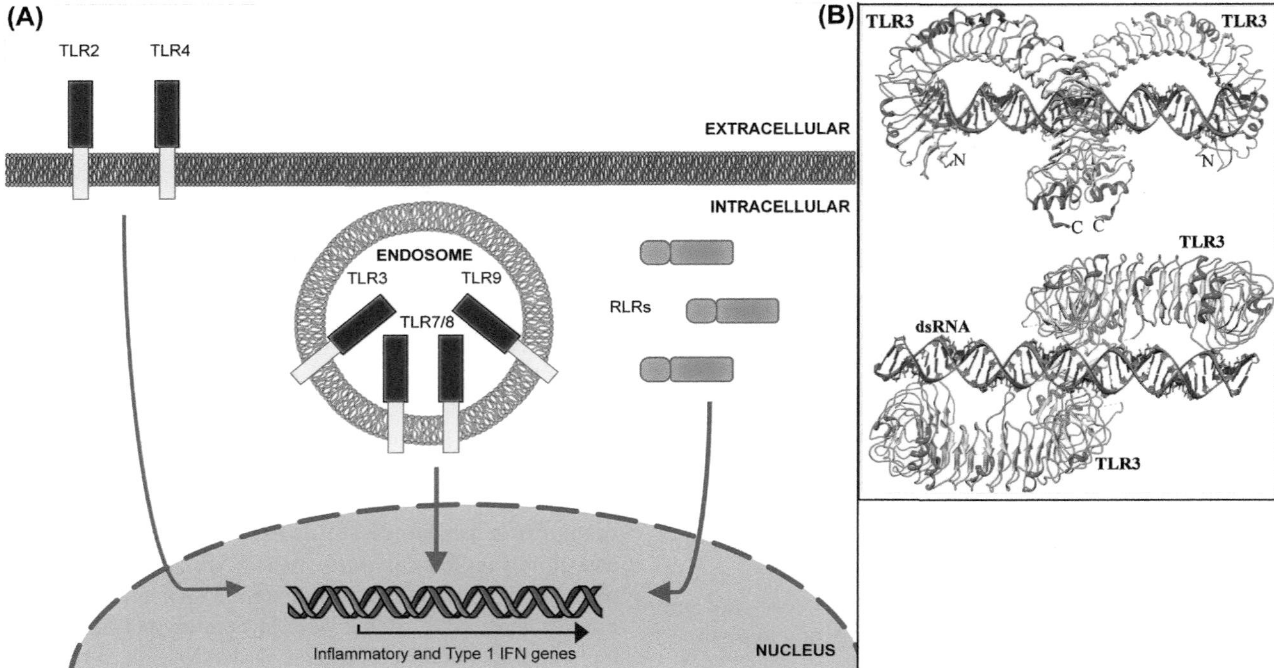

FIGURE 6.2 **Cellular pattern recognition receptors.** (A) Protein receptors are present in various locations of the cell that sense the presence of pathogens, including viruses. TLR2 and TLR4 are located at the plasma membrane, while TLR3, TLR7, TLR8, and TLR9 are found within endosomes. RLRs are in the cytosol. Binding of the PRR to its specific pathogen-associated molecular pattern (PAMP) sends off an intracellular protein signaling cascade that leads to the activation of inflammatory and cytokine genes, such as the type 1 interferons. (B) The TLRs are horn-shaped proteins that dimerize when they bind their cognate ligand. Shown here are two ribbon diagrams, from the side and from the top, of two TLR3 molecules binding to viral dsRNA. *(Image courtesy of Manavalan, B. et al., 2011. Similar structures but different roles—an updated perspective on TLR structures. Front. Physiol. 2, 41.* http://dx.doi.org/10.3389/fphys.2011.00041.*)*

recognize viral proteins in virions as they are attached to the surface of the cell for infection. The binding of the PAMP to the PRR activates a cascade of proteins within the cell that induces the transcription of specific genes.

Although there are several PRRs that recognize bacterial components, there are two main classes of PRRs that recognize viral PAMPs: the Toll-like receptors (TLRs) and RIG-I-like receptors (RLRs). These receptor classes are named after the initial proteins that were identified—Toll and RIG-I. Later, related protein receptors were discovered, so families of receptors were created based upon receptor similarity. These families are the TLRs and RLRs. Overall there are 10 members of the TLR family that recognize bacterial and viral PAMPs. Six of the TLRs recognize viral PAMPs: TLR2, TLR3, TLR4, TLR7, TLR8, and TLR9. Whereas certain TLRs also recognize bacteria, the RLRs recognize only viral and not bacterial RNA.

PRRs are situated within the cell at locations that are used for virus entry and replication (Fig. 6.2). TLR2 and TLR4 are transmembrane on the surface of the cell, where they are positioned to interact with the viral envelope proteins of viruses attaching to the plasma membrane. (TLR4 homodimerizes, while TLR2 forms heterodimers with one of two other TLRs, TLR1 or TLR6.) On the other hand, homodimers of TLR3, TLR7, TLR8, and TLR9 are sensors of viral nucleic acids, not proteins: they respond to dsRNA (TLR3), ssRNA (TLR7 and TLR8), or viral DNA sequences (TLR9). These TLRs are also transmembrane proteins but positioned within endosomes, where they come in contact with viral nucleic acids during the uncoating of virions. RLRs are unique in that they are found within the cytosol rather than at the surface of the cell or within intracellular compartments. RLRs recognize the ssRNA or dsRNA of actively replicating viruses. The RNA that RLRs recognize can be genomic RNA from viruses or transcribed RNA from DNA or RNA viruses (Table 6.1).

One of the ways that the body is alerted that an infection is taking place is through these pattern recognition receptors. Binding of the specific PAMP to the PRR causes a signal to be sent to the nucleus that results in the production of important proteins that influence the subsequent innate and adaptive immune responses.

6.1.2 Cytokines

TLRs and RLRs induce the production of **cytokines**, small proteins that are secreted by cells and cause effects within target cells through a **cytokine receptor** (Fig. 6.3). One of the most important classes of cytokines produced during viral infection through the PRRs are the **type 1 interferons**,

TABLE 6.1 Pattern Recognition Receptors That Recognize Viruses

Pattern Recognition Receptor	Ligand	Virus Recognized
Toll-like Receptors		
TLR2	Envelope proteins	Measles virus
TLR3	Viral dsRNA	Respiratory syncytial virus, reovirus, West Nile virus
TLR4	Envelope proteins	Respiratory syncytial virus, mouse mammary tumor virus
TLR7	Viral ssRNA	Influenza A virus, vesicular stomatitis virus
TLR8	Viral ssRNA	Human immunodeficiency virus, vesicular stomatitis virus, influenza A virus
TLR9	Viral dsDNA	Herpes simplex virus-1, herpes simplex virus-2, mouse cytomegalovirus
RIG-I-like Receptors		
RIG-I	Recognizes virally transcribed ssRNA and short dsRNA	–ssRNA viruses: Influenza A virus, influenza B virus, Sendai virus, respiratory syncytial virus, measles virus, rabies virus, vesicular stomatitis virus +ssRNA viruses: Hepatitis C virus, Japanese encephalitis virus dsDNA viruses: Epstein–Barr virus, herpes simplex virus-1, adenovirus
MDA5	Long dsRNA	+ssRNA viruses: Picornaviruses (encephalomyocarditis virus, Theiler's virus, murine norovirus-1, murine hepatitis virus)
LGP2	Regulates other RLRs	

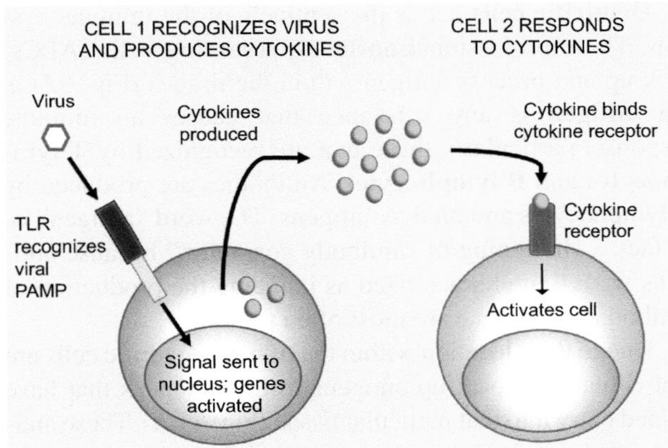

FIGURE 6.3 Cytokines and cytokine receptors. Cytokines are small proteins that are produced by a cell and act upon any cells that have the specific receptor for the particular cytokine. There exist over 30 different cytokines, and each one has a unique receptor that recognizes it. The binding of a cytokine to its receptor activates proteins within the cell bearing the receptor. The effects that ensue within the cell depend upon the type of cytokine. In this example, a TLR recognizes a viral PAMP, which signals to the cell to activate cytokine genes. The cytokines are secreted and act upon a neighboring cell that possesses the cytokine receptor, in a paracrine fashion. Cytokines can also act upon the cell that produced them, in an autocrine fashion, if the cell has the specific cytokine receptor.

IFN-α and IFN-β. The type 1 IFNs were the first cytokines to be discovered, named for their potent ability to "interfere" with influenza replication in cells of embryonated chicken eggs (see *In-Depth Look*). They bind to the IFN-α/β receptor, which is found on the surface of most cells within the body, and activate proteins within the cell that turn on hundreds of genes. Certain immune system cells, called **plasmacytoid dendritic cells**, produce high amount of IFN-α/β after sensing viral nucleic acid through TLRs. In this way, the type 1 IFNs function as a warning system for uninfected cells.

Many of the IFN-activated genes code for proteins that activate immune system cells, but others encode proteins that interfere directly with the process of viral replication. The two best characterized pathways to an antiviral state induced by the type 1 IFNs involve the dsRNA-dependent protein kinase (PKR) and the 2′–5′ oligoadenylate synthetase–ribonuclease L (OAS–RNase L) pathway (Fig. 6.4). PKR is a cytoplasmic protein that is activated when it binds to dsRNA, which is an intermediate of RNA virus replication found in the cytoplasm during infection. PKR then binds to eukaryotic translation initiation factor 2α (eIF-2α) and inactivates it. As its name implies, eIF-2α is normally involved in initiating protein translation. Without its activity, the synthesis of both viral and host proteins by ribosomes is halted.

OAS is also activated by dsRNA. Activated OAS generates short strings of nucleotides, or **oligomers**, from ATP. They are called 2′–5′ oligoadenylates because the 2′- and 5′-carbons of the nucleotides are bonded together in the molecule, which is how OAS gets its name: it synthesizes 2′–5′ oligoadenylates. These bind to and activate a cellular enzyme called RNase L that degrades single-stranded RNA. The result is that viral mRNAs and ssRNAs are cleaved, as well as host mRNA.

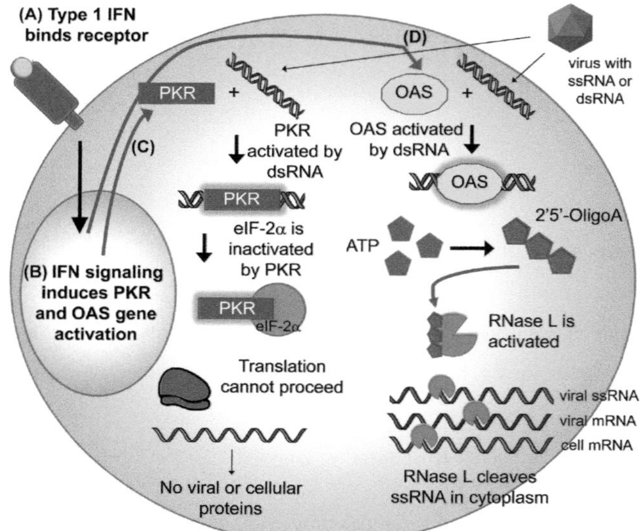

FIGURE 6.4 **Intracellular antiviral pathways induced by type 1 IFN.** When type 1 IFNs bind their receptor on the surface of a cell (A), intracellular proteins are activated that induce the transcription of PKR and OAS mRNA (B), which leaves the nucleus and is transcribed into the PKR and OAS proteins. This sets up an antiviral state within the cell. If the cell becomes infected with a ssRNA or dsRNA virus, dsRNA is produced during replication and activates these proteins. When PKR is activated (C), it inactivates eIF-2α, a translation initiation factor that assists in recruiting the ribosome complex to mRNA. Without eIF-2α, translation does not proceed and proteins are not produced. When OAS is activated by dsRNA (D), it synthesizes 2′5′-OligoAs from ATP. These activate RNase L, which cleaves any ssRNA within the cytoplasm. With no viral mRNAs or proteins, replication cannot occur within the cell.

In-Depth Look: Discovery of the Type 1 Interferons

Alick Isaacs and Jean Lindenmann discovered the first cytokine in 1957 while studying influenza virus. The virus is able to infect and replicate within the cells of the chorioallantoic membrane of fertilized hen eggs. (The chorioallantoic membrane is the bird equivalent of the mammalian placenta.) Isaacs and Lindenmann created a stock of virus that was unable to replicate by inactivating the virus with high heat or UV light. They noted that when they treated chorioallantoic membrane cells with inactivated virus, a substance was produced by the cells that would protect other cells from becoming infected with infectious virus. They named the proteins "interferons" after their demonstrated ability to interfere with viral replication. Biotechnological advances allowed recombinant human IFN-α and IFN-β to be produced by bacterial cells in the early 1980s, allowing the use of type 1 IFNs as treatments for medical conditions. Currently, IFN-α or IFN-β are approved to treat chronic hepatitis B and C infections, AIDS-related Kaposi's sarcoma, multiple sclerosis, and several cancers including hairy cell leukemia, malignant melanoma, and follicular lymphoma.

If an infected cell binds the secreted IFN-α/β, these pathways will be activated and virus replication will be slowed because viral mRNAs will be cleaved by RNase L and viral proteins will not be translated, due to the inactivated eIF-2α. If noninfected cells bind the cytokine, they set up an intracellular environment that is not conducive to infection because no proteins can be produced.

As part of the innate immune system, the production of the type 1 IFNs helps to reduce viral replication until the adaptive immune response is activated. These and other cytokines produced during viral infection also act as signals that tailor the developing adaptive response for the particular pathogen encountered.

6.1.3 Macrophages and Dendritic Cells

Two important innate cell types, macrophages and dendritic cells, are found within all tissues of the body. **Macrophages** (*macro* means "large," and *phage* is Greek for "eat" = "big eaters") are large immune cells that are specialized in phagocytosis, the ingestion and digestion of large particles (Fig. 6.5A and B). As mentioned in Chapter 5, "Virus Transmission and Epidemiology," antibodies circulating within the host can recognize viral capsid proteins and bind to them. Macrophages have antibody receptors that are used to bind and internalize virus/antibody complexes. Macrophages also phagocytose dead cells and remove them from the body, and they possess PRRs that recognize viral PAMPs.

Dendritic cells act as the sentinels of the immune system. These professional **antigen-presenting cells** (APCs) pick up and process antigen within the tissues (Fig. 6.5C). An **antigen** is any substance that causes an immune response, particularly those that are recognized by **T lymphocytes** and **B lymphocytes**. Antibodies are produced by B lymphocytes and bind to antigens. The word "antigen" is, in fact, a shortening of "antibody generator," because antigens were first characterized as inducing the production of antibodies. Antigens are most often proteins.

Due to their location within the tissues, dendritic cells are well situated to pick up antigens from pathogens that have gained entry into that particular tissue (Fig. 6.6A). These antigens can be derived directly from the phagocytosis of virions or dead infected cells, or they can enter the dendritic cell through endocytosis or pinocytosis of viral antigens found within the tissue. Antigens can also be acquired directly from viruses that have infected the dendritic cell. The dendritic cell processes the antigen and uses lymphatic vessels to travel to the regional lymph nodes. Cells within tissues are bathed in interstitial fluid, derived from blood plasma, that is collected in lymphatic vessels and drains to the lymph nodes, eventually returning to the bloodstream. Activated dendritic cells travel within the lymph in these lymphatic vessels to the nearest lymph node. The lymph node is the site of **antigen presentation** to T lymphocytes (Fig. 6.6B). The details of antigen presentation will be discussed in Section 6.2.1.

Viruses that are found within the blood, rather than within tissues, are filtered out by macrophages and dendritic

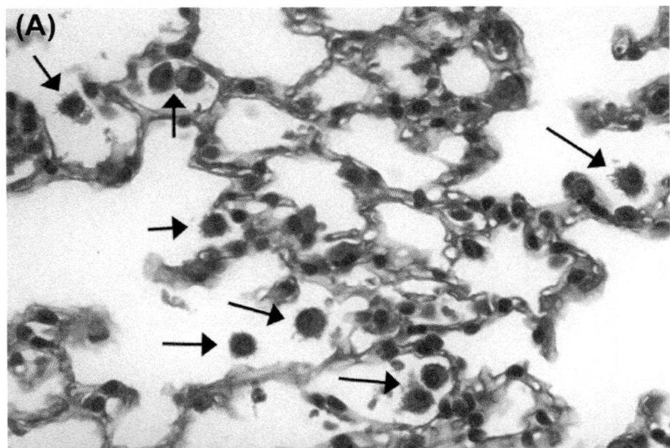

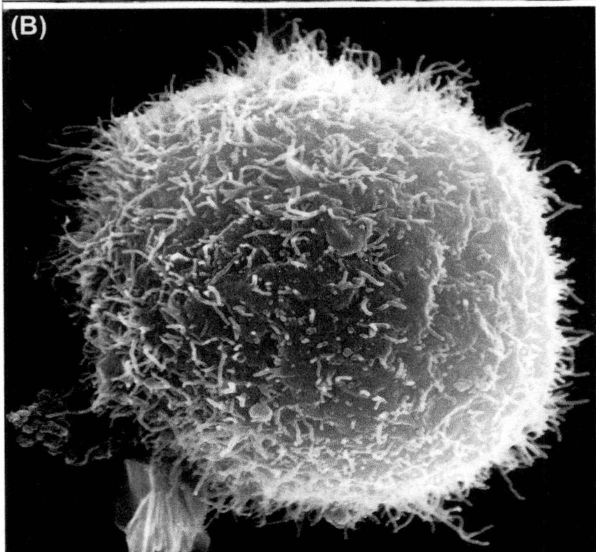

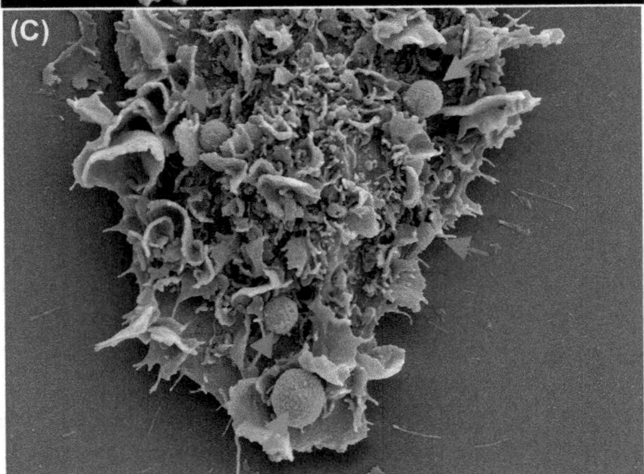

FIGURE 6.5 **Macrophages and dendritic cells.** Macrophages and dendritic cells are two important innate cells that are located within body tissues. (A) *Arrows* indicate macrophages, found here in the alveoli of the lung where they phagocytose inhaled pathogens and particles. (B) Scanning electron micrograph of a macrophage. *(Image courtesy of Dr. Timothy Triche, National Cancer Institute.)* (C) Scanning electron micrograph of a dendritic cell (gray) phagocytosing *Aspergillus* fungal spores. *(Image courtesy of Behnsen J., et al., 2007. Environmental dimensionality controls the interaction of phagocytes with the pathogenic fungi Aspergillus fumigatus and Candida albicans. PLoS Pathog. 3(2), e13.* http://dx.doi.org/10.1371/journal.ppat.0030013.*)*

cells within the spleen of the body. Here, dendritic cells can also pick up and present antigen to T lymphocytes found within the spleen.

Dendritic cells possess many of the pattern recognition receptors, as discussed above, that are used to recognize viruses. During infection, recognition of viral PAMPs by PRRs on the dendritic cell results in the maturation of the dendritic cell to one that is specialized in presenting antigen to T lymphocytes.

6.1.4 Natural Killer Cells

Natural killer cells (NK cells) are an important innate cell type in the defense against viruses. They possess lytic granules that induce **apoptosis** of virally infected target cells. Apoptosis, also known as programmed cell death, is carried out as an orderly process within cells. The proteins that are delivered to the target cell by NK cells set off a cascade of events that activates enzymes called caspases to fragment DNA within the nucleus of the cell. The contents of the target cell are degraded and shed within membrane-bound vesicles that are released to other cells. Having completed the process of apoptosis, the cell is eventually removed from the tissue through macrophage phagocytosis.

Type 1 IFNs, in addition to their role in activating antiviral proteins within the cell, significantly increase the killing activity and proliferation of NK cells during viral infection. The specifics of NK cell killing during viral infection will be discussed in more detail later in this chapter.

6.2 THE ADAPTIVE IMMUNE SYSTEM

In contrast to innate immune responses, the adaptive immune response is customized to respond to a pathogen specifically, rather than generally. This response must be properly stimulated by innate immune components, however.

The adaptive immune system is composed of white blood cells known as T lymphocytes and B lymphocytes, more often referred to as T cells and B cells. T cells and B cells are located within the lymph nodes or spleen of the body. It is at these locations that T cells will first be presented antigen by dendritic cells. In the case of infection within tissues, the dendritic cells mature and travel within lymphatic vessels to the lymph nodes in order to present antigen to the T cells located there.

The immune system has two broad classes of T cells that are important in defense against viruses, **cytotoxic T lymphocytes (CTLs)** and **helper T lymphocytes**. As their names imply, CTLs kill virally infected cells via apoptosis, and helper T cells activate other cell types, particularly B cells. Both types of T cells express a transmembrane **T cell receptor for antigen (TCR)** that specifically recognizes the antigen presented by the dendritic cell.

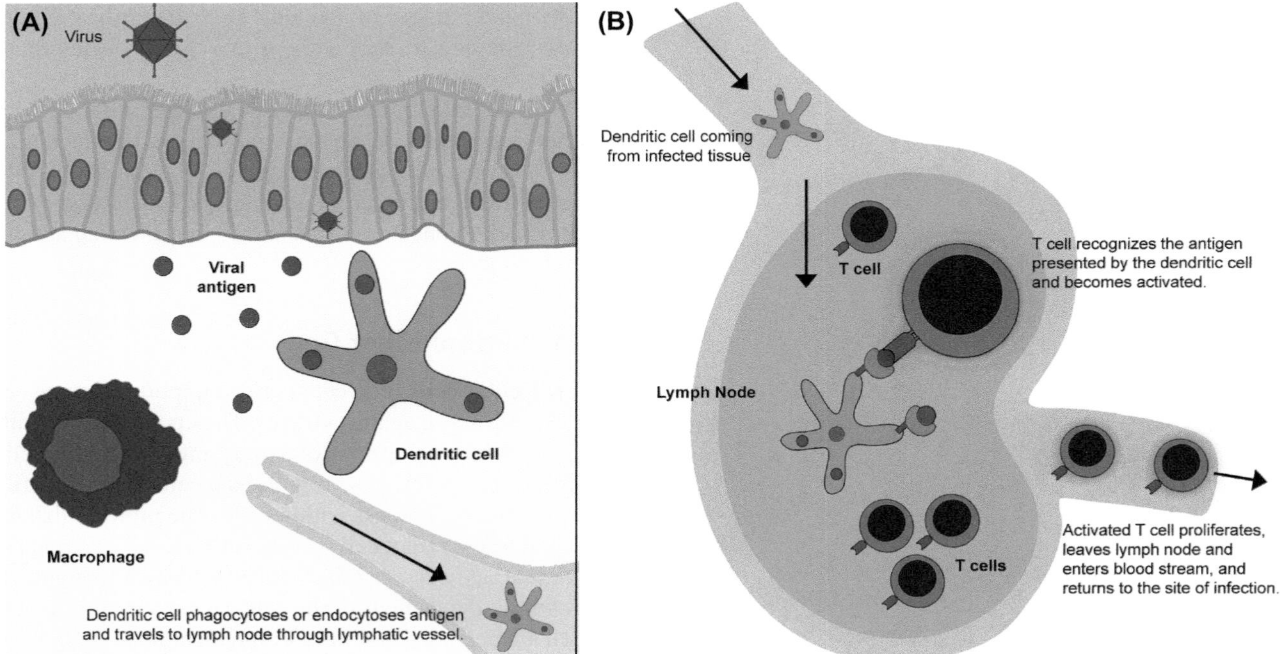

FIGURE 6.6 **Dendritic cell processing and presentation of antigen to T lymphocytes.** (A) Macrophages and dendritic cells are found in all tissues of the body, situated to respond to invading pathogens. (B) When a pathogen enters the tissue, macrophages can phagocytose antibody-bound virions or dead cells. Dendritic cells endocytose viral antigens, enter the lymphatic vessels located in the tissue, and travel to the regional lymph node. Dendritic cells present the antigen to the T cells in the lymph node, and a T cell becomes activated and proliferates when it recognizes the antigen. The activated T cells leave the lymph node and return to the site of infection to kill infected cells.

6.2.1 Cytotoxic T Lymphocytes

T cells are unable to recognize antigen unless it is presented to the T cell within specific proteins. Within the dendritic cell, the protein antigen is processed into short pieces known as peptides that are loaded onto cell surface molecules called **major histocompatibility complexes (MHC)**. MHC molecules allow for the display of antigens that are normally only found inside the cell, including viral antigens, on the outside of the cell. There are two different types of MHC, MHC class I and MHC class II ("class one" and "class two"), that are loaded with antigen and displayed on the surface of the dendritic cell (Fig. 6.7A).

Different subsets of T cells survey the antigen in the two classes of MHC molecules, using their TCR to do so. CTLs have a TCR that recognizes antigen presented in the MHC class I molecule. CTLs are also known as "CD8 T cells" because they have a protein, called CD8, that associates with the TCR and assists in the binding of the TCR to the MHC class I molecule (Fig. 6.7B).

The TCR of each T cell is slightly different and therefore recognizes slightly different antigen. Within the human body, there are an estimated 10^{18} different T cells—a billion billion T cells, each with a different TCR! The dendritic cell presents antigen, but only the T cell with the TCR that recognizes that specific antigen will bind and become activated by it. This is how the adaptive immune response is *antigen specific*: different viruses have different protein antigens, and the T cells specific for that viral antigen are those that respond (Fig. 6.8A). Lymphocytes that have not previously encountered their antigen are known as **naïve** lymphocytes.

The dendritic cell presents the antigen in the MHC class I molecule to naïve CTLs, and upon recognizing its **cognate** (specific) antigen, the CTL proliferates to create trillions of clones that all recognize the same antigen. The cells leave the lymph node and return to the site of infection, searching for infected cells expressing the cognate antigen.

All nucleated cells within the body express MHC class I, and like dendritic cells, each cell loads its MHC class I molecules with the various antigens found within the cell. Cells infected with a virus will display MHC class I molecules that contain antigens from the virus, a metaphorical "red flag" to alert CTLs that a virus has infected the cell. The activated CTLs survey each cell within the infected tissue with their TCRs. When the correct match is found, the CTL secretes proteins that cause the infected target cell to undergo apoptosis (Fig. 6.8B), similarly to how NK cells induce apoptosis. The CTL will repeat the process throughout the tissue until all infected cells have been given the "kiss of death."

As will be discussed in more detail later in this chapter, several viruses have evolved mechanisms to interfere with the presentation of antigen within MHC class I on the surface of the cell. Without the MHC molecule, the CTL is

Focus on a Viral Immunologist

Christine A. Biron, PhD, Esther Elizabeth Brintzenhoff professor of Medical Sciences, Brown University.

1. What is viral immunology? What aspect of it does your lab work on?
   Viral immunology is simply the study of immune responses to viruses. Within the discipline, however, there many different aspects. My laboratory's research has been dedicated to innate immune responses during viral infections, particularly the NK cells and cytokines that contribute to the earliest defense mechanisms after challenge.
2. How did you first become interested in viral immunology?
   It may seem unbelievable, but I developed interests in immunology and microbiology while in elementary school and made my way through biochemistry to these areas of study. The specific interest in viruses came during graduate school because of the exciting things that were (and continue to be) going on in this field: virus–host relationships, viral latency, viruses and cancer, antivirals drugs, and immune responses mediating viral defense as compared to those contributing to pathology.
3. What was your most exciting discovery or result?
   The one paper that many have said had the biggest impact on the field was a report with John Sullivan characterizing the lack of NK cells in an adolescent with unusual sensitivities to three different herpesviruses. It was the first time a link between NK cells and resistance to viral infections was established in humans. Overall, however, our work has made a number of discoveries on the ways NK cells and cytokines respond to and function during viral infections. Most recently, the role of relative intracellular signaling molecule concentrations in shaping responses to particular cytokines has provided important new insights for explaining how the host can be prepared with a limited number of genes to defend against an almost limitless number of infectious organisms.
4. What do you think have been the biggest breakthroughs in viral immunology since you joined the field?
   There are so many! I started graduate school shortly after Zinkernagel and Doherty did their ground-breaking work showing how CD8 T cells of the adaptive immune system recognize virus-infected host target cells. (They were given the 1996 Nobel Prize in Physiology or Medicine for this discovery of one way the immune system distinguishes *self* from *nonself*). More recently, workers following up on a model proposed by Charlie Janeway to explain how the innate immune system can distinguish between self and nonself led to the discovery of pattern recognition receptors in place to stimulate innate cytokine production. Beutler and Hoffmann shared the 2011 Nobel Prize for their key discoveries on the molecular mechanisms of innate immune activation with Steinman for his discovery of the specialized dendritic cell and its role in activating adaptive immunity. These are just a few very important highlights, but there have been so many others. It has been an exciting and stimulating time, with advances in knowledge seeming to be escalating.
5. What do you think the biggest challenge is in the field of viral immunology right now?
   Right now, attempts to make vaccines are coming up against the realization that we have so much more to learn about how to optimize immune responses against viruses while limiting their potentially damaging effects. The work will never end because new viruses come along to surprise and threaten humans. They rapidly adapt to avoid immune detection and control, and do so in unique and unexpected ways. As the mechanisms of avoidance are revealed, different features of the immune system are appreciated for the first time. Viral immunology will continue to be an exciting and rewarding career for the next generation of scientists.

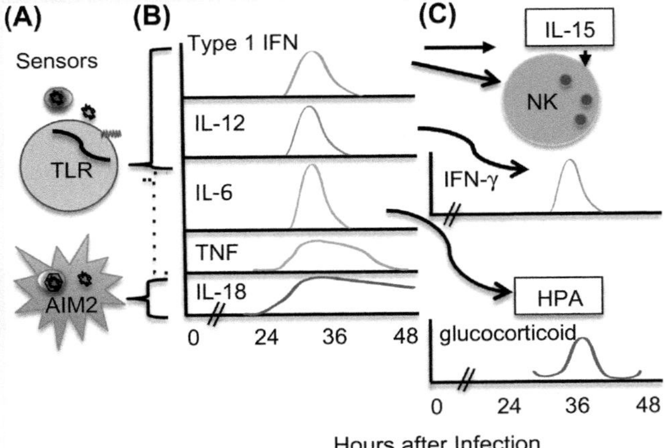

When mice are infected with mouse cytomegalovirus, pattern recognition receptors recognize viral components (A), inducing the production of several cytokines (B) that promote a proinflammatory response and downstream events (C). *Reprinted from Biron C.A., Tarrio M.L., 2015. Immunoregulatory cytokine networks: 60 years of learning from murine cytomegalovirus. Med. Microbiol. Immunol. 204 (3). Fig. 1, with kind permission from Spring Science and Business Media.*

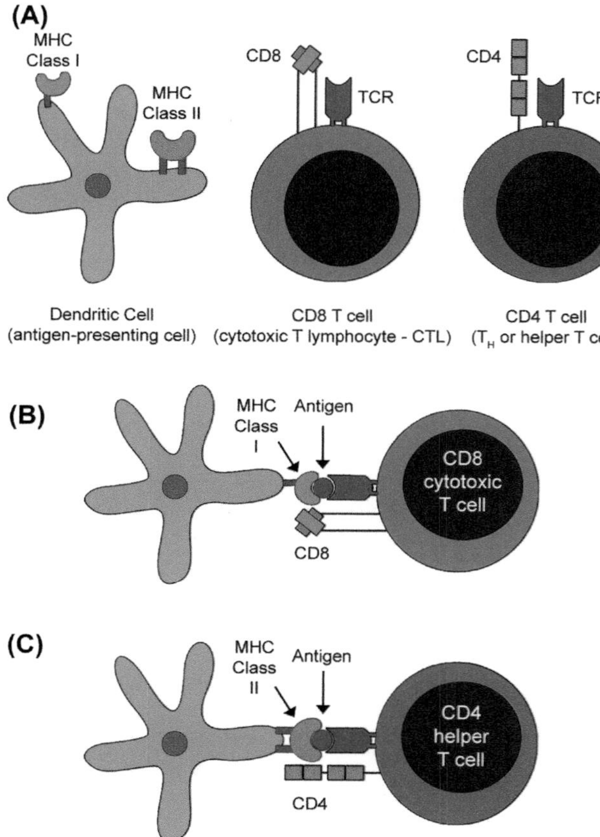

FIGURE 6.7 Important cell surface molecules found on immune cells. (A) Dendritic cells are specialized antigen-presenting cells. They possess MHC class I and MHC class II molecules, which they use to present antigen to T cells. (B) CD8 T cells (CTLs) recognize antigen presented by MHC class I, (C) while CD4 T cells (helper T cells) respond to antigen presented by MHC class II molecules. The TCR cannot recognize antigen unless it is presented by one of these two molecules. When a match is made, the T cell becomes activated.

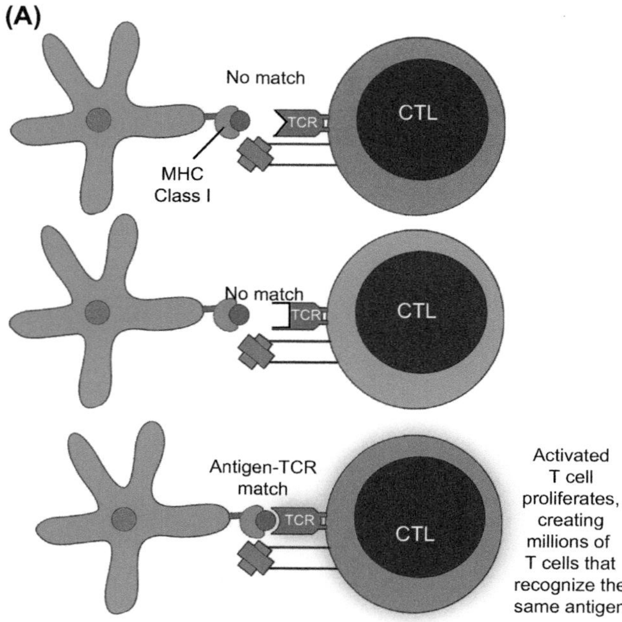

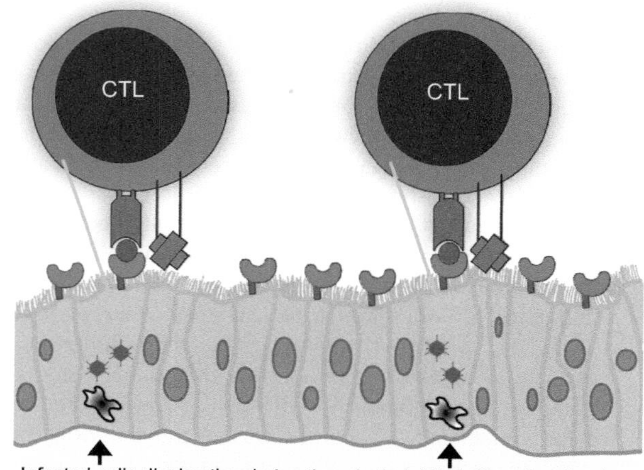

FIGURE 6.8 Antigen specificity and CTL effector functions. (A) Each T cell possesses a unique T cell receptor (TCR) that recognizes only one specific antigen. To become activated, the TCR has to recognize its cognate antigen. For cytotoxic T lymphocytes, the antigen must be presented by MHC class I. For helper T cells, the antigen must be presented by MHC class II. (B) Once activated, the antigen-specific CTL proliferates and returns to the site of infection. All cells display MHC class I molecules at the cell surface, and the MHC class I molecules of infected cells will contain viral antigens. The activated CTL recognizes the antigen and releases proteins that induce apoptosis of the infected cell. Uninfected cells do not display viral antigen and so are untouched by the CTLs.

unable to recognize that the cell is infected and does not kill the infected cell. In these situations, natural killer cells are critical for defense.

Natural killer cells are innate cells because they do not specifically recognize antigen. In contrast to CTLs, which recognize their antigen in an MHC class I molecule, NK cells recognize the *absence* of MHC class I molecules. This might occur if a virus interfered with the process in an attempt to escape from CTL detection. An NK cell has activating receptors and inhibitory receptors on its surface (Fig. 6.9A). When a body cell is in duress or under stress, such as during infection with a virus, an activating ligand is displayed on the surface of the cell. Binding of the NK activating receptor to an activating ligand signals the NK cell to kill the target cell. However, the NK cell also has an inhibitory receptor to safeguard against the unnecessary killing of cells. If the inhibitory receptor binds its ligand, which is MHC class I, then the cell receives an inhibitory signal that overrides the activating signal (Fig. 6.9B). In this case, a CTL, rather than an NK cell, will kill the infected cell since the cell still has MHC class I molecules. But if MHC class I is absent,

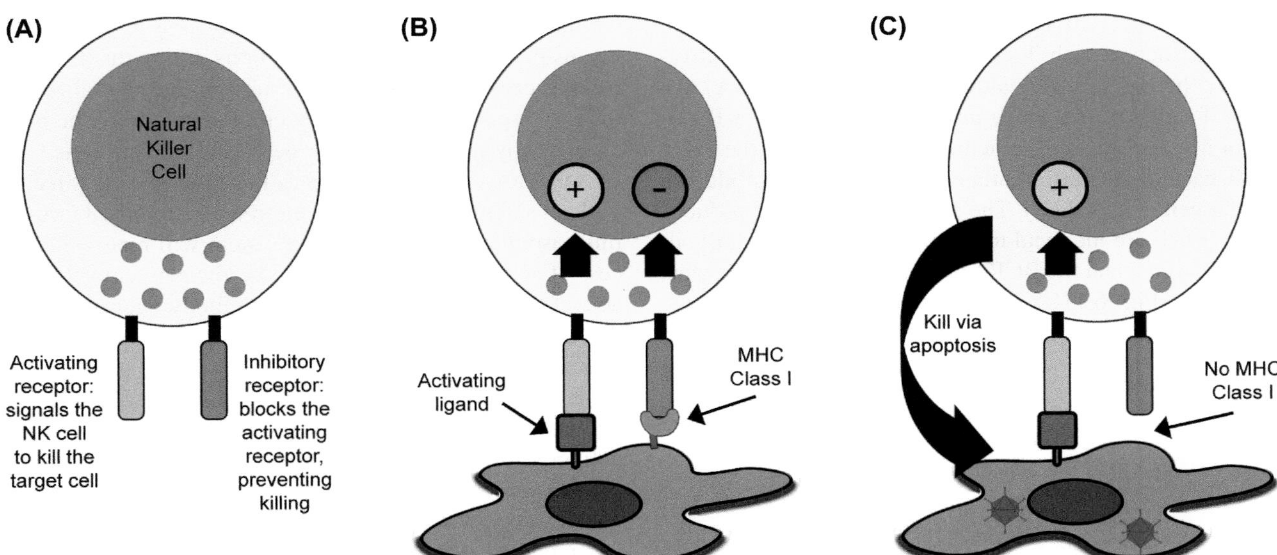

FIGURE 6.9 **Activation of natural killer cells.** (A) A natural killer cell has two receptors on its surface, an activating receptor and an inhibitory receptor. (B) The activating receptor binds a stress-induced activating ligand found on the target cell. This sends a positive "killing" signal to the NK cell. However, if MHC class I is present on the cell surface, it binds the inhibitory receptor. This sends a negative signal to the NK cell, preventing it from killing the target cell. (C) If the virus has interfered with the MHC class I molecule, the inhibitory receptor will not be engaged. Without the negative signal, the activating receptor sends the killing signal to the NK cell, which induces apoptosis of the target cell.

such as when it is interfered with by a virus, then the activating receptor is not inhibited. Without this inhibitory signal from MHC class I, the NK cell will kill the target cell (Fig. 6.9C). In this way, NK cells protect the host when CTLs cannot recognize virally infected cells. As an innate cell, NK cells are active while CTLs are in the process of being activated and proliferating, and so help to reduce the viral load until the major virus-eliminating cell is primed for action.

Study Break
Describe how CTLs become activated by dendritic cells and how they identify infected cells.

6.2.2 Helper T Lymphocytes

Like CTLs, helper T cells also recognize antigen presented by dendritic cells. In the same way they present antigen to CTLs, dendritic cells acquire antigen in the infected tissue and travel to the regional lymph nodes. However, instead of using MHC class I molecules, they present the antigen within MHC class II molecules to helper T cells (Fig. 6.7C). Helper T cells are also known as Th cells or CD4 T cells, because they have a protein called CD4 that associates with the TCR and assists in the binding of the TCR to the MHC class II molecule. Just as with CTLs, the TCR of each helper T cell recognizes a different antigen, so only the T cell with the corresponding TCR will recognize each specific antigen.

After binding to the antigen and MHC class II molecule presented by the dendritic cell, the naïve helper T cell becomes activated and proliferates. Instead of gaining the ability to kill infected cells, helper T cells acquire the ability to activate cells that bear MHC class II molecules. All cells in the body express MHC class I and are susceptible to cytolysis by CTLs, but only select immune cells express MHC class II: dendritic cells, macrophages, and B cells. These are the only cells capable of being activated by helper T cells. Several classes of helper T cells exist that are associated with immune responses tailored to certain pathogens.

6.2.3 B Lymphocytes and Antibody

Since they express MHC class II, B cells are activated by helper T cells. The primary function of B cells is to produce **antibody**, also called **immunoglobulin**. This is known as the **humoral** response, a term that originates as a tribute to the four humors (fluids) of ancient medicine, one of which was blood—the location of antibodies. (The T cell response is known as the cell-mediated response.) In the same way that each T cell has a unique TCR, each B cell bears a **B cell receptor (BCR)** that is able to bind antigen (Fig. 6.10). It is a Y-shaped transmembrane protein that attaches to free antigen or antigen found on the external surfaces of a pathogen, such as capsid proteins or viral envelope proteins. Dendritic cells endocytose antigen and break it down for display within MHC class II, while B cells use the BCR to internalize antigen and display it on their MHC class II molecules, where it is recognized by T cells (Fig. 6.10).

B cells are also found within the lymph node and closely associate with T cell areas. Following the activation of a helper T cell by a dendritic cell (Fig. 6.11A), the helper T cell can recognize antigen presented by a B cell within MHC class II. The helper T cell activates the naïve B cell, causing it to differentiate into an antibody-producing **plasma cell** (Fig. 6.11B). The antibodies that are produced by the B cell are identical to the BCR, except that antibodies are secreted and the BCR remains tethered to the plasma membrane (Fig. 6.11C). The antibodies made by a single B cell are specific to a particular viral antigen. Antibodies circulate in the blood for months following infection. This means that subsequent viral infections during this time will be unsuccessful because the virus will be neutralized by the virus-specific antibodies in circulation within the body. This is why a person doesn't get the same cold twice in a season. However, the antibodies made during the infection will not safeguard against infection with a different virus. In this case, the adaptive immune system will repeat the process of activating T cells specific for the new virus that will activate other B cells to produce antibody against the new virus. This is why the adaptive immune system is antigen specific.

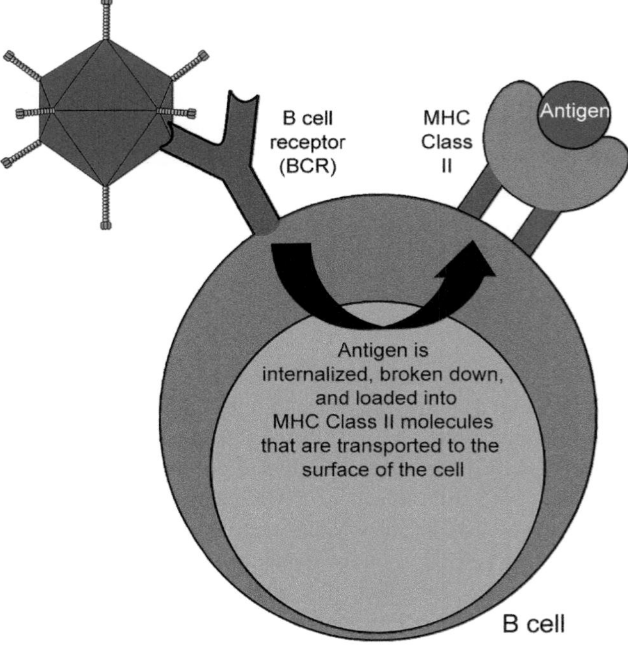

FIGURE 6.10 **The B lymphocyte.** B cells possess a BCR that allows them to bind antigens, including those on the surface of virions. The antigen is internalized within endosomes, broken down, and loaded onto MHC class II molecules. B cells interact with helper T cells through the antigen they display on the B cell MHC class II.

B cells produce different antibody classes, or **isotypes**, that have slightly differing sizes and properties. The first antibody isotype made by plasma cells is IgM. Soon after, the plasma cell undergoes internal gene modifications to create antibodies of the IgG class that have higher **affinity**, meaning that the antibody binds stronger to the antigen. Other cells, such as macrophages, also have receptors for IgG. This allows macrophages to bind antibody-coated viruses and phagocytose them for digestion (Fig. 6.12A). Whereas IgM is too large to escape from the bloodstream, IgG is smaller than IgM and can travel nearly anywhere in the body, including through the placenta to a developing child within the womb. When born, a child has circulating IgG derived from the mother that helps to protect it from pathogens.

Diagnostic tests can differentiate between IgM and IgG. This can be used to indicate whether a patient has a recent infection with a virus (and produces IgM) or a chronic infection or second exposure to the virus (and produces IgG).

Like macrophages, NK cells also possess receptors that bind to the stalk portion, or constant region, of antibodies. This is another way to induce apoptosis of infected cells. Many enveloped viruses assemble at the plasma membrane and must introduce their proteins into the plasma membrane as they assemble at this location. Antibodies

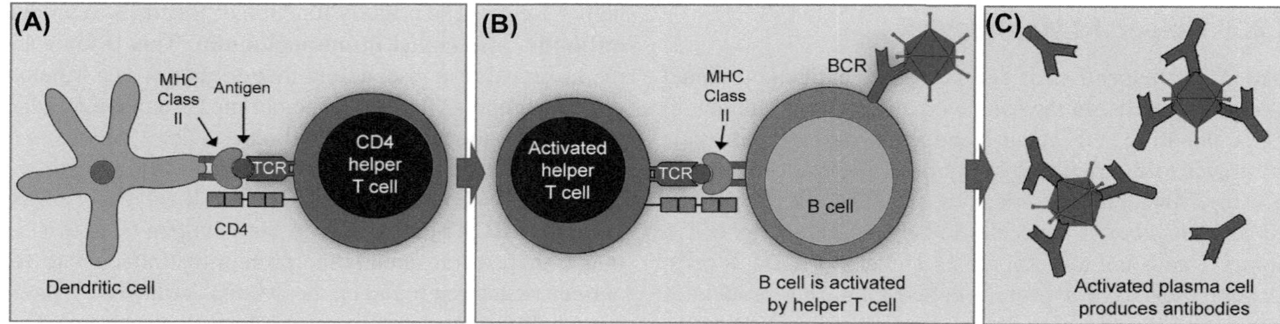

FIGURE 6.11 **Activation of helper T cells and B cells.** (A) Dendritic cells retrieve antigen in tissues, travel to the lymph nodes, and present antigen in MHC class II molecules. The helper T cell that recognizes the antigen with its TCR will become activated. (B) Activated helper T cells recognize the same antigen when it is presented in MHC class II by B cells. (C) Once this occurs, the T cell activates the B cell through cell–cell contact and release of cytokines. The activated B cell differentiates into an antibody-producing plasma cell. The antibodies produced recognize the same antigen as the BCR. They neutralize the virus by preventing virus attachment proteins from interacting with cell surface receptors.

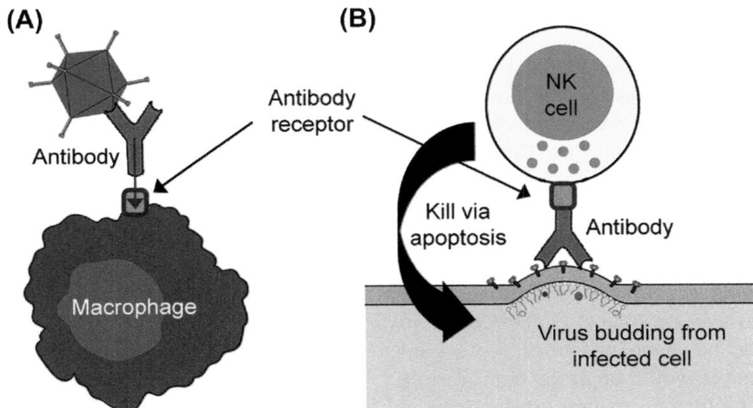

FIGURE 6.12 **The function of antibody receptors.** Both macrophages and NK cells express a receptor that binds antibody constant regions. (A) Macrophages use this receptor to attach to pathogen-bound antibody and phagocytose it. (B) In the process of antibody-dependent cell cytotoxicity, the NK cell's antibody receptor engages with antibody bound to virions budding from the surface of a cell. This activates the NK cell to kill the target cell.

that recognize these viral proteins are able to bind to them while they are within the cell's plasma membrane. The NK cell attaches to the antibodies using its antibody receptors; this triggers the NK cell to deliver its apoptosis-inducing granules to the target cell, thereby killing the infected cell. This process is known as **antibody-dependent cell cytotoxicity** (Fig. 6.12B).

As we learned in Chapter 5, "Virus Transmission and Epidemiology," IgA is produced by B cells at mucosal epithelial surfaces, such as in the lung, nasal cavity, or gastrointestinal tract. IgA is also produced within breast milk and ingested by a breastfeeding child. Both IgG and IgA are effective at neutralizing viruses by binding to the external surface of the virion and preventing the interaction of the virus attachment protein with its cell surface receptor, thereby preventing infection of the target cell.

Study Break
How do antibodies participate in eliminating viral infection?

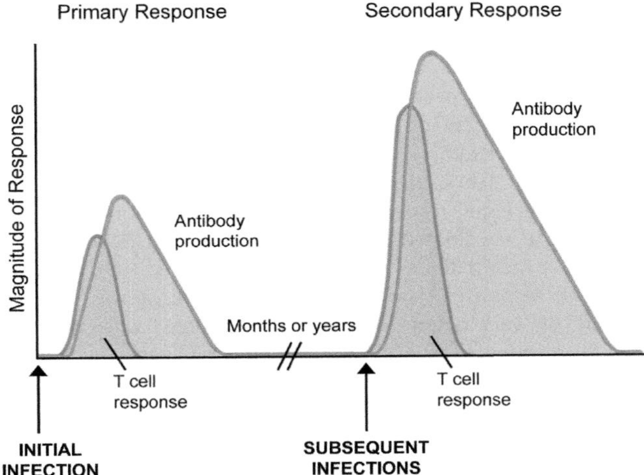

FIGURE 6.13 **Immunological memory.** The first encounter of the immune system with a pathogen is known as the primary response. T cell responses and B cell production of antibody take weeks to develop during the primary response. The immune response will be immediate and of greater magnitude upon subsequent infections because there will be more lymphocytes present and they will require less stimulation. This is known as immunological memory.

6.2.4 Immunological Memory

During the first immune response against a pathogen, it takes days for the adaptive immune response to begin. Dendritic cells must process antigen, travel to the lymph nodes, and present it to CTLs and helper T cells, which proliferate and activate other cells, such as B cells that produce antibody. After the infection has been cleared, many of these T cells, now no longer needed, will undergo apoptosis. The plasma cells continue to make antibody for a few months after infection, but antibody responses gradually slow down as the extra plasma cells also undergo apoptosis. Some of these T and B cells form long-lived **memory cells**, however, that remain within the body. These memory cells remain at frequencies that are 100- to 1000-fold higher than T cells that have not yet been stimulated by antigen. Memory cells are the basis for **immunological memory**: upon recognizing a pathogen that has infected the body previously, the adaptive immune system responds stronger and more rapidly. Memory T cells circulate through the tissues and require less activation by dendritic cells, which no longer have to travel to the lymph nodes to activate the lymphocytes, and the memory B cells already produce IgG of high affinity rather than low-affinity IgM. Taken together with the increased total number of memory cells, the **secondary response** against a pathogen is much faster and more effective than the **primary response** (Fig. 6.13). In many cases, the infected individual will not realize they have been reinfected.

In-Depth Look: How Long Does Immunological Memory Last?

Variola virus is the causative agent of smallpox. The vaccine against variola virus uses a related virus, known as vaccinia virus. Antibodies produced against the vaccinia virus recognize variola virus and protect against infection. **Cross reactivity** occurs when antibodies produced against one antigen also recognize a similar antigen, such as in the case of vaccinia and variola viruses.

Smallpox was successfully eradicated from the world in 1978 because of worldwide vaccination efforts, and vaccination against the virus ceased in the United States in 1972. In the early 2000s, however, questions arose concerning whether or not people would be protected if variola virus were to be used as biological warfare. Researchers at the Oregon Health and Science University examined the immune responses of people who had previously been vaccinated against smallpox to determine whether immunological memory still existed against the virus in these individuals. They separated study volunteers into four groups: people who were vaccinated 20–30 years prior, 31–50 years prior, 51–75 years prior, or those who had never been vaccinated. As expected, unvaccinated people did not have detectable T cell responses or antibodies against the vaccinia virus. When they examined helper T cell memory responses, they found that 100% of those vaccinated 20–30 years ago still had CD4 T cell responses (Fig. 6.14). This dropped to 89% for those vaccinated 31–50 years prior, and 52% of individuals vaccinated over 50 years ago still had CD4 memory T cell responses. Looking at CD8 CTL responses, around half of vaccinated individuals had detectable memory CTL responses, regardless of when they were vaccinated. Similarly, 50% of vaccinated individuals had neutralizing antibody levels that were considered high enough to provide immunity against the virus. Interestingly, the researchers found that receiving more than one vaccination against smallpox was not more effective than receiving a single dose of the vaccine in generating memory responses.

This study indicates that the vaccinia vaccine provides long-term immunity against smallpox. It also verifies that individuals that have not been vaccinated do not have memory responses and would be susceptible to infection with smallpox.

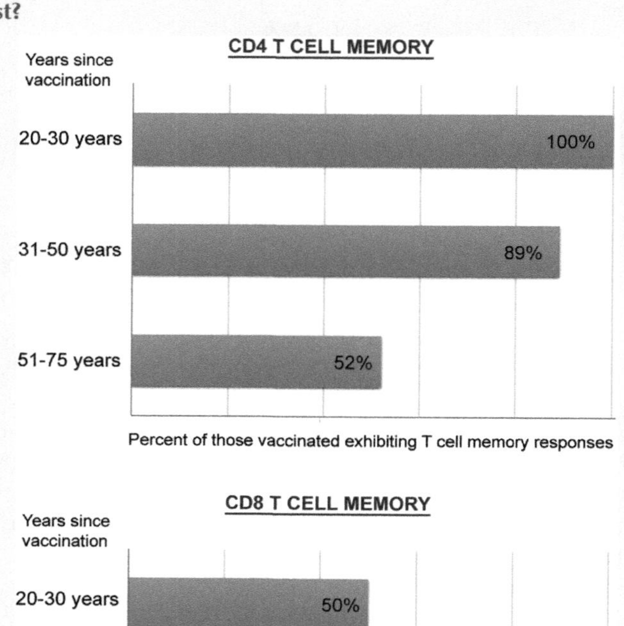

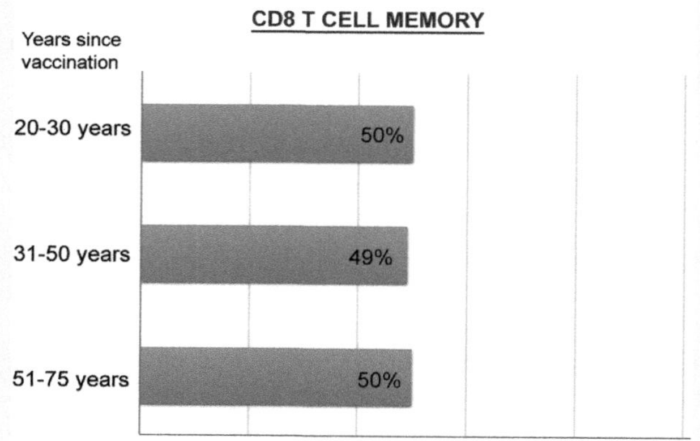

FIGURE 6.14 **Immunological memory provided by the smallpox vaccine.** A study carried out by the Slifka lab at Oregon Health and Science University examined memory T cell responses in individuals that had previously been vaccinated for smallpox. They tested the helper T cell and cytotoxic T cell responses in unvaccinated individuals or those vaccinated 20–30 years, 31–50 years, or 51–75 years before. *Data derived from Hammarlund, E., Lewis, M.W., Hansen, S.G., et al., 2003. Duration of antiviral immunity after smallpox vaccination. Nat. Med. 9, 1131–1137.*

The duration of immunological memory provided by a vaccine is dependent upon several factors that include the characteristics of the pathogen and the design of the vaccine. The vaccines against measles, mumps, polio, rubella, and yellow fever are all considered to provide long-term immunity. However, viruses that mutate quickly, such as influenza A virus, require a unique formulation each year that contains different strains of the viruses that are currently circulating.

The establishment of immunological memory is one of the most important functions of the adaptive immune system. **Vaccination** is the intentional inoculation of a person or animal with a harmless form of a pathogen. It activates the immune system much in the same way that the actual infection would and leads to the generation of memory T and B cells that are prepared to initiate a memory response should the individual encounter the actual pathogen.

6.3 VIRAL EVASION OF THE IMMUNE RESPONSE

The immune system is a dynamic collection of cells and proteins that safeguards the host from pathogens, including viruses. Over evolutionary time, however, viruses have also evolved mechanisms to interfere with the host immune response. Nearly every facet of the immune system is thwarted by some virus. RNA viruses, with error-prone RNA polymerases, mutate quickly and can escape immunological memory in this fashion, whereas large DNA viruses, such as herpesviruses and poxviruses, have large genomes that encode immune evasion proteins.

6.3.1 Antigenic Variation

Viruses contain antigens, the small portions of viral proteins that are picked up and processed by dendritic cells and presented to T cells. The human body has an estimated 10^{18} different T cells, each with a slightly different TCR, and the T cell that specifically recognizes the viral antigen will be the only T cell able to respond to it. Similarly, there are around 5×10^{13} B cells with different B cell receptors, and only those that specifically recognize the viral antigen will respond. The memory T cells and B cells that are generated will have the same antigen specificity as the initial effector T cells and B cells. The error-prone RNA polymerases of RNA viruses often introduce point mutations into viral genes when they place an incorrect nucleotide while copying or transcribing its RNA. As we learned in Chapter 3, "Features of Host Cells: Cellular and Molecular Biology Review," the host ribosome reads the RNA in 3-nucleotide codons and creates a protein out of amino acids based upon the sequence of nucleotides. If the viral mRNA has a mutation, this can result in an amino acid different than the original one being incorporated into the protein. This leads to a slightly different protein being translated, and as a result, a slightly different antigen is created. This is known as **antigenic variation** (Fig. 6.15A). The T cell that originally responded to the initial antigen may be unable to recognize the new antigen because of this slight difference, and the memory T cells that were generated will also not be responsive. Consequently, a new T cell with a TCR that recognizes the new antigen will need to be activated. The entire process of antigen presentation and T cell activation will need to be repeated. The same phenomenon occurs with B cells, and so antibodies created against the initial antigen may not recognize the mutated antigen. This constant acquisition of small mutations in viral proteins is called **antigenic drift**. Viruses with RNA polymerases lacking proofreading ability create small mutations that result in the creation of new strains of the virus that contain antigens unrecognized by previously activated T and B cells. This happens frequently with influenza viruses, which introduce point mutations into their hemagglutinin and neuraminidase proteins, which become unrecognizable by previously generated antibodies. Similarly, there are over 100 different strains of rhinovirus, so the immune response against the cold you catch one year will not provide immunological memory against a different strain of rhinovirus the following year.

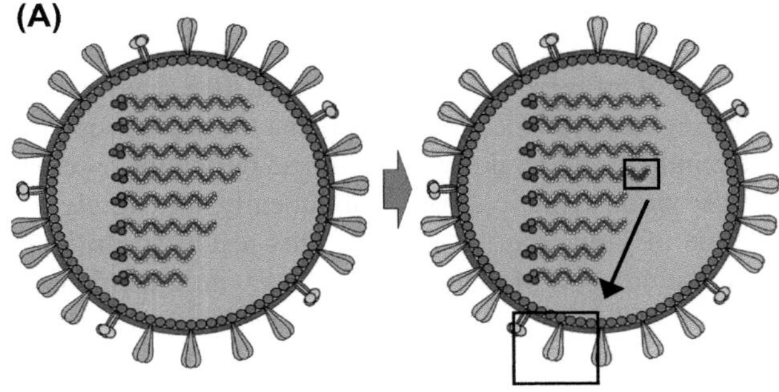

(A)
Small genome mutation leads to a modified antigen that is no longer recognized by lymphocytes that reacted to the initial antigen.
Immunity against first strain does not provide immunity against new **strain**.

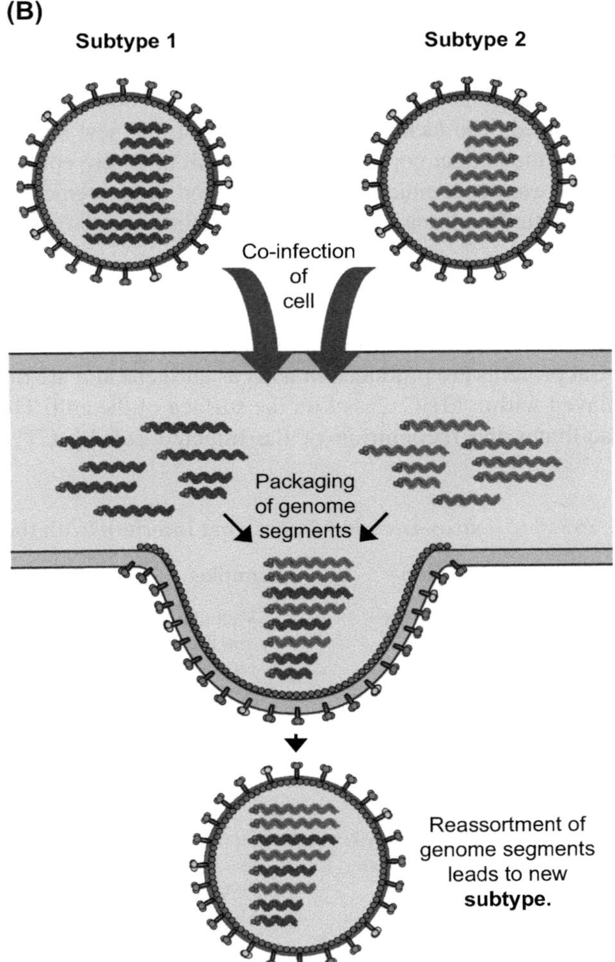

FIGURE 6.15 **Antigenic drift and antigenic shift.** (A) Error-prone viral RNA polymerases introduce point mutations into viral genes that result in the translation of a slightly different protein antigen, creating a new strain of virus. The result is that the T and B lymphocytes that specifically responded to the initial antigen may be unable to recognize the new antigen. A new primary response by other T and B lymphocytes would have to occur upon infection with the new strain as no memory response would be available. The constant acquisition of small genome mutations is known as antigenic drift. (B) If two different subtypes of the same segmented virus infect a cell, reassortment of the viral segments can occur during assembly of new virions. This results in the formation of a new subtype. Reassortment is of particular concern when an animal and human reassortant is created.

Segmented viruses, like influenza, encode their genetic information in several pieces of nucleic acid. If two different subtypes of influenza infect the same cell, then **reassortment** of the segments can occur, meaning that the copied segments of the initial virus strains may be mixed during the assembly of new virions (Fig. 6.15B). Consequently, an entirely new subtype of virus may be formed. This results in **antigenic shift**, the creation of a subtype of virus with a combination of antigens that a species has not seen before, and therefore, no one within the population has any immunity against the virus. This is a major concern with influenza, because certain subtypes of influenza circulate within bird (avian) or pig (porcine) populations but do not infect humans. However, several human influenza strains infect birds or pigs, and so if one of these animals was infected simultaneously with a human *and* animal strain, reassortment could occur within the animal, and an animal virus that infects humans could be created. The lack of immunity against this new subtype within the human population could result in severe epidemics, or even pandemics. The phenomenon of antigenic shift will be further discussed in Chapter 10, "Influenza."

6.3.2 Latency

When a virus infects and replicates within target cells, viral proteins are produced that act as antigens and are displayed within MHC class I on the surface of the cell. This facilitates the recognition of the infected cell by CTLs. Some viruses, however, enter a state, known as **latency**, where they no longer replicate within the cell but remain dormant until the immune system is weakened. Viral replication does not occur during latency, and so there are no viral proteins produced to act as antigen and alert the immune system of the infected cell.

Viruses can become latent in the initial cell type they infected or in a different cell type near the initially infected cell. For instance, Epstein–Barr virus (EBV) infects and establishes latency within B lymphocytes; on the other hand, varicella zoster virus (VZV), which causes chickenpox, infects epithelial cells of the skin but establishes latency within a sensory nerve at the posterior (dorsal) root ganglion, where the nerve joins the spinal cord. VZV replicates locally within skin epithelial cells but then infects nearby sensory nerves within the skin. The virus travels up the neuron to the dorsal root ganglion, where it remains dormant within the nucleus. When the immune system is weakened, the virus reactivates, travels down the nerve, and begins replicating again within the skin epithelium, causing the disease shingles. Latency is a hallmark of herpesviruses and will be discussed in more detail in Chapter 13, "Herpesviruses."

6.3.3 Virus-Encoded Evasion Mechanisms

Viruses encode the genes necessary for their replication, but some viruses also encode genes whose protein products interfere with the host immune response (Table 6.2). For example,

TABLE 6.2 Virus-Encoded Genes That Interfere With the Host Immune Response

Viral Gene Function	Examples
Interferes with innate cell functions	1. Vaccinia virus, Ebola virus, HSV-1, measles virus, and human cytomegalovirus infect dendritic cells and prevent them from functioning properly 2. Hepatitis C virus, Kaposi's sarcoma-associated herpesvirus, and vaccinia virus interfere with PRR pathways 3. Human cytomegalovirus expresses a "dummy" MHC class I molecule that binds to NK cell inhibitory receptors, preventing killing 4. West Nile virus increases expression of host MHC class I to avoid NK cell lysis in order to finish replicating before T cells are fully activated
Interferes with MHC class I (preventing CTL activation)	1. Human cytomegalovirus US3 and adenovirus E19 protein prevents peptide loading onto MHC class I 2. HSV-1 ICP47 and human cytomegalovirus U6 proteins prevents peptides from being loaded onto MHC class I 3. Human cytomegalovirus US2 and US11 proteins dislocate MHC I to the cytosol, where it is degraded 4. EBV has proteolysis-resistant proteins that are difficult to generate peptides for MHC molecules 5. HIV Nef protein causes the MHC class I molecule to be taken in from the surface of the cell and degraded (also occurs with MHC class II molecules)
Inhibits inflammatory response	1. Vaccinia encodes secreted cytokine receptors to soak up host cytokines 2. EBV prevents the infected cell from expressing adhesion molecules so cytotoxic cells can't easily bind the infected cells
Inhibits humoral response	1. HSV-1 and human cytomegalovirus encode their own antibody receptors that soak up antibodies 2. Measles virus prevents the activation of B cells

several viruses infect dendritic cells and prevent them from properly presenting antigen and activating T cells. Many viruses also interfere with the loading of peptides into MHC class I or the presentation of this molecule on the surface of the infected cell, ensuring that the infection is not recognized by CTLs. Other viruses interfere with various stages of host defense and inflammatory responses by interfering with cytokines or antibodies. Poxvirus and herpesvirus family members, with their large dsDNA genomes, each encode several host evasion mechanisms.

In-Depth Look: The Effects of Severe Combined Immunodeficiency

Severe combined immunodeficiency (SCID) is a term used to describe the genetic disorder that results in major defects of the immune system. This can be caused by a variety of mutations, including those that affect genes encoding cytokine receptors, enzymes, and transcription factors important for the development and differentiation of immune system cells. The result is that important immune cells such as T cells, B cells, and/or NK cells are absent or nonfunctional. Without these major players of the immune system, the person is left extremely vulnerable to infectious agents, including viruses.

One of the most famous cases of SCID was that of David Vetter, the "bubble boy" who lived his entire life in a sterile plastic bubble in order to avoid the infectious agents that could kill him. He ended up dying after a bone marrow transplant designed to cure his disease inadvertently also transferred latent EBV within the bone marrow cells. In David's body, a host with no immune system, the virus replicated uncontrollably and caused hundreds of Burkitt's lymphoma tumors that eventually overtook his body.

SUMMARY OF KEY CONCEPTS

Section 6.1 The Innate Immune System

- The two arms of the immune system, the innate and adaptive immune system, function together to provide long-term immunity against pathogens. The innate immune system, which is activated immediately after infection, controls viral replication and provides signals that stimulate the adaptive immune system, which provides an antigen-specific response that controls infection and creates immunological memory.
- Pattern recognition receptors are cellular proteins that recognize pathogen-associated molecular patterns. TLRs and RLRs are two major classes of PRRs.
- The binding of a PAMP to a PRR induces signaling pathways within the cell that lead to cytokine production. Type 1 interferons, IFN-α and IFN-β, are cytokines that activate antiviral PKR and OAS-RNase L pathways that inhibit protein translation and degrade RNA, respectively.
- Macrophages are specialized in phagocytosing pathogens. Dendritic cells are specialized antigen-presenting cells that acquire viral antigens within body tissues and travel to lymph nodes to present antigen to T cells. This initiates the adaptive immune response.
- Natural killer cells induce apoptosis of virally infected cells that lack MHC class I molecules on the cell surface.

Section 6.2 The Adaptive Immune System

- The innate immune system uses general effects to control infection, while the adaptive immune system is customized to respond specifically to an individual pathogen.
- Antigen must be presented to T cells in order for them to recognize it with their TCR. CTLs (CD8 T cells) respond to antigen in MHC class I molecules, and helper T cells (CD4 T cells) respond to antigen presented in MHC class II molecules.
- Once activated in the lymph node by antigen presentation, a CTL proliferates and returns to the site of infection, where it surveys the antigen that is displayed by the tissue cells. If the CTL binds its cognate antigen, it will induce apoptosis of the target cell.
- A helper T cell activated by antigen presentation proliferates and acquires the ability to activate cells presenting its antigen in MHC class II.
- The BCR on the surface of a B cell recognizes antigen, which is internalized, broken down, and then presented on MHC class II molecules. This allows for helper T cells that recognize the antigen to activate the B cell, which differentiates into an antibody-producing plasma cell.
- Antibodies are secreted proteins that bind to the same antigen against which the B cell was activated. Antibodies can neutralize viruses by preventing the interaction of the virus with its cell surface receptor. Antibody-bound viruses are phagocytosed and destroyed by macrophages. NK cells can induce apoptosis of infected cells by binding antibodies attached to viral proteins on the cell surface.
- Immunological memory is the phenomenon by which T cells and B cells that have previously encountered antigen respond faster and in higher magnitude upon subsequent encounters with the same antigen. Effective relies upon generating immunological memory.

Section 6.3 Viral Evasion of the Immune Response

- Viruses have evolved a diverse set of mechanisms to interfere with the host immune response. Error-prone RNA polymerases introduce antigenic variation through point mutations in genes, leading to antigenic drift. Coinfection with two different subtypes of a segmented virus can lead to reassortment of genome segments and antigenic shift.
- Certain viruses, particularly herpesviruses, are able to undergo latency. In this dormant state, the viruses do not activate the immune system.

FLASH CARD VOCABULARY

Immune response	Apoptosis
Virulent	Cytotoxic T lymphocyte
Immunity	Helper T lymphocyte
Innate immune system	T cell receptor
Pattern recognition receptors	Major histocompatibility complex
Pathogen-associated molecular patterns	Naïve lymphocyte
Cytokine/cytokine receptor	Cognate antigen
Type 1 interferons	Antibody/immunoglobulin
Macrophages	Humoral response
Dendritic cells	B cell receptor
Antigen presentation	Plasma cell
Natural killer cells	Isotype
Affinity	Cross reactivity
Antibody-dependent cell cytotoxicity	Antigenic variation
Immunological memory	Antigenic drift
Memory cells	Reassortment
Primary response/secondary response	Antigenic shift
Vaccination	Latency

CHAPTER REVIEW QUESTIONS

1. Explain the difference between the innate and adaptive immune system.
2. How does type 1 IFN lead to the establishment of an intracellular antiviral state?
3. What would be the consequence of a person having a genetic defect such that he/she does not develop natural killer cells?
4. Explain the process by which natural killer cells induce apoptosis of virally infected cells.
5. Make a list of the pattern recognition receptors, their PAMP ligands, and where each PRR is located in the cell.
6. How are T lymphocytes, which are located in lymph nodes or the spleen, alerted of an infection in the body tissues?
7. Create a chart that lists the types of T lymphocytes. For each subset, list the molecule that the T cell uses to recognize antigen, the TCR-assisting molecule found on its cell surface, and the function of the cell type.
8. From beginning to end, describe the steps required for the immune system to recognize a viral infection is taking place and create antibodies against the virus.
9. What is immunological memory and why is it important in combatting viruses?
10. Explain the difference between antigenic drift and antigenic shift. How could each occur? Which is more likely to lead to a pandemic?
11. Describe what the effect would be on the host immune response of the immune evasion mechanisms listed in Table 6.2.

FURTHER READING

Hammarlund, E., Lewis, M.W., Hansen, S.G., et al., 2003. Duration of antiviral immunity after smallpox vaccination. Nat. Med. 9, 1131–1137.

Isaacs, A., Lindenmann, J., 1957. Virus Interference: I. The interferon. Proc. R. Soc. London B. Biol. Sci. 147, 258–267.

Iwasaki, A., Medzhitov, R., 2015. Control of adaptive immunity by the innate immune system. Nat. Immunol. 16, 343–353.

Jensen, S., Thomsen, A.R., 2012. Sensing of RNA viruses: a review of innate immune receptors involved in recognizing RNA virus invasion. J. Virol. 86, 2900–2910.

Mempel, T.R., Henrickson, S.E., Von Andrian, U.H., 2004. T-cell priming by dendritic cells in lymph nodes occurs in three distinct phases. Nature 427, 154–159.

Røder, G., Geironson, L., Bressendorff, I., Paulsson, K., 2008. Viral proteins interfering with antigen presentation target the major histocompatibility complex class I peptide-loading complex. J. Virol. 82, 8246–8252.

Silverman, R.H., 2007. Viral encounters with 2′,5′-oligoadenylate synthetase and RNase L during the interferon antiviral response. J. Virol. 81, 12720–12729.

Chapter 7

Detection and Diagnosis of Viral Infections

Viruses have evolved alongside humans for as long as both have existed. "Filterable viruses" were classified as a separate pathogen upon their identification in 1898, but infectious diseases have been characterized throughout history by the clinical conditions they have caused, despite that the doctors of the time had no idea of the existence of microscopic pathogens. Hippocrates (460–377 BC), the father of modern medicine, described several illnesses characteristic of viral diseases, including influenza and poliomyelitis. The ability to definitively identify a specific virus as the cause of an illness has only become possible within the last 100 years.

As described in Chapter 1, "The World of Viruses," bacteria can be viewed under a light microscope, but viruses are too small to be visualized with light microscopes. Consequently, the first efforts to identify specific viruses relied upon **serology**, the analysis of the protein antibodies found in blood that the immune system synthesizes against pathogens. **Tissue culture**, the ability to grow tissues and cells outside of a living organism in a controlled environment, was invented and refined in the first half of the 20th century. This led to the propagation of viruses using cell culture and the detection of the pathogenic effects that viruses exert upon cells. Both serology and tissue culture have been refined and are still vital techniques for the diagnosis of viral infections. Advances in molecular biology have also accelerated our ability to conclusively identify a virus.

The detection of a virus as the cause of an illness is important for many reasons. Several viral infections result in similar clinical symptoms, and viruses with serious effects need to be identified early in order to prescribe the best treatment. Likewise, the infection of high-risk groups, such as transplant recipients, pregnant women, or immunocompromised individuals, needs to be monitored so that critical sequelae can be addressed. The development and availability of antiviral drugs is increasing, and proper diagnosis ensures an effective treatment is prescribed. The typing of viruses is also effective in determining subtypes or strains of viruses, including those that are resistant to certain drugs or are more likely to cause cancer. In the field of epidemiology, most case definitions rely upon laboratory confirmation of the specific virus to confirm a case. Proper diagnosis ensures that accurate surveillance takes place and adequate control measures are instituted during epidemics. It also ensures the safety of transplanted human tissues and safeguards the blood supply. This chapter discusses commonly used techniques for the detection and diagnosis of viruses in clinical samples. Many of these methods are staples in virology research laboratories, as well.

7.1 COLLECTION AND TRANSPORT OF CLINICAL SPECIMENS

For identifying a specific virus, the type of specimen obtained depends upon the type of virus. The specimen will be isolated from the location of infection for viruses that establish localized infections (Table 7.1). For instance, influenza virus is readily detected from nasopharyngeal swabs, and herpes simplex viruses can be isolated from the oral or genital lesions that these viruses cause (Table 7.2). Viruses that establish systemic infection may be isolated from several different sources, depending upon the virus. The site of pathology is often a good place to start, although the virus may be present in the blood as well. For example, hepatitis B virus and hepatitis C virus infect hepatocytes (liver cells) but are detectable in serum.

The choice of diagnostic test will also depend upon the stage of infection. A person's viral load is highest during acute infection but may drop to undetectable levels as the infection is cleared. On the other hand, it takes weeks for antibodies to develop during the primary response against a virus (Fig. 7.1). As described in Chapter 6, "The Immune Response to Viruses," IgM is the antibody isotype that is first produced by plasma cells against a pathogen, but the higher-affinity IgG isotype begins being secreted by plasma cells later during infection and during secondary responses. Therefore, the choice of diagnostic test will depend upon the patient's stage of infection. Tests for the virus itself are best performed before or while symptoms are present. The levels of IgM versus IgG antibodies against a virus can be used to help determine if the infection recently occurred, but neither of these will be present at the beginning of a primary infection.

Care must be taken in the collection, storage, and transport of clinical specimens. Blood is collected into appropriate tubes, depending upon whether cells, serum, or plasma is required for the diagnostic test. Tubes containing sodium heparin or EDTA as an anticoagulant block the clotting of blood and are used to obtain white blood cells (**leukocytes**) or **plasma**, the liquid fraction that remains when blood is

Essential Human Virology.

TABLE 7.1 Types of Specimens Collected for Viral Diagnosis

Site (or type) of illness	Possible viral cause	Types of specimens collected
Respiratory tract	Adenovirus	Nasopharyngeal swab, nasal aspirate, nasal swab, nasal wash, throat swab
	Cytomegalovirus	
	Enterovirus	
	Herpes simplex virus	
	Influenza virus	
	Parainfluenza virus	
	Respiratory syncytial virus	
Gastrointestinal tract	Adenovirus	Stool, vomit
	Rotavirus	
	Norwalk virus	
Skin (rash)	Coxsackie A virus	Biopsy, Tzanck smear
	Herpes simplex virus	
	Varicella zoster virus	
Eye	Adenovirus	Conjunctival swab, corneal swab
	Cytomegalovirus	
	Enterovirus	
	Herpes simplex virus	
	Varicella zoster virus	
Central nervous system (Meningitis, encephalitis)	Arboviruses (many)	Cerebral spinal fluid, stool, biopsy (or autopsy), throat swab, blood
	Coxsackie A virus	
	Coxsackie B virus	
	Dengue virus	
	Enterovirus	
	Herpesviruses	
	Lymphocytic choriomeningitis virus	
	Measles virus	
	Mumps virus	
	Poliovirus	
	West Nile virus	
Genital infections	Herpes simplex virus	Cervical swab, urethral swab, vesicle fluid, Pap smear, Tzanck smear
	Human papillomavirus	

centrifuged to pellet blood cells. **Serum** is obtained by allowing blood to clot and then centrifuging the clot, leaving behind the liquid portion of the blood (Fig. 7.2). The difference between plasma and serum, therefore, is that plasma contains clotting factors that are part of the clot when serum is obtained. Antibodies and virus/virus antigen will be found in serum and plasma, although virus could also be found in the leukocytes, if they are a target of the virus. For collecting fluid from skin lesions, a sterile swab is used to collect the fluid and cells from a lesion that has been opened, and then the swab is placed in a special transport medium. The same transport medium is used for nasopharyngeal swabs, cervical swabs, rectal swabs, or throat swabs. Stool is collected into a clean, leak-proof container.

TABLE 7.2 Specimens Collected for Select Human Viruses

Virus	Type of specimen collected for identification	Tests for virus?	Tests for antibodies?
Influenza	Nasopharyngeal swab, nasal aspirate, nasal swab, nasal wash, throat swab	Yes	No
Norwalk virus	Stool, vomit	Yes	No
Hepatitis viruses	Serum	Yes (HBV, HCV)	Yes
Herpes simplex virus	Scraping from site of infection: oral mucosa, genital mucosa, conjunctiva or cornea	Yes	No
	Serum	No	Yes
Human immunodeficiency virus	Serum	Yes	Yes
	Saliva	No	Yes
Human papillomavirus	Pap smear or cervical swab	Yes	No
Rabies virus	Cerebral spinal fluid, serum	No	Yes
	Saliva	Yes	No
Ebola virus	Serum	Yes	Yes
West Nile virus	Serum, cerebral spinal fluid	Yes	Yes

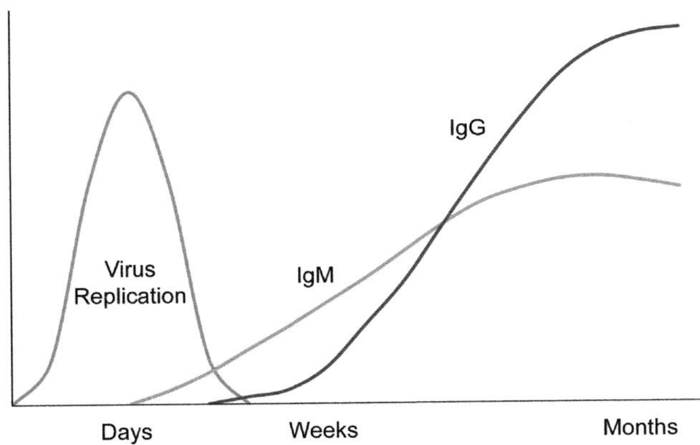

FIGURE 7.1 **Considerations for choice of diagnostics.** Diagnostic tests are available that test directly for the presence of virus, which occurs during active infection, or antiviral antibodies, which take weeks to develop and continue for months following infection. IgM is produced during a primary response, while IgG takes additional time to develop and will be produced during secondary responses. The choice of test depends upon the state of the infection and whether virus or antibody is likely to be present at that time.

The susceptibility of viruses to environmental factors can be an issue for diagnostic tests that rely upon "live" virus. As mentioned in Chapter 5, "Virus Transmission and Epidemiology," most viruses will become noninfectious after being exposed to extended periods of heat. Although nucleic acids may be able to be recovered from these samples, infectious virus will not be present. To prevent virus inactivation, samples other than blood that must be transported to diagnostic laboratories are refrigerated during shipment, or frozen at −80°C and shipped on dry ice if the transport will take 3 days or more. However, some viruses are not stable when frozen, including varicella zoster virus, respiratory syncytial virus, measles, and human cytomegalovirus. These viruses must be frozen in a special transport medium to prevent their inactivation. Much emphasis is placed upon the efficacy of the diagnostic test itself, but no assay can provide meaningful results if the specimen has not been collected, stored, and transported with care.

7.2 VIRUS CULTURE AND CELL/TISSUE SPECIMENS

Methods for detecting viruses are either direct or indirect methods. **Direct methods** assay for the presence of the virus itself, while **indirect methods** observe the effects of the virus, such as cell death or the production of antibodies by the infected individual. Tissue culture is a way to identify a virus based upon the effects of the virus upon the cells. It is also a way to amplify virus if a larger sample is needed for other diagnostic tests, since viruses require cells to replicate.

Tissue culture, also known as **cell culture** when cells are grown specifically, involves maintaining living cells or tissues in a controlled environment outside a living organism. The cells are housed in plastic flasks or bottles and bathed in a liquid growth medium that contains nutrients and supplements (Fig. 7.3A). These cultures are grown in

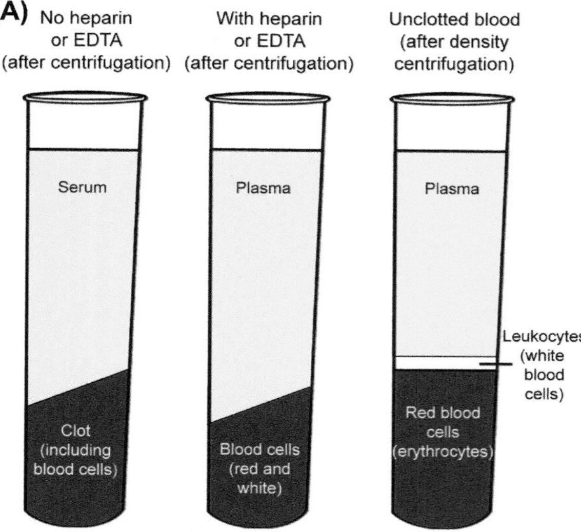

FIGURE 7.2 **The constituents of blood.** Different diagnostic tests require different components of blood. (A) In a tube without anticoagulants, a clot will form that can be centrifuged to the bottom of the tube. The acellular liquid fraction is serum. A tube with heparin or ETDA as coagulants prevents clotting factors from working. The cellular fraction is centrifuged, and the liquid fraction—with clotting factors—is known as plasma. If unclotted blood is centrifuged slowly in a special medium to separate the blood by density, a thin white line of leukocytes, called the buffy coat, separates the plasma and red blood cell layers. In this case, about 55% of the total blood is plasma, 45% are erythrocytes, and <1% are leukocytes. Antibodies are found in serum or plasma. (B) A researcher examines two blood density centrifugations. Note the thin layer of leukocytes in between the plasma and red blood cell layers. *Photo courtesy of FDA / Michael J. Ermarth.*

an incubator set to body temperature (37°C). A **cell line** is a set of cells that have been isolated from a tissue or organ fragment (Fig. 7.3B and C). Cell lines can be **finite** or **continuous.** Finite cell lines will only undergo mitosis a limited number of times, while continuous cell lines are immortal and will proliferate indefinitely. This characteristic is usually a result of genetic mutations. Cell lines derived from tumors, which have lost control of regulating the cell cycle and proliferate indefinitely, can also result in continuous cell lines. The choice of a finite versus continuous cell line for propagating viruses in culture will depend upon the virus that is needed to be isolated. Additionally, the cell line must be one that expresses the cell surface receptor specific for the virus and be permissive to infection; otherwise, the virus will not be able to attach and replicate within the cell line.

Cell cultures must be grown using aseptic technique and sterile conditions, otherwise bacteria and fungi that get into the cell culture will grow profusely in the rich growth medium and contaminate the culture. A sterile environment is provided by a **biological safety cabinet (BSC)**, which is different from a fume hood that is used when working with chemicals. A BSC uses a fan to filter air through **high-efficiency particulate air (HEPA)** filters, which filter out bacteria, fungi, spores, and viruses to generate sterile air. Particles of 0.3 μM in size are the most penetrating through the filters but are still removed with 99.97% efficiency. There are three different classes of BSCs, designated class I, class II, and class III, that provide varying levels of protection to the worker and to the biological material being manipulated. Class I BSCs act much like chemical fume hoods, except that the air is filtered through a HEPA filter before it is released to the environment. The worker and environment are protected, but the biological material is exposed to nonsterile air from the environment so these are not used for cell culture (Fig. 7.4A). Class II BSCs are most often used, as they afford protection to the worker, the biological material, and the environment. The air entering the BSC from the front is sucked into a grille to prevent it from contaminating the working surface, which is constantly bathed in HEPA-filtered air to ensure a sterile working environment (Fig. 7.4B and D). Class II BSCs rely on the **laminar flow** of air, which means that the BSC creates an uninterrupted flow of air in a consistent, uninterrupted stream. As long as the work is performed within the stream of air, then the material will remain sterile. The air is also sent through a HEPA filter before being released to the environment. Class II BSCs are sufficient to work with cells and the majority of viruses. Some dangerous pathogens must be manipulated within a class III BSC, which is air-tight to prevent any exposure of the virus to the worker, who must use the heavy-duty rubber gloves that are built into the BSC to work with the pathogen (Fig. 7.4C). The air leaving a class III BSC is passed through two HEPA filters, or a single HEPA filter and incinerator, to ensure the pathogen does not enter the environment.

Appropriate levels of safety must be taken when working with viruses. Rhinoviruses cause colds, and so the effects of accidentally being exposed to rhinovirus are minimal and self-limiting. Other viruses, such as viruses that cause hemorrhagic fevers, can lead to deadly effects if someone is

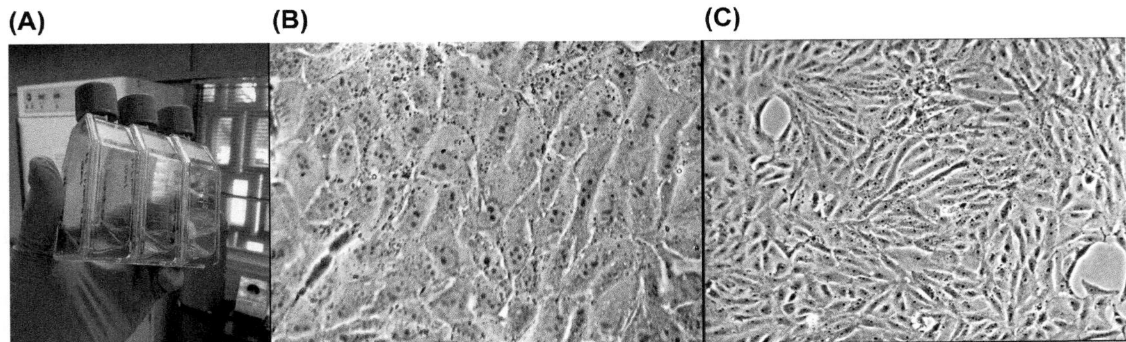

FIGURE 7.3 **Cell culture.** (A) Three cell culture flasks containing living cells and culture medium (pink). (B) MDCK cells, the preferred cell line for isolating influenza A and B viruses. (C) Vero cells, which are susceptible to infection with herpes simplex viruses, poliovirus, Coxsackie B virus, respiratory syncytial virus, mumps virus, rubella virus, SARS-CoV, and lymphocytic choriomeningitis virus, among others.

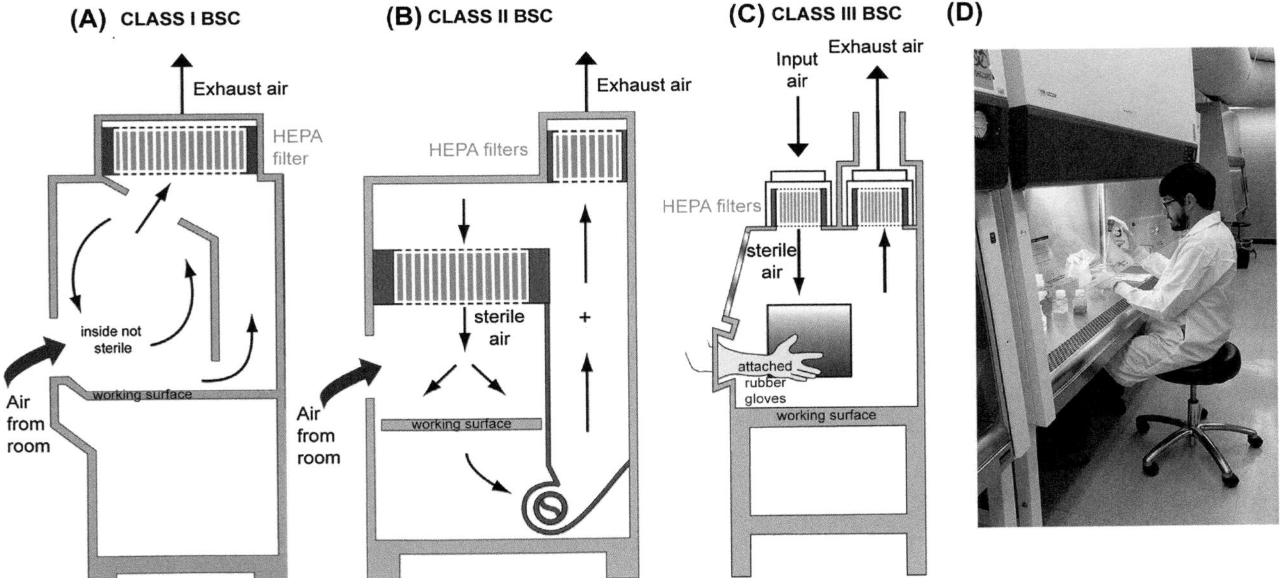

FIGURE 7.4 **Classes of biological safety cabinets.** (A) Class I BSCs protect the environment by filtering contaminated air through a HEPA filter, but the cabinet does not provide a sterile environment within it. Class I BSCs are mainly used to house equipment that might generate aerosols but are not for use with sterile cultures. (B) Class II BSCs filter the environmental air through a HEPA filter before entering the cabinet to provide a sterile working surface. Environmental air is sucked into a grille at the opening of the cabinet to prevent it from contaminating the sterile working area. (C) Class III BSCs are completely sealed and air-tight, and all air entering or exiting the hood is filtered. Work in the cabinet must be performed by using the attached rubber gloves. (D) This researcher is performing cell culture in a class II BSC. Notice the grille at the front that prevents environmental air from entering the working area. The great majority of research and clinical laboratory work involving viruses is performed in a Class II BSC.

exposed. Therefore, there exist four **biosafety levels (BSLs)** that specify what precautions must be taken with different pathogens (Table 7.3). Some notable differences between the BSL levels are summarized below:

Biosafety level 1 (BSL1) is for work involving well-characterized agents not known to cause disease in healthy adult humans. These pathogens present minimal potential hazard to laboratory personnel and the environment. All material must be handled in an appropriate way and decontaminated after use, and workers must use gloves for protection, along with a lab coat and safety glasses, if warranted. Work with BSL1 agents does not require a BSC, unless cell cultures require the use of one to maintain sterility.

Biosafety level 2 (BSL2) is for work with agents that are known to pose moderate hazards to personnel and/or the environment. It includes all the precautions of BSL1 but also requires that laboratory personnel are supervised, receive specific training in handling the pathogenic agents, and conduct any work that may generate infectious aerosols or splashes in a BSC.

Biosafety level 3 (BSL3) is for work that could cause serious or potentially lethal disease through inhalation. BSL3 work includes all the precautions of BSL2 but also requires a special BSL3-level laboratory that

TABLE 7.3 Biosafety Levels Required for Work With Certain Viruses

Biosafety level	Examples of viruses worked with at this level
BSL1	Bacterial viruses
	plant viruses
	Nonhuman insect viruses
BSL2	Hepatitis A virus
	Hepatitis B virus
	Hepatitis C virus
	Hepatitis E virus
	Human herpesviruses
	Seasonal influenzas
	Poliovirus
	Hantaviruses (for potentially infected serum)
	Lymphocytic choriomeningitis virus
	Rabies virus
	Human immunodeficiency virus
	Severe acute respiratory syndrome–associated coronavirus (SARS-CoV)
BSL3	Hantavirus propagation
	1918 influenza virus
	Highly pathogenic avian influenza viruses
	Lymphocytic choriomeningitis virus strains lethal to nonhuman primates
	Human immunodeficiency virus (for large-scale volumes or concentrated virus)
	SARS-CoV propagation
	West Nile virus animal studies and infected cell cultures
	Eastern equine encephalitis virus
	Western equine encephalitis virus
	Venezuelan equine encephalitis virus
	Rift Valley fever virus
BSL4	Hendra virus
	Nipah virus
	Variola virus (smallpox)
	Crimean–Congo hemorrhagic fever virus
	Ebola virus
	Guanarito virus
	Junin virus
	Lassa virus
	Machupo virus
	Marburg virus

is entered through two self-closing doors and is under negative pressure so that contaminated air is drawn to another area and HEPA-filtered before leaving the room. Laboratory personnel must wear additional protective laboratory clothing, such as a gown, scrub suit, or coveralls, that is only worn while in the laboratory and then decontaminated or disposed of upon exit of the lab. Work must be performed in class II or class III BSCs, depending upon the virus.

Biosafety level 4 (BSL4) is for work with dangerous and exotic agents that pose a high individual risk of aerosol-transmitted laboratory infections and life-threatening disease that is frequently fatal and for which no vaccines or treatments are available. In addition to BSL3 standards, personnel must change out of their normal clothing and into laboratory clothing in a specific entry room. Some BSL4 labs use BSL3-protective laboratory clothing but class III BSCs are required for manipulating any infectious materials, which must be decontaminated when passing out of the BSC. Other BSL4 labs require the use of a one-piece protective plastic suit with a separate air supply to keep the suit under positive pressure, which would push out air if any cuts or tears were to occur (Fig. 7.5).

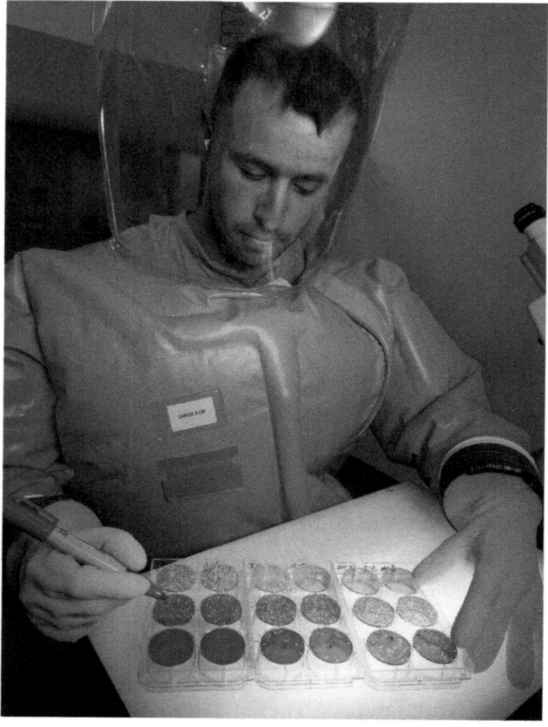

FIGURE 7.5 **BSL4 protective suit.** This researcher at the CDC is using a light box to count viral plaques in a BSL4 laboratory. Note that the suit is inflated because of the air supply, which keeps it under positive pressure in case of a leak in the suit. The researcher also must wear laboratory clothing under the suit (scrubs in this case). Upon exiting the lab, the worker will pass through a bleach shower to decontaminate the orange plastic suit, change out of his laboratory clothing, and then take a personal shower before changing back into his street clothes. *Photo taken by James Gathany, courtesy of the CDC and Dr. Scott Smith.*

This suit is decontaminated with a bleach shower upon leaving the lab. In either BSL4 lab scenario, workers must leave through a separate exit room, decontaminate their laboratory clothes, and shower before changing into street clothes. BSL4 labs must be in a separate building or clearly demarcated isolated zone within a building, have emergency power, be sealed from the environment, and provide multiple means of decontaminating the workspace and all infectious materials. These labs are reserved for the most dangerous infectious agents.

The majority of diagnostic cell cultures take place at BSL2. Generally, three to six cell lines are selected based upon the test being performed and cell cultures are inoculated with the clinical specimen. Although viruses cannot be seen under the light microscope, they biochemically affect the cells in which they are replicating, which sometimes leads to visible **cytopathic effects (CPEs)** that are distinguishable using a light microscope (Fig. 7.6A and B). For example, some viruses may cause cells that normally attach to the bottom of their culture vessel to round up and detach. Large bubblelike vacuoles are sometimes observed in the cytoplasm of infected cells, which may also swell or shrink, depending upon the virus. Other viruses cause adjacent cell membranes to fuse together, creating a **syncytium,** or giant multinucleated cell, that can have up to 100 nuclei within the cell. Some viruses cause lysis of the cells in which they replicate. Other viruses cause cell death or damage by hijacking the cell's transcription and translation machinery, leaving the cell at a deficit to translate its own proteins, including enzymes that are required for metabolic pathways.

Observation of CPEs—and how long it takes the virus to cause them—can provide clues for the diagnosis of the virus. For instance, adenoviruses cause cells to form grapelike clusters; herpesviruses cause cells to round up; and respiratory syncytial virus induces syncytia, as its name suggests. Other times, cell culture is used in conjunction with other assays. Electron microscopy can be used to identify the morphology of the virus, while **immunofluorescence assays (IFAs)** or **immunohistochemistry (IHC),** described in more detail below, can be used to definitely identify a specific virus within cells by recognizing viral antigens. For these assays, cells can be grown on slides or coverslips for easy removal from the culture.

Some types of clinical specimens bypass the need to infect cells in culture because they are infected cells or tissues taken directly from the patient. **Cytology** is the examination of cells, while **histology** refers to the examination of tissues. Blood, lung washings, and cerebral spinal fluid contain cells, and cells would also be collected during a cervical swab, which brushes some cells from the cervix and places them in a liquid preservative. The related Papanicolaou (pap) smear scrapes cells from the cervix but instead smears them across a slide that is then sent for analysis. A Tzanck smear uses a similar idea but is used to smear cells from skin lesions onto a slide for diagnosis of herpesvirus infections. These specimens would be subject to cytology. On the other hand, tissues are collected when a biopsy is taken. The tissue is sectioned into thin slices and undergoes histological examination at the diagnostic lab.

Most cells and tissue are devoid of color and therefore require colored stains for viewing under a microscope. Just as CPEs can be visualized in infected cell cultures, they can also be seen in cell or tissue specimens from a patient (Fig. 7.6C). Another type of CPE observed in infected cells or tissues is called an **inclusion body,** which is a visible site of viral replication or assembly within the nucleus or cytoplasm that can be observed with infection by some viruses (Fig. 7.6D; see also Fig. 13.9). Just as with infected cell cultures, cell or tissue specimens can be used for IFA or IHC.

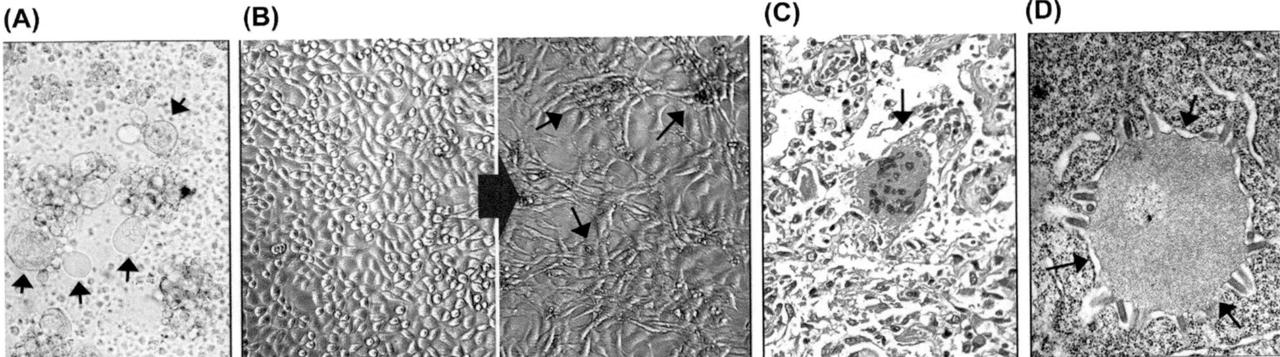

FIGURE 7.6 **Cytopathic effects of viral infection.** Viruses cause a variety of visible effects upon cells. (A) Human herpesvirus-6 infection causes ballooning of infected cells. *(Image courtesy of Zaki Salahuddin and the National Cancer Institute.)* (B) These photos show the cytopathic effects upon cells before (left) and after (right) infection with murine cytomegalovirus. Note the differences in morphology and organization of the cells. (C) A multinucleate syncytium caused by SARS-CoV in a histological section of stained human lung. *(Image courtesy of the CDC/Dr. Sherif Zaki.)* (D) A cytoplasmic inclusion body composed of viral proteins and nucleic acid is visible in this electron micrograph of a cell infected with Lagos bat virus. Note the bullet-shaped virions assembling from the inclusion body. *(Image courtesy of Dr. Fred Murphy and Sylvia Whitfield at the CDC.)*

7.3 DETECTION OF VIRAL ANTIGENS OR ANTIVIRAL ANTIBODIES

Although cell culture, cytology, and histology are valuable in providing visible clues as to the identity of a virus, some viruses cause similar CPEs or no visible CPEs at all. In these cases, it is necessary to use assays that detect viral antigens to prove the presence of a virus. This is accomplished by using antibodies that recognize specific viral antigens. Due to the advents of biotechnology, antibodies can be produced in large amounts and are commercially available. These antibodies recognize different antigens from a wide range of pathogens, including viruses. Several widely used and relatively fast assays make use of antibodies to identify the cause of a viral infection. "Immuno" is usually found in the name of the assay to indicate that it uses antibodies, which are produced by plasma cells of the immune system.

IFAs are performed on cells or tissues that have been affixed to slides and exposed to a fixative. The cells can be from patient specimens or they can be cell cultures that have been infected with patient samples. IFAs use antibodies that are conjugated to **fluorophores,** or fluorescent dyes, that give off a certain color when they are excited by a particular wavelength of light. The best-known fluorescent dye is **fluorescein isothiocyanate**, better known as **FITC**, which is excited by light of 490 nm (blue) and gives off light in the 519 nm range (green). In an IFA, FITC-conjugated antibodies specific for a certain virus are added in a liquid buffer onto the cells or tissue on the slide (Fig. 7.7A). If the viral antigen that is recognized by the antibody is present on the cell surface, perhaps because the virus was assembling at the plasma membrane or was in the process of infecting cells when the section was fixed, then the antibodies will bind to the viral antigen present. Alternatively, the cells can be permeabilized with a detergent to allow the antibodies to enter inside the plasma membrane and bind viral antigens there (Fig. 7.7B). In either case, the antibodies will not bind if the specific antigen is not present. After a sufficient period of time to allow binding of the FITC-labeled antibodies, the slides are washed with a buffer. Any antibody that is bound will remain bound, whereas unbound antibody will be rinsed off. The slides are examined under a fluorescence microscope, which contains a special lightbulb that can provide the wavelengths of light able to excite the fluorophores and filters that allow the viewer to see one emitted color at a time. In this case, if any green cells are seen, then the antibodies bound and the virus was present.

Because the antibody was directly bound to the specific antibody, this IFA is known as a **direct IFA,** or **direct fluorescent antibody** staining. However, sometimes conjugated antibodies are not available; in these cases, a second antibody must be used that is fluorescently conjugated and recognizes the primary antibody (Fig. 7.7C). This is a way to get around not having a conjugated primary antibody. It can also be used to amplify a weak signal when not much viral antigen is present. Because this assay requires a secondary antibody and the fluorochrome is not attached to the primary antibody, it is known as an **indirect IFA.** IFA is used for the identification of a variety of viruses, including several different herpesviruses, influenza, measles, mumps, and adenovirus (Fig. 7.7D).

IHC relies upon the same principles as an IFA except that the antibody is conjugated to an enzyme instead of a fluorescent molecule. After the tissue is exposed to the enzyme-bound primary antibodies, a liquid substrate is added to the slide. If the enzyme-linked antibodies have bound to the tissue, the enzyme will cleave the substrate and a visible colored precipitate will be deposited on the slide. IHC is visible using a normal light microscope and does not require the use of a fluorescence microscope. IFAs or IHC can take as little as a few hours to perform.

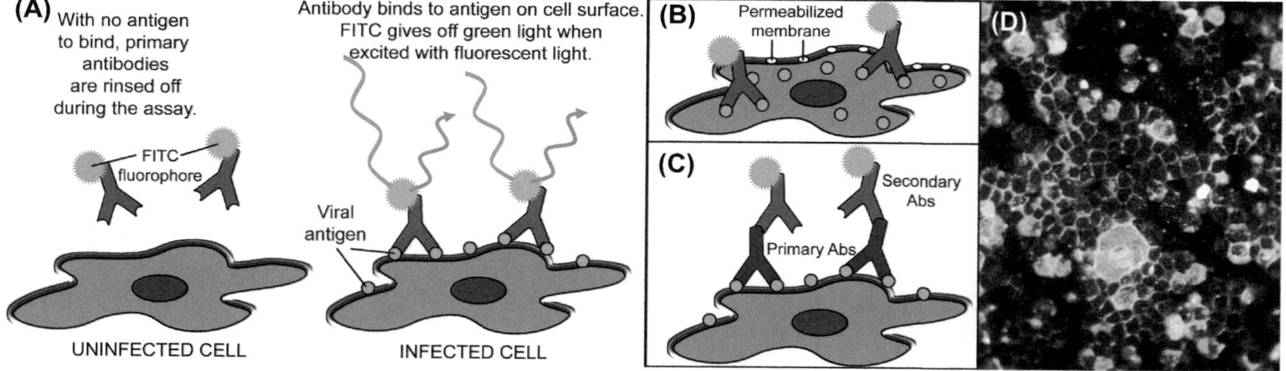

FIGURE 7.7 **Immunofluorescence.** Immunofluorescence uses virus-specific antibodies to verify infection. (A) A buffer containing FITC-conjugated antibodies is added to fixed cells or tissues (from cell culture or cell/tissue specimens). The antibody binds to cognate viral antigens expressed on the surface of infected cells, which prevents it from being rinsed off the slide after incubation. The "stained" cells are examined under a fluorescence microscope to excite the FITC dye, which fluoresces green. This process can also be used to detect intracellular antigen by permeabilizing the plasma membrane to allow the antibodies to enter the cell (B). Some antibodies are not commercially available as already conjugated, so a FITC-labeled secondary antibody that recognizes the first antibody must be used (C). (D) is an example of an IFA performed on cells to verify the presence of respiratory syncytial virus. The green cells are therefore infected. *Photo courtesy of the CDC and Dr. Craig Lyerla.*

An extension of this concept takes place in an **enzyme immunoassay (EIA)**, also known as an **enzyme-linked immunosorbent assay (ELISA)**. Like IFA or IHC, EIAs/ELISAs also detect viral antigens using antigen-specific antibodies. Unlike IFAs or IHC, however, they do not use cells or tissues. Instead, they assay for viral antigens in liquid samples, such as serum or urine. A **sandwich ELISA**, starts by adding antibodies in a buffer to a special plastic plate that has been treated to bind proteins—including antibodies—to it (Fig. 7.8). These are known as "capture antibodies" because they will be used to capture the antigen from the patient sample. Once the antibodies have bound, the wells of the plate—usually there are 96 of them—are rinsed with a buffer to remove any unbound antibody. The plate is **blocked,** meaning that a buffer is added that contains nonspecific proteins, which attach to the plastic wells wherever there is no antibody bound. The wells are rinsed again, and then the patient's sample is added. If the antigen is present in the sample, either because the virus or pieces of the virus are present, then the antigen will bind to the capture antibody. In contrast, if no viral antigen is present in the patient sample, then the capture antibodies will not capture any antigen.

The wells are again rinsed to remove any leftover patient sample. Next, an antibody is added that is conjugated to an enzyme, just as with IHC. (This is where the "enzyme-linked" part of the ELISA name is derived; the "immunosorbent" part indicates that antibodies, the "immuno" part, absorb the antigen.) The enzyme-linked antibody, called the detection antibody, also specifically recognizes the antigen and will create a "sandwich" with the antibodies as the bread and the antigen as the meat—hence the name "sandwich ELISA." After sufficient time for binding to occur, the wells are again rinsed. The detection antibody will remain bound to the plate if the antigen was present, but if no antigen has bound the capture antibody, then there will be nothing for the detection antibody to bind and it will be rinsed out of the well.

The final step is to add the liquid enzyme substrate. If the detection antibodies are present—meaning that the capture antibody bound antigen because the antigen was present in the patient sample—then the enzyme attached to the detection antibodies will cleave the substrate, producing a visible color in the well. If no antigen was present, then there will be no detection antibody to catalyze the substrate reaction. No color will occur.

Spectrophotometers measure the intensity of light, including colored light. A special 96-well spectrophotometer (Fig. 7.9B) measures the color in each well and provides an optical density value that indicates the relative amount of antigen present in each well. If a standard set of samples with known concentration is also tested in the same ELISA plate, then the value obtained with the patient sample can be compared to the standards to obtain a quantitative value. Positive and negative controls are always performed in each ELISA to verify that the assay was performed correctly.

Each step of an ELISA usually allows around an hour or two for binding to occur. Therefore, an ELISA takes several hours to perform all together, but results are usually available the same day the clinical specimen was received. Testing for the hepatitis B surface antigen is performed using a sandwich ELISA.

ELISAs can be used to not only detect viral antigens, but to determine the presence of antibodies against a virus, known as **antiviral antibodies.** Assaying for antiviral antibodies is an indirect way of seeing if a patient has been exposed to a virus, and ELISAs that differentiate between IgM and IgG can provide clues as to whether a patient has been newly exposed. If a person has IgM and becomes positive for IgG over the course of days, then the person is experiencing a primary infection. If the person has high levels of IgG, then it is a secondary or recurrent infection. Antibody isotypes are also used to diagnose congenital infections. IgG crosses the placenta from the mother to the child and is therefore not helpful in diagnosis, but IgM is too large to cross the placenta. Therefore, if IgM is present in the infant's blood, it is made by the infant's immune system and indicates an infection in the child.

To assay for antibodies, the viral antigen is coated onto the bottom of the ELISA plate (Fig. 7.8). The patient sample is added, and if the patient possesses antibodies that recognize the viral antigen, then the antibodies will attach to the antigen on the plate, thereby immobilizing the antibody to the plate, as well. As with a sandwich ELISA, the next step is the addition of an enzyme-linked detection antibody. In this case, the detection antibody recognizes the patient's antibodies. Finally, the wells are rinsed and a substrate is added. If the patient possesses antibodies against the viral antigen, the patient's antibodies and subsequent detection antibodies would have bound to the antigen coated in the wells, and the presence of the enzyme would cleave the substrate, producing color (Fig. 7.9).

A **western blot** utilizes a similar process, except that the protein antigens are immobilized in a gel first. In the same way that DNA can be separated using agarose gel electrophoresis, proteins can be separated by size using polyacrylamide gel electrophoresis (PAGE)(Fig. 7.10A). The proteins embedded into the polyacrylamide gel are transferred onto a membrane made of nitrocellulose. At this point, the membrane functions like the wells of the ELISA plate. Enzyme-linked antibodies against the proteins that have been separated are added in a liquid buffer to the membrane and will bind specifically to the bands of protein antigens (Fig. 7.10B). A substrate is added, and the enzymes present produce detectable color, light, or fluorescence, depending upon the enzyme and substrate used (Fig. 7.10C).

Western blots are often used to confirm positive ELISA results. For example, if a person takes an oral HIV test (that uses saliva) and receives a positive result, the test must be confirmed using a western blot. In this scenario, HIV antigens are separated through PAGE, transferred to

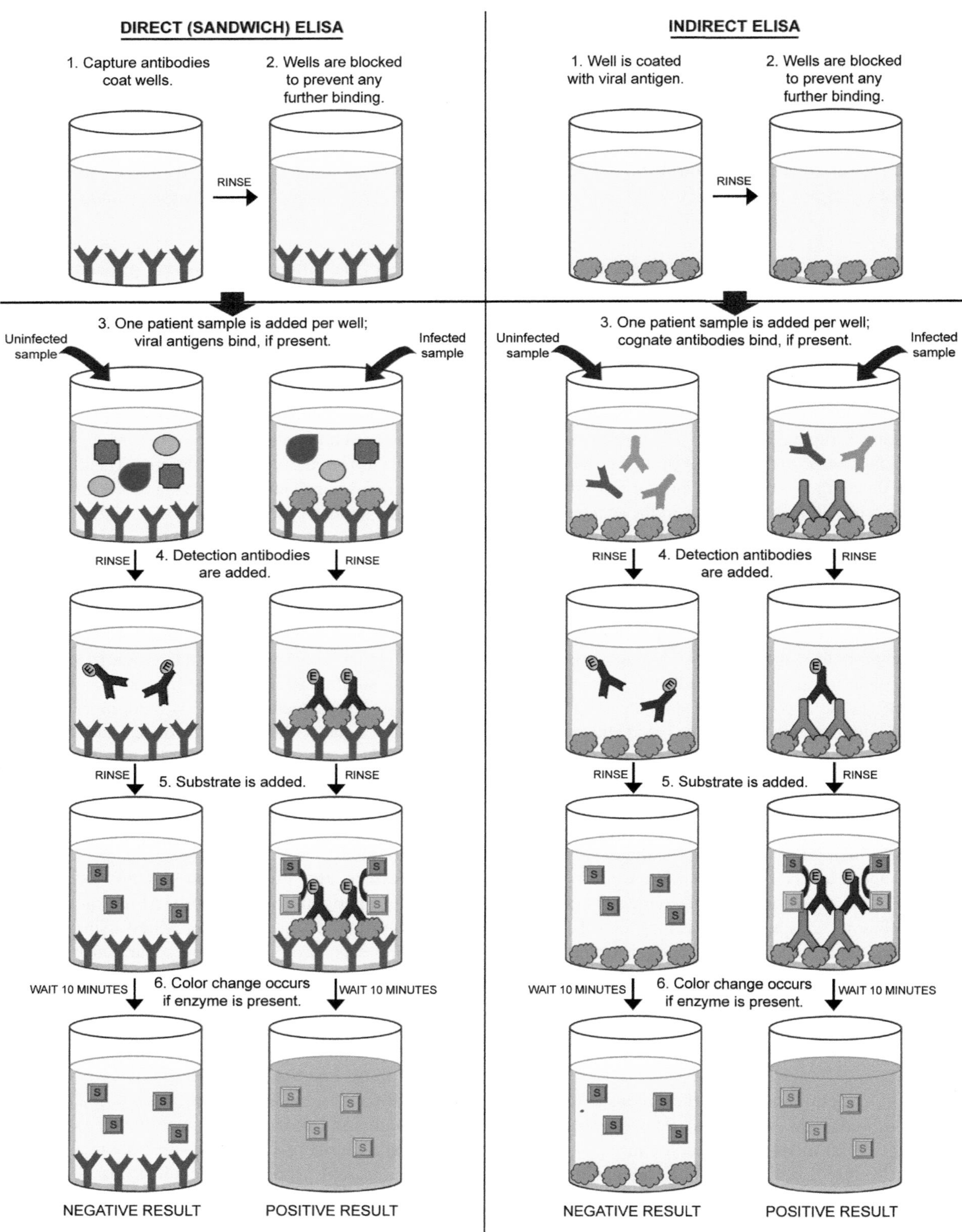

FIGURE 7.8 **Enzyme-linked immunosorbent assay (ELISA).** An ELISA measures viral antigens (direct "sandwich" ELISA) or antiviral antibodies (indirect ELISA). In a sandwich ELISA (left), capture antibodies specific for the viral antigen are coated on the bottom of the wells, which are then blocked to prevent nonspecific binding. The wells are rinsed out in between each step to remove unbound molecules. The patient samples are added, one sample per well, at which point any cognate viral antigens will be bound by the capture antibodies. Next, enzyme-linked detection antibodies are added to each well and also bind the viral antigen, if present. When the substrate is added, it will be cleaved by the enzyme, if present, to form a visible color that is read by a spectrophotometer. In an indirect ELISA (right), the viral antigen is coated to the bottom of the wells and blocked. When the patient samples are added, any patient antibodies that recognize the viral antigen will bind to the plate, as well. Secondary enzyme-linked antibodies are added that recognize human antibodies, and the enzyme cleaves the substrate when added to produce color.

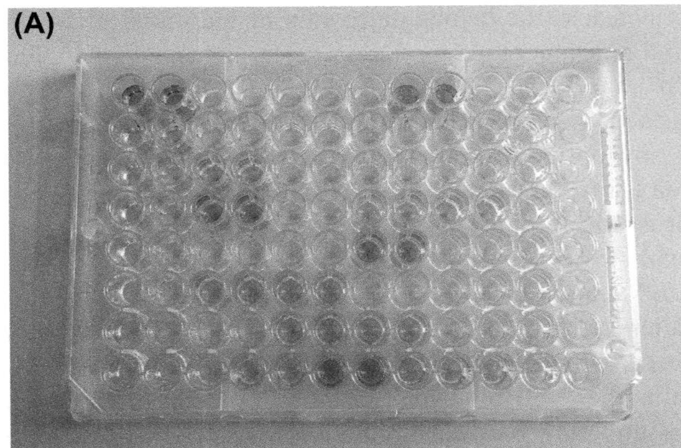

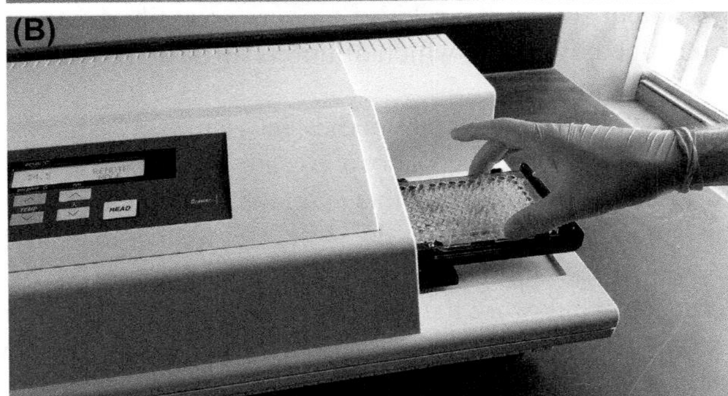

FIGURE 7.9 **ELISA.** (A) Note the variations in the color in the wells of the 96-well ELISA plate. A darker color indicates a higher amount of antigen or antibody was present in the patient sample. (B) An ELISA about to be read in a spectrophotometer.

a nitrocellulose membrane, and exposed to the patient's serum. If the patient has anti-HIV antibodies, they will bind onto the membrane where the HIV antigens are found. A secondary, enzyme-linked antibody will provide the enzyme that will cause a colored band to be produced when exposed to a substrate. (Western blots are still recommended for confirmation of oral HIV tests, although nucleic acid testing is now the choice of confirmatory test for positive HIV blood tests.)

Agglutination reactions take place when the binding of antibodies to antigen causes a visible clumping, or agglutination. Latex agglutination tests use the same principles as ELISAs or western blots, except the antigen or antibody is bound onto small latex beads (Fig. 7.11A). To test for patient antibodies, the viral antigen is coated onto the latex beads. When mixed with patient serum, the antibodies will bind to the antigen-coated latex beads. Because antibodies have two antigen-binding sites, each arm of the antibody is able to bind a different bead. The result is that the beads are splayed out in a lattice formation—the beads have agglutinated, and a visible "clump" has formed. If the patient does not have antibody against the particular viral antigen, then the latex beads will not agglutinate. Agglutination reactions can take place in tubes, 96-well plates, on slides, or using cardboard cards. Like ELISAs, latex agglutination tests can

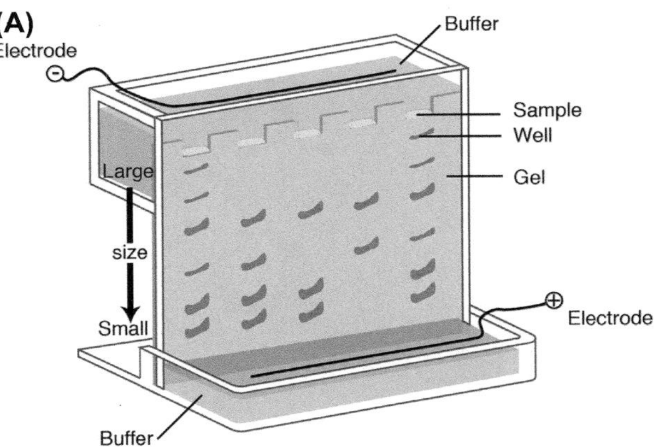

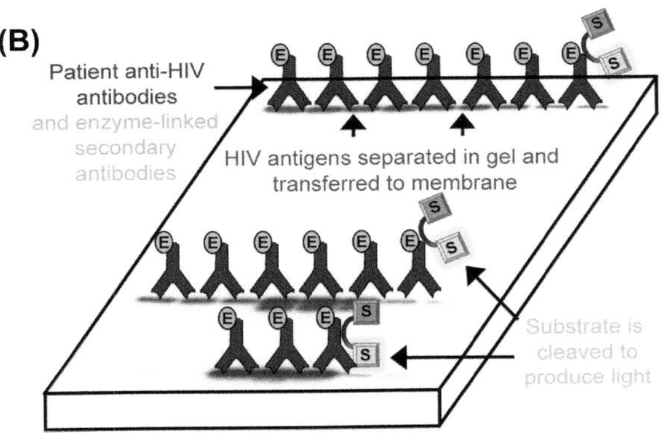

FIGURE 7.10 **Western blot.** Western blots work on the same principles as an ELISA, except the proteins are separated in a gel and transferred to a membrane instead of using wells. (A) To confirm a positive HIV result, a variety of different HIV antigens are separated using PAGE. The samples are loaded in the wells at the top of the gel, a charge is applied to the gel, and the proteins separate by size, with the smallest proteins traveling farthest in the gel. (B) Following PAGE, the HIV proteins are transferred to a nitrocellulose membrane. The patient serum sample is incubated with the membrane, and if the patient has anti-HIV antibodies, they will bind to the HIV proteins on the membrane. An enzyme-linked secondary antibody is added that recognizes human antibodies (just the enzyme part is shown here to save room). A substrate is then added that produces a detectable signal; in this case, the substrate gives off light that will create a dark mark when placed against X-ray film (C). In this scenario, the patient is positive for anti-HIV antibodies and thus has been infected with HIV. *Illustration in (A) by Darryl Leja and photo in (C) taken by Maggie Bartlett, National Human Genome Research Institute.*

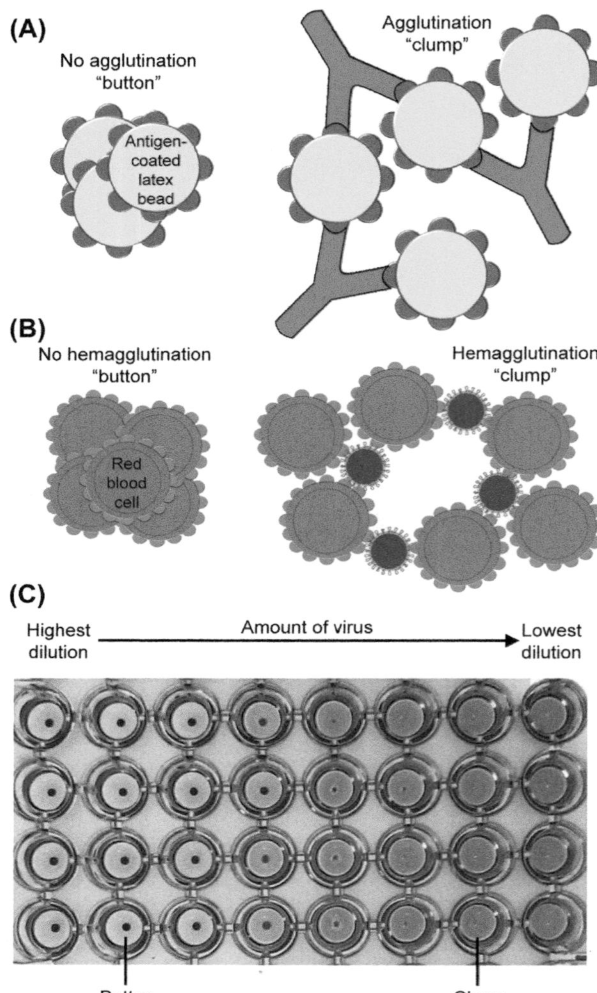

FIGURE 7.11 **Agglutination/hemagglutination.** (A) Latex agglutination assays coat viral antigen onto latex beads. If antibodies are present that recognize the antigen, they will agglutinate the beads. The same procedure can be performed to assay for the presence of viral antigens by coating antibodies on the beads. (B) Hemagglutination refers to the agglutination of red blood cells specifically. Several viruses, including influenza virus and measles virus, possess hemagglutinin proteins that bind to red blood cell surface glycoproteins. Hemagglutination can be used to identify these viruses. (C) The results of a hemagglutination reaction that show hemagglutination occurring as increasing amounts of virus are titrated into the wells. The "button" on the left and the "clump" on the right show the visible change that occurs with hemagglutinated blood (on right). *Image courtesy of Liu, Y., et al., 2013. Poly-LacNAc as an Age-Specific Ligand for Rotavirus P[11] in Neonates and Infants. PLoS One 8(11), e78113.*

also test for viral antigen in a patient sample by coating antibodies onto the beads that recognize the antigen. Several antibody-coated beads will bind to one antigen, resulting in visible agglutination.

Hemagglutination refers to the agglutination of red blood cells. A handful of viruses, including influenza virus, measles virus, mumps virus, rubella virus, and rabies virus, cause the hemagglutination of red blood cells. The viral attachment glycoproteins (aptly named "hemagglutinin" in influenza, measles, and mumps viruses) bind to the surface of red blood cells and agglutinate them, similarly to how antibodies do (Fig. 7.11B and C). Hemagglutination assays have been used to show the presence of these viruses in samples. A similar assay tests for the antibody levels in a patient sample against one of the viruses. A fixed amount of virus is added to each well, and then the patient serum is titrated into the wells at different dilutions. If the patient has antibodies against the virus, the antibodies will bind the virus and prevent it from hemagglutinating the red blood cells. By using dilutions of the patient serum, the relative amount of antibody can be determined by noting which dilutions do or do not prevent hemagglutination.

An assay that uses antibodies in a similar fashion as to the assays described above is called the **lateral flow immunoassay (LFIA)**. In this case, the presence of a virus (or antibodies against a virus, if the test assays for antiviral antibodies) will result in a colored band appearing in a particular window on the test. Pregnancy tests are the best-known LFIAs, but LFIAs are available that test for HIV antibodies, dengue virus antibodies, rotavirus, respiratory syncytial virus, and influenza virus. They can be in the form of a dipstick, like a pregnancy test, or they can be a small plastic test that requires a drop of sample be placed into a sample well (Fig. 7.12A).

In the LFIA, the test reagents are added onto a nitrocellulose membrane that draws the liquid along the length of the strip using capillary action (Fig. 7.12B). The test strip contains three major zones: a reaction zone, test zone, and control zone. In the case of the influenza LFIA, the patient specimen (usually a nasal or nasopharyngeal swab) is mixed with a buffer that disrupts any virions that are present, releasing the viral antigens. A drop of the liquid sample is placed in the sample well on one end of the LFIA test, and the viral antigens begin to move through the membrane through capillary action to the reaction zone of the test. The reaction zone contains free antibodies that are bound to gold beads or blue latex beads, which will be the basis for the formation of color later in the test. As the viral antigens pass through the reaction zone, the bead-linked antibodies bind to the antigens, and the antigen–bead complex continues moving through the membrane. The test zone is the next zone encountered. In the test zone, other antibodies are immobilized to the membrane and recognize the viral antigens that are bound to the reaction zone bead–linked antibodies. The binding of the antigen–bead complex to the test zone antibodies immobilizes the antigen–bead complexes to the test zone area, creating an antigen sandwich. As the beads begin accumulating, the color of the

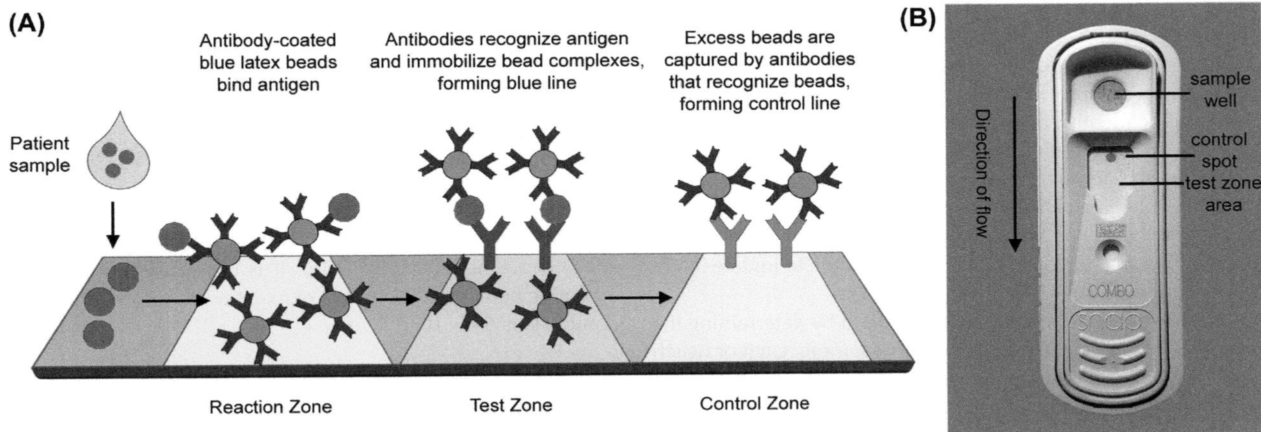

FIGURE 7.12 **Lateral flow immunoassay (LFIA).** (A) Lateral flow immunoassays have three major zones: a reaction zone, a test zone, and a control zone. The patient's sample is added to the sample well of the assay and begins moving though the membrane via capillary action. In the reaction zone, antigens found in the patient's sample bind to free antibody-coated beads. The bead-antigen complexes continue traveling to the test zone, where membrane-bound antibodies bind to the antigen, thereby immobilizing the beads. As the beads accumulate, a colored line is formed. Excess bead complexes continue moving through the membrane and are eventually bound by control zone antibodies that recognize the bead complex, rather than antigen. As excess beads accumulate here, a colored control line is formed. (B) LFIAs are useful for diagnosis of viral infection in animals, as well. This LFIA tests for two feline viruses, feline immunodeficiency virus and feline leukemia virus. Note that the control spot is present, but there are no other colored spots in the test zone area, indicating that the animal was negative for both viruses.

beads becomes apparent. Blue latex beads provide a blue band, and gold beads appear as a red band. This indicates a positive result—that the person's specimen contained the virus.

The final zone, the control zone, contains antibodies that will bind the bead complex. If the reaction zone beads are not bound to antigen, they will continue flowing through the test zone and be captured by the control zone antibodies. In fact, there are many more reaction zone beads than antigen, so some beads will be free even in the presence of antigen and make it to the control zone. The capture of the beads in the control zone produces a visible line in this section of the test. This is the positive control to show that the test worked.

The HIV or dengue virus LFIAs use the same principles, except they assay for the presence of antibodies against the viruses, rather than detecting the viruses themselves.

7.4 DETECTION OF VIRAL NUCLEIC ACIDS

Nucleic acid testing (NAT) has replaced many of the traditional, slower assays in diagnostic labs. Detecting viral nucleic acids is a sensitive and specific way to screen for viruses within a patient sample, and new methods allow for the screening of many viruses simultaneously. Additionally, NAT is able to detect the presence of viruses for which no other test currently exists. NAT results are available the same day the sample is processed.

NAT assays rely upon the principle of **polymerase chain reaction (PCR)**, which recapitulates the process of DNA replication in the laboratory by providing the molecules necessary to copy DNA (see Chapter 3, Features of Host Cells: Cellular and Molecular Biology Review for a review). In the process of PCR, DNA (including any viral DNA present) is isolated from the clinical specimen, generally blood cells or tissue, and added to a tube containing primers, DNA polymerase, and nucleotides (Fig. 7.14). The tube is placed in a **thermocycler,** a bench-top machine that simply changes the temperature of the tube, as its name suggests. In the *denaturation* stage of PCR, the thermocycler heats the DNA, usually to 95°C, which breaks the hydrogen bonds holding the two DNA strands together and so they separate from each other. In the *annealing* stage, the temperature is reduced to allow the binding (annealing) of two **primers** to the separated DNA. The primers are complementary to the sequence to which they bind, and they flank the region to be amplified, one primer on each strand. The annealing temperature is determined by the composition of the nucleotides in the primers, but is usually around 55°C.

In the final stage, *extension*, the thermocycler raises the temperature to the optimal temperature for the DNA polymerase enzyme. In this case, a special polymerase from the bacterium *Thermus aquaticus*, called Taq polymerase, is used. *Thermus aquaticus* was discovered to live in the hot springs of Yellowstone National Park, and its DNA polymerase is evolved to withstand high temperatures, such as those found in the hot springs. Taq polymerase is used for PCR because a human DNA polymerase would be denatured by the high temperatures required for the denaturation stage. The thermocycler holds the temperature at 72°C, and Taq polymerase extends

In-Depth Look: Counting Viral Particles Using a Plaque Assay

It is often necessary to know how many infectious virions are present in a sample. The diagnostic techniques described in this chapter identify the presence of a virus in a sample, or even the amount of viral nucleic acid, but these assays cannot determine the amount of virus present that is capable of productively infecting cells. A very common virology technique to determine this is known as the **plaque assay**, which measures the number of virions in a sample that are able to initiate infection of target cells.

Microbiologists measure viable bacteria by determining the number of individual bacterial cells in a sample, each of which will form a distinct bacterial colony. This results in the number of colony-forming units, or CFUs, in a sample. Plaque assays measure the number of individual cells that were infected by a single virion, each of which forms a **plaque**, or clearing, as the virus spreads among neighboring cells. This results in the number of **plaque-forming units (PFUs)** in a sample, the indication of viral infectivity.

To perform a plaque assay, the first step is to use cell culture to plate cells into several cell culture dishes or a multiwell plate—6-well or 24-well plates are often used for this purpose—in a liquid medium to support their growth (Fig. 7.13). The cell line used must be permissive to infection with the virus that is being studied. The cells are allowed to grow to near **confluency,** meaning that they have grown to completely cover the bottom of the cell culture vessel. At this point, the medium is removed from the wells. Tenfold serial dilutions of the initial sample are made, in case the initial sample has too much infectious virus and ends up harming the entire well of cells, and 0.1 mL of the experimental samples are added to individual wells. The plate of cells is gently shaken on a flat surface, moving the plate in the motion of a "plus" sign, to ensure that the virus that was just added is equally distributed over all the cells.

The virus is given time to bind and enter the cells (<1 h), at which point the cells are covered with a layer of cooled agarose that has been mixed with medium. Having entered random cells in the well, the virus will replicate over the course of the following days, and the agarose ensures that released virions are only able to infect cells immediately adjacent to the infected cell.

As the virus replicates and infects more cells, it begins forming plaques, or clearings, that are caused by the cytopathic effects upon the cluster of infected cells. In some cases, the virus is lysing infected cells, and in other cases, the virus has interfered with enough cellular processes to cause damage or death to the infected cells. Eukaryotic viruses typically take 3–5 days to form plaques that are large enough to see. Because cells are clear, a dye is used to stain the living cells, which leaves the clear plaques visible when the well is inspected.

The number of plaques are counted, giving the number of PFU per well. However, this needs to be converted into PFU/**mL** in the original sample, so the number of plaques must be divided by the dilution of the sample and by 0.1 mL, since that was the amount added to the cells. Therefore, 0.1 mL of a 0.001 dilution (1/1000) that resulted in 54 plaques means that the starting sample had 540,000 PFU/mL in the initial sample—54/0.1 mL/0.001 = 540,000. It is best to write this using scientific notation: the amount of infectious virus in the starting sample was 5.4×10^5 PFU/mL.

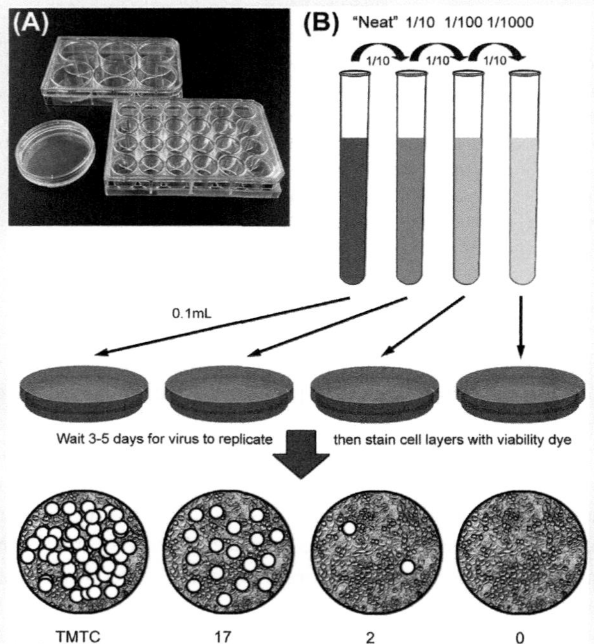

FIGURE 7.13 **A plaque assay determines the amount of infectious virus in a sample.** (A) Plaque assays are performed in a variety of different types of culture vessels, depending upon the size of the plaques that are formed. Shown here are a tissue culture dish, 6-well plate, and 24-well plate. (B) To perform a plaque assay, the undiluted "neat" sample is diluted several times using 10-fold dilutions. 0.1 mL of each dilution is added to an individual dish of cells. After allowing for time for the virus to enter the cells (about 1 h), a layer of agarose is used to overlay the cells to ensure lateral infection by infected cells. Depending on the virus, it generally takes 3–5 days for the visible sites of replication, called plaques, to be visible. At this point, the live cells are stained, and the clearings are counted (see Fig. 7.5). A dilution that has multiple but distinct plaques should be chosen to determine the PFU/mL in the original sample. In this case, the undiluted sample had 1.7×10^3 PFU/mL.

the 3′-end of the primer using the available nucleotides, creating a double-stranded piece of DNA from the single-stranded template. In this way, one double-stranded piece of DNA was separated and replicated to create two copies.

The stages of PCR are repeated, usually 30–35 times, to create billions of copies of the target DNA segment. In the case of viral diagnosis, the primers would be specific for a piece of the viral genome. Good primers will bind to just the

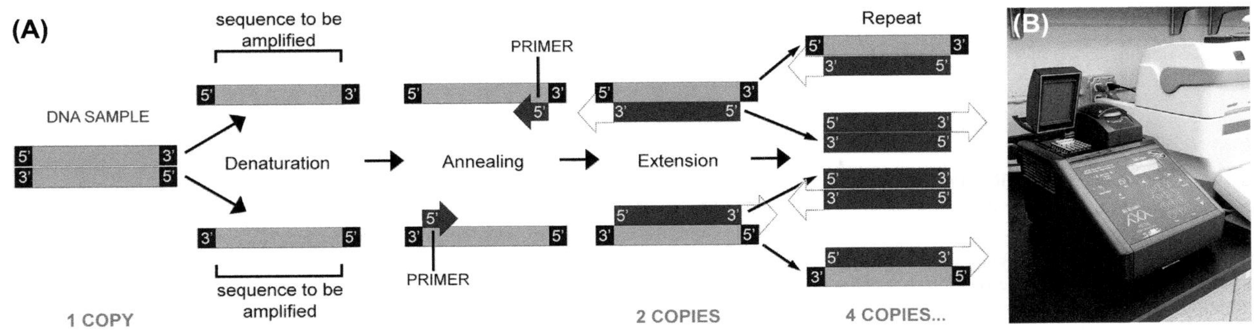

FIGURE 7.14 **Polymerase chain reaction (PCR).** PCR amplifies a specific sequence of DNA, based upon the location of primers. (A) In *denaturation*, high heat separates the two strands of the DNA sample. In *annealing*, the temperature is lowered, which allows the primers to hybridize to the strands. In *extension*, DNA polymerase extends the primers, creating the new strand. One DNA strand has now been copied into two. The three stages are repeated 30–35 times to generate billions of copies of the target sequence. If amplicons are created, then it means viral nucleic acid was present in the patient's DNA sample. (B) A thermocycler, the machine that controls the reaction temperature during PCR. (The white machine on the right is another brand of thermocycler.)

viral DNA being amplified and not to any isolated cellular DNA or genomes of other viruses.

What about viruses with an RNA genome? In this case, **reverse transcriptase PCR (RT-PCR)** is carried out. It involves the same steps as those performed with PCR, but because the viral genome is RNA, it must be reverse transcribed into cDNA first. This is accomplished using reverse transcriptase, which transcribes cDNA from an RNA genome. The reverse transcriptase used in RT-PCR is derived from retroviruses, generally Moloney murine leukemia virus or avian myeloblastosis virus. This additional step creates cDNA that is then amplified using PCR as above.

The amplified DNA fragment, known as an **amplicon,** can be detected in several different ways, but agarose gel electrophoresis remains the simplest and cheapest method. Agarose gel electrophoresis uses electricity to separate DNA fragments in an agarose gel. The distance traveled by the DNA is based upon the fragment size, with smaller fragments traveling farther in the gel than larger fragments. After separation is complete, the gel is stained with a chemical known as ethidium bromide that intercalates in between the base pairs of the DNA and fluoresces when exposed to UV light. A fluorescent band, therefore, indicates that DNA is present and was amplified by the PCR reaction. Because smaller fragments travel farther through the gel, the relative location of the band can verify the band that was produced is the anticipated size when compared to a known DNA ladder.

Multiplex PCR allows for the amplification of several different pieces of DNA at the same time in the same tube. This technique involves the same process as normal PCR or RT-PCR, but multiple primer pairs are added to the reaction tube. Each primer pair is designed to recognize the nucleic acid from a distinct virus, and the size of each amplicon produced is different for each virus. This allows for the simultaneous amplification of DNA from several viruses at one time, and analyzing the size of the amplicon reveals which viruses were present in the initial starting sample.

PCR was invented by Kary Mullis in 1983, and this molecular biology technique has been built upon and adapted to produce numerous assays of great utility. A modification of PCR that is ubiquitously used is known as **real-time PCR**, so named because the user can monitor the amplicon amplification in real time using a special thermocycler. A great advantage of monitoring the reaction as it proceeds is that the rate of amplification can be noted. Traditional PCR machines analyze the product after the completion of all cycles, which makes it impossible to know if the reaction had plateaued at an earlier cycle because of limitations in reagents, such as nucleotides. Real-time PCR machines address this concern by providing a visual graph of the amplification. An advantage of being able to see the amplification in real time is that the quantity of the PCR product can be used to back calculate the amount of starting template nucleic acid that was present, since we know that each cycle doubles the amount of DNA. For this reason, real-time PCR is also known as **quantitative PCR.** Therefore, the standard abbreviation for real-time PCR, is **qPCR.** (Recall that RT-PCR stands for reverse transcriptase PCR, *not* real-time PCR!) RT-PCR can also be performed using qPCR; this technique is known as RT-qPCR.

How is DNA amplification monitored in real time during qPCR? There are two major methods, and both involve

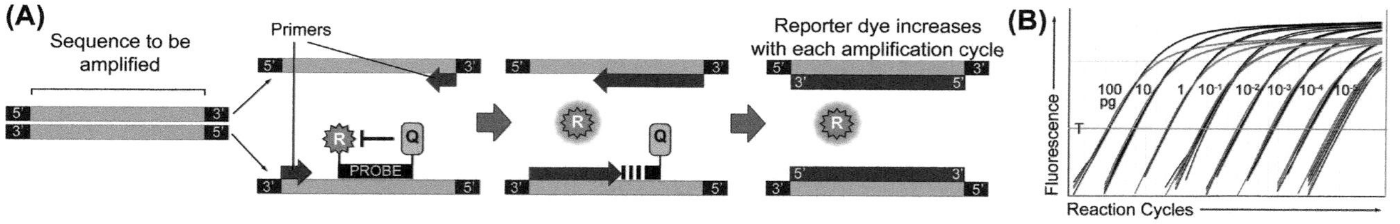

(A) Sequence to be amplified; Primers; Reporter dye increases with each amplification cycle; (B) Fluorescence; Reaction Cycles

FIGURE 7.15 **Real-time PCR (qPCR).** Real-time PCR works on the same principles as PCR, except that it measures the fluorescence of a reporter dye in real time to monitor the amplification reaction. DNA-binding dyes are used, as are probes attached to a reporter dye and quencher molecule (A). The probe hybridizes to a section of the target DNA in between the two primers. As long as the reporter and quencher are physically close, the quencher absorbs any fluorescence and none is given off. As the primer is extended by DNA polymerase, the probe is broken down, releasing the reporter dye and quencher. Free from the quencher, the reporter dye fluoresces. The fluorescence increases as each cycle is repeated. (B) A real-time PCR amplification plot showing the amplification of eight different concentrations of JC polyomavirus (JCPyV) DNA. The number of reaction cycles is on the X axis, while the fluorescence generated in the reactions is on the Y axis. A few important things to note: The samples that start with more DNA are amplified more quickly (require fewer cycles to reach a set threshold, T). Also, each of the cycles eventually plateaus around the same fluorescence, even though we know some samples contain more DNA than others. This is why the samples are compared at the threshold limit (green line). This was multiplex qPCR: black lines are the amplification of a JCPyV protein–coding genomic sequence, while the red lines are the amplification of a noncoding genomic region. *Amplification plot reprinted from Ryschkewitsch, C.F., et al., 2013. Multiplex qPCR assay for ultra sensitive detection of JCV DNA with simultaneous identification of genotypes that discriminates non-virulent from virulent variants. J. Clin. Virol. 57(3), 243–248, Copyright 2013 with permission from Elsevier.*

the use of fluorescence. In the first method, a fluorescent dye is used that intercalates into double-stranded DNA, in the same way that ethidium bromide does. An example of a fluorescent DNA-binding dye is SYBR Green, which is excited with light of 488 nm and emits light at 522 nm (green). The real-time thermocycler is able to provide the excitation wavelength and has detectors to measure the emitted wavelength. Another method involves fluorescent reporter probes, such as Taq-Man® probes (Fig. 7.15A). Like a primer, the probe recognizes a sequence of the DNA target, but it is located in the middle of the amplified sequence, in between the primers. Attached to the probe are a fluorescent **reporter dye** at one end and a **quencher** at the other. As long as the reporter dye and quencher are attached to the probe, the quencher absorbs the fluorescence emitted by the excited reporter dye (through the process of fluorescence resonance energy transfer, or FRET). Taq polymerase breaks down the nucleotides of the probe as it amplifies the DNA to which the probe is bound, releasing the reporter dye and quencher from the probe. When this occurs, the two are physically separated and the quencher molecule can no longer inhibit the reporter's fluorescence. As with the fluorescent double-stranded DNA dyes, reporter dyes are excited and detected by the thermocycler.

Using either method, an increase in fluorescence indicates an increase in amplified product. A **DNA amplification plot** is created as the measurements of fluorescence are taken in real time (Fig. 7.15B). This allows the DNA product to be quantified at a cycle number where the amplification is in exponential phase, which can be used to back calculate the amount of starting template DNA. Multiplex qPCR assays have also been developed that allow for the simultaneous detection of several fluorescent probes in one tube, one for each amplified viral DNA segment.

PCR and qPCR amplify specific sequences of viral DNA. On the other hand, **DNA microarrays** rely upon the **hybridization** of viral nucleic acid segments to a synthesized piece of DNA. "Hybridization" refers to the complementary binding of two nucleic acid pieces to each other; for instance, primers hybridize with their target DNA during PCR. With DNA microarrays, thousands of short single-stranded pieces of DNA, called **oligonucleotides (oligos),** are immobilized onto a small silicon chip or glass slide (Fig. 7.16B). Oligos, like primers, are able to be synthesized in the laboratory using a specialized piece of equipment that bonds individual nucleotides together into a strand of nucleic acid. When detecting the presence of viral nucleic acid, the sequences of the synthesized oligo probes spotted on the chip are complementary to known sequences within the viral genome of interest.

Since viral nucleic acids may be in low abundance in a sample, microarray experiments begin with PCR to amplify the sample DNA (Fig. 7.16A). Primers are used that amplify randomly, rather than specifically, so that all DNA in the sample is amplified. Fluorescent nucleotides are incorporated into the amplified DNA. Following amplification, the fluorescent DNA is added to the microarray, and hybridization is allowed to occur between the sample DNA and the oligos on the microarray. Any hybridized DNA will remain bound to the oligos on the chip, while unhybridized DNA is rinsed away. The microarray is read by a machine that measures fluorescence. If fluorescence is present at a particular oligo spot, then the DNA from the sample is bound there and the machine will report a positive result (Fig. 7.16C). The identity of each oligo is known, and so the sequences present in the initial DNA sample can be determined.

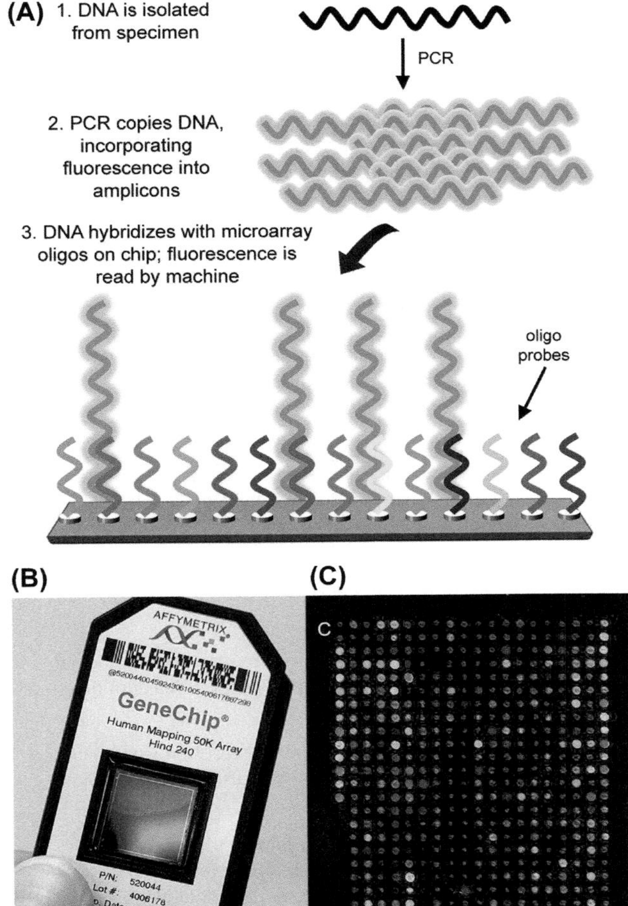
(A)
1. DNA is isolated from specimen
PCR
2. PCR copies DNA, incorporating fluorescence into amplicons
3. DNA hybridizes with microarray oligos on chip; fluorescence is read by machine
oligo probes
(B)
(C)
AFFYMETRIX
GeneChip®
Human Mapping 50K Array Hind 240
C

FIGURE 7.16 **DNA microarrays.** (A) Microarrays work on the principle of hybridization. For microarrays like the ViroChip that identify viruses within a sample, the starting DNA is copied using PCR. Fluorescent nucleotides are incorporated in the process. Then, the fluorescent amplicons are added to the microarray, which has spotted on it thousands of different oligonucleotide probes that each recognize a different viral sequence. If hybridization occurs with the sample DNA, then the virus's nucleic acid was present in the starting sample. A machine measures the fluorescence at each spot. (B) An example of a microarray, found on the blue chip in the middle of the photo. Note the size of the chip, compared to the size of this researcher's thumb. This chip contains 50,000 probes that hybridize to single nucleotide differences within the human genome. (C) Just a small portion (400 oligo spots) of an actual readout of a chip, showing in which spots fluorescence is present. This result used two different samples of cDNA, one labeled green and one labeled red. *(B) and (C) courtesy of Maggie Bartlett and the National Human Genome Research Institute.*

Microarrays can be used for virus identification, but by modifying the oligos on the chip, microarrays can also identify strains of viruses that are particularly virulent or are genetically resistant to certain therapeutics. As long as the genome sequences are known that correspond to these attributes, then an oligo can be made that corresponds to the genetic trait.

Microarray principles can be employed for identification of RNA transcripts or proteins, as well. For example, instead of using DNA oligos to capture nucleic acids, protein microarrays use antibodies to capture different proteins.

An exciting application of the DNA microarray is the **ViroChip,** developed by the DeRisi and Ganem laboratories at the University of California, San Francisco. The ViroChip uses long, 70-nucleotide oligos ("70-mers") as probes that are complementary to known viral sequences. The ViroChip contains over 60,000 probes that detect over 1000 viruses. The researchers included oligos for conserved and novel sequences found in related viruses. In this way, the ViroChip can identify similar viruses based upon conserved genomic regions and can also identify individual subtypes or strains based upon novel nucleotide sequences not possessed by related viruses. The ViroChip can also be used to characterize novel viruses. For example, it was used in 2003 to assist the CDC in identifying a novel virus that was causing deaths in Southeast Asia. The microarray suggested that the unknown virus was a previously unrecognized coronavirus, later to be known as the severe acute respiratory syndrome–associated coronavirus (SARS-CoV).

To create the oligo probes used on the ViroChip, researchers used the genome sequences of any viruses that had been sequenced. **High-throughput sequencing** methods allow for the determination of the nucleotide sequence of potentially any biological entity, including viruses. In the earlier days of viral diagnosis, CPEs were used to classify differences in subtypes or genotypes of viruses. This was replaced by serology and the use of antibodies to determine the viral subtype, and now sequencing of viral genomes provides a definitive differentiation at the nucleic acid level. Currently, over 4600 viral genomes have been completely sequenced, and public databases exist for the genomes of HIV, influenza viruses, dengue virus, and hepatitis C virus, among others. Sequencing also allows us to track the genetic differences between related subtypes and strains of viruses, and it assists scientists in identifying novel viruses by comparing genome sequences with those of known viruses. **Bioinformatics** is the field of study that uses computers to analyze and compare biological data, including genome and protein sequences. Bioinformatics is used to compare viral genome sequences, monitor viral evolution over time, and track virus mutations that appear during epidemics.

There are advantages and disadvantages to every method of diagnosis (Table 7.4), which is why there are a range of tests available for the diagnosis and confirmation of commonly encountered viral infections (Table 7.5).

TABLE 7.4 Advantages and Disadvantages of Various Viral Diagnostic Techniques

Technique	Advantages	Disadvantages
Cell culture	Can be used to detect viruses of unknown identity, cheaper than molecular assays, can distinguish infectious versus noninfectious virus, can be used to amplify a small amount of virus, can be a starting point for other tests	Slow (takes time for virus to replicate), requires specialized technicians, relatively expensive, living cells lead to variability, requirement for specialized equipment (BSCs) and laboratories (BSLs), not all viruses will replicate in cell culture, requires infectious virus
Cytology/histology	Quick because cells/tissues are sent as the specimen, staining assays are fast	May not provide definitive identification of virus, cannot differentiate between strains or subtypes based upon visual examination, requires medically trained personnel, often pathologists
Immunofluorescence assays	Relatively fast to perform, does not require specialized technicians, can definitively identify virus	Requires specialized equipment (fluorescence microscope), antibodies must exist for the virus being detected
Enzyme immunoassay/Enzyme-linked immunosorbent assay	Can be performed within a few hours, easy to perform, can assay for viral antigen or antiviral antibodies, can distinguish between primary and recurrent infection, can be used to diagnose congenital infections	Not available for all viruses, requires specialized equipment (spectrophotometer), low levels may not be detected by the assay
Western blot	Can confirm other tests, relatively quick to perform	Requires specialized reagents and equipment, low sensitivity
Agglutination/hemagglutination reactions	Quick to perform, no specialized equipment necessary, visible read-out	Can be difficult to interpret results, requires specialized reagents, only certain viruses hemagglutinate red blood cells
Lateral flow immunoassay	Can be performed at home or in a clinic, does not require specialized equipment or training, fast, relatively inexpensive	Can result in an indeterminate result, must be verified by other tests, may not distinguish between strains
Polymerase chain reaction	Very little starting material is required, fast, specific, can be performed on nonliving tissues	Requires specialized equipment (thermocycler), knowledge of sequence required for primers, expensive reagents, results are semiquantitative
Real-time PCR	Fast, extremely sensitive, specific, quantitative, can be performed on nonliving tissue, provides immediate results	Requires expensive reagents and specialized equipment, knowledge of sequence required for primers and probe
DNA microarrays	Fast, can test for thousands of different viruses at one time, can be used on nonliving samples, can identify novel viruses	Expensive, requires specialized equipment, knowledge of sequences are required to create oligo probes, does not indicate infectious versus noninfectious virus
High-throughput sequencing	Relatively fast, can provide entire viral sequence, does not require living tissues, can be used to differentiate between closely related viruses, can track viral mutations	Expensive reagents and equipment, requires specialized technicians to interpret results, amount of data generated can be overwhelming and not easily interpreted

TABLE 7.5 Types of Diagnostic Tests Available for Well-Known Viral Infections

Virus	Diagnostic tests available
Adenovirus	Cell culture, IFA
Cytomegalovirus	Cell culture, IFA, qPCR, ELISA (IgM/IgG), DNA sequencing
Dengue virus	ELISA (IgM/IgG)
Enteroviruses	Cell culture, IFA, qPCR
Hepatitis A virus	ELISA (IgM/total)
Hepatitis B virus	ELISA (IgM against surface or core antigen), ELISA (HBV surface and core viral antigens), qPCR
Hepatitis C virus	qPCR, genotyping using RT-qPCR of HCV genome portions, RNA genome amplification and probe hybridization
Herpes simplex virus	Cell culture, IFA, ELISA (IgG)
Human immunodeficiency virus	ELISA (IgM/IgG antibodies, p24 viral antigen), western blot (confirmation), LFIA, RT-PCR then sequencing, genome amplification and probe hybridization
Human papillomavirus	Cytology, genome amplification and probe hybridization
Influenza virus	Cell culture, hemagglutination, IFA
Measles virus	Cell culture, IFA, ELISA (IgM/IgG)
Rotavirus	Electron microscopy, latex particle agglutination
Rubella virus	Cell culture, IFA, ELISA (IgM/IgG)
Varicella zoster virus	Cell culture, IFA
West Nile virus	ELISA (IgM/IgG)

SUMMARY OF KEY CONCEPTS

Section 7.1 Collection and Transport of Clinical Specimens

- Diagnostic tests are paramount in determining the etiology of viral infections.
- Specimens can be collected from a variety of body fluids. The choice of specimen will depend upon the site and stage of infection and whether it is best to test for virus or antiviral antibodies.
- Heparin or EDTA prevents the clotting of blood. Plasma is the liquid portion of nonclotted blood (and so contains clotting factors), while serum is the liquid portion of clotted blood.
- Specimens must be carefully acquired, stored, and transported to ensure integrity of the samples. This is critical to ensure a meaningful and accurate test result.

Section 7.2 Virus Culture and Cell/Tissue Specimens

- Direct diagnostic methods assay for the presence of the virus, while indirect methods test for effects of the virus.
- Cell culture is the process of growing cells or tissues in the laboratory, using an incubator and special culture medium. Cells must be manipulated in at least a class II BSCs to maintain the sterility of the cultures.
- Biological safety cabinets rely upon filtering air through HEPA filters, which filter out 99.97% of particles 0.3 μM in size. Class II BSCs are most often used in research and clinical diagnostic laboratories.
- Certain viruses can cause severe effects and must be handled with caution. BSLs specify which precautions should be taken and are determined by the type of pathogen. They range in stringency from BSL1 to BSL4.
- Cell lines can be infected with patient samples to allow viral replication within the cells. CPEs, such as morphological changes, ballooning, syncytia formation, or inclusion bodies, can help to identify the virus identity. Infected cells can also be used for IFAs.
- Cytology and histology are the staining and microscopic examination of cell and tissue specimens, respectively.

Section 7.3 Detection of Viral Antigens or Antiviral Antibodies

- There are a variety of immunoassays that use antibodies to identify viruses or antiviral antigens. The commercial availability of manufactured antibodies has revolutionized diagnostics.
- Immunofluorescence assays are performed on fixed cells or tissue. Fluorescently labeled antibodies bind to viral antigens present in infected cells. A fluorescence microscope is used to excite the fluorophores so they give off colored light. IHC works in the same way except a colored precipitate is deposited at the site of the antibody.
- ELISAs can be used to verify the presence of viral antigens or antiviral antibodies in liquid patient specimens. In a direct sandwich ELISA, capture antibodies that specifically recognize the viral antigen are coated on the bottom of an ELISA plate with 96 wells. If the patient sample contains the virus, then it will bind to the antibodies and become immobilized to the well. Detection antibodies are conjugated to an enzyme that will cause color change when a substrate is added. An indirect ELISA uses the same principles but coats viral antigen on the plate bottom to detect antiviral antibodies in a patient sample.
- Western blots are sometimes used as a confirmatory test. They are analogous to indirect ELISAs, except the viral antigens are separated by PAGE and transferred to a nitrocellulose membrane before patient samples are added.
- Agglutination reactions use antigen- or antibody-coated latex beads. If the patient sample contains the complementary antibody or antigen, the beads will agglutinate and form a lattice clump. Some viruses naturally hemagglutinate red blood cells.
- LFIAs work like an ELISA in a stick. They rely upon a liquid patient sample traveling through a membrane and encountering antibody-coated beads that accumulate to cause a visible line.
- Plaque assays measure the amount of infectious virus in a sample. It measures the number of plaques formed by allowing a single virion to infect a cell and laterally infect neighboring cells, forming clearings that become visible when the cells are stained. To determine the PFU/mL, the number of plaques is divided by the sample dilution and the volume added to the cells.

Section 7.4 Detection of Viral Nucleic Acids

- Nucleic acid testing is a sensitive and specific way to identify viruses and viral subtypes/strains.
- PCR recapitulates DNA replication in a test tube. Following the isolation of nucleic acid from the clinical specimen, a thermocycler uses heat to separate the two DNA strands. Primers anneal to a target sequence on each strand, and Taq polymerase extends the primer to create the complementary strand. The process is repeated 30–35 times to generate billions of copies of the amplified sequence.
- Real-time PCR uses fluorescence to monitor PCR reactions in real time. It is quantitative because the rate of the reaction can be used to determine the initial starting material.
- DNA microarrays rely upon the hybridization of fluorescently tagged DNA or cDNA to oligo probes coated on a glass slide or silicon chip. These are currently used for research purposes but have great potential for viral diagnosis.
- High-throughput sequencing allows for the rapid determination of nucleotide sequences, including viral

genotypes. It generates nucleic acid sequences that can be analyzed using bioinformatics.

FLASH CARD VOCABULARY

Serology	Histology
Tissue culture	Inclusion body
Cell culture	Immunofluorescence assay
Leukocytes	Fluorescein isothiocyanate
Plasma/serum	Direct versus indirect fluorescent antibody
Direct versus indirect diagnostic methods	Staining
Biological safety cabinet	Immunohistochemistry
HEPA filter	Enzyme immunoassays/enzyme-linked immunosorbent assays
Laminar flow	
Biosafety level	Direct (sandwich) ELISA
Cytopathic effects	Indirect ELISA
Syncytium	Antiviral antibodies
Cytology	Western blot
Agglutination/hemagglutination reactions	Multiplex PCR
Lateral flow immunoassay	Real-time PCR
Plaque assay	Reporter dye and quencher
Plaque-forming units	Amplification plot
Confluency	DNA microarrays
Nucleic acid testing	Hybridization
Polymerase chain reaction	Oligonucleotides (oligos)
Thermocycler	ViroChip
Primer	High-throughput sequencing
Reverse transcriptase PCR	Bioinformatics
Amplicon	

CHAPTER REVIEW QUESTIONS

1. You are a doctor. A patient shows up in your office that appears to have shingles, which is a reactivation of varicella zoster virus, the virus that causes chickenpox. How would you suggest going about definitively diagnosing her infection?
2. In the 2009 H1N1 influenza pandemic, specimens from patients with potential influenza infections were tested to verify the influenza subtype. What types of specimens might have been collected for such purposes?
3. Describe the observable CPEs that viruses induce in cells.
4. Explain how a class II BSC works to maintain a sterile working environment.
5. You are working with samples that contain human respiratory syncytial virus, which causes coldlike symptoms in adults. Which BSL is most likely required for work with this virus?
6. You have a serum sample, and you would like to verify whether it contains antibodies against a certain virus. Which of the following assays could be used for this? Explain why each is or is not appropriate to use: Immunofluorescence, ELISA, LFIA, cell culture.
7. List the steps involved in performing an indirect ELISA and direct sandwich ELISA. What is each used to measure?
8. You perform a plaque assay with different dilutions of virus, plating 0.1 mL per well of your cells. Your 1/10,000 dilution has no plaques, your 1/1000 dilution has 59 plaques, and your 1/100 dilution has too many plaques to count—they are not distinctive. How many pfu/mL are in your undiluted sample?
9. Describe what happens at each stage of PCR. How is real-time PCR performed differently?
10. You perform qPCR on two patient samples. Both show amplification of viral DNA. One sample crosses the threshold limit at 25 cycles, and the other patient's sample crosses at 32 cycles. Which patient sample had more viral DNA in it?
11. Which type of test would be most effective in determining the entire nucleic acid sequence of a new strain of influenza virus?

FURTHER READING

Centers for Disease Control and Prevention, December 7, 2012. Infectious Diseases Pathology Branch (IDPB). http://www.cdc.gov/ncezid/dhcpp/idpb (accessed 15.06.15.).

Choo, Q.L., Kuo, G., Weiner, A.J., Overby, L.R., Bradley, D.W., Houghton, M., 1989. Isolation of a cDNA clone derived from a blood-borne non-A, non-B viral hepatitis genome. Science 244, 359–362.

Espy, M.J., Uhl, J.R., Sloan, L.M., et al., 2006. Real-time PCR in clinical microbiology: applications for routine laboratory testing. Clin. Microbiol. Rev. 19, 165–256.

Jerome, K.R., Lennette, E.H., 2010. Lennette's Laboratory Diagnosis of Viral Infections, fourth ed. Informa Healthcare USA, Inc., New York, NY.

Kumar, S., Henrickson, K.J., 2012. Update on influenza diagnostics: lessons from the novel H1N1 influenza A pandemic. Clin. Microbiol. Rev. 25, 344–361.

Mahony, J.B., 2008. Detection of respiratory viruses by molecular methods. Clin. Microbiol. Rev. 21, 716–747.

Mendelson, E., Aboudy, Y., Smetana, Z., Tepperberg, M., Grossman, Z., 2006. Laboratory assessment and diagnosis of congenital viral infections: rubella, cytomegalovirus (CMV), varicella-zoster virus (VZV), herpes simplex virus (HSV), parvovirus B19 and human immunodeficiency virus (HIV). Reprod. Toxicol. 21, 350–382.

Methods, T.A., Chevaliez, S., Rodriguez, C., Pawlotsky, J.-M., 2012. New virologic tools for management of chronic hepatitis B and C. Gastroenterology 142, 1303–1313. e1.

Miller, M.B., Tang, Y.W., 2009. Basic concepts of microarrays and potential applications in clinical microbiology. Clin. Microbiol. Rev. 22, 611–633.

Quan, P.L., Briese, T., Palacios, G., Ian Lipkin, W., 2008. Rapid sequence-based diagnosis of viral infection. Antivir. Res. 79, 1–5.

Storch, G.A., Wang, D., 2013. Diagnostic virology. In: Knipe, D.M., Howley, P.M. (Eds.), Fields Virology, sixth ed. Wolters Kluwer | Lippincott Williams and Wilkins, pp. 414–451 (Chapter 15).

Tenorio-Abreu, a, Eiros, J.M., Rodríguez, E., et al., 2010. Influenza surveillance by molecular methods. Expert Rev. Antiinfect. Ther. 8, 517–527.

United States Centers for Disease Control and Prevention and Association of Public Health Laboratories, June 27, 2014. Laboratory Testing for the Diagnosis of HIV Infection: Updated Recommendations. http://stacks.cdc.gov/view/cdc/32447 (accessed 15.06.15.).

U.S. Department of Health and Human Services, 2009. Section IV—Laboratory biosafety level criteria. In: Biosafety in Microbiological and Biomedical Laboratories, fifth ed. U.S. Department of Health and Human Services, pp. 30–59.

Wang, D., Coscoy, L., Zylberberg, M., et al., 2002. Microarray-based detection and genotyping of viral pathogens. Proc. Natl. Acad. Sci. U.S.A. 99, 15687–15692.

Wong, R., Tse, H., 2009. Lateral Flow Immunoassay. Springer, New York, NY.

Chapter 8

Vaccines, Antivirals, and the Beneficial Uses of Viruses

It is a constant evolutionary battle between humankind and the viruses that infect it. And yet, the existence of viruses is a double-edged sword. On the one hand, experts have proposed that viruses are responsible for the deaths of more people over time than all other infectious diseases combined. Viruses are still running the evolutionary race alongside humans, and some viruses are clearly ahead. On the other hand, the molecular biology revolution that we are currently experiencing has revealed the great potential for viruses to be used for advantageous purposes. The very attributes that make them difficult to overcome may be the properties that allow us to cure genetic diseases or cancer in the not-so-distant future.

This chapter discusses our efforts to prevent infection of the host through vaccination, treat the host once infection has been established, and finally, turn the tables on viruses and use them for the benefit of humankind.

8.1 VACCINE DEVELOPMENT

As discussed in Chapter 6, "The Immune Response to Viruses," the development of immunological memory is one of the most important functions of the adaptive immune system. Following primary infection, a person's immune system forms memory T lymphocytes and memory B lymphocytes that are long-lived and remain in the body at a higher frequency than naïve lymphocytes. These cells require less stimulation to become activated, and memory B cells produce higher-affinity IgG, rather than IgM. This immunological memory forms the basis for a secondary response that is faster and of greater magnitude than the primary response (see Fig. 6.13). During a secondary response, a person will experience reduced symptoms of the infection, if the individual even realizes he/she was infected at all.

Vaccination is the intentional inoculation of a person or animal with a harmless form of a pathogen. This activates the immune system much in the same way that a virulent pathogen would, and generates memory T and B cells that are prepared to initiate a memory response should the individual ever encounter the actual pathogen. Vaccination is the only way to prevent viral infection upon exposure to a virus. It goes without saying that it is important to avoid malnutrition and extreme stress—factors that affect the proper functioning of the immune system—but studies have shown that vitamins or supplements are not sufficient to consistently prevent infection. Websites that claim otherwise should be carefully scrutinized, because they often misquote scientific data—if they reference it at all—and may have an objective of selling the products they are supporting.

In the United States, both vaccines and antivirals must be tested extensively in cells and animals before they are allowed by the Food and Drug Administration (FDA) to be used in clinical trials of human volunteers. Several stages of clinical trials must statistically show that the vaccine/drug is effective and safe for its indication. All vaccines and drugs have side effects, and these are considered by the FDA before it makes a decision whether or not to approve the vaccine or drug for marketing (sale) to the general population. Clinical trials contain thousands of volunteers, yet not all side effects may initially appear during trials. By law, health care providers are required to report adverse events that occur following vaccination, and the FDA can require large-scale studies to be performed by the company when the drug is available for use in the public at large.

8.1.1 A Brief History of Vaccination

The history of vaccination begins with variola virus, the cause of smallpox, which will be discussed in detail in Chapter 15, "Poxviruses." Smallpox claimed the lives of around 30% of those it infected—a total of over 500 million people in the 20th century alone. Attempts to thwart infection trace back to ancient China, where pulverized dried smallpox scabs were inhaled or injected into uninfected persons in a process known as **variolation** (Fig. 8.1A). This led to a milder form of the disease, but the resulting infection still resulted in a 2–3% case fatality rate. Infected people could also transmit the virus to others, but in the absence of a better solution, the practice of variolation spread throughout China, India, and Africa, before being introduced into Europe and the Americas in the 1700s.

A notable advancement against smallpox occurred in 1774 in England, when in an act of desperation, farmer

Essential Human Virology.

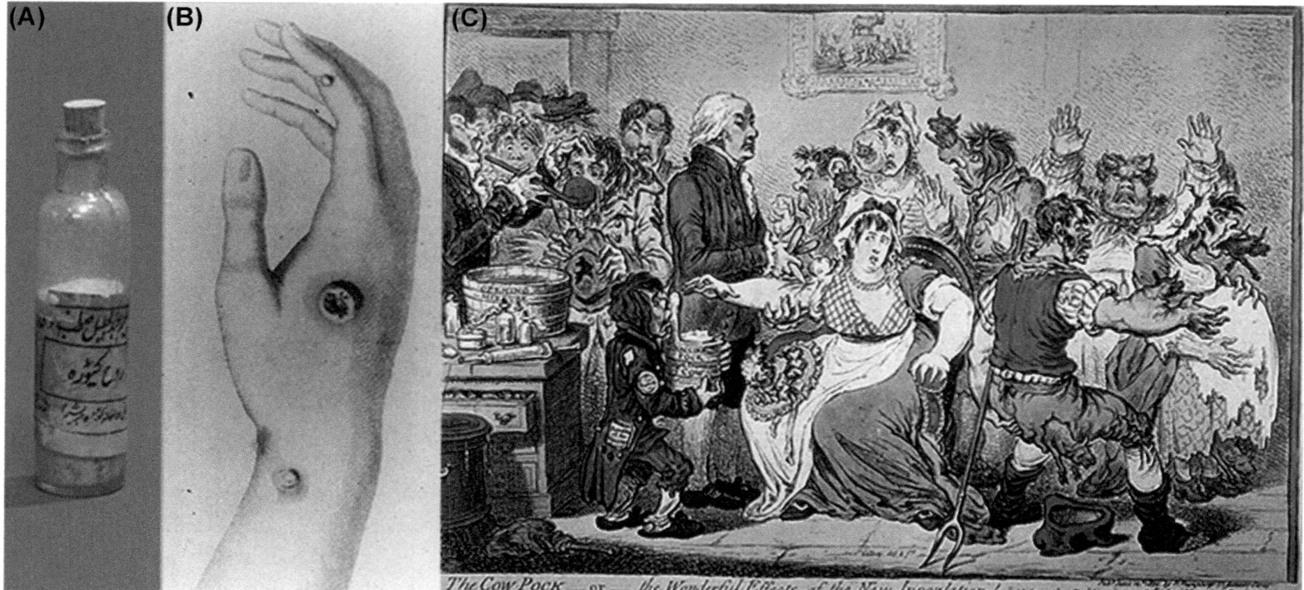

FIGURE 8.1 **Smallpox variolation and vaccination.** (A) A variolation vial, thought to be one of the last dosages from India. *(Photo courtesy of CDC/Brian Holloway.)* (B) A cowpox lesion on the hand of milkmaid Sarah Nelmes, used as a source of Edward Jenner's first vaccine material. *(From Jenner, E., 1923. An inquiry into the causes and effects of the variolae vaccinae.)* (C) A color etching by James Gillray, 1802. This satire piece shows Edward Jenner vaccinating a woman against smallpox and the claims alleged by the antivaccine society that receiving the cowpox vaccine induces the sprouting of various cow parts.

Benjamin Jesty used material from cowpox lesions to successfully prevent his family from contracting smallpox. At the time, country doctors were also taking note that milkmaids who contracted cowpox from the cows they were milking seemed to be protected from smallpox. In 1796, country doctor Edward Jenner made the same observation and came up with an idea to test it. He collected the fluid from a cowpox sore on the hand of milkmaid Sarah Nelmes (Fig. 8.1B) and injected it into a young boy, James Phipps. When later exposed to people with smallpox, James Phipps appeared to be protected from contracting the virus. Jenner initiated larger-scale tests, and found them to be successful in preventing infection. Jenner promoted the procedure, and vaccination gained widespread support over the following 100 years—although there was no shortage of critics and those concerned about the ramifications of transferring animal material into humans (Fig. 8.1C). Although likely not the first to attempt vaccination, Jenner receives deserved credit for vaccination by promoting the widespread use of his **vaccine**, a word derived from the Latin word *vacca*, meaning "cow."

You will recall from Chapter 1, "The World of Viruses," that germ theory only gained acceptance in the mid- to late-1800s, and viruses were first discovered in 1898. As such, Edward Jenner had no knowledge that infectious diseases were caused by microscopic biological entities but tested a hypothesis based upon the presentation of the disease. In 1885, Louis Pasteur created the second vaccine against viruses, although he did not know of the existence of viruses at the time, either. Pasteur and colleagues had discovered that the causative agent of rabies could be transmitted from dog to dog by transferring spinal cord or brain tissue from an infected dog into an uninfected dog. They infected rabbits with the infected tissue and removed their spinal cords, which were dried and used as a successful vaccine in dogs. In July, a 9-year old child, Joseph Meister, was bitten 14 times by a rabid dog. Knowing the fatal consequences of this, Pasteur reluctantly convinced physicians to administer his rabies vaccine to the child. The child lived, and within the following year Pasteur had successfully treated 350 people with his vaccine—only one child died who had been bitten 6 days before the treatment. Although rudimentary in their manufacturing, these two vaccines provided the foundation and scientific proof that vaccination could prevent the occurrence of infectious diseases.

8.1.2 Types of Current Vaccines

As performed with smallpox and rabies, the first vaccines were created from infected animal tissues as this was the only known method to propagate viruses at the time. In 1931, Alice Miles Woodruff and Ernest Goodpasture discovered that embryonated hen's eggs (*eggs containing a developing embryo*) would support the growth of fowlpox virus, and soon other viruses were found to be cultivable in embryonated eggs. This became the basis for creating vaccines against yellow fever virus in 1935 and influenza virus in 1945. After the discovery of antibiotics in 1928 and development of standardized cell growth media in 1959, cell culture replaced the use of eggs for propagation of virus. This has remained the standard method of propagating virus for vaccine use, although embryonated eggs are still used to grow virus for the influenza vaccine (Fig. 8.2)—which is why those with egg allergies are discouraged from taking the vaccine. New molecular biology techniques have also facilitated the creation of new ways to effectively create vaccines. Currently, 14 immunizations are recommended for children under 18; of these, 8 are for viral infections (Table 8.1).

To be effective, vaccines must be **immunogenic**: able to stimulate an immune response that establishes immunological memory. Vaccines are most effective when they induce the activation of both the innate and adaptive arms of the immune system. For protection against infection, most vaccines ultimately rely upon the production of antibodies, known as the **humoral response**, since antibodies can bind to the envelope or capsid proteins of viruses to prevent attachment to host cells. On the other hand, the immune response against certain viruses requires T cell activity, known as the **cell-mediated response**, to prevent infection. Knowledge of the virus, its pathogenicity, and how the immune system effectively responds against it is extremely helpful in designing an effective vaccine. Other factors can influence the immune response against vaccination, too, including how the vaccine is delivered, the type and dose of antigen within the vaccine, and whether a mother's antibodies are still circulating within the vaccinated child and could bind to vaccine antigens, thereby lessening the child's own response against them. Many vaccines also contain an **adjuvant** that boosts the immunogenicity of the vaccine (adjuvant is from the Latin word *adjuvare*, meaning "to help"). Some adjuvants are molecules that sequester the antigen within the tissue for longer, thereby allowing more time for it to be recognized and endocytosed by the dendritic cells that will present it to T cells. Other adjuvants are composed of molecules that can stimulate pattern recognition receptors. As discussed in Chapter 6, "The Immune Response to Viruses," activation of pattern recognition receptors on innate immune cells leads to the production of cytokines that customize the type of T cell response that ensues.

FIGURE 8.2 **Cultivating influenza virus for vaccine manufacturing.** This laboratory technician is using a Class II BSC to inoculate embryonated eggs, where the virus replicates before being harvested and purified for the influenza vaccine. *Photo courtesy of the FDA.*

There are three different formulations of viral vaccines that are currently being used and several that are still in experimental stages (Table 8.2). **Live attenuated virus**

TABLE 8.1 Recommended Viral Immunizations for People Aged 0 Through 18 Years—United States, 2015

Vaccine	Type	Schedule of Immunizations
Hepatitis B	Recombinant subunit	First dose at birth, second dose between 1 and 2 months, third dose between 6 and 18 months
Rotavirus	Live attenuated	First dose at 2 months, second dose at 4 months
Poliovirus	Inactivated	First dose at 2 months, second dose at 4 months, third dose between 6 and 18 months, fourth dose 4–6 years
Influenza	Subunit	Annually for seasonal influenza
Measles, mumps, rubella	Live attenuated	First dose at 12–15 months, second dose at 4–6 years
Varicella	Live attenuated	First dose at 12–15 months, second dose at 4–6 years
Hepatitis A	Inactivated	2 doses spaced 6–18 months apart starting at 1st year
Human papillomavirus	Recombinant subunit	First dose at 11–12 years, second dose 1–2 months later, third dose 6 months later

From United States Centers for Disease Control and Prevention, 2015. Appendix A. In: Epidemiology and Prevention of Vaccine-Preventable Diseases, thirteenth ed. Public Health Foundation, Washington, DC.

TABLE 8.2 Types of Viral Vaccines Used in Humans

Vaccine Type	Examples	Advantages	Disadvantages
Live attenuated	Measles, mumps, rubella, vaccinia (for smallpox), varicella, zoster, yellow fever virus, rotavirus, influenza (intranasal)	Highly immunogenic, relatively easy to create, well-established process, cheaper	May revert to virulent virus in immunocompromised, needs to be refrigerated
Inactivated	Poliovirus, hepatitis A virus, rabies (human), Japanese encephalitis virus, seasonal influenza (purified subunit)	More stable, can be lyophilized for easy transport, cheaper, can be used in immunocompromised	Often requires booster shots, must ensure proper inactivation, not all viruses are immunogenic after inactivation
Recombinant subunit	Hepatitis B virus, human papillomavirus	Can be created for viruses that do not propagate well in the laboratory, no chance of live virus reversion	Requires specialized expertise to create, is not as immunogenic as whole virus preparations, expensive, more difficult to produce
DNA vaccine (experimental)	Influenza, HIV, West Nile virus, herpesviruses	Does not require live virus, can be created for viruses that do not propagate well in the laboratory, no chance of viral reversion	Not as immunogenic as whole virus preparations, requires expertise and fine-tuning to create, difficulties in delivery, may require prime-boost strategy
Recombinant vector (experimental)	Rabies (approved for animal use), HIV, West Nile virus, measles, Ebola virus, avian influenzas	Can be created for viruses that do not propagate well in the laboratory, viral vector strains are well characterized	Viral vector can cause immune response, requires expertise to create, difficulties in delivery

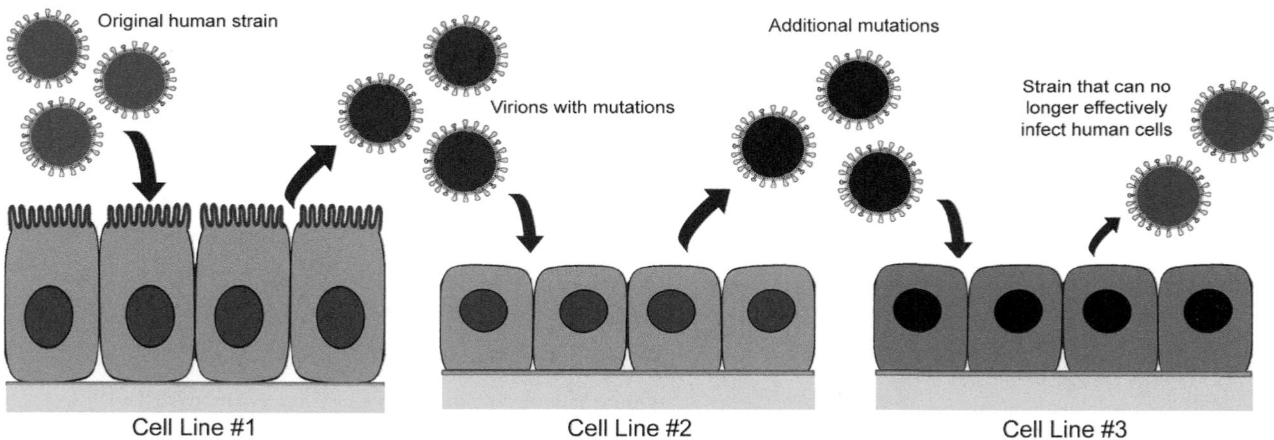

FIGURE 8.3 **Virus attenuation.** Live attenuated virus vaccines reply upon the attenuation, or weakening, of a viral strain. One way of doing this is to passage the initial virus in cell lines that are progressively different than the normal target human cells. Because of the sheer number of virions created during replication, some virions will have point mutations that allow them to replicate better in the new cell line. Eventually, a strain of the virus is created that does not effectively replicate within the human cells. Upon infection, the immune system quickly controls the infection, creating memory cells and immunity in the process.

vaccines are the most commonly used for immunizations. As the name suggests, this type of vaccine is composed of infectious virus, but the virus has been attenuated, or weakened, so that it no longer replicates efficiently in humans. One way of doing this is by repeatedly growing—or **passaging**—the virus in nonhuman cells, or cells for which the virus does not have optimal tropism (Fig. 8.3). Because so many virions are made during infection, mutations occur and create slightly different genetic versions of the virus. The ones that are better able to replicate in the new cell line will replicate faster and produce a greater proportion of progeny virions. After many passages, the resulting virus strain has genetically evolved to replicate in the new cell line rather than in the normal human cellular targets. The strain still produces an immunological response in humans, but the virus can no longer replicate well in humans and so the immune system quickly controls the inapparent infection. For instance, the Oka strain of varicella zoster virus used in the vaccines against chickenpox or shingles was attenuated by passaging the human strain 11 times in human embryonic lung cells, 12 times in guinea pig fibroblasts, and 8 times in two other strains of human fibroblasts for a total of 31 passages. Other viruses are attenuated by passaging them in human cells at a nonoptimal temperature. For example, the attenuated rubella virus used in the measles–mumps–rubella (MMR) vaccine has adapted to replicate best at 30°C (86°F), whereas human body temperature is 37°C (98.6°F).

A general rule of thumb for vaccine development is that the more similar a vaccine is to the pathogenic virus, the better the immunological response will be. As such, live attenuated virus vaccines generally stimulate strong T and B cells responses that lead to long-lasting immunity. For instance, measles antibodies develop in 95% of children vaccinated at 12 months of age and 98% of children vaccinated at 15 months of age with a single dose of the vaccine; 99% of people who receive two doses of the vaccine develop lifelong immunity.

Despite that live attenuated vaccines are considered to produce the most biologically similar immunity, there are important limitations that must be considered. Live attenuated vaccines usually need to be refrigerated to remain immunogenic, which may be of concern in countries that are unable to provide constant refrigeration. It also means that care must be taken during the shipment and storage of the vaccine to prevent its inactivation. Another major limitation of live attenuated virus vaccines is that, being a replicating virus, the possibility always exists that the attenuated virus may revert back to a wild-type strain that is fully virulent. This is of particular concern in immunocompromised individuals, for whom most live attenuated vaccines are not recommended. Without a fully functional immune system, the virus may replicate enough to mutate back to a strain that effectively causes illness. This is the reason that the live attenuated polio vaccine was discontinued in the United States. In the 20 years prior to its discontinuation, 154 cases of vaccine-associated paralytic polio (VAPP) were reported as a result of reversion.

In these situations, **inactivated virus vaccines** are a better choice. In these preparations, the virus is completely inactivated by high heat or low amounts of formaldehyde. Although they do not contain live virus and therefore generate less of an immune response, they can be used in people without fear of reversion to a wild-type strain. "Booster" shots of inactivated vaccine help to maintain immunity against the virus. Since the discontinuation of the live attenuated polio vaccine in 2000, the inactivated polio vaccine has been administered exclusively. It is a mixture of all three serotypes of polio virus, inactivated with formaldehyde. Although 99% of individuals are immune following three doses of the vaccine, it produces less gastrointestinal immunity than the live vaccine. As such, those who receive the inactivated polio vaccine are more readily infected with wild poliovirus strains.

A drawback of inactivated virus vaccines is that some viruses are not immunogenic following inactivation. An inactivated mumps vaccine only produced short-lived immunity, and an inactivated respiratory syncytial virus vaccine caused worse symptoms than actual infection when it was tested in a group of children, resulting in the deaths of two toddlers who had received it. (Upon inactivation with formalin, the immunogenicity of the antigen was altered to produce a proinflammatory immune response in the lung.) However, most inactivated vaccines do not require refrigeration and can be easily transported in a **lyophilized** (freeze-dried) form, which makes distribution easier in areas of the world without stable refrigeration.

Study Break

Compare the advantages and disadvantages of live attenuated virus vaccines and inactivated virus vaccines.

The great majority of vaccines that are currently administered or have existed in the past were based upon live attenuated viruses or inactivated virus. As a result of these vaccine formulations, one virus has been eradicated (variola virus, discussed in Chapter 15, "Poxviruses"), and the morbidity associated with several common childhood viruses has dropped precipitously (Fig. 8.4 and Table 8.3). However, a major limitation of both vaccine types is that the virus must be able to replicate in our current systems, and there are several clinically important viruses that we are still unable to propagate well in the laboratory. For these viruses, researchers have employed new biotechnology techniques to create effective vaccines. Specifically, they have made use of **recombinant protein expression** systems (Fig. 8.5).

Recombinant DNA refers to DNA from one organism that is placed into another organism. Most often, this is done to create medications, like human insulin for diabetics. In this process, the human gene for insulin is inserted into a small circular piece of DNA called a **plasmid**. This is known as **gene cloning**, and most often the cDNA sequence is used because it lacks introns (see Chapter 3, "Features of Host Cells: Cellular and Molecular Biology Review," for a review of introns). Plasmids are found normally in bacteria and yeast, and the genes contained on the plasmids are transcribed into mRNA and translated into proteins. When the modified gene-containing plasmid is inserted into bacteria, the bacteria transcribe and translate the gene as if it were their own, creating recombinant insulin that is purified, bottled, and prescribed to diabetics. The insulin protein used to be isolated from the pancreases of pigs for diabetics, but the pig insulin was recognized

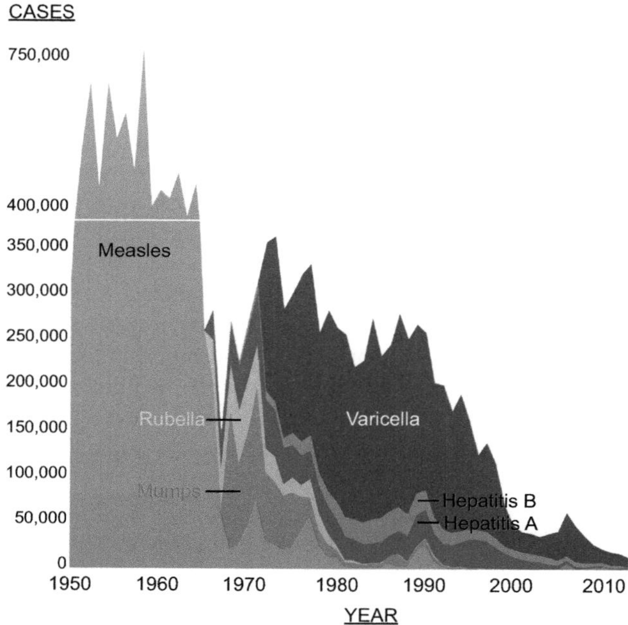

FIGURE 8.4 **Reduction in serious childhood diseases due to vaccination.** Childhood viral diseases and their consequences have been seriously reduced due to the development of vaccines. Shown here are the reductions in the number of cases of measles, mumps, rubella, varicella, hepatitis B, and hepatitis A viruses since 1950. *Data taken from Center for Disease Control and Prevention (CDC), 2015. Appendix E: Reported Cases and Deaths from Vaccine Preventable Diseases, United States, 1950–2013. In: Haborsky, J., Kroger, A., Wolfe, S. (Eds.), Epidemiology and Prevention of Vaccine-Preventable Diseases, thirteenth ed. Public Health Foundation, Washington, DC.*

TABLE 8.3 Reductions in Childhood Morbidity Due to Vaccination

Disease	20th Century Annual Morbidity	Cases Reported in 2013	Percent Decrease (%)
Smallpox	29,005	0	100
Polio	16,316	1	>99
Measles	530,217	187	>99
Mumps	162,334	584	>99
Rubella	47,745	9	>99
	Prevaccine Era Annual Estimate	2013 Estimate	Percent Decrease (%)
Hepatitis A	117,333	2890	98
Hepatitis B	66,232	18,800	72
Rotavirus	62,500	12,500	80
Varicella	4,085,120	167,490	96

From United States Centers for Disease Control and Prevention, 2015. Appendix A. In: Epidemiology and Prevention of Vaccine-Preventable Diseases, thirteenth ed. Public Health Foundation, Washington, DC.

by the human immune system as a foreign antigen and eventually did not work well. Now, biotechnology has facilitated the manufacturing of pure, human insulin using recombinant protein expression systems.

The same process can be used to create viral proteins, which can be purified and injected as a vaccine. Hepatitis B virus (HBV) is one of the viral causes of **hepatitis**, or inflammation of the liver. Chronic HBV infection can lead to cirrhosis (scarring of the liver), and the virus is responsible for 80% of the cases of hepatocellular carcinoma, the second most common cause of cancer death in the world. HBV does not replicate well in cell culture, however, and so recombinant protein expression systems were used to create a vaccine against the virus. HBV encodes seven viral proteins, one of which is displayed on the surface of the virion, called the hepatitis B surface antigen (HBsAg). Antibodies against HBsAg are sufficient to provide immunity, and so this was the target for vaccine development. The HBsAg gene sequence was identified through sequencing, and the recombinant HBsAg was produced by inserting the gene for HBsAg into a plasmid that was introduced into *Saccharomyces cerevisiae*, common baker's yeast (Fig. 8.5). Just as bacteria do with insulin, the yeast produce the recombinant HBsAg protein, which is harvested, purified, and injected with an adjuvant as the vaccine preparation. This type of vaccine is known as a **recombinant subunit vaccine**: it is composed of only certain viral proteins—not the entire virus, as with live attenuated or inactivated vaccines—and it is created using recombinant DNA technology. The challenge is determining which and how many viral proteins should be included in the vaccine formulation to ensure adequate immunity against the virus. Currently, the HBV and human papillomavirus (HPV) vaccines are created in this manner.

Subunit vaccines do not exclusively have to be created through recombinant protein expression. If the virus is able to be propagated in tissues or cell culture, then chemicals can be used to inactivate the virus, and only certain immunogenic viral antigens are isolated and injected as the vaccine. The most common version of the yearly flu vaccine propagates the virus in embryonated eggs, inactivates the virus, and purifies the hemagglutinin (HA) proteins. The yearly flu vaccine contains the HA proteins from at least three different influenza strains, the identity of which are selected each year based upon the viruses that are anticipated to circulate.

As with the other types of vaccines, subunit vaccines have advantages and disadvantages. The greatest advantage of recombinant subunit vaccines is that the virus does not need to be successfully propagated to create the vaccine. As long as the DNA sequences of the immunogenic antigens are known—and the genomes of most common pathogenic human viruses have all been sequenced—then a recombinant subunit vaccine can be developed. Without a live virus present, there is also no concern of infection with subunit vaccines.

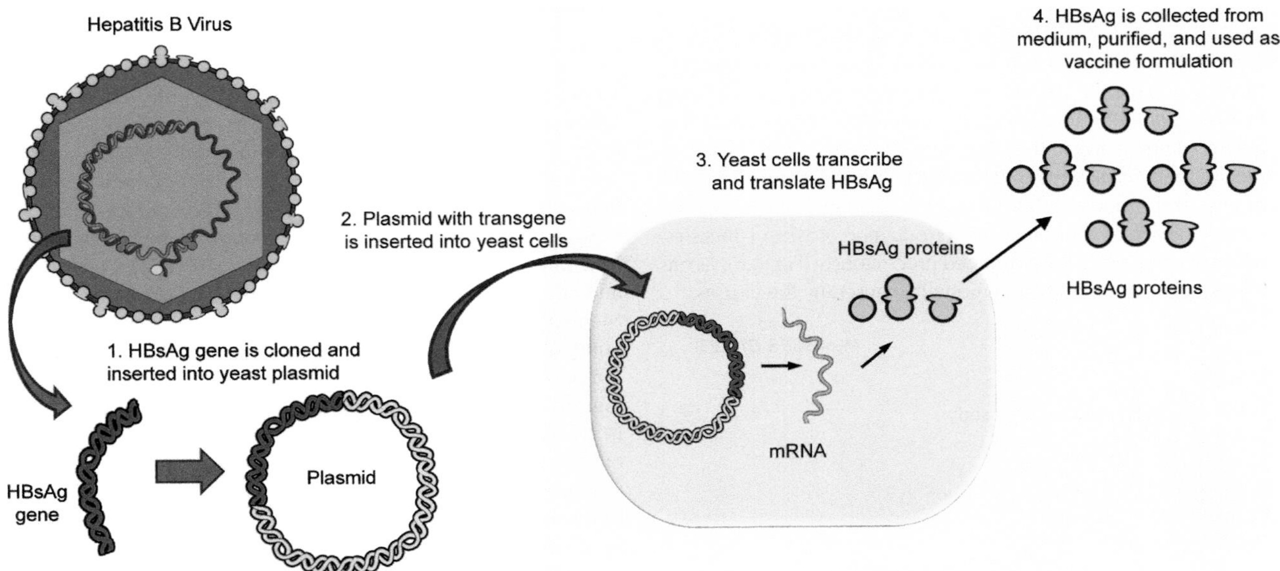

FIGURE 8.5 **Generation of recombinant subunit vaccines.** In the creation of the HBV vaccine, recombinant DNA technology is used to insert the HBsAg gene into a yeast plasmid, which is transferred into yeast cells. The yeast transcribes the gene into mRNA that is translated into the HBsAg protein and secreted. The HBsAg protein is harvested, purified, and injected as the HBV vaccine. Because only a portion of the virus is used for immunization, it is known as a subunit vaccine.

Although several different protein antigens can be included within a subunit vaccine, this type of immunization frequently requires booster shots to achieve or maintain immunity. Without the entire virus present, the immune response is not as robustly developed. In addition, because this type of vaccine preparation has to be performed in a laboratory using living bacterial or yeast cells, it is very expensive and requires expertise in recombinant DNA and protein technologies.

8.1.3 Experimental Vaccines

Other types of vaccines are currently in development that rely upon biotechnology and the manipulation of DNA or biological systems. **DNA vaccines** bypass the production of protein antigens by delivering viral DNA directly into the cells of an individual. Within the cell, the DNA is transcribed into mRNA and translated into viral proteins to which the host immune system responds. DNA vaccines are created in the same way that recombinant DNA plasmids are created: the viral genes of interest are inserted into a plasmid under the control of a promoter that induces their transcription once in human cells. The DNA can be injected into the muscle of a person or delivered using a "gene gun" that coats minute gold beads with the DNA and uses compressed air to "shoot" them into the skin cells of an individual. This process is known as **particle bombardment** and requires direct contact with the tissue of interest because the injection only penetrates a few layers of skin. DNA vaccines can also be delivered with a needle-free injector that uses compressed air to force the liquid through the skin, or by **electroporation**, the delivery of an electric pulse, to create temporary pores in the plasma membrane through which the DNA can pass.

Like recombinant subunit vaccines, DNA vaccines could theoretically be created for viruses that are unable to be propagated in cell culture, or for viruses that would be dangerous to propagate in cells. However, the immunological response to the virus needs to be understood in order to know which genes should be included on the injected plasmid. Other technical issues also need to be addressed, such as which promoter should be used for the viral genes so that a sufficient amount of viral protein is produced once in the human cell.

There is no DNA vaccine that has yet been approved by the FDA for use in humans, although clinical trials are taking place for human DNA vaccines against West Nile virus (WNV), influenza virus, HPV, HBV, HIV, and dengue virus, among several others. However, a DNA vaccine against WNV has been created for use in horses. The premembrane (prM) and envelope (E) genes of WNV were inserted onto a plasmid that was delivered intramuscularly to horses in two doses. The vaccine demonstrated protection against WNV and was the first DNA vaccine ever approved by the FDA. It is no longer on the market because it was discontinued following the acquisition of its manufacturer, Fort Dodge Animal Health, by Pfizer, another pharmaceutical company.

A very similar human DNA vaccine is currently in clinical trials. It also contains the E protein, which is the major surface protein that is used by the virus for attachment and

In-Depth Look: The MMR and Autism Controversy

The MMR vaccine is a combination of live attenuated measles, mumps, and rubella viruses. The triple virus vaccine was licensed in 1971, and a formulation is now available, called MMRV, that also contains a live attenuated varicella virus strain. Children receive their first MMR immunization between 12 and 15 months of age, and a booster shot of vaccine is administered between 4 and 6 years of age. Since its introduction, cases of measles, mumps, and rubella have dropped precipitously (Fig. 8.6), as has the associated morbidity and mortality caused by the viruses.

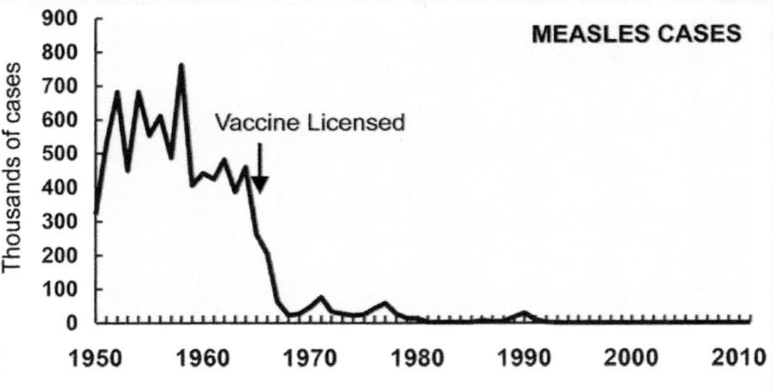

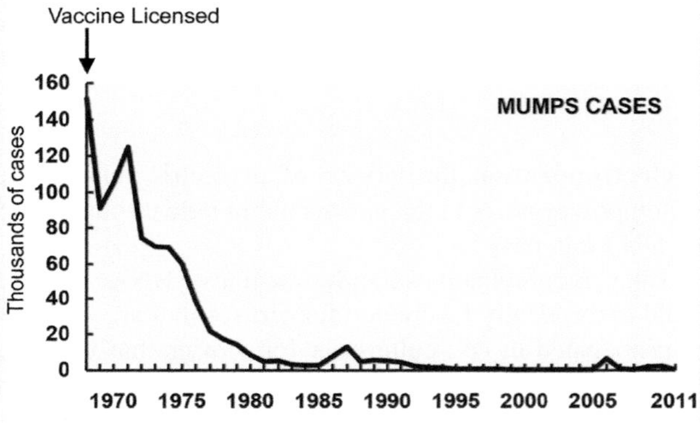

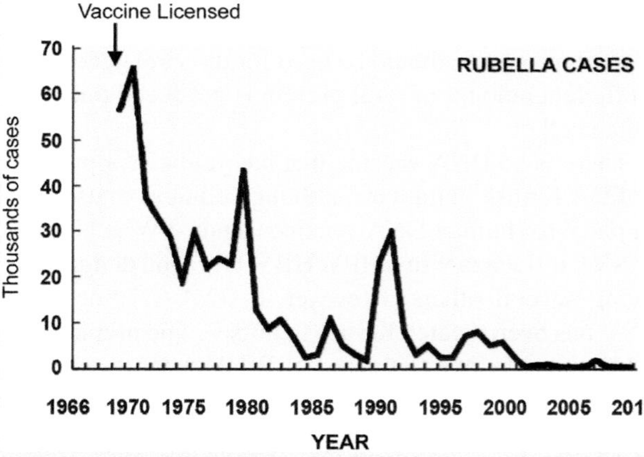

FIGURE 8.6 **Reductions in measles, mumps, and rubella cases following vaccine licensure.** Vaccines have significantly reduced the cases of measles, mumps, and rubella cases. The MMR triple virus formulation was licensed in 1971 and has been the basis of providing immunity against these three viruses for the past 45 years. *From Center for Disease Control and Prevention, 2015. In: Haborsky, J., Kroger, A., Wolfe, S. (Eds.), Epidemiology and Prevention of Vaccine-Preventable Diseases, thirteenth ed. Public Health Foundation, Washington, DC.*

The vaccine has caused much controversy in the past 15 years due to a research paper that was published in 1998. This paper, led by former doctor Andrew Wakefield, claimed to have found a link between the triple virus MMR vaccine and **autism**, a spectrum of childhood developmental disorders. He later recommended at a press conference that the viruses within the MMR vaccine should be given individually, rather than all at once. The study had serious technical and ethical issues, and an investigation by the British General Medical Council (GMC) found in 2010 that Wakefield was guilty of several wrongdoings in the research he conducted for the paper. They found he had subjected 11 children to a barrage of unnecessary invasive tests, such as lumbar punctures in the spine and colonoscopies, without obtaining required ethical approvals for doing so. They also found Wakefield had a financial interest in the results of the study, as he had been hired to advise a lawyer representing parents who believed their children had been harmed by the MMR vaccine. In addition, Wakefield had filed a patent the year before for a single virus measles vaccine, and at the time of the research publication, was setting up a company to manufacture and sell it. Data from the paper had also been deliberately falsified, and the journal that had published the paper, *The Lancet*, retracted the paper. The journal's medical editor, Richard Horton, spoke of the paper, saying that "it was utterly clear, without any ambiguity at all, that the statements in the paper were utterly false." Wakefield was removed by the GMC from its list of registered medical doctors.

Since that time, many carefully performed, ethically conducted scientific studies have found no link between the MMR vaccine and autism, including a study that followed over 500,000 children over 7 years and found no association between the MMR vaccine and autism. The American Academy of Pediatrics agrees that the MMR vaccine is not responsible for recent increases in the cases of autism in children. There has also been no link found between autism and thimerosal, a mercury-based preservative used to prevent contamination of multidose vials of vaccines. Thimerosal was removed or reduced to trace amounts in all childhood vaccines as a precautionary measure before the research had been completed.

As a result of this fallacious report, these viruses that were nearly eradicated from the country are now reappearing in outbreaks because of reduced vaccination rates and subsequently ineffective herd immunity. The CDC reports that 140 measles cases were reported in the United States in 2008, imported from travelers from countries with active outbreaks, and 91% of those infected were unvaccinated because of personal or religious beliefs or of unknown vaccination status. They concluded the increase in measles cases was due to the increased transmission of measles after it was imported, not a greater number of imported cases. It is difficult to correct an erroneous statement once it is released into the public, especially with support of antivaccine groups, but it is important to continue protecting children from potentially deadly infectious diseases while keeping the focus on the research being performed to find a data-substantiated cause of autism.

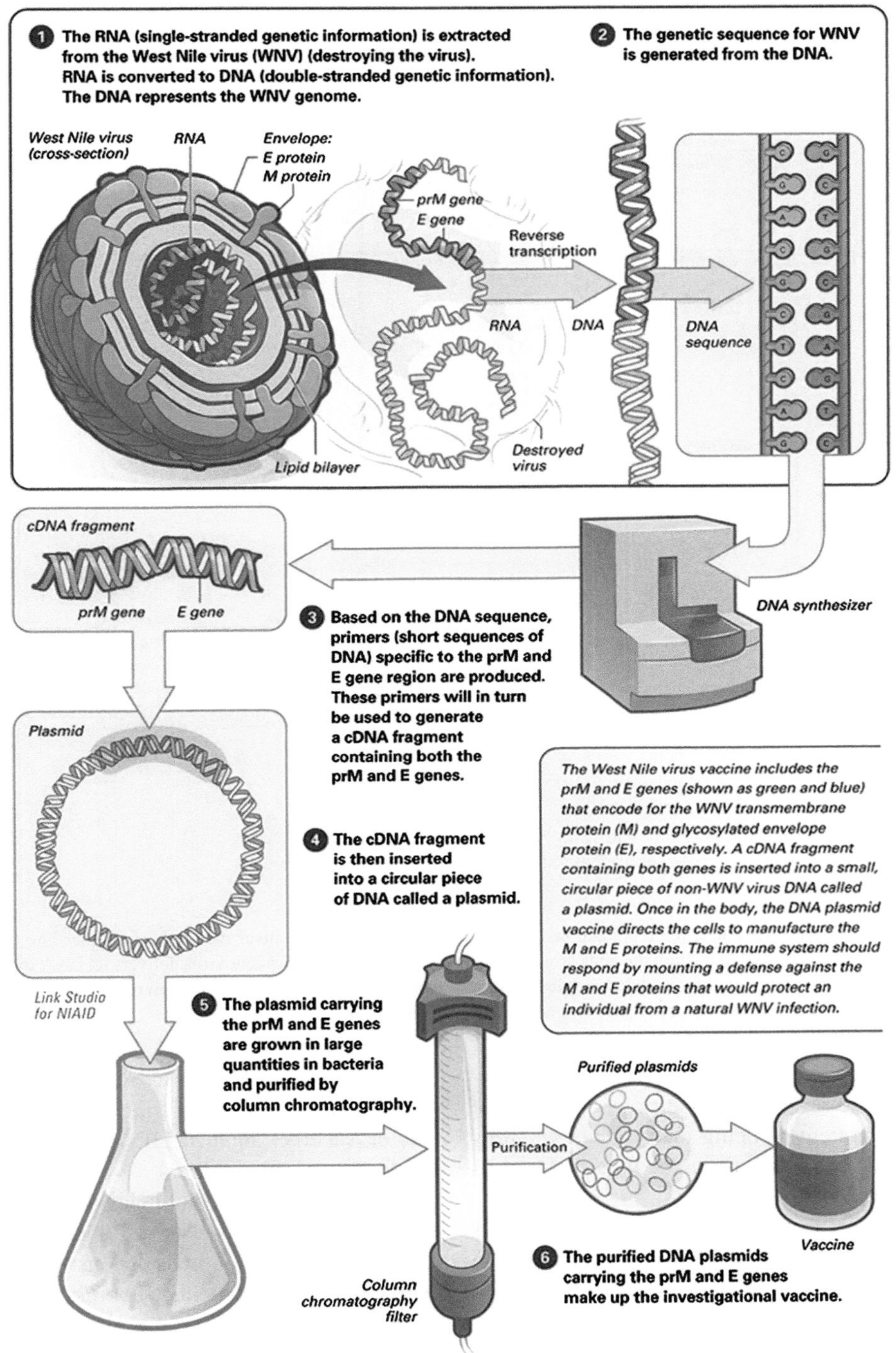

1 The RNA (single-stranded genetic information) is extracted from the West Nile virus (WNV) (destroying the virus). RNA is converted to DNA (double-stranded genetic information). The DNA represents the WNV genome.
2 The genetic sequence for WNV is generated from the DNA.
West Nile virus (cross-section)
RNA
Envelope: E protein M protein
prM gene
E gene
Reverse transcription
RNA
DNA
DNA sequence
Destroyed virus
Lipid bilayer
cDNA fragment
prM gene
E gene
DNA synthesizer
3 Based on the DNA sequence, primers (short sequences of DNA) specific to the prM and E gene region are produced. These primers will in turn be used to generate a cDNA fragment containing both the prM and E genes.
Plasmid
4 The cDNA fragment is then inserted into a circular piece of DNA called a plasmid.
The West Nile virus vaccine includes the prM and E genes (shown as green and blue) that encode for the WNV transmembrane protein (M) and glycosylated envelope protein (E), respectively. A cDNA fragment containing both genes is inserted into a small, circular piece of non-WNV virus DNA called a plasmid. Once in the body, the DNA plasmid vaccine directs the cells to manufacture the M and E proteins. The immune system should respond by mounting a defense against the M and E proteins that would protect an individual from a natural WNV infection.
Link Studio for NIAID
5 The plasmid carrying the prM and E genes are grown in large quantities in bacteria and purified by column chromatography.
Purified plasmids
Purification
Vaccine
6 The purified DNA plasmids carrying the prM and E genes make up the investigational vaccine.
Column chromatography filter

FIGURE 8.7 **DNA vaccines.** This graphic shows how DNA vaccines are created, using the experimental West Nile virus vaccine as an example. The viral RNA corresponding the E and prM genes is identified and cloned into a plasmid that is copied to large numbers by bacteria. The plasmid is purified and injected as the vaccine. In host cells, the viral genes are transcribed and translated, producing viral proteins to which the immune system responds. *Image courtesy of the National Institute of Allergy and Infectious Diseases.*

fusion, and the prM protein, which is cleaved into a membrane protein during the maturation of the virus (Fig. 8.7). In the DNA vaccine, these two genes are placed under the control of a cytomegalovirus promoter and a regulatory element from the human T-lymphotropic virus-I to ensure their transcription within cells.

Despite promising results in animals and their reduced cost compared to recombinant subunit vaccines, a major

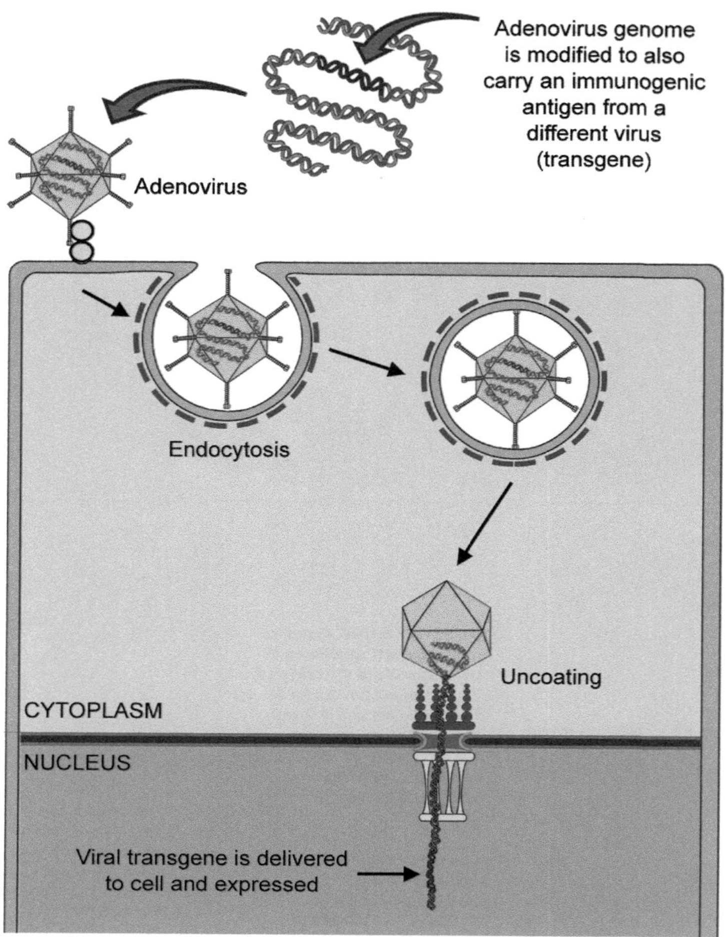

FIGURE 8.8 **Recombinant vector vaccines.** Recombinant vector vaccines use one virus to deliver the DNA of another into the cell. In this example, the adenovirus genome has been modified to contain the gene from another virus. Upon infection, adenovirus delivers its DNA into the nucleus of the cell. The viral transgene will also be transcribed and translated, producing viral proteins to which the immune system responds. Recombinant vector vaccines rely upon a similar process as DNA vaccines, except that a virus delivers the transgene, not a plasmid.

limitation of DNA vaccines has been their inability to generate a protective immune response on their own. Because of this, a **prime-boost** strategy is being investigated whereby the DNA vaccine is followed by a vaccine that uses a virus to deliver the same or different antigens.

A **recombinant vector vaccine** uses viral **vectors** as carriers to deliver DNA. In this procedure, the DNA of an attenuated virus—or a virus rendered harmless through modification of its genome—is modified to include DNA that encodes the viral antigen of interest (Fig. 8.8). It is related to the creation of a DNA vaccine, except that instead of modifying plasmid DNA, the vector DNA is modified to include the vaccine antigen, and the virus delivers the DNA to the cell instead of a gene gun or needle-free injection device. Commonly used attenuated viral vectors include adenovirus strains that have been used for military immunizations; poxviruses, such as canarypox or an attenuated version of the vaccinia virus used for the smallpox vaccine; and a specific type of retrovirus known as a lentivirus, primarily derived from an HIV backbone stripped of any virulence genes. There are clinical trials ongoing to test recombinant vector vaccines against HIV, avian H5N1 influenza, and Ebola virus, among others.

Although no recombinant vector vaccine has been approved for human use, one has been successfully used to vaccinate wild animals against rabies. This vaccine is a live vaccinia vector that expresses the rabies virus glycoprotein and is delivered orally. The vaccine is enclosed within a plastic wrapper that looks like a ketchup packet and wrapped in food that would attract the animal of interest—such as fishmeal and oil for raccoons—to create a bait unit (Fig. 8.9). The easiest way to distribute these is to drop them out of helicopters or low-flying planes. The goal is for the animal to bite into the bait, which releases the vaccine onto the food that the animal ingests. As the animal becomes infected with the rabies glycoprotein–expressing virus, its immune system would respond and provide immunity against subsequent infections with the actual rabies virus.

FIGURE 8.9 **Two examples of live rabies baits for immunizing wildlife.** Live recombinant vector vaccines that use adenovirus or vaccinia virus to deliver rabies antigens are used to immunize wildlife against rabies. Sachets containing liquid virus are coated with fishmeal (left) or inserted into fishmeal blocks (right) so that animals will find and bite into them, thereby ingesting the virus. *Photo by John Forbes, USDA-APHIS-Wildlife Services.*

In 2013, 92% of reported rabid animals were wildlife, so control of rabies in the wildlife population is an important task. Animal bites of humans by rabid animals can transmit the disease, which is always fatal without prompt intervention. A human vaccine against rabies exists, but is only given to those that may have been infected with the virus in an attempt to boost the person's immune response against it. Because rabies transmission to humans occurs very infrequently, the widespread use of the rabies vaccine in the general population is not warranted. However, people who are exposed to potentially rabid wildlife because of their occupation may receive the vaccine preventatively.

Recombinant vector vaccines hold great potential for immunizations. It is challenging to deliver DNA into cells, and even more so to deliver DNA to specific types of cells. Viral vectors can be modified to contain the virus attachment proteins that will allow them to specifically bind to cells of interest. They can also be modified to create **replication-defective viruses** that can infect cells but are unable to replicate within them, thereby preventing the generation of additional infectious vector virus. It is also a big advantage that several viral vectors have been used for human immunizations so we already know they are safe in humans and can be propagated on a large scale. On the other hand, recombinant vector vaccines are very expensive, and the very proteins that allow the vector virus to enter cells are those that are usually immunogenic, so the immune system responds to both the vector virus and the vaccine antigen. This can be problematic if booster shots of the vaccine are necessary but the immune system has protective immunity against the vector virus. The use of recombinant viral vectors for other purposes will be discussed later in this chapter, in Section 8.3.

Taken together, there are a variety of effective techniques to create vaccines against human viruses, and novel advancements hold promise for the development of future vaccines. However, this does not mean that it is easy to create a vaccine for every virus. HIV was discovered in 1983, and over 30 years later, scientists have yet to be able to create a safe and efficacious vaccine to prevent against its infection. There have been many challenges encountered along the way, which will be discussed more in Chapter 11, "Human Immunodeficiency Virus," after the biology of the virus has been discussed.

Study Break
What are the major challenges of using a live recombinant vector vaccine?

8.1.4 Passive Immunity

The types of vaccinations described above all provide **active immunity**, meaning that the person's own immune system responds to the pathogen by generating memory cells that will be reactivated upon subsequent exposures to the virus. On the other hand, **passive immunity** is the transfer of immune system components, primarily antibodies (immunoglobulins), into a person. Without plasma cells to continue producing them, transferred antibodies will eventually break down and become nonfunctional, but they serve several purposes. First, they are an immediate response. If a person is bitten by a rabid bat, the rabies vaccine is given but takes weeks to generate plasma cells that produce antibody. Transferred human rabies immunoglobulin immediately neutralizes virions to prevent continued infection. Antibodies can also be given to immunocompromised individuals who may not be able to be vaccinated but are exposed to virus that could be of concern in their state. Transferred immunoglobulin may partially or completely protect the individual.

Several immunoglobulins have been approved for use as passive therapy in the United States. Often, they can be recognized because their names end in "IG," the abbreviation for "immunoglobulin." "IVIG" refers to those immunoglobulins that are administered intravenously. Immunoglobulin preparations are available against HBV, varicella zoster virus, vaccinia virus, and respiratory syncytial virus. (They are also used against venoms from scorpions, black widow spiders, vipers, and rattlesnakes.)

8.2 ANTIVIRALS

It is always preferable to prevent infection through vaccination or control measures that interrupt the chain of infection. After infection has taken place, however, **antivirals** are a possible way to interfere with the replication of certain viruses. At this point in time, antivirals are only available for a limited number of viruses, and they reduce viral load and symptoms but rarely cure the infection. This is in contrast to antibiotics—which are different from *antibodies*—that work by interfering with bacterial processes, such as the generation of a cell wall, and are hopefully successful in killing all bacteria present. Antibiotics are specific for bacteria and do not work against viruses. It is a challenge to discover antiviral drugs that do not interfere with cell activities; bacteria have biological characteristics that eukaryotic cells do not possess and therefore can be targeted, but viruses are largely dependent upon cellular processes to create nascent virions.

The antivirals that have thus been discovered work by blocking one of the seven stages in the viral life cycle (attachment, penetration, uncoating, replication, assembly, maturation, and release) (Fig. 8.10). The majority of antivirals either inhibit a viral enzyme involved at one or more of these stages or prevent genome replication. Unlike some antibiotics that inhibit a wide range of bacteria, antivirals generally only work against a specific virus or closely related viruses.

Antivirals can be manufactured from several different types of components. Most medical drugs are composed of chemicals that are bonded together to create a molecule that binds to either the virus or to a cellular protein to prevent its function. Other drugs are made of one of the types of biological molecules—proteins, nucleic acids, carbohydrates, or lipids—which are much larger than the "small molecule" chemical drugs. For comparison, hundreds of amino acids form a protein, and the molecular weight of a single representative amino acid (tryptophan) is 204 g/mol. The molecular weight of the antiinfluenza drug oseltamivir is 410 g/mol, showing that small molecule drugs are only the size of a few amino acids put together. "Large molecule" drugs are composed of biological molecules, usually peptides (chains of amino acids) and the proteins (polypeptides) they form. These are generally made using biotechnology and living systems. Antibodies are an example of a large molecule

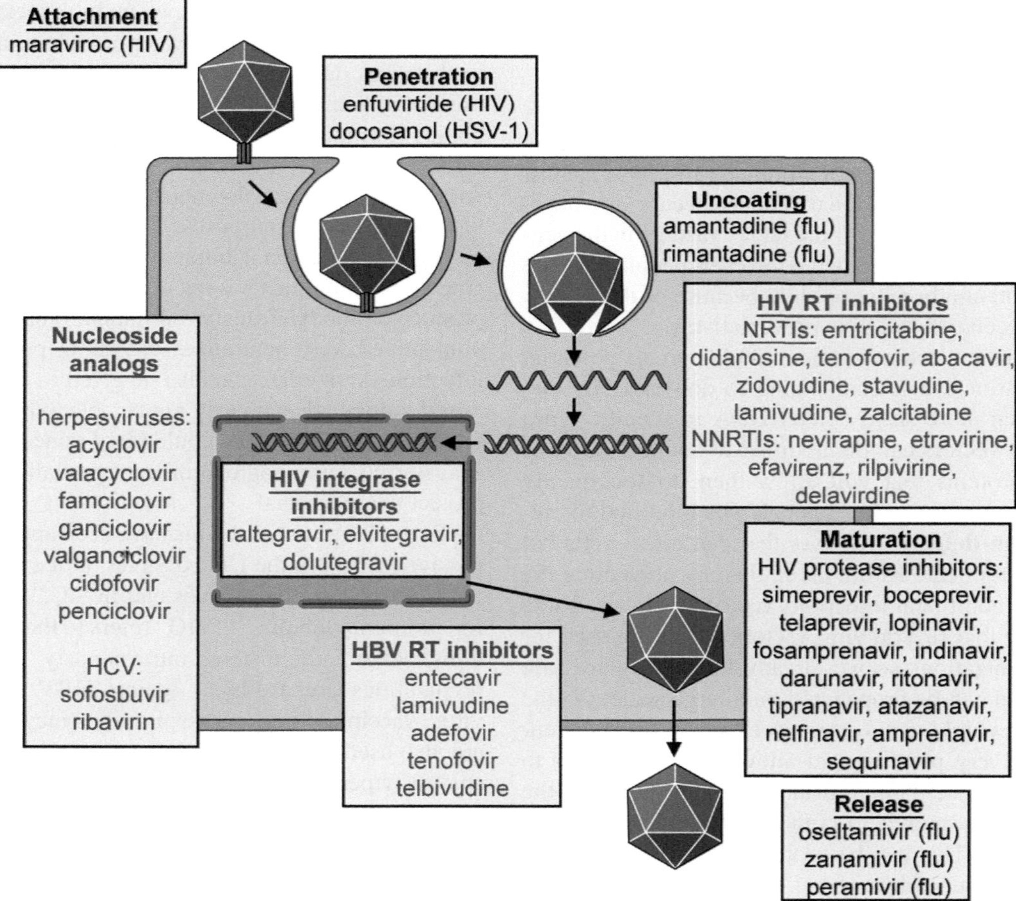

FIGURE 8.10 **Antivirals interfere with the stages of viral replication.** Listed are the stages of viral replication and the antivirals that interfere with each step. Antivirals listed in green text interfere with an aspect of genome replication, integration, or reverse transcription. This is not an exhaustive list, and many others are in clinical trials.

drug. Antibodies are proteins, and they are thousands of times larger than a small molecule drug, typically around 150,000 g/mol.

For certain viruses, it is necessary to use more than one antiviral agent at a time to prevent the virus from quickly mutating to a strain that can replicate in the presence of the antiviral. This is absolutely the case with HIV, which requires treatment with a "drug cocktail" of at least three antiviral medications, often called **highly active antiretroviral therapy (HAART)**. A single point mutation in the HIV protease leads to resistance against a protease inhibitor drug that appears within days of starting it, emphasizing the importance of HAART. Viruses, particularly those with error-prone polymerases, introduce mutations during replication, so evolution of drug-resistant viruses is to be anticipated.

8.2.1 Inhibition of Entry, Penetration, and Uncoating

Biologically, inhibitors of viral attachment and fusion are an optimal way to inhibit viral replication. Preventing the virus from entering a cell inhibits it from replicating and creating mutations that could translate into drug resistance. It also prevents **proviruses** from integrating their genomes into host chromosomes.

Currently, only a few entry and penetration inhibitors exist. The immunoglobulin preparations described in Section 8.1.4 are inhibitors of entry because they bind to viruses and prevent their attachment. Maraviroc is a small molecule drug that inhibits the binding of HIV-1 to one of its coreceptors, CCR5, to prevent attachment and subsequent entry into the cell. However, it does not work against HIV-1 strains that use CXCR4 as a coreceptor, and the drug will not work on a person infected with both because the CXCR4 strain will simply replace the CCR5 strain blocked by maraviroc. Drugs that block the human proteins that viruses use for entry are promising avenues to explore, although all cellular proteins have functions of their own for the cell, so it is important that antivirals do not also inhibit necessary cellular functions. In the case of CCR5, cells possess other proteins that carry out the activities of CCR5 and so inhibiting this receptor does not cause irreparable damage to the cell.

Enfuvirtide is another anti-HIV drug, a peptide that binds to and inhibits the HIV protein that fuses the viral envelope with the plasma membrane of the cell. Docosanol is thought to also inhibit the fusion of the HSV-1 envelope with the plasma membrane.

Some viruses gain entry into the cytosol by fusing their envelopes with the endosomal membrane. The influenza A virus M2 protein facilitates this process by forming pores in the viral envelope that allow the hydrogen ions within the endosome to enter the virion and disrupt the virion structure. Amantadine and rimantadine are small molecule drugs that are thought to inhibit the M2 protein, thereby preventing endosomal fusion and uncoating.

8.2.2 Inhibition of Genome Replication

Once the viral genome is released into the cell, viral genes are transcribed, and viral proteins are translated. The viral genome must also be copied. The largest number of antiviral drugs that have been developed interfere with genome replication, and the great majority of them have been designed to interfere with the replication of human viruses that reverse transcribe, namely HIV and HBV. As a retrovirus, HIV possesses an RNA genome that is reverse transcribed into cDNA upon entry into the cell. HBV uses reverse transcription at the end of its replication cycle to create copies of its DNA genome.

Nucleoside analogs are one class of small molecule drugs that terminate reverse transcription. A **nucleoside** is a DNA or RNA nucleotide without the phosphate group. As such, it is composed of a nitrogenous base (adenine, guanine, cytosine, thymine, or uracil) attached to a sugar (deoxyribose for DNA, or ribose for RNA). During normal DNA or RNA synthesis, nucleosides are converted into nucleotides that are incorporated into the replicating DNA or RNA. A nucleoside analog is a drug that resembles a nucleoside in its structure but is unable to bond with other nucleotides on its 3′-end. Nucleoside analogs compete with normal nucleosides during DNA or RNA replication and are converted into nucleotide analogs. When they are incorporated into a growing strand, they are unable to bond to subsequent nucleotides, thereby terminating the growing strand and replication of the genome (Fig. 8.11A). Due to their structure, viral polymerases are more likely to incorporate the nucleotide analogs than are cellular polymerases. This type of drug is referred to as a nucleoside/nucleotide reverse transcriptase inhibitor (NRTI) when it is used to prevent reverse transcription, such as with HIV and HBV infection. NRTIs are available that function as competitive analogs for all four bases of DNA: adenine, thymine, cytosine, and guanine. See the *In-Depth Look* for the generic names of these compounds, and how their names were determined.

Nucleoside inhibitors also exist against hepatitis C virus (HCV), which is an RNA virus that uses an RNA polymerase to copy its genome. For example, the nucleoside inhibitor sofosbuvir is an analog of uracil, which is incorporated into the virus RNA to stop replication.

A class of purine nucleoside analogs is used to treat infection with herpesviruses. Herpesviruses are double-stranded DNA viruses with large genomes. Because of this, they encode enzymes that increase the pool of nucleotides in the cell so that they do not become a limiting factor during replication of the DNA genome. One enzyme

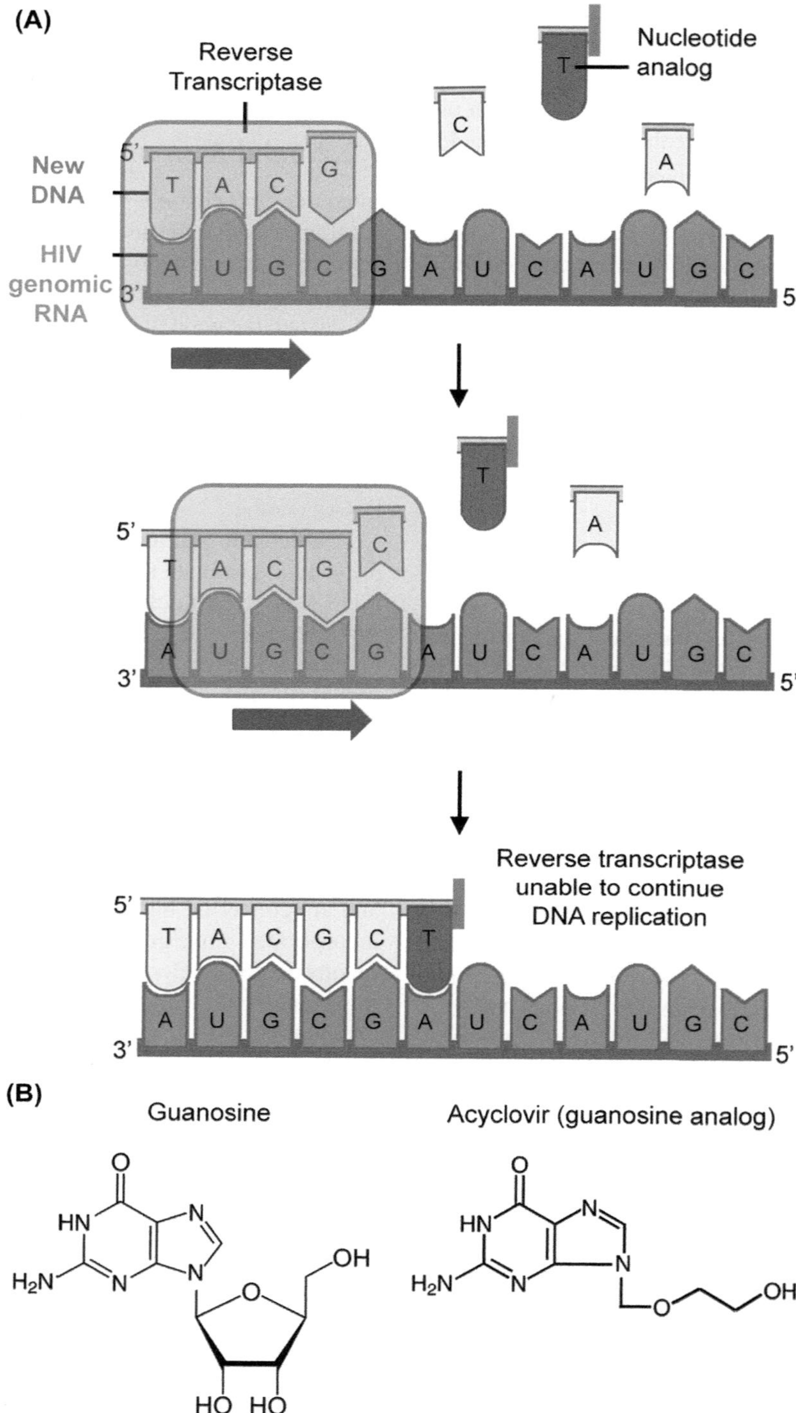

FIGURE 8.11 **Nucleoside analogs as antivirals.** A nucleoside refers to a DNA or RNA nucleotide without its attached phosphate groups. Nucleosides are converted into nucleotides before becoming part of a DNA strand. (A) Nucleoside analogs are drugs that are converted to nucleotide analogs. They compete with the normal cellular nucleotides and are unable to bond subsequent nucleotides, thereby halting new strand synthesis. (B) Guanosine is a normal nucleoside. Acyclovir is a guanosine analog used to treat herpesvirus infections.

is thymidine kinase (TK). TKs are enzymes that catalyze the first step in creating a nucleotide, the addition of a phosphate group to a nucleoside. Human cells have TK enzymes, too, but they only add phosphates to nucleosides containing thymidine or uridine. The herpesvirus TK has a much broader specificity—it binds to nucleosides containing thymidine or uridine, but also the purines adenine and guanine. Since human TK does not act upon purines,

purine nucleoside analogs have been developed that specifically bind to herpesvirus TK but not human TK (Fig. 8.11B). As nucleoside analogs, they are converted into nucleotides that end replication when incorporated into a growing strand of replicated DNA. The herpesvirus TK will be mentioned in Section 8.3.2, as it is exploited for beneficial use, as well.

Not all drugs that interfere with genome replication are nucleoside analogs (although the majority are). For example, foscarnet inhibits a binding site on the herpesvirus DNA polymerase. Ledipasvir is a drug that is thought to directly interfere with the HCV RNA polymerase. In combination with nucleoside analog inhibitors such as sofosbuvir, ledipasvir treatment has been shown to cure over 90% of certain genotypes of chronic HCV infection. **Nonnucleoside reverse transcriptase inhibitors (NNRTIs)** are small molecule drugs that bind directly to the active site of HIV-1 reverse transcriptase, disrupting its RNA-dependent and DNA-dependent DNA polymerase activities.

Following reverse transcription from the viral genome, the HIV proviral DNA integrates into a host chromosome. This is a necessary step for transcription of vmRNA and copying of the genome. **Integrase inhibitors** block the active site of integrase, the HIV enzyme that performs this step.

8.2.3 Inhibition of Assembly, Maturation, and Release

Thus far, no antivirals inhibit assembly. This may be because the nucleocapsid and capsid proteins involved in assembly are often the most numerous viral proteins translated. As a consequence, a high concentration of drug would be necessary to effectively inhibit assembly. However, when viral enzymes facilitate maturation of the virion, then they are potential drug targets. As described in Chapter 4, "Virus Replication," the HIV protease cleaves the *gag* gene product to form the capsid, matrix, and nucleocapsid proteins of the HIV virion. This is blocked by many different **protease inhibitors** that have been developed. An infectious virion is not formed if this maturation step does not happen.

The final step in the replication of viruses, *release*, is also a target if the virus-specific proteins facilitate the process. An example of when this occurs is with the influenza A virus neuraminidase, an enzyme that cleaves the sialic acids from the cell surface and from new influenza particles as budding is occurring to prevent the attachment of nascent virions to the cell from which they were produced. Oseltamivir, zanamivir, and peramivir are all neuraminidase inhibitors.

8.2.4 Boosting the Immune Response

As described in the previous sections, many antiviral drugs are available that inhibit various stages of viral replication. The majority of these drugs have been designed for use against HIV, influenza, herpesviruses, HBV, and HCV. Because viruses have distinct properties, most antivirals are specific for only one or a few viruses, which makes the process of drug discovery a slow and difficult one. Another approach is to develop drugs that boost the immune response to viruses to tip the scales in favor of the host. Although still in experimental stages, the hope is that this type of drug could be used against larger categories of viruses based upon the type of immune response that would be successful in overcoming the virus.

One immune-modulating drug already in existence for viral infections involves the type 1 interferons. Chapter 6,

In-Depth Look: Naming Antivirals

Drugs that are approved for clinical use generally have three names: a chemical name, a generic name, and a brand or trademark name. The chemical name is the molecular formula that describes the **active pharmaceutical ingredient (API)**, the molecule that causes the biological effect upon the body. This API has a single generic name associated with it. The brand name is a proprietary name for specific company's version of the API. Another company may have a different brand name for the exact same API, but the generic name associated with both drugs will be the same. For example, Copegus, Ribasphere, and Ribavarin are three brand names for the same antiviral drug with the generic name of ribavirin. The brand name is always capitalized, whereas the generic name never is.

The generic name of a drug is known as the International Nonproprietary Name (INN) (or United States Adopted Name (USAN) in the United States). Generic drug names can seem very bizarre, but the origin of the generic drug name follows an established convention that classifies the type of drug. The generic name contains a drug *stem* that indicates the type of drug it is. For example, the stem "vir" means the drug is an antiviral. The *infix* further subclassifies the drug. For example,

"amivir" refers to a neuraminidase inhibitor, such as oseltamivir, peramivir, zanamivir

"navir" refers to an HIV protease inhibitor, such as amprenavir, atazanavir, tipranavir

"virine" refers to a NNRTI, such as capravirine, dapivirine, etravirine

Generic names may also use stems that refer to the structure of the molecule, rather than the clinical use of it. For example, the influenza M2 inhibitor amantadine has a "mantadine" stem that classifies adamantine derivatives.

The prefix at the beginning of the generic name is chosen by the company and means nothing specifically. The name must be distinctive in sound and spelling to prevent errors in dispensing medication, should not be inconveniently long, and should facilitate easy pronunciation in other languages (avoiding the use of *th*, *ae*, or *ph*, for example).

Now check out the drugs listed in Fig. 8.10 to see if you can better understand the origin of their generic names!

"The Immune Response to Viruses," described the discovery of IFN-α/β and how the cytokines bind to the IFN-α/β receptor to produce an antiviral state within cells, through the activation of dsRNA-dependent protein kinase (PKR) and the 2'-5'-oligoadenylate synthetase-ribonuclease L (OAS-RNase L) pathways (Fig. 6.4). However, they have pleotropic effects upon the immune system and activate several immune cell types, so they have not proven to be the "magic antiviral bullet" that was hoped upon their discovery. Nonetheless, IFN-α has been shown to be useful alone or in combination for the treatment of chronic hepatitis C infection, as well as for several different cancers. (IFN-β is used for the treatment of multiple sclerosis.) These protein therapeutics are produced by *E. coli*, engineered using the recombinant DNA technologies described above in Section 8.1.2 for the production of recombinant subunit vaccines.

8.3 THE BENEFICIAL USES OF VIRUSES

Viruses are a unique biological entity. Not even "alive," they infect all types of living organisms and cause a range of effects, from subclinical infections to persistent diseases that last indefinitely. We attempt to overcome infection through vaccination and other control measures, but new viruses evolve and emerge from animal sources. Viruses have unique properties that have ensured their continued existence alongside unicellular organisms and humans alike.

Nearly since the discovery of viruses, persistent and ingenious scientists have realized the potential of viruses and their unique properties for the benefit of humankind. Although many more beneficial applications exist for viruses, gene therapy and its associated use to create anticancer therapies will be discussed in this section.

8.3.1 Gene Therapy

Gene therapy is the process of modifying an individual's DNA for therapeutic purposes. Gene therapy is currently being pursued for genetic defects, primarily for those disorders that can be traced back to the function of a single gene. These individuals lack a functional copy of a gene, and gene therapy seeks to insert a functional copy into a person's genome to lessen or cure the person's genetic condition.

Gene therapy requires that the genetic cause of the defect is known so that the proper gene ends up being inserted. A vector will be required to deliver the gene into the person's cells, and the vector must be designed to have promoter and enhancer elements that will allow proper expression of the gene, once inside the patient's cells. Gene therapy can be performed ex vivo, meaning that the person's cells are removed, modified, and then reinserted back into the person, or it can be performed in vivo, meaning that the cells will be modified while still within the person. In either case, a vector is required to deliver the functional gene. Because of their natural propensity to gain entry and deliver their genomes into cells, several types of viruses are being investigated for their potential as gene therapy vectors.

The most common virus used for gene therapy has been human adenovirus, specifically serotype 5. Adenoviruses are one cause of mild respiratory and gastrointestinal illnesses. They are relatively simple dsDNA viruses with a broad tropism. A protein "fiber" that facilitates attachment extends from each vertex of the icosahedral capsid (see Fig. 8.8). The adenovirus fiber initially binds to a receptor known as the Coxsackie adenovirus receptor on the surface of cells, and then the base of the virus attaches to integrins on the cell surface. It is taken into the cell via endocytosis and disassembles at the nuclear envelope, transporting its DNA into the nucleus.

One of the first adenovirus genes transcribed is known as E1A, which encodes a transcription factor that turns on several other early genes that eventually lead to the production of nascent virions. For gene therapy, the E1A gene is replaced with the therapeutic human gene. A few other genes can also be removed to make more space for the insert. Upon entry into the nucleus, the therapeutic gene is transcribed and translated into a functional protein, thereby compensating for the genetic deficiency. This works the same way as a recombinant vector vaccine, except a functional copy of the human gene is carried by the adenovirus, rather than a viral gene.

Adenoviruses are an attractive choice for gene therapy for several reasons. Most strains are relatively harmless, and they have a wide cellular tropism. They infect dividing and nondividing cells, and their life cycle is well understood. They also do not integrate their DNA into the host genome. This safeguards against an insertion that interrupts an important gene, but it also means that the effects of the gene will be short-lived and multiple injections of the virus will likely be necessary. And this is one of the biggest drawbacks of using adenoviruses: repeated challenges with the virus cause an immune response against it, which reduces the efficiency of gene delivery. Some individuals will have naturally encountered adenoviruses and will already have strong immunity against them. In fact, a human gene therapy trial in 1999 ended up causing the death of 18-year-old Jesse Gelsinger due to his body's robust response against the virus. This unfortunate event led researchers, the public, and government regulatory agencies alike to take a step back to reassess the challenges of gene therapy. The maximum adenovirus DNA insert is currently around 8 kb, which is large but will exclude many important human genes.

The viruses that are used in gene therapy need to be able to infect their target cells, but they need to be modified so that they do not form nascent virions in vivo that contain the human transgene and could be released to infect other individuals. It is for this reason that **replication-defective** viruses, also called replication-incompetent viruses, are created. Because the E1A gene has been removed from adenovirus, it is unable to replicate when it enters the host cell. This seems like a catch 22, however, because how can the adenovirus vectors be made before injection if they are unable to replicate within cells? The answer lies in the use of a **packaging cell line**, often the HEK-293 cell line, that has been engineered to express the E1A protein (Fig. 8.12). When the adenovirus vector containing the therapeutic human gene is inserted into the packaging cell line, new virions are made because the cell line has the E1A protein that initiates transcription of the viral genes to create new virions. The E1A gene is not encoded within the adenovirus genes, however, and so the virions that are created do not contain the gene for E1A. This means that the virions will infect but will be unable to replicate inside the cells.

Adeno-associated virus (AAV) is also used for gene therapy. It is a single-stranded DNA virus in the *Parvoviridae* family that has low immunogenicity, and it is not associated with any human diseases. AAV is a **satellite virus**, which means that it can infect cells but cannot replicate without a "helper virus" present to provide certain proteins. This is most often adenovirus itself, which is where the virus derives its name. Different AAV serotypes have unique receptors that determine their tissue specificity.

Like adenovirus, AAV can infect dividing and nondividing cells. AAV additionally possesses integration sequences within its genome that match a location within human chromosome 19, aptly called *adenovirus-associated virus integration site 1* (*AAVS1*). Integration is a low-frequency event, but when it occurs, the human transgene is stably integrated into the host's chromosome. Integration is a double-edged sword: on the one hand, it translates into a more stable delivery (and less frequent injections of the virus), but on the other hand, an integrated gene is currently impossible to remove, if ever the need arose. Although the virus has many positive characteristics for gene therapy, because of its small size the maximum insert is around 5 kb, the smallest of any virus used for gene therapy.

One of the more promising types of viral vectors for gene therapy is a type of retrovirus known as a lentivirus. As described in Chapter 4, "Virus Replication," retroviruses reverse transcribe their single-stranded RNA genomes into cDNA that is flanked at either end with a long-terminal repeat region. All retroviruses have at minimum three genes– *gag*, *pol*, and *env*– that encode polyproteins that are cleaved into the viral proteins. The *gag* and *pol* genes

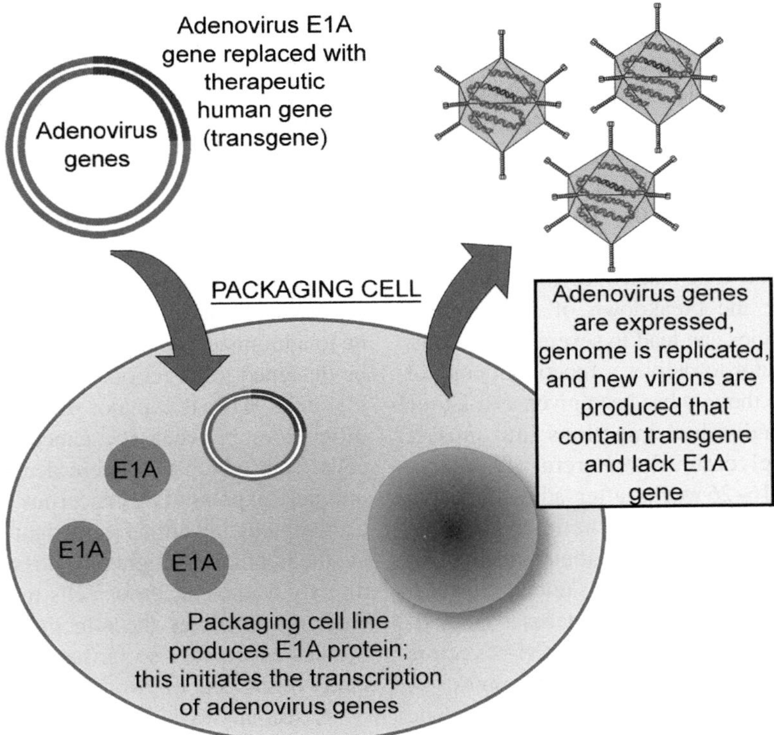

FIGURE 8.12 **The creation of replication-defective adenovirus.** Adenovirus possesses a gene called E1A that encodes a transcription factor that activates gene transcription and the subsequent creation of nascent virions. In replication-defective viruses, the E1A gene has been replaced with the therapeutic human gene; in its absence, new virions cannot be made. To create a stock of replication-defective viruses, a packaging cell line is used that expresses the E1A protein. When the adenovirus genes are inserted into the cell (either on a plasmid or from a previous stock of replication-defective virus), the cellular E1A initiates adenovirus gene transcription, leading to the assembly of new virions.

encode the structural and packaging genes, among others, and the *env* gene encodes the viral envelope proteins. In addition to those already listed, lentiviruses contain three to six more proteins that are responsible for replication and persistence, but most of these nonessential viral genes are removed from the lentiviral vector. Lentiviruses generally have a narrow tropism, but modifications have been made to the vector such that the *env* gene is replaced with genes from other viruses that encode viral receptors that target the virus instead to liver, lung, neurons, or muscle, among others. The maximum insert is around 8 kb, similar to adenoviruses, and the viruses are made in packaging cell lines such that they are replication defective.

Lentiviruses infect dividing and nondividing cells, and as retroviruses, they randomly integrate into the host chromosome to provide long-term expression of the therapeutic transgene. The optimal result would be that the gene integrates into an area of noncoding DNA, but the random integration can be cause for concern if the insertion of the transgene disrupts an essential gene, such as a tumor suppressor gene. In support of this possibility, development of leukemia has been an adverse event noted in some human gene therapy trials, and scientists are working to design better systems to avoid such effects.

There are over 1500 clinical trials for applications of gene therapy currently ongoing in the United States. No gene therapies have yet been approved for use in the United States, but two have been approved elsewhere in the world. In 2004, China approved the world's first gene therapy, called Gendicine, that is discussed in more detail in Section 8.3.2. The other, known as Glybera, was approved in the European Union in 2012. It is used to treat familial lipoprotein lipase (LPL) deficiency, a rare deficiency that interferes with the breakdown of fats and oils (triglycerides). People with LPL deficiency have a mutation that inactivates the LPL gene, which is responsible for the breakdown of triglycerides. High circulating triglycerides can lead to severe pancreatitis. Glybera uses an AAV vector to deliver a functional copy of the human LPL gene. The therapy has been given as a 1-time treatment of up to 60 individual injections into muscle tissue, and although triglyceride levels returned to pre-treatment levels within 16–26 weeks after administration, patients continued to show improvement in triglyceride metabolism, translating to reductions in their pancreatitis. The transgene sustained expression in the skeletal muscle.

Although adenoviruses, AAV, and retroviruses (including lentiviruses) are the most common types of viral vectors being explored for gene therapy, herpesviruses, poxviruses, and several other types of viruses are also being used. Gene therapy is currently only being performed on **somatic cells**, or body cells, rather than **germ cells**, which form the sperm or eggs that come together in the process of fertilization to create a new individual. This means that any genetic defects that are corrected in an individual through gene therapy may still be passed along to future generations, since the original defective gene is not modified in the germ cells.

8.3.2 The Use of Viruses in Anticancer Therapies

Cancerous cells are derived from normal human cells that have lost regulation of the cell cycle, leading to uncontrolled proliferation of cells that invade nearby tissues. The goal of cancer therapy is to eliminate cancerous cells while leaving healthy cells unharmed, but cancerous cells are extremely difficult to specifically target because the immune system is trained to not attack its own cells. Thus, the majority of treatments for cancers so far have been chemotherapeutics that become incorporated faster into rapidly proliferating cells and radiation to damage the cancerous DNA permanently.

One therapy in development involves the use of **oncolytic viruses**, meaning that the virus kills the cancerous cells it infects. This type of therapy is known as **virotherapy** and can use naturally occurring or modified strains. In comparison to the replication-defective viruses that are used in gene therapy, oncolytic viruses are meant to replicate and infect nearby tumor cells until they have all been destroyed through viral replication. Immune-stimulating genes are also being inserted into these viruses to boost the local immune response against the tumor cells.

Virotherapy has huge potential, but major challenges need to be overcome for it to be a safe, consistent, and effective therapy. The first challenge involves the delivery of the virus. Thus far, most oncolytic viruses have been injected directly into the tumor. **Metastasis**, the movement of cancerous cells from their initial location to additional sites, is a major concern with cancer, and not all locations of metastases are necessarily known. A solution to this problem would be to administer the oncolytic virus systemically, but it must be designed to target only the cancerous cells and not normal cells. This is a major difficulty because it relies upon differences between the cancerous cells and the normal cells from which they were derived. **Cancer antigens** are uniquely expressed by cancerous cells and not normal cells. Cancer antigens are not abundant and not all tumors generate them, but they might be a possible way to deliver viruses directly to the cancerous cells that express them. A second challenge involves the effectiveness of the tumor killing. Studies carried out so far have had mixed results, and certainly this aspect of virotherapy needs to be further studied.

A second way that viruses are being investigated as a means to target cancer cells is through gene therapy. In one situation, viral vectors are used to introduce copies of genes that interfere with the proliferation of the cancerous cells. One such gene is a tumor suppressor named p53, which blocks the cell from being able to continue through the cell cycle. Over 50% of cancers have mutations in p53 that allow them to replicate uncontrollably, so normal copies of these so-called "suicide genes" are being engineered into viral vectors and delivered into cancerous cells. An example of this is Gendicine, approved in 2004 in China to treat head and neck squamous cell carcinomas, which are cancers of the cells that line the mucosal epithelium inside the head and neck. Researchers designed an adenovirus vector containing the p53 gene and injected 10^{12} virus particles directly into each tumor once a week for 8 weeks, given along with normal radiation therapy. The only reported side effect was self-limited fever, and the rate of tumor clearance was 2.31-fold higher with the viral vector than with radiation therapy alone. The virus itself is not oncolytic; the p53 transgene interferes with the proliferation of the cancerous cells by providing a functional copy of the gene (Fig. 8.13A). Clinical trials using the adenovirus-p53 approach are being performed in the United States and Europe to examine if these results can be recapitulated elsewhere.

Virus-directed enzyme prodrug therapy (VDEPT) is another manifestation of gene therapy where the viral vector contains genes for certain enzymes. The viruses are delivered to target cells, where the enzyme gene is transcribed and translated to create the functional enzyme, which is able to modify an inactive **prodrug** to its active drug form. An example of when this is used against cancerous cells involves the insertion of the herpesvirus thymidine kinase into an adenovirus or AAV vector (Fig. 8.13B). As described in Section 8.2.2, the herpesvirus TK converts nucleosides into nucleotides by adding phosphate groups to the nucleoside. Human TK only does this to nucleosides containing thymidine or uridine, but the herpesvirus TK can act upon any nucleoside, and so purine nucleoside analogs (of adenine and guanine) only interact with the herpesvirus TK. In this case, VDEPT uses a viral vector containing the herpesvirus TK enzyme gene, which is transcribed and translated when the virus enters the cancerous cell. The purine nucleoside analog is administered to the patient and diffuses into his/her cells. In the virus-infected cancerous cells, the herpesvirus TK enzyme converts the nucleoside analog into a nucleotide analog that is incorporated into the DNA and terminates replication of new strands. The cancerous cell's proliferation ceases. Without the herpesvirus TK enzyme, normal cells are unaffected because they are unable to convert the purine analog into nucleotides.

All of the above uses of viruses are promising and yet have significant hurdles to overcome before they could be regularly used in the clinic. The most pressing issues of investigation are how to deliver oncolytic viruses or transgene-containing viral vectors specifically to tumor cells, and how therapeutic viruses could avoid initiating an immune response, the very system designed to keep the host free of viruses.

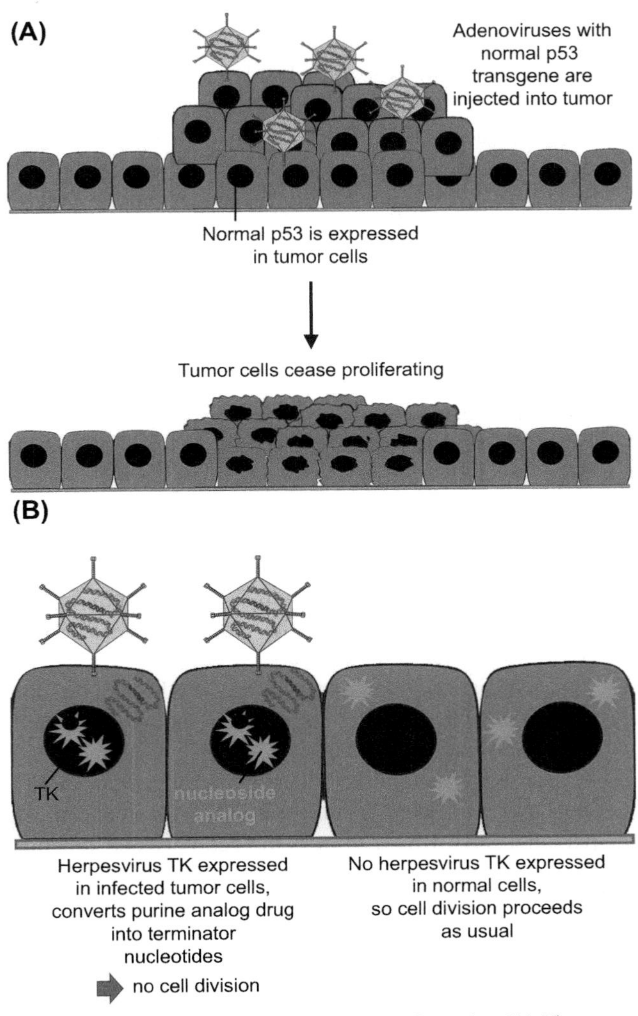

FIGURE 8.13 **Viruses as anticancer gene therapies.** (A) Viruses can be engineered to contain normal copies of tumor suppressor genes. In this case, adenoviruses carrying the normal p53 gene infect tumor cells. The expression of p53 blocks the cell cycle and inhibits the proliferation of the tumor cells. (B) In virus-directed enzyme prodrug therapy, a patient is treated with a drug that will only become activated by the gene product delivered by the virus. In this case, adenoviruses delivered to tumor cells carry the herpesvirus thymidine kinase gene, the product of which is able to convert purine nucleoside analog drugs into terminator nucleotides that cease DNA replication. The drug diffuses into normal cells, too, but they are unable to modify it into terminator nucleotides so DNA replication is unaffected.

SUMMARY OF KEY CONCEPTS

Section 8.1 Vaccine Development

- Vaccination is the intentional inoculation of a person with a harmless form of a pathogen. The purpose of vaccination is to establish an immune system memory response that provides immunity upon actual exposure to the pathogen. The immunogenicity of the vaccine is highest when the vaccine closely resembles the actual pathogen.
- Ancient vaccination efforts began with variolation, the intentional inoculation of a person with dried smallpox scabs. In 1796, Edward Jenner tested and promoted the use of cowpox as a vaccine that provides immunity against smallpox.
- The majority of current vaccines are based on live attenuated or inactivated virus vaccines. Viruses are gradually passaged in suboptimal replication conditions to create an attenuated vaccine strain. Inactivated viruses have been treated with high heat or low amounts of formaldehyde to render the viruses unable to infect cells.
- Molecular biology techniques are useful in creating vaccines for viruses that are not easily propagated. HBV and HPV vaccines are created using recombinant protein expression systems to express viral proteins, which are purified and injected as the vaccine material. These are known as recombinant subunit vaccines.
- The fabricated link between the MMR vaccine and autism has been soundly disproven by many large, well-controlled scientific studies.
- Several experimental vaccines are being tested, including DNA vaccines that inject a DNA transgene into a person's cells, which express the gene product to which the person's immune system responds. Recombinant vector vaccines use the same principles but rely upon benign viruses to deliver the gene into cells.
- Passive immunity relies upon the transfer of temporary immune system components, mainly antibody, to provide short-lived immunity against a virus.

Section 8.2 Antivirals

- Antivirals are not antibiotics. Antibiotics affect bacteria and have no effect upon viruses.
- Antivirals prevent virus replication by blocking one or more of the stages of the virus's replication. The majority of antivirals block virus proteins from working or interfere with genome replication. Other antivirals interfere with virion attachment, penetration, uncoating, maturation, or release.
- A nucleoside is composed of a nitrogenous base and a sugar—it is the DNA/RNA nucleotide without the phosphate group(s) attached. Nucleoside analogs are similar to nucleosides in that they are converted into nucleotides that become incorporated into DNA/RNA, but nucleotide analogs are unable to bond subsequent nucleotides and thus terminate replication.
- Nucleoside analogs that interfere with reverse transcription are known as NRTIs. Other drugs that directly bind reverse transcriptase are known as NNRTIs. These are effective against HIV or HBV, two human viruses that possess reverse transcriptase activities.
- Thymidine kinase is an enzyme that creates a nucleotide by adding a phosphate group to a nucleoside. Human TK can only do this to nucleosides containing thymidine or uridine, but herpesviruses encode a TK that is able to convert any nucleoside into a nucleotide. This fact is exploited with purine nucleoside analog drugs, which are only converted into terminator nucleotides by cells infected with herpesviruses.
- Numerous HIV protease inhibitors prevent the maturation of HIV virions. Several anti-influenza drugs target neuraminidase, leaving the virions unable to be released from the infected cell.
- Antiviral drugs have three names: the chemical name, a generic name, and a brand/trademark name. The chemical name and generic name will always be the same for one specific active pharmaceutical ingredient, although companies may market the API under different brand names. The stem of the generic name is determined based upon the drug class.

Section 8.3 The Beneficial Uses of Viruses

- Gene therapy is the process of modifying an individual's DNA for therapeutic purposes. This can be used to insert a functional copy of a gene in a person that has a genetic defect.
- Gene therapy can be performed ex vivo on cells that have been removed from the person, modified, and reinjected into the person, or it can be performed in vivo by modifying the cells while still in the person.
- Several viruses are used as vectors to deliver the therapeutic gene into a person's cells. Adenovirus, adeno-associated virus, and retroviruses are among those most-commonly used for such a purpose.
- In virotherapy, oncolytic viruses cause the destruction of the tumor cells they infect. Gene therapy can also be used to deliver functional copies of genes into tumor cells to inhibit their proliferation or boost the immune response against the tumor.
- VDEPT uses a viral vector to deliver an enzyme gene into tumor cells. The expressed enzyme modifies an inactive prodrug to its active form, which interferes with the tumor cell. The inactive prodrug has no effect on normal cells, which would not contain the virus-delivered enzyme gene.
- All beneficial uses of viruses struggle with the same challenges: ensuring targeted delivery to the desired cells, preventing uncontrolled viral replication and dissemination, and avoiding the immune response against the vector virus.

FLASH CARD VOCABULARY

Vaccination	Gene cloning
Variolation	Hepatitis
Vaccine	Recombinant subunit vaccine
Immunogenic	Autism
Humoral response	DNA vaccines
Cell-mediated response	Particle bombardment
Adjuvant	Electroporation
Live attenuated virus vaccine	Prime-boost
Passaging	Vector
Inactivated virus vaccine	Recombinant vector vaccine
Lyophilized	Active immunity
Recombinant protein expression	Passive immunity
Recombinant DNA	Antivirals
Plasmid	Highly active antiretroviral therapy (HAART)
Provirus	Packaging cell line
Nucleoside/nucleotide	Satellite virus
Nucleoside analog	Somatic cell/germ cell
Non-nucleoside reverse transcriptase inhibitors	Oncolytic virus
	Virotherapy
Integrase inhibitors	Metastasis
Protease inhibitors	Cancer antigens
Active pharmaceutical ingredient	Virus-directed enzyme prodrug therapy (VDEPT)
Gene therapy	Prodrug
Ex vivo/in vivo	Replication-defective virus

CHAPTER REVIEW QUESTIONS

1. What are the characteristics of a good vaccine?
2. Describe how a pathogenic virus is attenuated for vaccine use. Which individuals should not receive live attenuated virus vaccines? Why?
3. You want to design a vaccine against Ebola virus. Which type of vaccine formulation do you think would be safest but effective?
4. You want to design a recombinant subunit vaccine that induces neutralizing antibodies. Thinking about the architecture of a virion, which viral proteins are likely the best ones to include in your vaccine?
5. Why is it recommended that people with egg allergies do not receive the normal seasonal influenza vaccine?
6. You have a friend who is wrestling with whether he should avoid vaccinating his infant for fear of autism. As an informed scientist, how would you respond to his concerns with sensitivity?
7. How are DNA vaccines and recombinant vector vaccines similar? How are they different?
8. Why is it necessary to treat HIV infections with a cocktail of several antiviral medications?
9. Create a table with the seven stages of viral replication, and list the types of drugs that inhibit each stage.
10. How specifically do the major anti-HIV drugs reduce viral replication?
11. Spell out in detail how a nucleoside analog is processed and prevents the copying of DNA or RNA. For which viruses are nucleoside analog drugs currently available?
12. Consult Fig. 8.10. Which type of drug do you think uses the generic stem "tegravir"?
13. For gene therapy, why is it important to create replication-defective viral vectors?
14. Create a table that lists the major viruses that are used for gene therapy. In your table, list the major advantages and disadvantages of each virus type.
15. How are viruses used in anticancer therapies?

FURTHER READING

Amanna, I.J., Slifka, M.K., 2014. Current trends in West Nile virus vaccine development. Expert Rev. Vaccines 13, 598–608.

Anderson, L.J., 2013. Respiratory syncytial virus vaccine development. Semin. Immunol. 25, 160–171.

Bartlett, D.L., Liu, Z., Sathaiah, M., et al., 2013. Oncolytic viruses as therapeutic cancer vaccines. Mol. Cancer 12, 103.

Deval, J., Symons, J.A., Beigelman, L., 2014. Inhibition of viral RNA polymerases by nucleoside and nucleotide analogs: therapeutic applications against positive-strand RNA viruses beyond hepatitis C virus. Curr. Opin. Virol. 9, 1–7.

El Garch, H., Minke, J.M., Rehder, J., et al., 2008. A West Nile virus (WNV) recombinant canarypox virus vaccine elicits WNV-specific neutralizing antibodies and cell-mediated immune responses in the horse. Vet. Immunol. Immunopathol. 123, 230–239.

Kamimura, K., Suda, T., Zhang, G., Liu, D., 2012. Advances in gene delivery systems. Pharmaceut. Med. 25, 293–306.

Kennedy, R.B., Ovsyannikova, I.G., Lambert, N.D., Haralambieva, I.H., Poland, G.A., 2014. The personal touch: strategies towards personalized medicine and predicting immune responses to them. Expert Rev. Vaccines 13, 657–669.

Ledgerwood, J.E., Pierson, T.C., Hubka, S.A., et al., 2011. A west nile virus DNA vaccine utilizing a modified promoter induces neutralizing antibody in younger and older healthy adults in a phase I clinical trial. J. Infect. Dis. 203, 1396–1404.

PhRMA, 2013. Vaccine fact book 2013. Pharm. Res. Manuf. Am. 1–97.

Plotkin, S., 2006. The history of rubella and rubella vaccination leading to elimination. Clin. Infect. Dis. 43 (Suppl. 3), S164–S168.

Poulet, H., Minke, J., Pardo, M.C., Juillard, V., Nordgren, B., Audonnet, J.C., 2007. Development and registration of recombinant veterinary vaccines. The example of the canarypox vector platform. Vaccine 25, 5606–5612.

Pulendran, B., Oh, J.Z., Nakaya, H., Ravindran, R., Kazmin, D.A., 2013. Immunity to viruses: learning from successful human vaccines. Immunol. Rev. 255, 243–255.

Riedel, S., 2005. Edward Jenner and the history of smallpox and vaccination. Proc. Bayl Univ. Med. Cent. 18, 21–25.

Samulski, R.J., Muzyczka, N., 2014. AAV-mediated gene therapy for research and therapeutic purposes. Annu. Rev. Virol. 1, 427–451.

United States Centers for Disease Control and Prevention, 2015. Epidemiology and Prevention of Vaccine-preventable Diseases, thirteenth ed. Public Health Foundation, Washington, D.C.

World Health Organization, 2006. State of the Art of New Vaccines: Research and Development. WHO Document Production Services, Geneva, Switzerland.

Chapter 9

Viruses and Cancer

In the United States, an estimated 1.66 million people were diagnosed with cancer in 2015. It is the second most common cause of death in the United States, accounting for nearly 1 of every 4 deaths: 1600 people die of the disease every day, for a total of over half a million people each year. Globally, cancer is the cause of over eight million deaths each year. Viruses are a contributing cause of ~15% of total cancer cases and are a major cause of cervical, liver, and oral cancers. This chapter will discuss the molecular mechanisms behind cancer and the role viruses play in **oncogenesis**, the development of cancer (*onco* means "tumor").

9.1 PROPERTIES OF CANCEROUS CELLS

Cancer is a disease caused by uncontrolled cell division that results in proliferating cells that invade nearby tissues and spread to other areas of the body. About 5–10% of cancers involve inherited mutations that predispose an individual to developing cancer, while the rest are associated with the acquisition of somatic mutations that lead to aberrant cellular processes.

The development of a tumor is a multistep process. Within a single cell, mutations in the genes that control cell division can lead to dysregulated proliferation. If the cells stop dividing once they form a mass of a certain size, this is known as a **benign** tumor. On the other hand, if other mutations are acquired as the cell's replication and cell cycle control processes become more unstable, then the cell may proliferate to form a **malignant** tumor that invades nearby tissues. **Angiogenesis**, the growth of new blood vessels, occurs to feed the tumor, which eventually **metastasizes**, or spreads to other sites in the body. Malignant tumors become life threatening as they compete with normal tissues for limited resources and interfere with the required functions of organs.

Carcinogens are substances that induce mutations in DNA that lead to the development of a cancerous cell. Carcinogens can be categorized as physical carcinogens, such as ionizing radiation or ultraviolet light, or chemical carcinogens, such as asbestos or the components of cigarette smoke. Infectious pathogens can be classified as carcinogens, too. In the 1960s, researchers began performing viral infection of cells in culture and noted that certain viruses were able to induce changes within the cells that resembled those involved in malignancy. This phenomenon, known as **transformation**, is characterized by changes in the morphology and growth of cells (Fig. 9.1A and B). Transformation is a phenotypic change in the properties of cells that results from genotypic alterations. Like cancerous cells, transformed cells increase their rate of proliferation and become immortal. They reactivate **telomerase**, the enzyme that rebuilds the telomeres of chromosomes that are shortened in normal cells when chromosome replication takes place. They rely less on the constituents of the cell culture medium and produce their own autocrine growth factors. Known as **contact inhibition**, normal cells cease proliferating when they make contact with nearby cells, but transformed cells continue dividing and will grow on top of one another, increasing their motility (Fig. 9.1C and D). Transformed cells also acquire **anchorage independence**, meaning that they are able to grow without having to attach to a substrate, which is a requirement of normal cells. This feature is also acquired by metastasizing tumor cells. Finally, transformed cells may be **tumorigenic** and cause tumors when implanted into an animal. Described in more detail below, seven human viruses are classified as carcinogens because of their ability to transform cells or induce tumor formation. These oncogenic viruses include hepatitis B virus (HBV), hepatitis C virus (HCV), human papillomavirus (HPV), Epstein–Barr virus (EBV), Kaposi's sarcoma–associated herpesvirus (KSHV), Merkel cell polyomavirus (MCPyV), and human T-lymphotropic virus type I (HTLV-I).

Word Note: Transformation

In virology, *transformation* refers to the phenotypic changes that occur in cells upon infection with certain viruses. In molecular biology or biotechnology, however, the word *transformation* refers to the uptake of DNA—most often plasmids—by bacterial or yeast cells. For example, in the process of creating the HBV-recombinant subunit vaccine, yeast are transformed with a plasmid containing the hepatitis B surface antigen gene. In comparison, the infection of cells with the polyomavirus SV40 causes their transformation.

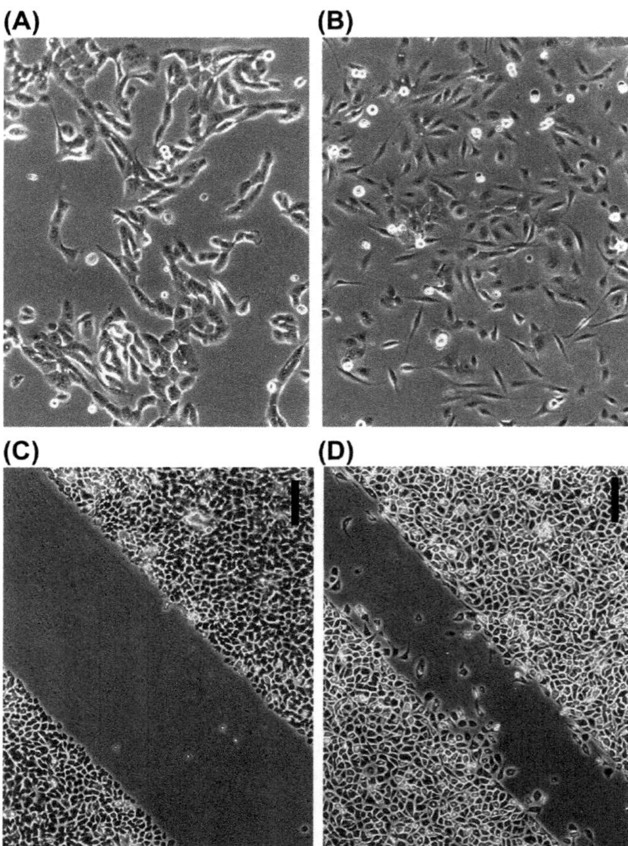

FIGURE 9.1 **Viral transformation of cells.** Transformation of normal cells is characterized by morphological and biological changes. Nasopharyngeal epithelial cells were transfected with a control vector (A) or vector containing the Epstein–Barr virus LMP1 protein (B). Note that the cells expressing LMP1 have a different morphology than the control cells, which are more cobblestone-shaped than elongated. Those expressing EBV LMP1 also exhibit reduced cell–cell contact. The control cells form smooth edges after being scraped with a plastic pipette (C), while the EBV LMP1-expressing cells migrate, forming uneven edges (D). *Reprinted with permission from Lo, A.K., et al., 2003. Alterations of biologic properties and gene expression in nasopharyngeal epithelial cells by the Epstein–Barr virus-encoded latent membrane protein 1. Lab. Invest. 83(5), 697–709. Macmillan Publishers Ltd., Copyright 2003.*

TABLE 9.1 The 10 Hallmarks of Cancer, as Stated by Hanahan and Weinberg

Hallmark	Description
Sustaining proliferative signaling	Normal cells carefully control mitosis and the production of growth-stimulating factors; cancer cells acquire mutations that ensure their proliferation.
Evading growth suppressors	Normal cells have tumor suppressors, including pRB and p53, that regulate the cell cycle. Cancer cells inactivate or interfere with the function of tumor suppressors.
Enabling replicative immortality	Normal cells have a finite life span; cancer cells have unlimited replicative potential, often through the reactivation of telomerase.
Resisting cell death	Normal cells respond to stimuli and undergo apoptosis; cancer cells become refractory to these proapoptotic signals.
Deregulating cellular energetics	Normal cells undergo anaerobic cellular respiration in low oxygen concentrations; cancerous cells switch their metabolism to perform mainly glycolysis even in the presence of oxygen.
Genome instability and mutation	Normal cells induce senescence or apoptosis if mutations occur; cancer cells successively acquire mutations that confer selective advantages, lending further instability to the genome.
Inducing angiogenesis	Tumors induce the sprouting of new blood vessels to ensure the supply of nutrients and oxygen and ability to shed metabolic waste products and carbon dioxide.
Activating invasion and metastasis	Most normal cells maintain their position within the body; cancer cells alter genes involved in cell–cell adhesion, acquiring mutations that lead to invasion and metastasis.
Avoiding immune destruction	The immune system kills off highly immunogenic cancer cells. The remaining cancer cells that have low immunogenicity proliferate and continue to acquire mutations that evade the immune system.
Tumor-promoting inflammation	Immune system cells normally eliminate cancer cells. However, cancer cells can acquire mutations that signal for the influx of innate cells that assist in fostering an oncogenic environment.

9.2 CONTROL OF THE CELL CYCLE

As with other carcinogens, infection with an oncogenic virus is not sufficient in and of itself to cause malignancy. Not all people who contract these viruses will develop cancer; in other words, additional cofactors and biological modifications are required for cancer to develop in a virally infected cell. In 2000, Douglas Hanahan and Robert A. Weinberg proposed six biological "hallmarks" that malignant cells progressively acquire in order to become a "successful" cancer. In 2011, they added four additional characteristics based upon more knowledge of oncogenic processes, for a total of 10 hallmarks of cancer (Table 9.1). Several of these hallmarks involve the dysregulation of cell cycle control mechanisms in cancerous cells. As described in Chapter 3, "Features of Host Cells: Cellular and Molecular Biology Review," the cell cycle is divided into four phases, Gap 1, Synthesis, Gap 2, and Mitosis (G_1, S, G_2, and M) (Fig. 3.10). G_1 is the period of normal cellular growth and metabolism, and if cell division is necessary, the cell will pass

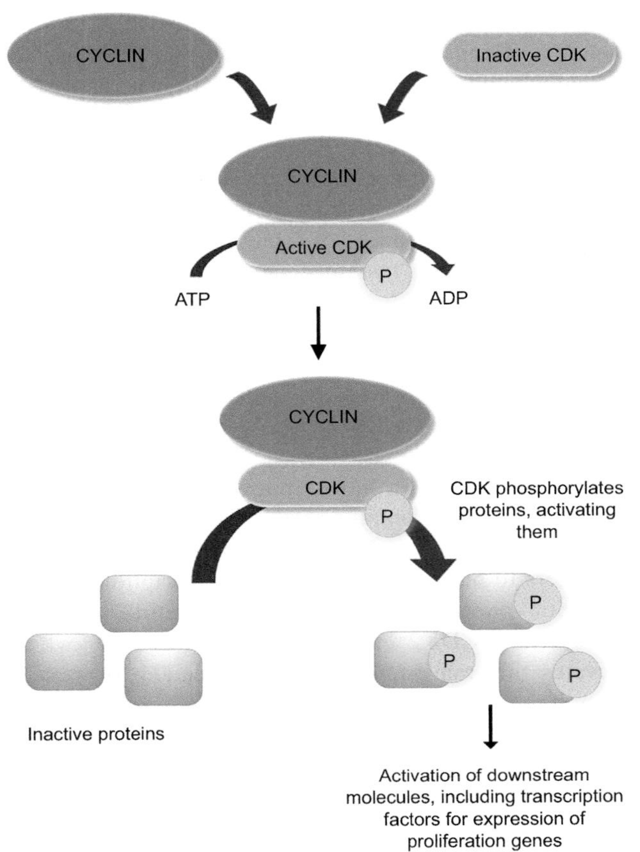

FIGURE 9.2 **Cyclins and CDKs.** The expression of various cyclins increases and decreases during the cell cycle as the cyclin proteins are expressed or degraded. When the cyclin level reaches a certain threshold, it binds to and activates its CDK by inducing a conformational change in the enzyme. This reveals the active site of the CDK, allowing the kinase to activate downstream proteins through phosphorylation. Some of the CDK targets are transcription factors that induce the genes that cause the cell to continue progressing through the cell cycle.

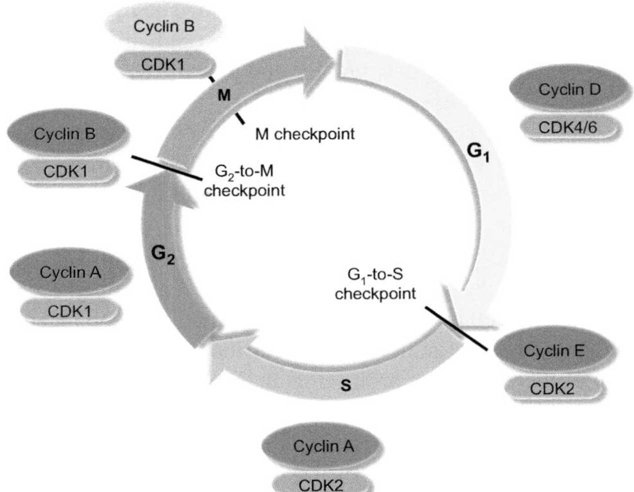

FIGURE 9.3 **Cyclins and CDKs involved in cell cycle progression and regulation.** A range of cyclins and CDKs regulate progression through the cell cycle. CDKs are constitutively expressed, only becoming activated when their corresponding cyclins reach a certain level. Cyclins and CDKs also regulate the progression of the cell through the cell cycle checkpoints.

into S phase, at which point the chromosomes are duplicated through DNA replication. The cell passes through G_2 before undergoing mitosis, which separates the duplicated chromosomes and splits the cytoplasm of the cell in half, forming two daughter cells (Fig. 1.6).

The cell has a network of regulatory proteins that determine whether it will progress through the stages of the cell cycle. Three cell cycle **checkpoints** either prevent the cell from entering the next stage or activate the transcription of genes that begin the next stage. The major checkpoints are known as the **G_1-to-S checkpoint** at the transition between G_1 and S phases, the **G_2-to-M checkpoint** at the transition between G_2 and Mitosis, and the **mitotic spindle (M) checkpoint** that occurs during metaphase of mitosis when the sister chromatids are all aligned on the metaphase plate. The M checkpoint initiates anaphase to divide sister chromatids into separate daughter cells. The cell cycle checkpoints function as gatekeepers of the cell cycle, preventing or inducing progression through the phases. DNA damage prevents progression through the G_1-to-S checkpoint, ensuring that damaged DNA is not replicated. Similarly, the G_2-to-M checkpoint prevents progression into mitosis if errors were introduced during DNA replication in S phase.

Cellular proteins called **cyclins** and **cyclin-dependent kinases (CDKs)** orchestrate the progression of the cell through these checkpoints. As their name suggests, the levels of cyclins fluctuate during the stages of the cell cycle. Cyclin expression increases as a response to external signals, such as through growth factors, nutrients, or hormones, for example. When a cyclin level reaches a certain threshold level, it activates its CDK by inducing a conformational change in the enzyme, revealing the CDK active site (Fig. 9.2). CDKs are **kinases**, enzymes that activate other proteins through **phosphorylation** (adding phosphate groups to them). Some of the CDK-activated proteins are transcription factors that induce the genes that continue the cell through the cell cycle.

CDKs are always present but are inactive until combined with a cyclin molecule, hence the name *cyclin-dependent* kinases. As the cell cycle proceeds, the cyclin is degraded through **ubiquitination**, which attaches a small regulatory protein known as ubiquitin to the cyclin protein. This targets it to the proteasome, where it is degraded.

Different combinations of cyclins and CDKs regulate the transition at the cell cycle checkpoints (Fig. 9.3). Cyclin D is the first cyclin to be synthesized and interacts with CDK4 and CDK6 to drive progression through G_1. This causes the upregulation of cyclin E, which activates CDK2 and pushes the cell through the G_1-to-S checkpoint. Cyclin A and CDK2 advance the cell through S phase, and cyclin

A and CDK1 function through G2 phase. The cyclin B and CDK1 complex drives progression through the G_2-to-M checkpoint. Cyclin B is broken down at the M checkpoint to allow mitosis to complete.

Cell cycle regulation ensures that cells with damaged DNA do not undergo cell division. In this situation, several regulatory proteins inhibit cyclins and CDKs, thereby arresting the cell cycle until DNA repair has occurred. Malignant cells acquire mutations in cyclins, CDKs, and their regulatory proteins that allow the cells to proceed through the cell cycle in situations when normal cells would be arrested. Oncogenic viruses contribute to the dysregulation of the cell cycle, promoting an environment within the cell that is more conducive to becoming malignant.

9.3 IMPORTANT GENES INVOLVED IN THE DEVELOPMENT OF CANCER

There are two general types of genes that play a major role in the development of cancer: oncogenes and tumor suppressor genes. **Proto-oncogenes** are normal cellular genes that promote cell division. When they are mutated or constitutively activated, however, proto-oncogenes become cancer-causing **oncogenes** that drive cell proliferation. Oncogenes are caused by **gain-of-function mutations** that increase the expression of the protein or increase the effectiveness of the protein. (You can remember this because *oncogene* begins with *on*—oncogenes are harmful when turned *on*.) The result is the same: the protein has increased activity, and the cell is forced to progress through the cell cycle. The majority of oncogene mutations, which can occur in the gene itself or in a regulatory region, are dominant; because the change results in a gain-of-function mutation, the normal, wild-type allele cannot make up for a mutated copy. *Myc* and *Ras* are two examples of cellular proto-oncogenes that are susceptible to mutations that cause an increase in their activity. Several viruses encode their own viral oncogenes to induce proliferation, while other viruses may integrate into a human chromosome in such a way that a strong viral promoter drives the expression of a human oncogene.

Note on Protein Nomenclature

You may notice that several proteins are named with a number following the letter "p." This convention originated with the functional identification of **p**roteins—the origin of the letter "p"—based upon their molecular weight. For example, the p53 protein was identified as a 53 kiloDalton (kD) protein. The p21 protein is a protein of 21 kD. This historical convention for naming proteins is generally now discouraged, as multiple proteins can be of the same molecular weight, and the molecular weight of a protein is affected by glycosylation and other factors, such as genetic polymorphisms.

A **tumor suppressor gene** is the opposite of a proto-oncogene: tumor suppressor genes encode proteins that normally regulate the cell cycle. If a tumor suppressor gene becomes inactivated, then the regulation of the cell cycle is lost and the cell proliferates in an uncontrolled manner. Several viral proteins are capable of interfering with tumor suppressor proteins. When caused by mutations, tumor suppressor genes are normally recessive mutations, so "two hits" are required—one inactivation of each allele in the genome—to prevent the expression of a functional protein with regulatory activity. Two important tumor suppressor genes encode the retinoblastoma protein (pRB) and the p53 protein. As a tumor suppressor, pRB is involved in preventing the transition through the G_1-to-S checkpoint. pRB binds to and inactivates E2F, a transcription factor that activates genes involved in promoting cell division (Fig. 9.4). pRB also inhibits proteins that remodel chromatin to allow promoters to become accessible to transcription factors. Sequential phosphorylation of pRB by cyclin D-CDK4/6 and cyclin E-CDK2 causes the release of E2F, which activates the transcription of genes needed to proceed through the G_1-to-S checkpoint. Therefore, when pRB is blocked, there is no regulation of E2F and the cell continues unchecked through the checkpoint. RB is considered a tumor suppressor gene because proliferation occurs when it is inactivated. In order to induce cell proliferation, several viruses possess proteins that block pRB.

Another important tumor suppressor gene encodes p53, which is the most frequently altered gene in human cancers. p53 is a DNA-binding protein (Fig. 9.5A). It becomes phosphorylated when the cell incurs DNA damage or stress stimuli, at which point it binds to DNA and stimulates the expression of its target genes, including a

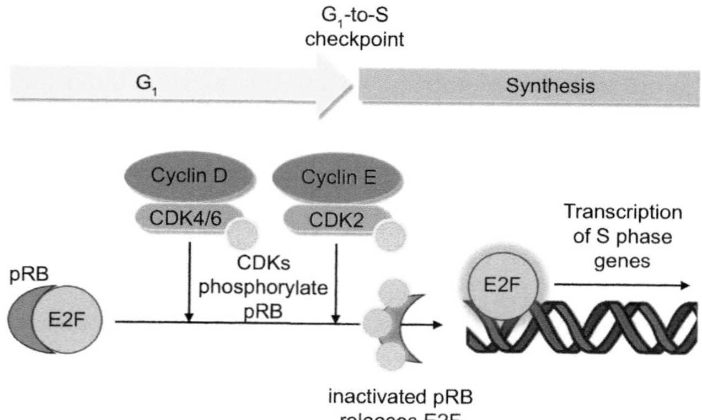

FIGURE 9.4 **Function of pRB.** pRB is a tumor suppressor that functions at the G_1-to-S checkpoint. pRB negatively regulates the G_1-to-S checkpoint by restraining E2F, a transcription factor that activates cell cycle genes. Sequential phosphorylation of pRB by cyclin D-CDK4/6 and cyclin E-CDK2 causes the release of E2F, which activates the transcription of genes needed to proceed through the checkpoint.

gene that encodes a protein known as p21 (Fig. 9.5B). The p21 protein binds to and inhibits CDK2 and CDK1, preventing progression through the cell cycle. p53 is considered a tumor suppressor gene because even though it becomes activated, the result is that the cell cycle is arrested.

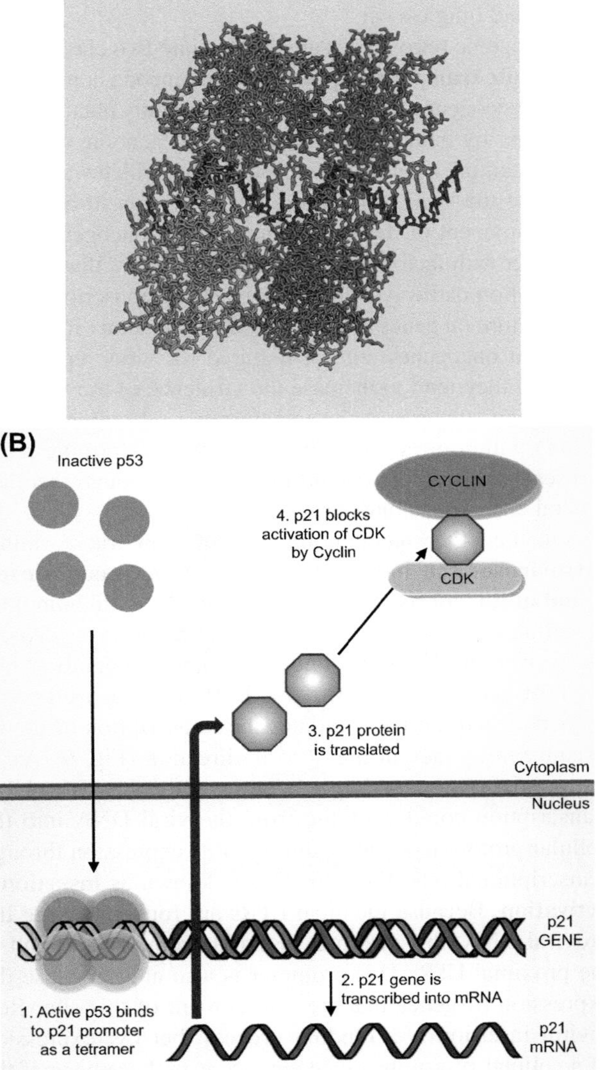

FIGURE 9.5 Tumor suppressor p53. (A) p53 is a DNA-binding protein that tetramerizes on the promoters of genes to activate their transcription. Here, four subunits of p53 (in blue and green) are bound to the promoter of the p21 gene (DNA double-helix in orange and red). *(Image created with QuteMol, 2006. IEEE Trans. Vis. Comput. Graphics 12, 1237–1244 using PDB 3TS8, 2011. Mol. Cancer Res. 9, 1493–1499.)* (B) p53 becomes phosphorylated in response to DNA damage or cell stress, at which point it translocates to the nucleus to induce gene transcription. One gene it activates encodes p21, a protein that blocks the interaction of CDKs with cyclins. In this way, p53 prevents progression through the cell cycle by interfering with CDK activity, including those that phosphorylate pRB.

Oncogene activation results in cell cycle progression, while tumor suppressor proteins prevent it. As with pRB, several oncogenic viruses interfere with the regulatory functions of p53, leading to uncontrolled cell division. The following section describes these viruses and how they induce a pro-oncogenic state within a cell.

Study Break

Draw out how all of the following proteins interact with each other: cyclin D, cyclin E, CDK4 and CDK6, CDK2, pRB, E2F, and p53. You should be able to connect all of them together in one network.

9.4 ONCOGENIC VIRUSES

Seven human viruses have been causally associated with the development of certain cancers (Table 9.2). Interestingly, the majority of the viruses are unrelated but target similar pathways involved in cell proliferation or evasion of the immune response. These viruses directly or indirectly contribute to the development of a cancerous cell, although this is a by-product, rather than a requirement, of infection.

TABLE 9.2 Cancers Associated With Human Oncogenic Viruses

Virus	Cancers
Retroviruses	
Human T-lymphotropic virus type I	Adult T cell leukemia
RNA tumor viruses	
Hepatitis C virus	Hepatocellular carcinoma
Small DNA tumor viruses	
Hepatitis B virus	Hepatocellular carcinoma
Merkel cell polyomavirus	Merkel cell carcinoma
Human papillomavirus	Cervical cancer, vulvar cancer, vaginal cancer, penile cancer, anal cancer, oropharyngeal cancer, mouth cancer
Large DNA tumor viruses	
Epstein–Barr virus	Burkitt's lymphoma, Hodgkin's lymphoma, non-Hodgkin's lymphoma, nasopharyngeal carcinoma, gastric carcinoma
Kaposi's sarcoma–associated herpesvirus	Kaposi's sarcoma, primary effusion lymphoma, multicentric Castleman's disease

It is simply that the growth-promoting properties of the viruses have oncogenic consequences. A common feature of human oncogenic viruses is that they do not initiate productive infection within tumor cells. In fact, some of these viruses become unable to replicate as a result of the changes that occur in the cancerous cell. There are four broad categories of oncogenic viruses: oncogenic retroviruses, RNA tumor viruses, small DNA tumor viruses, and large DNA tumor viruses.

9.4.1 Oncogenic Retroviruses

Retroviruses are RNA viruses that reverse transcribe their genome into DNA before integrating it into a host cell chromosome. Retroviruses share a common genome structure, with long-terminal repeats (LTRs) flanking at least three genes– *gag, pol,* and *env*– that encode polyproteins that are cleaved into structural proteins, enzymes, and glycoproteins (Fig. 9.6). The LTRs are required for integration into the host chromosome, which occurs in a random fashion. The first oncogenic viruses discovered, avian leukemia viruses and Rous sarcoma virus (RSV, not to be confused with the human pathogen with the same acronym, respiratory syncytial virus), are retroviruses. In 1911, Peyton Rous found that a filterable agent from a chicken **sarcoma**, a tumor of connective tissue, was able to cause tumors in healthy chickens. It was later discovered that the retrovirus RSV contains an additional nonessential gene, named

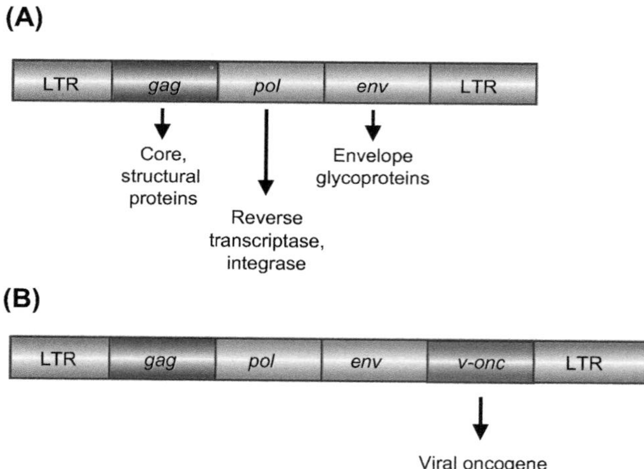

FIGURE 9.6 **Retrovirus gene structure.** All retroviruses contain at least three essential genes: *gag*, *pol*, and *env*. These genes respectively encode core and structural proteins, such as the matrix, capsid, and nucleocapsid proteins; reverse transcriptase and integrase enzymes; and the envelope glycoproteins used by the virus for attachment. (A) Shown here is the cDNA following reverse transcription, as evidenced by the two long-terminal repeats (LTR) segments that flank the genes. The LTRs are important for integration into the host chromosome. (B) Oncogenic retroviruses encode a viral oncogene that also becomes integrated into the host chromosome. No human retroviruses have been identified that contain oncogenes of cellular origin.

src (pronounced "sark") for its ability to induce sarcomas (Fig. 9.6). This was the first oncogene ever discovered, and research on viral *src* led to the discovery that humans also contain a *src* gene that functions as a proto-oncogene. To differentiate the two, the viral oncogene is known as *v-src,* while the cellular oncogene is referred to as *c-src*. Src is a kinase that promotes cell proliferation, invasion, and survival, and as an oncogene, its expression has been found to be dysregulated in several types of cancer, including colon, breast, and lung cancer.

Oncogenic retroviruses are divided into two classes. Like RSV, **acute transforming retroviruses** encode homologs of cellular oncogenes. These viruses inadvertently obtained their oncogenes by integrating into the cellular genome within an area containing a cellular proto-oncogene, which was copied alongside the viral genome during replication to become a new component of the viral genome. Viral oncogenes often encode growth factor receptors, kinases like *src* that activate proliferation pathways within the cell, and transcription factors that turn on genes involved in cell replication (Table 9.3). The viral oncogenes are not required for virus replication, although they tend to increase the virulence of the virus. As their name suggests, acute transforming retroviruses cause tumors within days. Thus far, no acute transforming retroviruses have been discovered in humans, although there are several that exist in birds or mice.

On the other hand, **nonacute** (or slow-transforming) **retroviruses** can induce tumor formation through **insertional mutagenesis**, which does not involve viral homologs of cellular oncogenes. The first way this occurs involves the activation of cellular oncogenes upon integration into the host genome. The retrovirus LTR contains a promoter sequence, and the LTR initiates the transcription of downstream viral genes in the 5′ to 3′ direction (Fig. 9.7A). If the provirus contains large deletions after the distal LTR, transcription could continue from the viral DNA into the cellular proto-oncogene, resulting in its expression through transcriptional read through. This is known as **insertional activation**. Because identical LTRs are found flanking the proviral DNA, this also means that a promoter is found in the proximal LTR. This promoter is also able to cause the expression of genes that are downstream of this site. Retroviral insertion is a random process, but overexpression of a cellular proto-oncogene can occur in these ways if the proviral DNA is integrated upstream of the gene. Similarly, cellular proto-oncogenes can be activated if strong viral enhancers in the LTR are integrated in such a way as to activate the promoters of proto-oncogenes (Fig. 9.7B). On the other hand, **insertional inactivation** could occur if a viral integration event disrupts a cellular tumor suppressor gene (Fig. 9.7C). This process is a slow one, taking years if it occurs at all, because the chances of such integration are low and would require many rounds of viral replication to occur within the person. As discussed in Chapter 8,

TABLE 9.3 Examples of Viral Oncogenes With Their Cellular Homologs

Oncogene	Function	Virus	Species
src	Kinase	Rous sarcoma virus	Chicken
mos	Kinase	Moloney murine sarcoma virus	Mouse
abl	Kinase	Abelson murine leukemia virus	Mouse
ras	G protein (GTPase)	Harvey murine sarcoma virus	Mouse
		Kirsten murine sarcoma virus	Mouse
myc	Transcription factor	Avian myelocytomatosis virus MH2	Chicken
		Avian myelocytomatosis virus MC29	Chicken
		Avian carcinoma virus MH2	Chicken
jun	Transcription factor	Avian sarcoma virus 17	Chicken
erbB	Growth factor receptor	Avian erythroblastosis virus ES4	Chicken
fms	Growth factor receptor	McDonough feline sarcoma virus	Cat

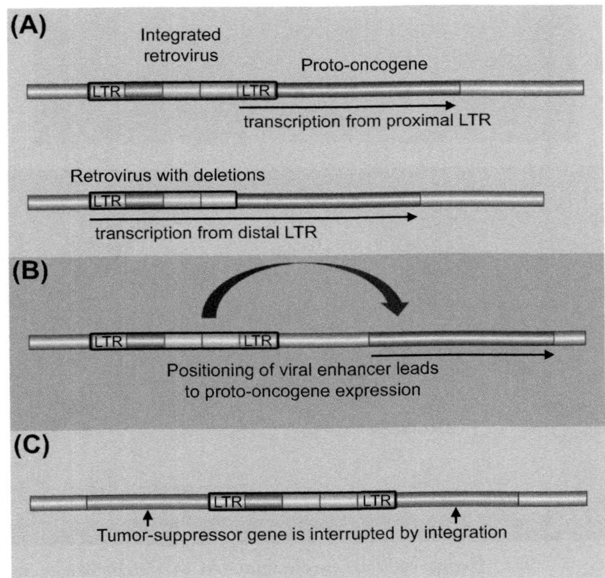

FIGURE 9.7 Retroviral insertion. Nonacute retroviruses can activate proto-oncogenes or inactivate tumor suppressor genes as a result of integration. (A) Insertional activation occurs when the promoters found in retrovirus LTRs induce transcription of proto-oncogenes after integration upstream of the proto-oncogene. Similarly, viral enhancer elements can induce transcription of proto-oncogenes from a distance (B). Insertional inactivation occurs when a retrovirus integrates within the sequence of a tumor suppressor gene (C), leading to the absence of a functional allele.

"Vaccines, Antivirals, and the Beneficial Uses of Viruses," the insertional mutagenesis outcomes described above are a consideration in using retroviruses (and lentiviruses) for gene therapy.

Although HIV is a retrovirus, compelling evidence has not been found to support that HIV reproducibly causes cancer through insertional activation. The rate of certain cancers is much higher in individuals with HIV, although this appears to be attributable to the immunosuppression that accompanies the disease that reduces the immune system surveillance of cancerous cells and oncogenic viruses. In support of this, the cancers that HIV-infected individuals develop more often than the general population are caused by oncogenic viruses: non-Hodgkin's lymphoma (caused by EBV), Kaposi's sarcoma (KSHV), and anal cancer (HPV). In this case, immunosuppression occurs before HIV multiplies enough times to randomly cause insertional activation of cellular proto-oncogenes.

One human virus, HTLV-I, has been found to be associated with adult T cell leukemia (ATL), a rare and highly aggressive malignancy of T cells. Approximately 15–25 million people worldwide are infected with HTLV-I, and 1–5% of these develop ATL, generally at least 20–30 years after initial infection with HTLV-I. In this case, a viral gene called *Tax* is responsible for transforming human T cells. Tax is a transcription factor that binds with host transcription factors to both viral and cellular promoter sequences. This leads to a plethora of intracellular events, including the phosphorylation of pRB, induction of cyclin D, and activation of CDK4 and CDK6, all events that progress the cell through cell cycle checkpoints. In addition, Tax prevents the phosphorylation of p53, leading to an inactive p53 that is unable to block cell cycle progression. HTLV-I oncogenesis is unique from those mentioned above for other retroviruses because although *tax* is considered a viral oncogene as its effects foster the development of malignancy, it has no human counterpart, which distinguishes it from the viral oncogenes mentioned above. Additionally, ATL is not associated with insertional activation of a cellular oncogene by the virus. Other HTLV-I proteins are also being investigated for their roles in reactivating cellular telomerase and evading immune responses.

9.4.2 RNA Tumor Viruses

In addition to retroviruses, one other human RNA virus has a strong association with oncogenesis: hepatitis C virus (HCV). HCV is an enveloped +ssRNA virus of the family *Flaviviridae* that infects hepatocytes, the cells of the liver. Roughly 80% of those infected are asymptomatic, but only 15–25% of initial infections are cleared by the host immune system. The remainder result in chronic infections, 15–30% of which eventually progress within 20 years to **cirrhosis**, irreversible scarring of the liver. There are an estimated 2.7–3.9 million chronic cases of HCV in the United States, and 170 million people are infected globally.

Chronic HCV infection is associated with up to 50% of the cases of hepatocellular carcinoma (HCC), a hepatocyte tumor that is the fifth most common cancer of men and seventh most common cancer of women worldwide (Fig. 9.8A). With a 5-year survival of <12%, HCC is globally the second leading cause of cancer-related mortality. Cirrhosis is definitively associated with the development of HCC, and HCV contributes to HCC indirectly by causing chronic hepatocyte inflammation that leads to cirrhosis (Fig. 9.8B). Reactive oxygen species that damage DNA are generated during chronic inflammation, introducing mutations that lead to the development of a cancerous cell. HCV is also thought to directly contribute to oncogenesis through its effects upon numerous cellular proteins to induce cellular transformation.

HCV does not integrate into the host genome. As an RNA virus, the HCV life cycle is entirely cytoplasmic. After attaching to hepatocytes, it enters the cytoplasm by endocytosis where the 9.6 kb genome is released from the capsid and is translated in its entirety as one large open reading frame by host ribosomes into a polyprotein of around 3000 amino acids in length. The polypeptide is processed by both viral and host proteases to produce a total of 10 viral proteins, 4 of which affect pathways that are involved in oncogenesis. Other than inhibiting antiviral type 1 IFN signaling, the HCV core protein promotes the expression of cyclin E and CDK2, necessary for progression through the G_1-to-S checkpoint. Core protein increases the expression of telomerase and inhibits apoptosis by interfering with caspase-8, an enzyme at the top of an intracellular cascade that leads to the induction of apoptosis. NS5A, an HCV protein implicated in resistance to type 1 IFN, also inhibits apoptosis by stabilizing proteins that prevent caspase activation. Both core and NS5A block the tumor suppressor p53 to induce proliferation. Similarly, core and NS5B, the viral RNA polymerase, interfere with tumor suppressor pRB, leading to E2F-dependent transcription, progression through the cell cycle, and cellular proliferation. Transfection of HCV NS3 DNA into normal cells induces their transformation, and these cells are tumorigenic when injected into immunocompromised animals. So although HCV induces a chronic inflammatory response that leads to cirrhosis, which is strongly associated with the development of HCC, the virus also encodes several genes that are likely to encourage an oncogenic state due to their influence upon cellular proliferation and apoptosis pathways.

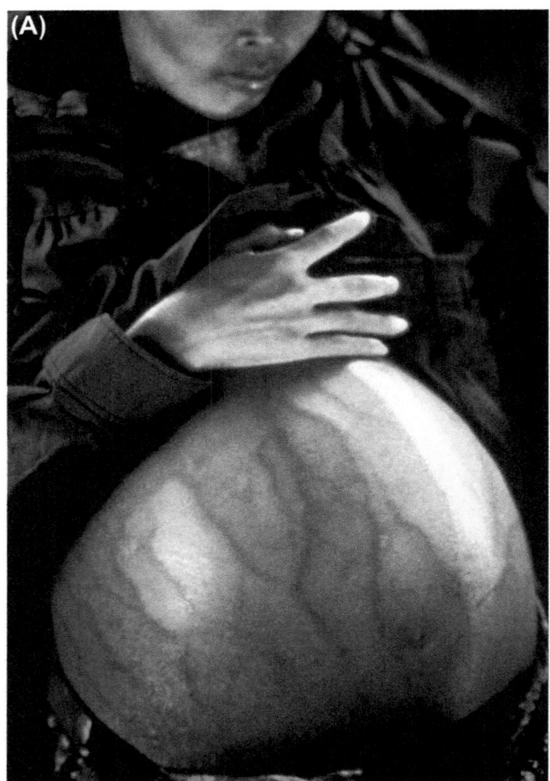

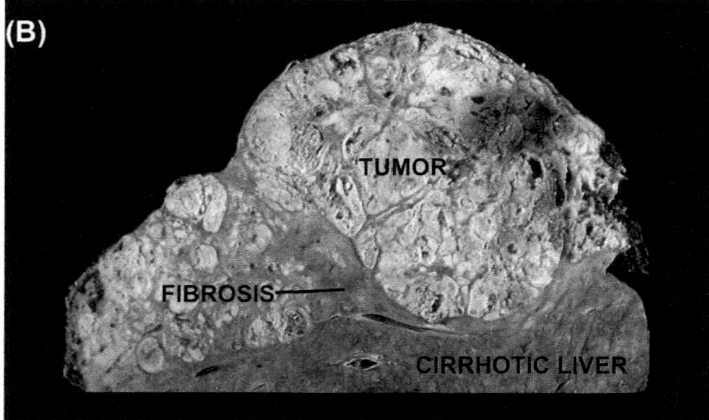

FIGURE 9.8 **Hepatocellular carcinoma.** (A) A Cambodian woman exhibits a visible protrusion from her abdomen, due to the presence of a large hepatocellular carcinoma. *(Photo courtesy of the CDC and Dr. Patricia Walker, Regions Hospital, Minnesota.)* (B) A cross section of a hepatocellular carcinoma extending from the cirrhotic liver of an HCV-positive individual. Fibrosis is visible in between the two. *(Photo courtesy of Dr. Ed Uthman.)*

Study Break

Describe which HCV proteins interfere with cell proliferation and apoptosis. Why is it necessary for oncogenesis that both these pathways are inhibited?

9.4.3 Small DNA Tumor Viruses

Three small human DNA viruses are associated with the development of cancers: HBV, MCPyV, and certain types of HPV, known as "high-risk" HPVs. DNA tumor viruses contain oncogenes that are of viral origin, not cellular origin as with acute transforming retroviruses. These viruses constitute a diverse group from the *Hepadnaviridae, Polyomaviridae,* and *Papillomaviridae* families, respectively, although they all target similar pathways within the cell to induce a malignant state.

Although it has many different properties than HCV, HBV shares some similarities with HCV, as well. HBV is a virus that also exhibits tropism for the liver, induces chronic hepatitis and cirrhosis, and causes up to 50% of the cases of HCC. However, whereas acute HCV infection progresses to chronic infection in the majority of cases, only about 5% of adults develop chronic HBV infection—the remainder of acute infections are cleared from the host. This number is significantly higher in children, where 30–50% of those infected between 1 and 5 years of age and >90% of those <1 year of age become chronically infected. Altogether, 350 million people in the world have chronic, life-long infections.

HBV is an unusual virus. It possesses a partially double-stranded 3.2 kb DNA genome, and is one of the few non-retroviruses to encode and utilize a reverse transcriptase to complete its life cycle. After attaching to hepatocytes, it is brought into the cell via endocytosis and the genome is transported to the nucleus. The genome is then repaired to a fully double-stranded circular form, known as the covalently closed circular DNA (cccDNA), that is transcribed by host RNA polymerase II. It encodes a total of seven proteins, one of which, protein X (HBx), has been implicated in oncogenic processes.

Like HCV, HBV is associated with fibrosis and cirrhosis that leads to a cancerous state due to the inflammation that induces hepatocyte DNA damage and mutations. However, there is also evidence that HBx plays a synergistic role in this process. HBx is an essential viral regulatory protein that has pleomorphic effects within the cell, stimulating cell proliferation pathways and interfering with DNA repair mechanisms. It also represses the p53 promoter and interferes with the binding of p53 to its targets.

Following repair of its single-stranded genome segment, the HBV cccDNA episome is usually maintained as a separate entity within the nucleus. However, in 80% of HCC cases associated with HBV, the genome is found to have been integrated into the host chromosome, which is not a requirement of viral replication. This certainly leads to genomic instability and insertional activation of cellular proto-oncogenes or disturbance of cellular tumor suppressor genes. Integration of HBV appears to play an important role in the development of HCC.

The HBV vaccine was effectively the first anticancer vaccine to be developed. Because it prevents infection with HBV, it also prevents HBV-mediated HCC. All children in the United States are vaccinated against HBV with a three-shot series, the first of which is administered at birth (see Chapter 8, "Vaccines, Antivirals, and the Beneficial Uses of Viruses").

MCPyV was discovered in 2008 and is the first human polyomavirus definitively associated with the development of human cancer (although other polyomaviruses, such as SV40, have been shown to induce cellular transformation and have been found in patients with certain cancers). MCPyV DNA is detected in around 80% of biopsies of Merkel cell carcinoma (MCC), a rare and aggressive type of skin cancer. MCC predominantly occurs in older and immunocompromised patients, although MCPyV is a common human virus, infecting up to 80% of adults by age 50.

Polyomaviruses are nonenveloped icosahedral viruses with circular dsDNA genomes of ~5000 bp. They are characterized by the expression of T (tumor) antigens, a subset of proteins of different sizes due to alternative splicing. Because polyomaviruses are completely dependent upon the host cell for replication, the large T antigen and small T antigen cooperate to drive the cell into S phase so that the enzymes required for viral replication are available. The large T antigen physically interacts with and inhibits tumor suppressors pRB and p53, allowing the cell to progress through the G_1-to-S checkpoint, and small T antigen induces transcription of E2F, necessary for turning on essential S phase genes. Large T antigen is a DNA-binding protein that possesses DNA helicase and ATPase activity, all of which are required for it to bind to and unwind viral DNA during its replication in S phase. It then interacts with cellular primase and DNA polymerase to stimulate replication of the viral genome. Because of these attributes, the large T antigen is indispensable for polyomavirus replication.

In MCCs, the MCPyV genome becomes randomly integrated into host chromosomes. In this state, the virus is unable to replicate because it has no way to excise its genome, although the large T antigen gene is still present. In MCCs, the large T antigen gene becomes mutated so it is no longer able to initiate DNA replication, although it is still able to interfere with pRB and drive cell proliferation. As a result, the low frequency and random integration of McPyV into human cells lead to a pro-oncogenic state that may lead to malignancy if other mutations are acquired. Because MCC is a skin cancer, it is possible that UV radiation may be a contributing factor in the acquisition of additional mutations. As with other viruses, tumor formation by

MCPyV is an unintended, chance event that does not promote viral replication.

The final small DNA tumor virus that is currently associated with cancer is HPV, the most common sexually transmitted infection in the United States. Papillomaviruses are small, nonenveloped icosahedral viruses (Fig. 9.9A) with circular dsDNA genomes that infect the basal layer of cells in the skin or mucosal epithelium and cause **papillomas**, or wartlike growths. Over 200 different types of HPV have been described and categorized into "low-risk" or "high-risk," based upon their oncogenic potential. Low-risk HPVs, such as types 6 and 11, cause genital warts (Fig. 9.9B), but high-risk HPVs, including types 16 and 18, act as carcinogens and are detected in 99% of cervical cancers (Fig. 9.9C). Cancers of epithelial cells are known as **carcinomas**. Type 16 alone causes 50% of cervical cancers, while 16 and 18 together account for 70%. Types 21, 33, 45, 52, and 58 are responsible for an additional 20% of cervical cancers. High-risk HPVs are also associated with other anogenital cancers, including cancer of the vulva, vagina, penis, and anus. It is also a major cause of head and neck cancers that are not caused by smoking, such as cancer of the oropharynx, the back of the throat (Fig. 9.9D). Most acute HPV infections are cleared by an effective immune response, but a small proportion result in persistent infections that, within years or decades, can form cervical intraepithelial neoplasias (CINs), transformations of cervical cells that can resolve spontaneously or progress to cervical cancer.

The HPV genome encodes eight genes, six that are encoded from the early region of the genome and two that are late genes. During infection of the cervix, HPV infects the basal cell layer of the stratified (multilayer) mucosal epithelium, and two early-encoded HPV genes are involved in transformation by HPV: E6 and E7. As with MCPyV, the purpose of these proteins is to cause infected cells to progress to S phase, when the enzymes needed for viral replication are available. HPV E7 shows sequence similarity to the polyomavirus large T antigen, and similarly, it binds to pRB, freeing E2F to activate transcription of genes necessary for cell cycle progression. This would normally signal p53 to prevent G_1-to-S progression, but the HPV E6 protein is targeted to the nucleus and degrades p53 by inducing its ubiquitination. E6 positively affects other transcription factors, such as Myc, that increase telomerase production within the cell. Both E6 and E7 inhibit type 1 IFN signaling, as well.

During normal replication processes, the circular HPV genome remains as an episome within the nucleus. In contrast, high-risk HPV types are found to be nonspecifically integrated into the host genome in malignant tumors. As with the other DNA tumor viruses, this acts as a terminal event for the virus, preventing further replication. Integration results in persistent expression of E6 and E7, leading to increased genomic instability and further mutations that result in transformation and tumor development.

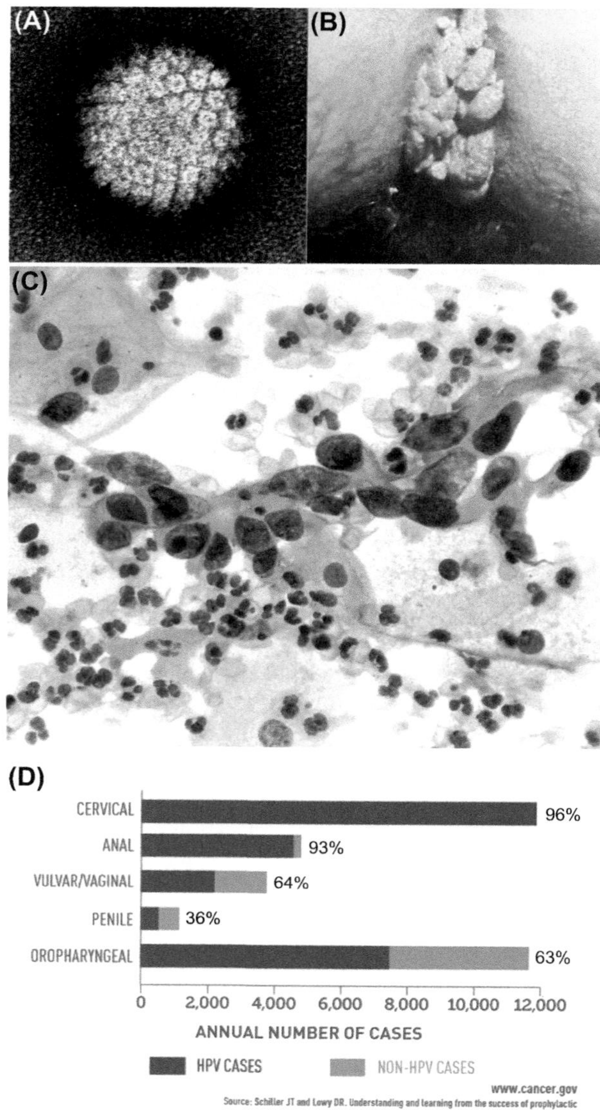

FIGURE 9.9 **Human papillomavirus.** (A) HPV is a small, nonenveloped icosahedral virus. *(Image courtesy of the National Cancer Institute, Laboratory of Tumor Virus Biology.)* (B) Low-risk HPV types, such as types 6 or 11, cause genital warts, such as those observed in the perineal region between the legs of this man. *(Photo courtesy of CDC/Dr. Wiesner.)* (C) In contrast, high-risk HPVs, such as types 16 and 18, are the cause of virtually all cervical cancers. This cytological cervical specimen reveals cervical cancer cells exhibiting HPV-associated changes. These include a large nuclear to cytoplasmic ratio, irregularity of the nuclear membrane, variations in nuclear size, and cells undergoing mitosis. The orange dye stains for keratin. *(Image courtesy of the National Cancer Institute.)* (D) HPV infection causes a substantial percentage of several types of cancer each year, including cervical, anal, vulvar/vaginal, penile, and oropharyngeal cancers. The HPV vaccine is anticipated to significantly reduce these numbers. *(Image modified from the National Cancer Institute.)*

Approved in 2006, the HPV vaccine is the second anticancer vaccine to be produced, and it is recommended for both female and male adolescents. The bivalent vaccine protects against HPV type 16 and type 18, the two high-risk types that are responsible for 70% of cervical cancer cases. The quadrivalent vaccine also protects against genital warts from low-risk types 6 and 11, as well as types 16 and 18. In December 2014, another HPV vaccine was approved that protects against nine types of HPV, both the low-risk (types 6 and 11) and high-risk (types 16, 18, 31, 33, 45, 52, and 58) types. The vaccines are recombinant subunit vaccines that contain the HPV L1 protein, which forms the capsid of the virus. The vaccines are highly immunogenic after the three-shot series: 99% of recipients develop an antibody response against HPV, and nearly 100% of precancerous cervical cell changes are prevented by the vaccine. The vaccines have also been proven to be hugely effective in preventing cervical, vulvar, vaginal, anal, and oral HPV infections and cancers. If administered consistently, the vaccine could drastically reduce the seven million new HPV infections that occur each year in the 15–24 year age group. Currently, 27,000 cases of cancer in the United States are attributable to HPV infection: 19,000 in women and 8000 in men. The HPV vaccine is expected to significantly diminish this number over time in those that are vaccinated. Vaccination efforts began less than 10 years ago and cancers take years to develop, so ongoing studies will provide definitive proof that inhibiting HPV infection prevents the development of the cancers that are currently associated with it.

9.4.4 Large DNA Tumor Viruses

EBV and KSHV are two viruses that are associated with the development of tumors in humans. As herpesviruses, EBV and KSHV are large, enveloped viruses with complex dsDNA genomes that each encodes over 50 genes. A hallmark of all herpesviruses is that they are never fully cleared from the host, establishing latency within cells after the initial infection.

EBV, which infects B lymphocytes and epithelial cells, causes 90% of the cases of infectious mononucleosis, or "mono." Additionally, the virus also plays a role in the development of B cell **lymphomas**, malignancies that arise in the lymph nodes or in lymphatic tissues. These include Burkitt's lymphoma (Fig. 9.10A and B) and Hodgkin's lymphoma, as well as non-Hodgkin's lymphoma in immunodeficient individuals. EBV also causes epithelial carcinomas of the stomach and nasopharynx, the back of the nasal cavity (see Fig. 10.1).

To become latent, EBV progresses through several gene expression stages, known as "latency programs," in infected cells (Fig. 9.10B). During productive ("lytic") infection of B cells, EBV undergoes the latency III program and expresses several viral genes that induce proliferation and prevent apoptosis, but this also activates immune recognition of the virus. This causes the virus to enter the latency II program, which leads to the differentiation of the proliferating cells into memory B cells. At this point, they undergo the latency

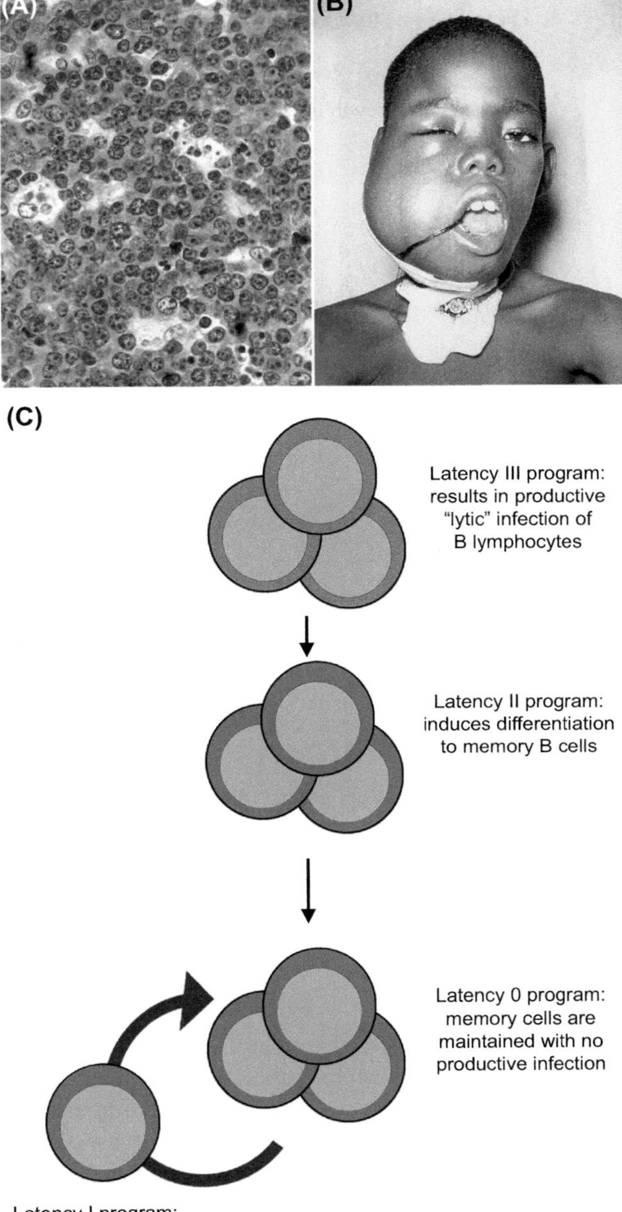

FIGURE 9.10 **Burkitt's lymphoma and the latency transcription programs of EBV.** EBV causes several cancers, including Burkitt's lymphoma, a malignancy of B cells (A). *(Photo by Louis M. Staudt/National Cancer Institute.)* (B) A Nigerian child exhibiting a Burkitt's lymphoma tumor on the right side of his face. *(Photo by CDC/Robert S. Craig.)* (C) EBV goes through several stages of gene expression, known as latency programs. Upon initial infection of B cells, EBV induces a productive infection through genes expressed during the latency III program. Due to immune recognition, the virus switches to the latency II program, which induces the differentiation of the cells into memory B cells. The virus then reduces its viral protein expression, undergoing latency 0 program, which is associated with a latent state. Occasionally, the memory B cells undergo reduced proliferation to maintain their populations, at which point EBV activates the latency I program to maintain the EBV episome within infected cells.

TABLE 9.4 EBV Oncogenes Expressed During Latency Replication Programs

Program	Oncogene	Involvement in Oncogenesis
Latency III (growth transcription program)		
	EBNA1	Prevents cell death, increases genomic instability through reactive oxygen species
	LMP1	Mimics cell receptor CD40 to induce proliferation and promote cell survival
	LMP2	Promotes differentiation, survival, and cell growth
	EBNA2	Upregulates cyclin D and E, downregulates CDK inhibitors, induces proliferation
	EBNA3A/B/C	Engages cellular ubiquitin ligase to target pRB for proteasomal degradation, drives proliferation
	EBNALP	Transactivator of viral and cellular genes, involved in cell transformation
	miRNAs	BHRF miRNAs promote cell cycle progression and inhibit apoptosis
Latency II (differentiation transcription program)		
	EBNA1	See above
	LMP1	See above
	LMP2	See above
	miRNAs	BART miRNAs inhibit apoptosis and type 1 IFN signaling
Latency I (memory proliferation program)		
	EBNA1	See above
	miRNAs	BART miRNAs inhibit apoptosis and type 1 IFN signaling

EBNA, EBV nuclear antigen; *LMP*, latent membrane protein.

0 program, whereby no viral proteins are expressed, thereby preventing immune detection. This is the gene program that exemplifies true viral latency. Occasionally, the memory B cells must undergo proliferation to maintain their populations. When this occurs, the virus activates the latency I program. Viral expression of EBV nuclear antigen 1 (EBNA1) at this point functions to maintain the EBV genome as an episome alongside the host chromosome.

During the latency III program, which is associated with cell proliferation, a large complement of EBV genes are expressed that induce the cell cycle and prevent apoptosis; these are listed in Table 9.4. Because the latency III program is very immunogenic, EBV-related cancers from this stage of cell only appear in immunocompromised individuals, such as AIDS-associated non-Hodgkin's lymphoma, which occurs in 4–10% of HIV/AIDS patients. A smaller subset of EBV genes are expressed in latency II, namely EBNA1, latent membrane protein 1 (LMP1), and LMP2, to induce cell proliferation and promote cell survival (Fig. 9.11A and B). Hodgkin's lymphoma and nasopharyngeal carcinomas are from cells in the latency II program. Burkitt's lymphoma, on the other hand, is associated with cells in the latency I memory cell proliferation program, when mainly EBNA1 is expressed. **MicroRNAs** (miRNAs) are small pieces of antisense RNA that bind to genes to prevent their activity, and EBV produces many of its own miRNAs during latency I, II, and III programs that inhibit cellular tumor suppressor regulatory genes, block apoptosis, and prevent immune responses, further contributing to an oncogenic cellular environment. miRNAs will be discussed in more detail in Chapter 13, "Herpesviruses."

Although EBV infection is associated with these cancers, it is not sufficient to induce malignancy—as with other oncogenic viruses, all EBV-induced cancers require other cellular mutations to occur, as well. This can occur through the genomic instability caused by viral infection or by infection with other pathogens, among other causes. For

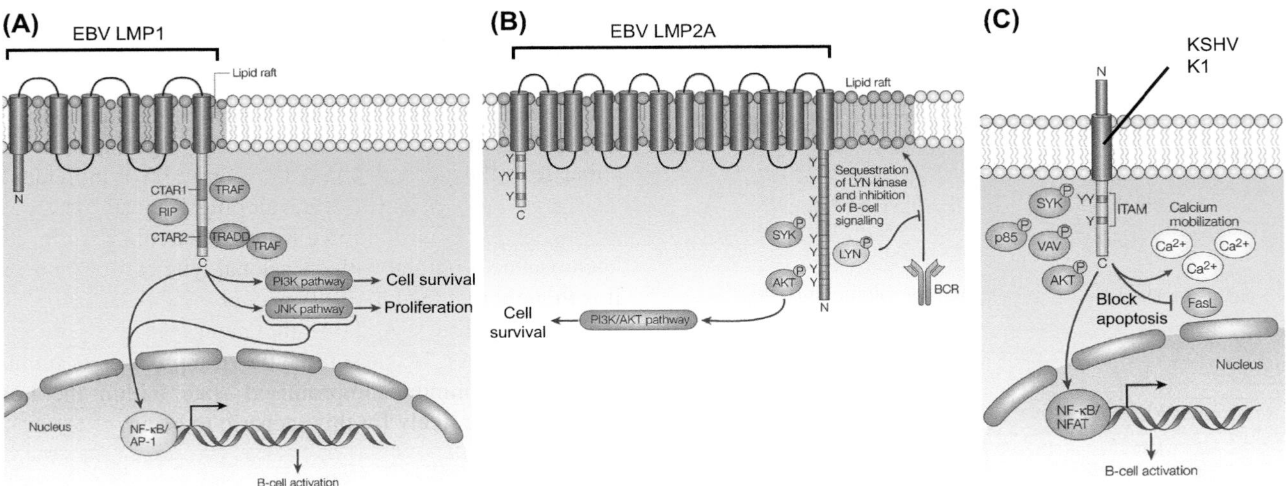

FIGURE 9.11 **EBV and KSHV oncoproteins.** EBV and KSHV each encode several genes that induce cell proliferation and prevent apoptosis. (A) The EBV LMP1 protein is a transmembrane protein that integrates into the plasma membrane of infected B cells. It constitutively activates signaling pathways within the cell that lead to cell survival, proliferation, and B cell activation. (B) LMP2A is another EBV oncoprotein that promotes cell survival of B cells. LMP2 transforms epithelial cells, inducing proliferation and preventing differentiation. (C) The KSHV K1 protein, also constitutively active, induces the phosphorylation of several signal transduction pathways within the cell that lead to B cell activation and the prevention of apoptosis. It is expressed in Kaposi's sarcoma, primary effusion lymphomas, and multicentric Castleman's disease. *Reprinted with permission from Damania, B., 2004. Oncogenic γ-herpesviruses: comparison of viral proteins involved in tumorigenesis. Nature Reviews Microbiology. 2(8), 656–668. Macmillan Publishers Ltd., Copyright 2004.*

example, 90% of Burkitt's lymphomas possess a chromosomal translocation whereby pieces of chromosome 8 and 14 have been translocated, inducing the activation of Myc, a cellular oncogene. Coinfection with *Plasmodium falciparum,* the parasite that causes malaria, is hypothesized to increase the likelihood of chromosomal translocations in EBV-infected B cells.

The other oncogenic herpesvirus, KSHV, usually causes subclinical infection, but the virus is also a contributing factor in Kaposi's sarcoma, which is an endothelial cell cancer that causes visible red or purple skin lesions (Fig. 9.12), as well as a rare non-Hodgkin's lymphoma known as primary effusion lymphoma. It is also associated with multicentric Castleman's disease, a lymphocyte proliferation disorder that acts much like a lymphoma. Like EBV, KSHV expresses a different subset of viral proteins during lytic or latent stages of infection. During latency, a reduced subset of genes is expressed (Table 9.5) that inhibits p53 and pRB tumor suppressor pathways, promote proliferation through oncogene activation, block apoptosis, and activate expression of telomerase. A full complement of KSHV genes are expressed during lytic replication that promote proliferation, inhibit cell death (Fig. 9.11C), modify the metabolism of the cell, and interfere with immune recognition. miRNAs also function during both latent and lytic stages to prevent apoptosis and induce mitosis.

A unique characteristic of herpesviruses, particularly KSHV, is that they encode several homologs of host genes that are involved in proliferation. KSHV encodes a viral homolog of cyclin D (called v-cyclin) that is constitutively active, forcing the activation of CDK4 and CDK6 and progression through G_1. Viral FLICE/caspase-8-inhibitory protein (vFLIP), the viral homolog of cellular FLIP, is a constitutively active protein as well, blocking caspase-8 and its proapoptotic signaling. KSHV also encodes several other viral homologs that induce the production of growth factors, block apoptosis, and prevent immune recognition. The virus even encodes viral homologs of cellular miRNAs!

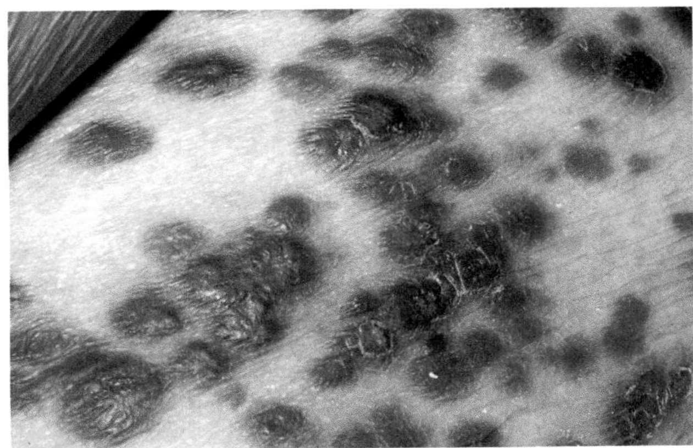

FIGURE 9.12 **Kaposi's sarcoma.** Kaposi's sarcoma on the skin of an AIDS patient. *Photo courtesy of the National Cancer Institute.*

TABLE 9.5 KSHV Oncogenes Expressed During Latency or Lytic Replication Stages

Stage	Oncogene	Involvement in Oncogenesis
Lytic		
	vIRFs	Prevent apoptosis, inhibit p53 activation, interfere with type 1 IFN response
	vGPCR	Activates immortalization, upregulates proangiogenic factors, induces reactive oxygen species, leading to mutations
	vBCL2	Promotes cell survival
	K1	Induces angiogenesis, mediates invasiveness and metastasis, switches metabolism toward aerobic glycolysis
	vIL-6	Stimulates angiogenesis, inflammation
	vCCL-1, -2, -3	Proangiogenic factors, attract inflammatory cells
	miRNAs	Inhibit type 1 IFN signaling, prevent apoptosis
Latent		
	LANA	Maintains genomic episome, inhibits pRB and p53, stimulates cell proliferation, activates telomerase
	v-cyclin	Constitutively active version of cyclin D that activates CDK4/6
	vFLIP	Activates antiapoptotic signaling
	miRNAs	Promote angiogenesis, inhibit p53, prevent apoptosis, inhibit immune responses

As with other viruses, KSHV-infected cells require additional cofactors to become malignant, and immunodeficiency is one means of achieving unregulated viral responses. Patients undergoing organ transplant have a 500–1000 times greater risk of developing Kaposi's sarcoma than the general population, and HIV+ individuals have a 50% lifetime risk of developing Kaposi's sarcoma. The majority of cells within a Kaposi's sarcoma exhibit the latency gene program, with occasional lytic cells appearing. It is thought that the lytic cells may provide the growth factors and signaling that allows the latent cells to proliferate within the tumor. Because the lytic viral program is immunogenic, an immunocompromised state within the host would be more likely for this scenario to occur.

SUMMARY OF KEY CONCEPTS

Section 9.1 Properties of Cancerous Cells

- Oncogenesis refers to the development of cancer, which causes over eight million deaths each year worldwide.
- Cancer is caused by uncontrolled cell division that results in proliferating cells that invade nearby tissues and metastasize to other areas of the body. Carcinogens are substances that cause cancer.
- Infection with certain viruses can lead to cell transformation. Transformed cells reactivate telomerase, proliferate indefinitely, and become immortal. They also lose contact inhibition and become anchorage independent.
- Inducing malignancy is not a requirement of viral replication. It is a by-product of the effect of viral proteins that cause cell proliferation, prevent apoptosis, and evade immune responses.

Section 9.2 Control of the Cell Cycle

- Viral infection is not sufficient to induce cancer. Other factors must occur to create proliferative cells capable of indefinite replication, invasiveness, and metastasis.
- Cyclins and CDKs regulate the progression through the cell cycle and through checkpoints that prevent or induce passage into the next phase. When a cyclin level reaches a certain threshold level, it interacts with its CDK, which activates other proteins through phosphorylation. Different combinations of cyclins and CDKs regulate progression through the cell cycle phases.
- Proto-oncogenes are cellular genes that induce oncogenic proliferation when mutated or modified. Tumor suppressor genes encode proteins that regulate the cell cycle. pRB and p53 are two major tumor suppressor genes, both of which are inhibited by many viral proteins.

Section 9.3 Oncogenic Viruses

- Acute transforming retroviruses, such as Rous sarcoma virus, encode viral homologs of cellular oncogenes that induce cancer. No human acute transforming retroviruses have been discovered, although they have been found in birds and certain mammals.
- Nonacute retroviruses induce tumor formation as a side effect of integration. Viral promoters or enhancers can activate cellular proto-oncogenes, and integration into a tumor suppressor gene can prevent its functionality.
- HTLV-I is associated with adult T cell leukemia. The viral gene *Tax* encodes a protein that leads to the phosphorylation of pRB, activation of cyclin D and CDK4/6, and inhibition of p53.
- HCV is associated with cirrhosis and progression to hepatocellular carcinoma. In addition to the proinflammatory effects of viral infection, the HCV core, NS5A, NS5B, and NS3 proteins are implicated in inducing proliferation and inhibiting apoptosis.
- HBV is also a major cause of HCC. The HBx protein stimulates cell proliferation and represses the activity of p53. A vaccine is available that prevents HBV and therefore the cancers it induces.
- MCPyV causes Merkel cell carcinoma, a skin cancer. MCPyV atypically becomes randomly integrated into the genome, and the large T antigen induces cell proliferation while inhibiting pRB and p53.
- High-risk HPV types are a major cause of cervical, anal, vulvar, vaginal, penile, and oropharyngeal cancers. HPV E6 protein degrades p53, and E7 inhibits pRB to support oncogenesis. An HPV vaccine is available that prevents both low-risk and high-risk HPV types.
- The herpesvirus EBV is associated with Burkitt's lymphoma, Hodgkin's lymphoma, non-Hodgkin's lymphoma, nasopharyngeal carcinoma, and gastric carcinoma. The virus progresses through several gene programs, known as latency programs, to become less immunogenic and finally establish latency. Viral proteins with pro-oncogenic properties are expressed during latency I, II, and III programs, and EBV-associated cancers express gene programs that are representative of one of these programs. miRNAs are also expressed by the virus that inhibit viral and cellular genes.
- The herpesvirus KSHV is associated with Kaposi's sarcoma, primary effusion lymphoma, and multicentric Castleman's disease. KSHV has many proteins that promote proliferation, inhibit apoptosis, and interfere with host immune responses, including several viral homologs of cellular genes.

FLASH CARD VOCABULARY

Oncogenesis	Malignant
Cancer	Angiogenesis
Benign	Metastasize
Carcinogens	Proto-oncogenes/oncogenes
Transformation	Gain-of-function mutations
Telomerase	Tumor suppressor gene
Contact inhibition	Acute transforming retrovirus
Anchorage independence	Nonacute retrovirus
Tumorigenic	Insertional mutagenesis
Checkpoints	Insertional activation
G_1-to-S checkpoint	Insertional inactivation
G_2-to-M checkpoint	Cirrhosis
M checkpoint	Papilloma
Cyclins	Carcinoma
Cyclin-dependent kinases	Lymphoma

Kinases	Sarcoma
Phosphorylation	MicroRNAs
Ubiquitination	

CHAPTER REVIEW QUESTIONS

1. Make a list describing the properties of transformed cells.
2. Consult the 10 Hallmarks of Cancer in Table 9.1. For each virus, identify which stages are affected by the viral proteins that were discussed.
3. Explain how cyclins are involved in the activation of CDKs.
4. Draw the phases of the cell cycle, and fill in which cyclin-CDK combinations are involved in the progression through each stage.
5. Why is pRB considered a tumor suppressor? How does it control the G_1-to-S checkpoint?
6. Describe four different ways that retrovirus integration can affect cell proliferation or apoptosis mechanisms.
7. Several oncogenic viruses integrate into host chromosomes, when normally they do not. How does this create a pro-oncogenic state within cells?
8. Describe how p53 works to prevent cell cycle progression.
9. How do vaccines against HBV and HPV prevent certain cancers?
10. Why do you think that cancers caused by EBV and KSHV are more common in HIV-infected individuals than in the general population?
11. How would you go about creating a drug that inhibits virally induced oncogenesis? Which of the viral proteins discussed would be easiest or most effective to target?
12. Several very different viruses support oncogenesis. What are the common cellular targets of these viruses?

FURTHER READING

American Cancer Society, 2015. Cancer Facts & Figures 2015. American Cancer Society, Atlanta, GA. http://dx.doi.org/10.3322/caac.21254.

Centers for Disease Control and Prevention, 2015. Human papillomavirus. In: Hamborsky, J., Kroger, A., Wolfe, C. (Eds.), Epidemiology and Prevention of Vaccine-Preventable Diseases, thirteenth ed. Public Health Foundation, Washington D.C, pp. 175–186.

Cutts, F., Franceschi, S., Goldie, S., et al., 2007. Human papillomavirus and HPV vaccines: a review. Bull. World Health Organ. 85, 719–726.

El-Serag, H.B., 2012. Epidemiology of viral hepatitis and hepatocellular carcinoma. Gastroenterology 142, 1264–1273.

Hanahan, D., Weinberg, R., 2011. Hallmarks of cancer: the next generation. Cell 144, 646–674.

Hariri, S., Dunne, E., Saraiya, M., Unger, E., Lauri Markowitz, M., 2011. Human papillomavirus. In: Roush, S.W., Baldy, L.M. (Eds.), Manual for the Surveillance of Vaccine-Preventable Diseases. Centers for Disease Control and Prevention, Atlanta, GA, pp. 5-1–5-11 (Chapter 5).

Kannian, P., Green, P.L., 2010. Human T lymphotropic virus type 1 (HTLV-1): molecular biology and oncogenesis. Viruses 2, 2037–2077.

Maeda, N., Fan, H., Yoshikai, Y., 2008. Oncogenesis by retroviruses: old and new paradigms. Rev. Med. Virol. 18, 387–405.

Martin, D., Gutkind, J.S., 2008. Human tumor-associated viruses and new insights into the molecular mechanisms of cancer. Oncogene 27 (Suppl. 2), S31–S42.

McLaughlin-Drubin, M.E., Munger, K., 2008. Viruses associated with human cancer. Biochim. Biophys. Acta 1782, 127–150.

Mesri, E.A., Feitelson, M.A., Munger, K., 2014. Human viral oncogenesis: a cancer hallmarks analysis. Cell Host Microbe 15, 266–282.

Moore, P.S., Chang, Y., 2010. Why do viruses cause cancer? Highlights of the first century of human tumour virology. Nat. Rev. Cancer 10, 878–889.

Pecorino, L., 2012. Molecular Biology of Cancer, third ed. Oxford University Press, Oxford, United Kingdom.

Robertson, E.S., 2012. Cancer Associated Viruses. http://dx.doi.org/10.1007/978-1-4614-0016-5.

Shlomai, A., de Jong, Y.P., Rice, C.M., 2014. Virus associated malignancies: the role of viral hepatitis in hepatocellular carcinoma. Semin. Cancer Biol. 26, 78–88.

Shuda, M., Feng, H., Kwun, H.J., et al., 2008. T antigen mutations are a human tumor-specific signature for Merkel cell polyomavirus. Proc. Natl. Acad. Sci. U.S.A. 105, 16272–16277.

Spurgeon, M.E., Lambert, P.F., 2013. Merkel cell polyomavirus: a newly discovered human virus with oncogenic potential. Virology 435, 118–130.

Vogt, P.K., 2012. Retroviral oncogenes: a historical primer. Nat. Rev. Cancer 12, 639–648.

Chapter 10

Influenza Viruses

Influenza, more commonly known as "the flu," is caused by an influenza virus. The word "influenza" stems from the 15th century Italian word *influentia,* meaning *influence,* from the belief that epidemics caused by the virus were attributable to the "influence of the stars." Influenza viruses circulate within human and animal populations. They can cause mild seasonal illnesses, but they have also been the cause of pandemics that have caused the deaths of millions of people within the course of only months. This chapter discusses the clinical course of infection, the intracellular replication strategy of influenza viruses, and how genetic changes in the virus can lead to serious pandemics.

10.1 INFLUENZA TAXONOMY AND TYPES

Influenza viruses are enveloped, –ssRNA viruses found within the *Orthomyxoviridae* family. Three genera exist with this family that pertain to influenza viruses, *Influenzavirus A, Influenzavirus B,* and *Influenzavirus C,* each of which contain a single species, or **type**: *Influenza A virus, Influenza B virus,* and *Influenza C virus,* respectively (Table 10.1). Influenza A viruses infect humans and a variety of other wildlife, including pigs and birds, which perpetuate the virus in nature throughout the world. Human influenza A viruses are thought to have originated from strains that infected wild aquatic birds. Influenza B viruses infect humans (and seals, interestingly) and primarily affect children. Infection with influenza C generally results in mild or subclinical infections. Influenza A and B cause seasonal epidemics, but only influenza A has caused worldwide pandemics. Generally, influenza A infections account for about 2/3 of human infections each year.

The **subtypes** of Influenza A virus are determined by the two viral proteins embedded into the envelope of the virion: **hemagglutinin** (HA or H) and **neuraminidase** (NA or N) (Fig. 10.3). Eighteen different HA subtypes (H1–H18) and eleven different NA subtypes (N1–N11) exist (Table 10.2), each subtype varying in amino acid sequence by at least 30%. These could theoretically combine in any combination, but currently only H1, H2, and H3 can be transmitted from person to person. In fact, H1N1, H1N2, H2N2, and H3N2 subtypes are the only combinations that have ever circulated in the human population. Influenza A (H1N1) and A (H3N2) are currently circulating in humans, while H2N2 ceased circulating in 1968.

In-Depth Look: Naming Influenza Strains

The World Health Organization (WHO) and the Centers for Disease Control and Prevention (CDC) follow an internationally accepted convention for naming influenza virus strains. It involves, in this order:

1. The antigenic virus type (A, B, or C)
2. The species of origin, only indicated if not a human strain
3. The geographic site where it was first isolated
4. The strain number
5. The year of isolation
6. For influenza A viruses, the HA and NA virus subtype, in parentheses

For example:

- A/duck/Alberta/35/76 (H1N1) for a virus from duck origin
- A/Perth/16/2009 (H3N2) for a virus from human origin

Currently circulating influenza A viruses are of H1N1 and H3N2 subtypes. Circulating influenza B viruses belong to one of two lineages: B/Yamagata and B/Victoria.

TABLE 10.1 Types of Influenza Viruses

	Influenza A	Influenza B	Influenza C
Hosts	Humans, waterfowl, poultry, pigs, horses, sea mammals, bats	Humans, seals	Humans, pigs, dogs
Gene segments	8	8	7
Proteins	11	11	9
HA/NA antigenic subtypes	18 HA, 11 NA	None	None
Clinical features	Moderate to severe illness	Milder disease than Influenza A	Largely subclinical
Epidemio-logical features	Causes pandemics	Less severe epidemics than Influenza A; no pandemics	Does not cause epidemics or pandemics

Essential Human Virology.

TABLE 10.2 Influenza A Virus HA and NA Subtypes and Their Hosts

	Birds	Humans	Swine	Horses	Bats
H1	X	X	X		
H2	X	X			
H3	X	X	X	X	
H4	X				
H5	X	X (avian)[a]	X (avian)[a]		
H6	X				
H7	X	X (avian)[a]		X	
H8	X				
H9	X	X (avian)[a]			
H10	X				
H11	X				
H12	X				
H13	X				
H14	X				
H15	X				
H16	X				
H17					X
H18					X
N1	X	X	X		
N2	X	X	X		
N3	X	X (avian)[a]			
N4	X				
N5	X				
N6	X				
N7	X	X (avian)[a]		X	
N8	X			X	
N9	X	X (avian)[a]			
N10					X
N11					X
Strains having circulated		H1N1	H1N1	H7N7	H17N10
		H1N2	H1N2	H3N8	N18N11
		H2N2	H3N2		
		H3N2			

[a]*indicates avian-to-human/animal transmission has occurred but no circulation within the latter population.*

10.2 CLINICAL COURSE OF INFECTION

Influenza is a respiratory virus that is transmitted in droplets that are generated when an infected person coughs or sneezes. These droplets are inhaled by a susceptible person less than 3 feet away into the respiratory tract, where the virus comes into contact with cells that express the cell surface receptor to which the virus binds. Transmission can also occur through direct or indirect contact with respiratory secretions that are transferred to the eyes, nose, or mouth.

As described in Chapter 5, "Virus Transmission and Epidemiology," the respiratory tract is subdivided into the upper respiratory tract, which consists of the nose, sinuses, pharynx, and larynx (voice box), and the lower respiratory tract, which consists of the trachea, bronchi, and lungs. Within the two lungs, the bronchi branch into bronchioles of smaller diameter that lead to the alveoli, where gas exchange occurs. However, not all epithelial cells of the respiratory tract are identical. The layered, squamous epithelium of the nose becomes a single, thicker layer of ciliated columnar epithelium within the sinuses (Fig. 10.1A). The pharynx comprises the nasopharynx, the oropharynx, and the hypopharynx (Fig. 10.1B). The nasopharynx is composed of ciliated columnar epithelium, which is susceptible to influenza infection, but the oropharynx and hypopharynx are squamous epithelium and involved with the gastrointestinal tract. The epithelium becomes ciliated and columnar again in the larynx, continuing through the trachea and bronchi. At the lower end of the respiratory tract, this ciliated columnar epithelium transitions into ciliated cuboidal epithelium within the smaller bronchioles. These cells become devoid of cilia as they transition into the thin alveolar epithelium, only one cell thick. Human influenza viruses exhibit tropism for the ciliated cells within the sinuses, nasopharynx, larynx, trachea, bronchi, and bronchioles (Fig. 10.1C). The cilia are destroyed as the virus replicates, affecting the clearance of environmental particles and potentially harmful microorganisms from the lung.

The incubation period for a typical, uncomplicated influenza infection ranges from 1 to 4 days with an average of 2 days. About 50% of infected people develop the classic symptoms of influenza:

- abrupt onset of fever (101–102°F) that lasts for 2–3 days
- **myalgia** (muscle aches, usually in the back and shoulders)
- generalized **malaise** (a feeling of discomfort)
- fatigue
- headache
- sore throat
- **rhinorrhea** (a runny nose)
- a dry, scratchy cough that is nonproductive, meaning that it does not work to clear mucus from the lungs (thought to occur as a result of epithelial cell damage)
- in children, nausea and vomiting

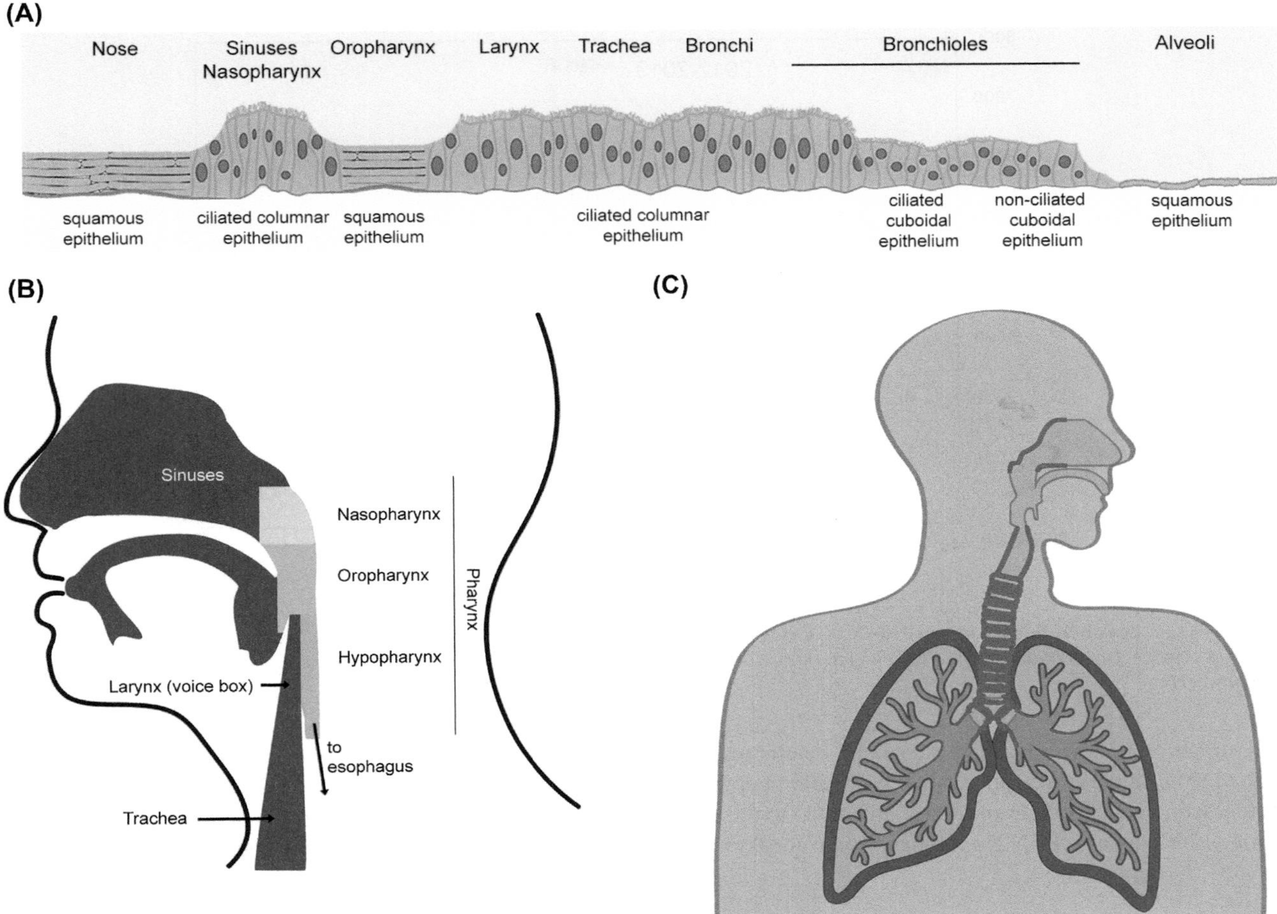

FIGURE 10.1 **Influenza-susceptible areas of the respiratory tract.** (A) The epithelium of the respiratory tract changes as it progresses through the different anatomical sections, from the nose to the alveoli. Human influenza viruses infect the ciliated columnar epithelium found in the sinuses, nasopharynx, larynx, trachea, bronchi, and bronchioles. (B) The pharynx is divided into the nasopharynx, oropharynx, and hypopharynx. The nasopharynx is composed of ciliated columnar epithelium. As a part of the digestive tract, the oropharynx and hypopharynx have squamous epithelium. (C) The areas in red highlight the locations of ciliated columnar epithelium susceptible to infection by human influenza viruses due to the cell surface expression of alpha-2,6-linked sialic acids, the receptor for human influenza viruses.

These symptoms usually last 3–5 days, although the cough, malaise, and fatigue can last several weeks. The presence of myalgia, extreme fatigue, and a fever that lasts several days help to differentiate influenza infection from the common cold.

Shedding of the virus in adults occurs from the day before symptoms appear to 5 days after they begin, although children can shed the virus for longer than 10 days. Viremia is uncommon.

Children ≤2 years old, people ≥65 years of age, immunocompromised individuals, and people with chronic conditions (such as asthma, diabetes, or heart disease) have the highest rates of hospitalizations from influenza-related complications. Influenza A infection is associated with four times more hospitalizations than influenza B, which affects mostly children. The most frequent complications arise from secondary bacterial infections from *Streptococcus pneumoniae, Staphylococcus aureus,* or *Hemophilus influenzae* that cause **pneumonia**, inflammation in the lungs that causes difficulty in breathing. The case fatality rate is around 7%. Viral pneumonia from influenza itself is rare but can progress within hours to a severe pneumonia that can cause death. Influenza can also exacerbate previously existing lung conditions, such as bronchitis and chronic obstructive pulmonary disease. Although not a direct result of infection, around 10% of people infected with influenza also experience **myocarditis** (inflammation of the heart), which can be mild or can lead to cardiac arrest.

Influenza usually peaks from December through March, although infections have occasionally peaked in November, April, and May (Fig. 10.2). Overall, the CDC estimates an average of 226,000 influenza-associated hospitalizations occurred each year from 1979 to 2001. From 1976 to 1990,

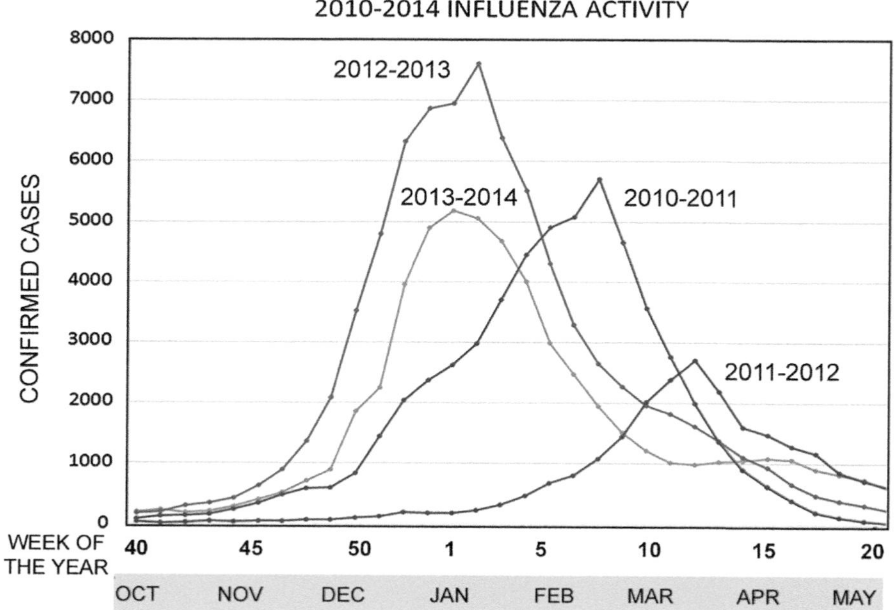

FIGURE 10.2 **Peak of influenza from 2010–2014.** Cases of influenza usually peak between December and March, as occurred in the most recent influenza seasons. *Data obtained from FluView (internet), Atlanta, GA. Centers for Disease Control and Prevention. Available at: http://www.cdc.gov/flu/weekly.*

an average of 19,000 influenza-associated deaths occurred each season; in contrast, approximately 36,000 people died each season between 1990 and 1999, attributed to the circulation of Influenza A (H3N2) subtypes, which are associated with higher mortality rates. Worldwide, the WHO estimates that seasonal influenza epidemics result in a quarter- to a half-million deaths every year. People over 65 years of age account for more than 90% of influenza-associated deaths.

In-Depth Look: Why Do Influenza Viruses Have Seasonal Patterns?

Cases of influenza tend to peak during the winter months in temperate climates, although the exact reason for this is unknown. Several hypotheses have been proposed to explain this phenomenon. One of the most commonly cited reasons is that people tend to congregate indoors during the winter months, which would increase contact rates between individuals and influence the rate of transmission. On a similar line of reasoning, students are out of school and family vacations are taken more often during the warmer months, contributing to reduced person-to-person contact. Weather has been suggested as another driver of influenza transmission, since influenza virions are more stable during months when the humidity is low, which occurs alongside dropping temperatures during the winter months. Respiratory droplets also remain airborne longer in lower humidity environments, increasing their chance of being inhaled. In addition, the influenza ribonucleic acid (RNA) genome is sensitive to UV radiation from the sun, which is lower in the winter due to reduced daylight. Finally, the aspects of the host immune system must also be considered. Studies carried out in the past two decades have shown an effect of vitamin D on the human immune system, and the synthesis of the vitamin in the body occurs in the skin after exposure to sunlight. Ongoing research is attempting to carefully dissect these hypotheses and reconcile the rates of infection in temperate versus tropical climates.

10.3 MOLECULAR VIROLOGY

Influenza virions (Fig. 10.3A) are enveloped and contain a helical −ssRNA genome, referred to as the viral RNA (vRNA). The virions are pleomorphic, meaning that they often have varying shapes (Fig. 10.3B). The particles are elongated upon initial isolation from a host, but after being propagated in cells, the virions are more spherical, and about 80–120 nm in diameter. The influenza genome is **segmented**, and each helical segment encodes a different viral gene. Influenza A and B have eight gene segments (that encode 11 proteins), while influenza C has seven (that encode 9 proteins).

Each single-stranded, negative-sense vRNA segment is encapsidated by a repeating protein, known as the **nucleocapsid protein** (NP), each of which binds approximately 24 nucleotides of influenza vRNA (Fig. 10.3C). This forms a helix of about 9 nm in diameter. Interestingly, the influenza genomic vRNA winds around the NP protein—like thread on a spool—rather than being

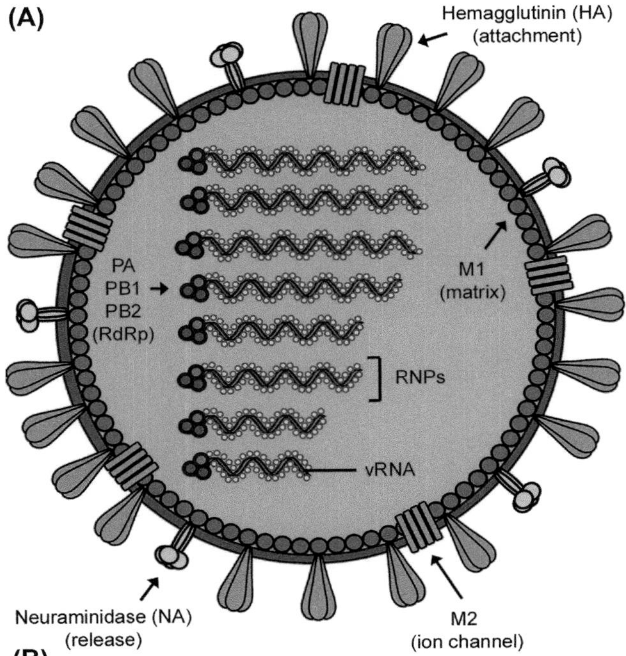

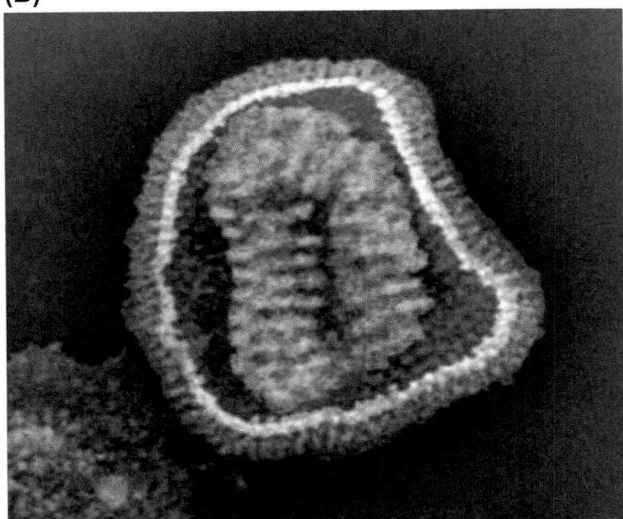

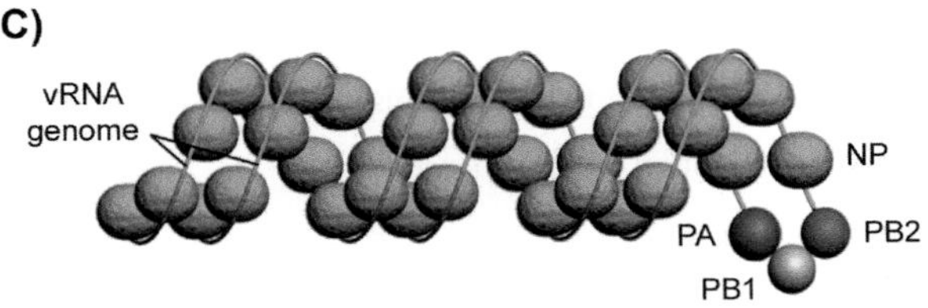

FIGURE 10.3 **The structure of influenza A virions.** (A) Influenza A virions are enveloped and contain eight gene segments. HA and NA proteins form spikes protruding from the envelope, while M2 is embedded into the envelope and M1 is directly beneath it. PA, PB1, and PB2 form the RNA-dependent RNA polymerase (RdRp), attached to the vRNA, which is wrapped around nucleocapsid protein (NP) subunits to form ribonucleoprotein (RNP) complexes. (B) Transmission electron micrograph of an influenza A virion. *(Photo courtesy of the CDC and Dr. Frederick Murphy.)* (C) The RNP complex of influenza A virus is composed of the –ssRNA genome, the protecting NP, and three associated viral proteins that form the RdRp. *(Graphic modified from Fig. 1 of Tao et al., 2010. PLoS Pathogens 6(7), e1000943.)*

encased within it. The NP protein is also important in the transport of the RNA segments into the nucleus of the host cell. Associated with the 5′ end of the NP-associated RNA segment is a complex of three viral proteins—PA (polymerase acidic protein), PB1 (polymerase basic protein 1), and PB2 (polymerase basic protein 2)—that together form the viral polymerase. Together, the NP-protected vRNA with its associated polymerase complex is known as the **ribonucleoprotein (RNP) complex** (a term that indicates the presence of RNA *and* proteins). Each segment of influenza RNA exists as an RNP complex.

The three proteins that form the RNA-dependent RNA polymerase (RdRp) complex (PA, PB1, and PB2) are the three largest influenza genes and proteins (Fig. 10.4 and Table 10.3). Within the PB1 gene of some influenza A viruses is an alternative reading frame that encodes a small protein, termed PB1-F2, that has been shown to induce the apoptosis of immune cells, thereby interfering with host immune responses.

Embedded within the influenza envelope are two integral glycoproteins, HA and NA (Fig. 10.3A). There are generally around 600 of these proteins projecting from the envelope of the virion; the HA protein is 4 times more abundant than NA. HA is spike shaped and forms a homotrimer, while NA is more umbrella shaped and forms a homotetramer of 4 NA proteins. As detailed previously, the identity of these two proteins determines the particular subtype of Influenza A viruses. The HA protein is translated and termed HA0, which is then cleaved into two subunits, HA1 and HA2, that remain covalently associated with each other. Whereas HA1 facilitates attachment, HA2 triggers fusion with the endosomal membrane. NA is an enzyme that is employed by the virus upon virion release.

Interestingly, the reason why influenza C has one less gene segment than influenza A or B is because it does not have separate HA and NA proteins. Instead, it has an envelope glycoprotein, known as **hemagglutinin–esterase fusion**, that has both HA and NA activity.

The smallest two gene segments each encode two proteins. In the influenza A genome, segment 7 encodes M1 and M2, a product of alternative viral mRNA (vmRNA) splicing. M1 is the matrix protein found immediately under the virion envelope. It is involved in the transport of new RNPs out of the nucleus and plays a role in the budding of the virion from the plasma membrane of the cell. M1 is the most abundant protein within the influenza virion: there are around 3000 found inside each virion. M2 is a transmembrane protein that acts as an ion channel and is involved in virion fusion within the endosome.

The smallest influenza gene segment encodes nonstructural protein 1 (NS1) and 2 (NS2). NS1 plays a crucial role during infection as it interferes the host cell's production of type 1 IFN and inhibits the activity of PKR,

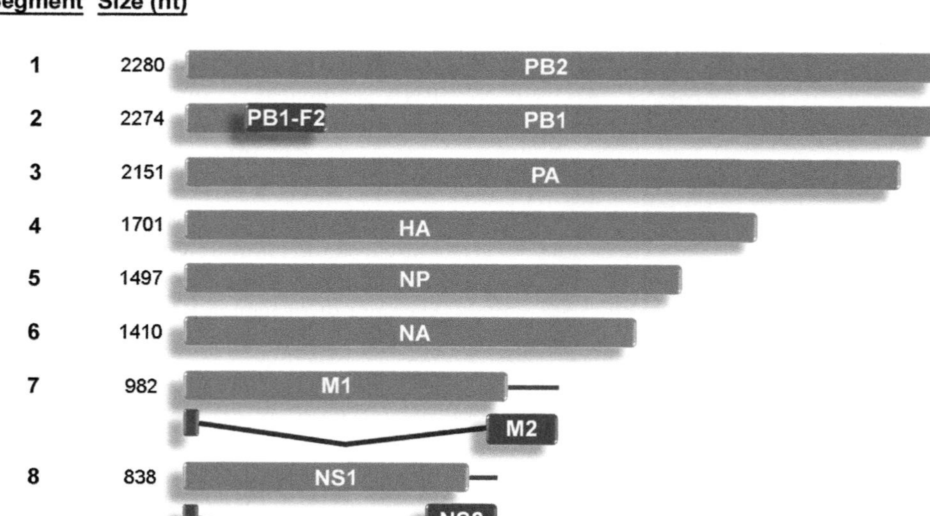

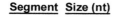

FIGURE 10.4 **Size and function of influenza genes.** The influenza A genome is composed of eight segments that encode 11 proteins. The gene segments are ordered from largest to smallest. Small protein PB1-F2 is derived from an alternative reading frame in some strains. M2 and NS2/NEP are products of alternative splicing. *Nucleotide sizes of influenza A genes were derived from GenBank accession numbers FJ966079-966086 and FJ969513 for influenza A/California/04/2009(H1N1).*

TABLE 10.3 Influenza A Genome Segments and Their Encoded Proteins

Gene segment	Protein name	Protein size (aa)	Function
1	PB2	759	Part of RdRp; binds 5′ cap of host mRNA
2	PB1	757	Part of RdRp; transcribes vmRNA and antigenome
2	PB1-F2	90	Induces apoptosis of host cells (not present in all strains)
3	PA	716	Part of RdRp; cleaves 5′ cap from host mRNA
4	HA	566	Binds to sialic acid on cell surface; induces fusion within endosome
5	NP	498	Encapsidates vRNA; contains nuclear localization signal for nuclear import of RNPs
6	NA	469	Cleaves sialic acid from cell surface during virion release (neuraminidase)
7	M1	252	Transport of RNPs out of nucleus; assists in budding from plasma membrane
7	M2	97	Ion channel protein that allows H+ ions to enter virion during uncoating
8	NS1	219	Inhibits antiviral response: Type 1 IFN antagonist, inhibits PKR
8	NS2 (NEP)	121	Transports nascent RNPs from nucleus to cytoplasm

Protein sizes derived from GenBank accession numbers FJ966079-FJ966086 and FJ969513 for strain A/California/04/2009(H1N1). Since this strain has a nonfunctional PB1-F2 protein, DQ415301 from strain A/Taiwan/3286/03(H3N2) was used for PB1-F2.

an enzyme that inactivates eukaryotic translation initiation factor 2α to prevent protein translation and stall virus replication (reviewed in Chapter 6, "The Immune Response to Viruses"). NS2 is also known as the nuclear export protein (NEP) for its role in exporting newly formed RNPs from the cell nucleus.

As described in Chapter 4, "Virus Replication," every virus completes seven stages of replication to interact with a host cell and create new, infectious virions: attachment, penetration, uncoating, replication, assembly, maturation, and release. The previously described proteins all play a role in the replication of influenza virions within target cells (Table 10.3).

10.3.1 Attachment, Fusion, and Uncoating

Influenza is a respiratory virus that attaches to the ciliated respiratory epithelium within the nasopharynx and trachea, bronchi, and bronchioles (the tracheobronchial tree). The HA1 subunit of the HA protein binds to carbohydrates extending from the plasma membrane of ciliated lung cells that are parts of various proteins and lipids. Specifically, the influenza HA protein binds to a sugar known as sialic acid when found at the very end of these carbohydrate chains (Fig. 10.5A). The penultimate sugar in these carbohydrates is galactose, and human influenza viruses bind with greater affinity to sialic acids linked to galactose with an alpha-2,6 linkage, meaning that the second carbon of the terminal sialic acid binds to the sixth carbon of the galactose sugar (Fig. 10.5B). Sialic acid is also known as neuraminic acid, hence why the influenza protein enzyme that cleaves this molecule from the cell surface is known as neuraminidase. After binding sialic acid, the influenza virions are internalized into the cell via endocytosis (Fig. 10.5A). The best characterized pathway involves clathrin-mediated endocytosis, but nonclathrin, noncaveolin pathways have also been described.

Fusion of the viral envelope with the endosomal membrane requires a drop in the pH of the endosome. The HA1 subunit facilitates attachment, but in the low pH of the endosome, the HA protein—which is anchored into the viral envelope—undergoes a conformational change that results in the fusion peptide within the HA2 subunit hooking into the endosomal membrane (Fig. 10.6B). The two membranes fuse at this spot, creating a pore (Fig. 10.6C). Meanwhile, the M2 protein, also embedded in the viral envelope, acts as a channel to allow H+ ions (which acidify the endosome) to flow into the virion from the endosome. This is thought to disrupt the internal virion structure so that the RNPs can be released through the newly formed pore into the cytoplasm, completing the process of uncoating. Antiinfluenza drugs amantadine and rimantadine target the M2 protein.

Unlike other RNA viruses, influenza viruses replicate within the nucleus of their target cell. The NP protein that encapsidates the vRNA has a nuclear localization signal that interacts with the machinery at the nuclear pores. The RNPs are imported into the nucleus, where all transcription and genome replication take place. The next section explains why influenza replication requires nuclear entry of the RNPs.

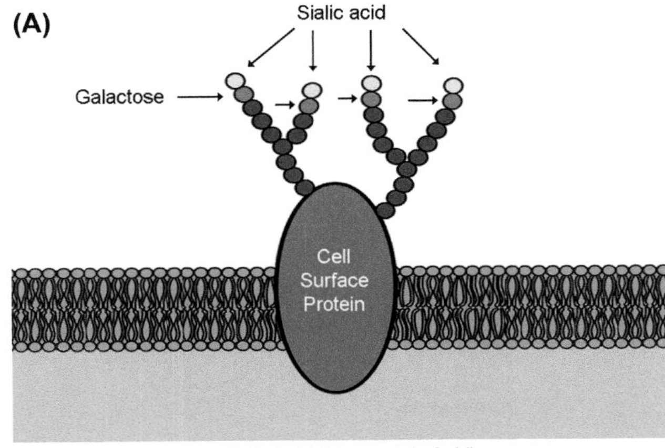

FIGURE 10.5 **Sialic acid, the receptor for influenza viruses.** (A) The influenza HA protein binds to terminal sialic acids (in yellow) of membrane glycoproteins when the penultimate sugar is galactose (in green). (B) The tropism of the virus for different species depends upon how the sialic acid and galactose are linked. Human influenza A viruses bind with greater affinity to sialic acids linked with an alpha-2,6 linkage, meaning that the second numbered carbon of sialic acid creates a bond with the sixth carbon of galactose (highlighted in red). On the other hand, avian influenza A viruses recognize alpha-2,3 linkages.

10.3.2 Replication

Upon entry of the RNPs into the nucleus, each negative-sense vRNA is used as a template for the creation of two positive-sense RNAs: the vmRNAs and a piece of RNA that is complementary to the genome. This complementary RNA (cRNA) is termed the **antigenome**, because it is the exact complement of the genome. During replication, the positive-sense cRNA acts as a template to create full-length copies of the negative-sense vRNA segments. The RdRp of influenza synthesizes both the vmRNA and cRNA. It is composed of three proteins: PA, PB1, and PB2.

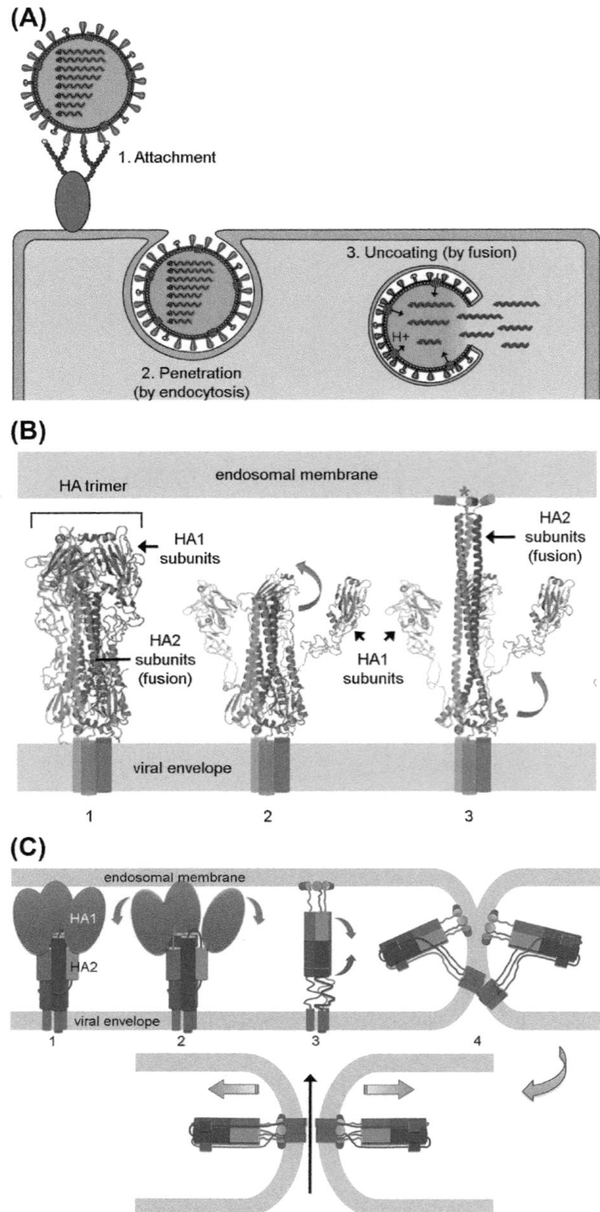

FIGURE 10.6 **Influenza virus attachment, penetration, and uncoating.** (A) The influenza HA protein attaches to terminal sialic acids of membrane glycoproteins, and the virion enters the cells via clathrin-mediated endocytosis. The HA protein changes configuration within the acidified endosome, leading to the fusion of the viral envelope with the endosomal membrane. The M2 protein acts as a channel to allow H+ ions to enter the virion, further weakening the virion structure. The eight RNPs are released into the cytoplasm. (B) The HA protein is composed of two covalently linked subunits, HA1 and HA2, that form a trimer protruding from the viral envelope (1). HA1 facilitates attachment to sialic acid, but in the acidified endosome, the HA1 undergoes a conformational change and peels back toward the envelope (2). The HA2 subunits are translocated toward the endosomal membrane, where the HA2 fusion peptides (asterisk) make contact (3). *(Reprinted with permission from Harrison, S.C., 2008. Nat. Struct. Mol. Biol 15(7), 690–698. doi:*10.1038/nsmb.1456, *Macmillan Publishers Ltd.)* (C) After the endosome is acidified and the HA2 fusion peptides engage the endosomal membrane, the molecule folds back upon itself, drawing the two membranes together (4) and creating a channel through which RNPs are exported (5). *(Reprinted from Harrison, S.C., 2015. Viral membrane fusion. Virology 479–480, 498–507, with permission from Elsevier.)*

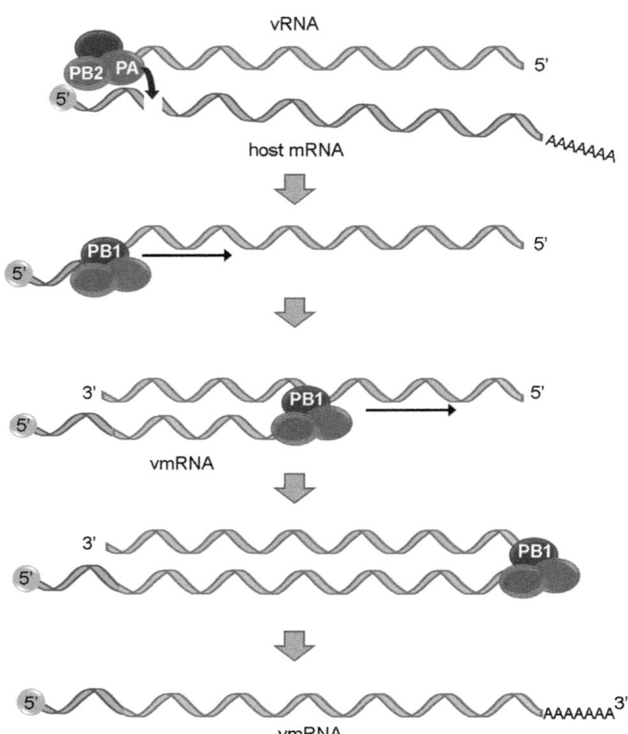

FIGURE 10.7 **Initiation of transcription via cap-snatching.** The influenza RdRp complex steals the 5′ caps of host mRNA transcripts for its own vmRNA transcripts. PB2 binds the host mRNA 5′ cap, which PA cleaves off, leaving the host mRNA to become degraded. PB1 extends the nucleotides of the capped sequence, using the vRNA as a template. When it reaches the end of the vRNA segment, it encounters a series of uridines on which it stutters, creating the poly(A) tail of the vmRNA transcript. This transcription process is repeated for all eight segments of the genome.

One reason that replication of influenza occurs in the nucleus is because the polymerase complex begins transcription of vmRNAs by stealing the 5′ caps of host mRNA transcripts (Fig. 10.7A). In this **cap-snatching** process, PB2 binds to the 5′ cap of one of the host pre-mRNAs. PA then cleaves the cap about 10–13 nucleotides from the host mRNA. PB1 adds to the capped strand of nucleotides as it begins transcribing the vmRNA from the vRNA template. PB1 transcribes until it encounters a stretch of five to seven uridines near the end of the vRNA. Here, it stutters, creating the poly(A) tail of the vmRNA transcript, which is then exported from the nucleus and translated by host ribosomes. Although transcription does not directly require host RNA polymerase II, it indirectly requires it because host mRNA transcripts—which are transcribed by RNA polymerase II—are required to start the process of vmRNA transcription.

The 5′ caps are not removed from the virus's own vmRNAs because the viral polymerase remains bound to the section where the cleavage would occur. The remaining cap-less cellular pre-mRNAs are retained in the nucleus and eventually degraded.

The two smallest vRNA segments each encode two proteins because the vmRNAs undergo alternative splicing.

This results in the M1 and smaller spliced M2 protein for segment 7, and NS1 and smaller spliced NS2/NEP for segment 8.

The absence of host mRNAs allows vmRNAs to be preferentially translated by ribosomes. In addition, the influenza NS1 protein binds to the 5′ ends of the vmRNA and recruits components of the translation initiation machinery, also leading to the preferential translation of the vmRNAs.

HA, NA, and M2 are synthesized by ribosomes on the rough endoplasmic reticulum. The translated HA protein is cleaved by host proteases into HA1 and HA2, which remain associated with each other through a covalent disulfide bond. This step is required for the virion to be infectious because otherwise, attachment and fusion would not occur. The proteins are folded within the rough ER, and HA and NA are glycosylated. HA is assembled into a trimer, and NA and M2 are each assembled into tetramers, before being exported to the Golgi complex. The other influenza proteins are translated by cytosolic ribosomes.

After sufficient viral proteins have been transcribed and translated, the polymerase complex in the nucleus switches from vmRNA transcription to genome replication. To do so, the positive-sense antigenome cRNA is synthesized. Unlike the shorter vmRNAs, cRNAs are full-length complementary copies of the vRNA that do not have a 5′ cap or a poly(A) tail. The positive-sense cRNAs are used repeatedly as templates for the synthesis of negative-sense genomic vRNA by the polymerase complex. Interestingly, the antigenome cRNA also becomes coated with NP.

Still within the nucleus, the newly synthesized vRNA segments are encapsidated by NP and associate with newly translated PA, PB1, and PB2 to form RNP complexes. The M1 protein assists in the formation of the RNPs and then binds to NS2/NEP, which associates with the nuclear export machinery to transport the RNPs from the nucleus into the cytoplasm. M1 is one of the later proteins to be translated, which prevents the premature assembly and export of RNPs before all the components have been synthesized.

10.3.3 Assembly, Maturation, and Release

Assembly of the influenza virion occurs at the plasma membrane (Fig. 10.8). Specifically, this occurs at the **apical** side of the epithelial layer, the side that faces the lumen of the airway rather than the underlying tissues. HA, NA, and M2 localize to the plasma membrane, and M1 recruits the RNPs, which were assembled in the nucleus, to the site of assembly. Influenza possesses eight genome segments, and how all eight are recruited into each virion is not completely understood. Although the process could be random and those virions that end up with one of each segment would be infectious, research suggests that each RNA segment contains specific packaging sequences that ensure all eight segments are incorporated into the virion.

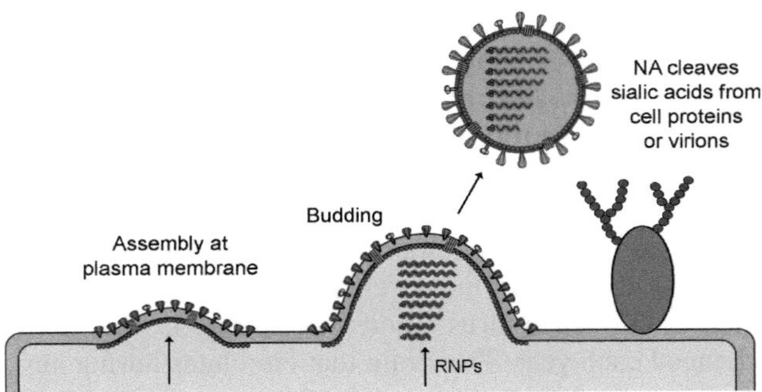

FIGURE 10.8 **Assembly, maturation, and release.** Following translation, HA, NA, and M2 localize to the plasma membrane. M1 recruits the RNPs to the budding virion. As the enveloped virion is released, NA cleaves sialic acids from plasma membrane proteins or from the HA protein to prevent aggregation of virions at the cell surface.

The interaction of the M1 proteins with each other causes the surface of the cell to bend, forming a bud in the membrane. As the virion buds from the plasma membrane, the NA protein cleaves sialic acid from the surface of the cell and from any HA proteins that have bound it; otherwise, the newly formed virions would immediately attach again to the cell from which they are budding. Antiinfluenza drugs oseltamivir and zanamivir prevent infection of new cells by inhibiting the NA protein, leading to the aggregation of virions at the infected cell surface.

Study Break
Create a list of the seven stages of viral replication and indicate how each step is accomplished by influenza virus.

10.4 GENETIC CHANGES IN THE INFLUENZA GENOME

Because the influenza RdRp complex is error-prone, mutations occur in the vRNA during replication of the genome. Some of these result in silent mutations that do not change the amino acid sequence of the viral proteins, but because of how the genetic code is read by ribosomes in three-nucleotide codons, some point mutations lead to a change in the amino acid that is placed at that position. For instance, the codon CUU encodes the amino acid leucine, but a point mutation in the first nucleotide that results in GUU now codes for the amino acid valine. These single-nucleotide point mutations lead to **antigenic drift**: the constant acquisition of small changes in the influenza genome that result in the creation of new *strains* (not *subtypes*).

As described in Chapter 6, "The Immune Response to Viruses," antibodies are created against a particular viral antigen in the capsid or envelope and neutralize the virus upon binding the antigen, thereby preventing the attachment of the virion. In the case of influenza, antibodies are created against the proteins on the outside of the virion, namely HA and NA. When antigenic drift occurs, however, point mutations in the viral genome may result in an HA or NA antigen that is no longer recognized by previous antibodies, and so the virus is not neutralized. These **escape mutants** have a selective advantage within the host: those mutations that result in virions that are not recognized by host antibodies will be the ones that are able to continue infecting new cells. Antigenic drift is the basis for seasonal influenza epidemics; it is for this reason that the influenza virus strains within the flu vaccine are changed each year. The strain that circulates during any 1 year may mutate into a strain that is not recognized by the antibodies created by previous vaccine strains.

Antigenic drift leads to minor antigenic changes that constantly result in new influenza strains. **Antigenic shift**, on the other hand, can occur in influenza A viruses and results in a major antigenic change leading to the creation of a new influenza A *subtype* (Fig. 6.15). Influenza can undergo antigenic shift through a process known as **reassortment** (often incorrectly referred to as recombination). As described in Section 10.1, 18 different HA subtypes (H1–H18) and 11 different NA subtypes (N1–N11) exist. Because the influenza genome is segmented, if a cell is infected simultaneously with two different subtypes of influenza, newly replicated vRNA segments could be mixed within forming virions, leading to the creation of a novel virus subtype.

Antigenic shift is a major concern for humans. Although only four subtypes of influenza A viruses have ever circulated in the human population (H1N1, H1N2, H2N2, and H3N2), the other subtypes regularly circulate in other animal reservoirs. Wild aquatic birds have been shown to transmit avian influenzas to poultry, pigs, horses, and sea mammals; coinfection of a pig, for instance, with a human influenza and an avian influenza could result in reassortment of viral segments, and the creation of a novel virus able to undergo human-to-human transmission but with a combination of antigens that has never before been seen by the human population at large. The result would be that no one within the human population would have any immunity against the virus, which could lead to a severe epidemic, or even a worldwide pandemic. An example of this occurred in 1968, when circulating influenza A (H2N2) was replaced by influenza A (H3N2) due to antigenic shift. Currently, influenza A (H1N1) and (H3N2) subtypes are the only two circulating in humans, and so an antigenic shift to another subtype would mean that the human population, at large, would be immunologically unprotected. This also means that even though influenza A (H2N2) circulated in the human population previously, anyone born after 1968 has no immunity against the virus. When immunity within the population has waned, a previous strain could reemerge as a pandemic virus. Antigenic shifts occur much less frequently than antigenic drifts, which occur constantly, but have the potential to cause pandemics of varying severity.

10.5 HISTORICAL INFLUENZA ANTIGENIC SHIFTS

In the 20th century, four major pandemics, caused by antigenic shifts, struck the world. The pandemic of 1918 is the single greatest infectious disease outbreak in history, killing an estimated 20–50 million people worldwide in the course of months (Fig. 10.9). It is difficult to assign a point of origin to the first wave of the pandemic, as it appeared nearly simultaneously in North America, Europe, and Asia in spring of 1918. One hypothesis is that it began at the US Army base at Camp Funston (now Fort Riley), Kansas, in March of 1918. Here, soldiers had been cleaning manure from pig pens, and then burned tons of the manure. Within a week, hundreds of the soldiers had been reported sick, and 48 died from an associated pneumonia. The virus spread with the troops, who had been called to serve in World War I, and soldiers in Europe of all nationalities contracted the virus. The press in Spain was the first to cover cases reported in its cities, which is how the flu got the nickname, the "Spanish Flu," while other European countries were concerned that reporting about the outbreak during the war might affect public morale. The virus became more virulent as it spread through Europe, and troops brought it back with them to the United States. In early September of 1918, soldiers began dying at Camp Devens, near Boston, and soon civilians began dying of pneumonia. It moved down the eastern seaboard to New York City and Philadelphia; war efforts crammed men into offices to sign up for the draft and enticed hundreds of thousands of people to parades to support Liberty Loan drives, which only increased the transmission of the virus. Once the virus began spreading, hundreds of people died every day: in New York City, nearly 900 people died in 1 day alone, and over 11,000 people died in Philadelphia in the month of October. Everyone lacked immunity because no one had been previously exposed to this subtype. Hospitals were overflowing with patients laying on stretchers or the floor, waiting for a bed to open up. People ran out of caskets, and mass graves were dug at the potter's field in Philadelphia, where unknown and poverty-stricken people were buried. Public gatherings were prohibited and the infected were quarantined in their houses, but the virus spread south and west, into small towns and large cities, and eventually made it to the west coast.

Those infected presented with a high fever that caused delirium, severe headache, cough, and myalgia, and death occurred from secondary bacterial pneumonia or from fluid

(A)

(B)

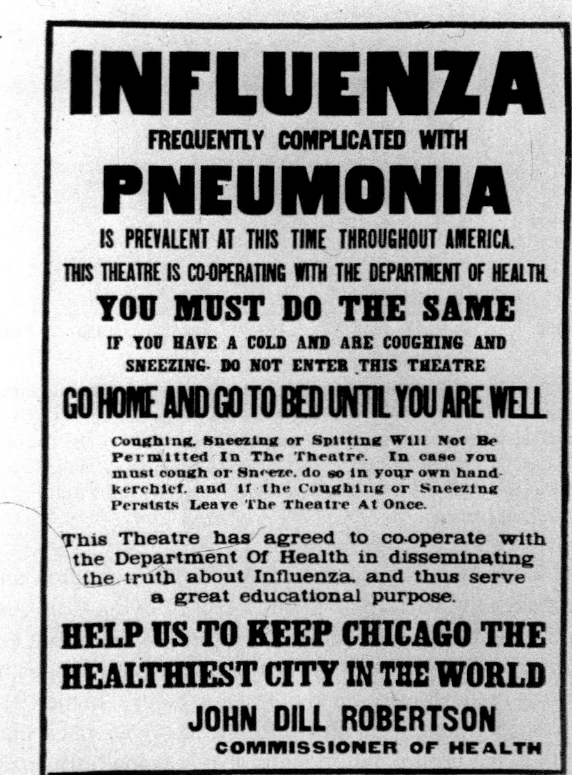

FIGURE 10.9 **Photos from the 1918 influenza pandemic.** (A) An emergency hospital at Camp Funston, Kansas, one of the possible origins of the 1918 H1N1 influenza. *(Courtesy of the National Museum of Health and Medicine.)* (B) A photograph of a poster displayed at a Chicago theater during the 1918 pandemic. *(United States National Library of Medicine.)*

and inflammation in the lungs. Patients would develop "heliotrope cyanosis," a term coined to describe the bluish color of the people afflicted. The case fatality rate was >2.5% of those infected, over 25 times more than a typical flu season.

As the virus ravaged cities and town, it eventually ran out of new hosts to infect, because those that did not die developed immunity to the virus. By the middle of November of 1918, only two months after it had started, the virus was disappearing in Boston, New York, and Philadelphia. The outbreak was not over, however. A third wave of influenza spread throughout the United States in the early months of 1919 with similar vigor as the second wave. Overall, the pandemic killed an estimated 675,000 people in the United States, more Americans than all the wars of the century combined.

Later examination of stored serum samples from the time or antibody analysis of survivors indicated that the virus was an influenza A (H1N1) subtype. Although the original outbreak in the United States was thought to have jumped from pigs, the virus was identified as an avian-like influenza. This was confirmed when the complete sequences of the vRNA segments from the 1918 virus were determined by amplifying viral genes from the preserved tissues of two soldiers who died in 1918 and from the frozen tissue of a woman who died of influenza and was buried in a mass grave in the permafrost of Alaska. Both the HA and NA proteins were derived from an avian-like influenza: an antigenic shift had resulted in an influenza virus not seen in the human population before.

Studies using **reverse genetics**, the creation of viral strains by artificially inserting influenza vRNAs into a cell to generate functional virions, were able to recreate the 1918 virus, which was used to infect mice and monkeys (Fig. 10.10). These experiments indicated that the 1918 HA and NA genes are more virulent than typical seasonal influenzas, and the 1918 virion, particularly the HA protein, induced a "cytokine storm" of inflammatory cytokines from host innate immune cells that caused tissue damage, hemorrhaging, and massive infiltration of additional immune cells into the lungs. In addition, the type 1 interferon genes, which normally induce antiviral pathways, were downregulated in mice infected with a contemporary H1N1 virus containing the 1918 NS1 gene. Taken together, the 1918 influenza viral proteins culminated in a more virulent strain that inhibited antiviral immune responses while overstimulating immune cells that cause lung damage.

Although the virus was of avian origin, the precise path it took to get into humans is still under investigation. The virus could have jumped directly from birds into humans, as has been shown to occur since that time, and then evolved to efficiently infect humans. In support of this hypothesis, there was an outbreak of a respiratory syndrome during World War I at a British army base in northern France in 1916 that also caused severe symptoms, including the "heliotrope cyanosis" that was seen in the 1918 outbreak, and a similar syndrome with a high fatality rate struck England in 1917.

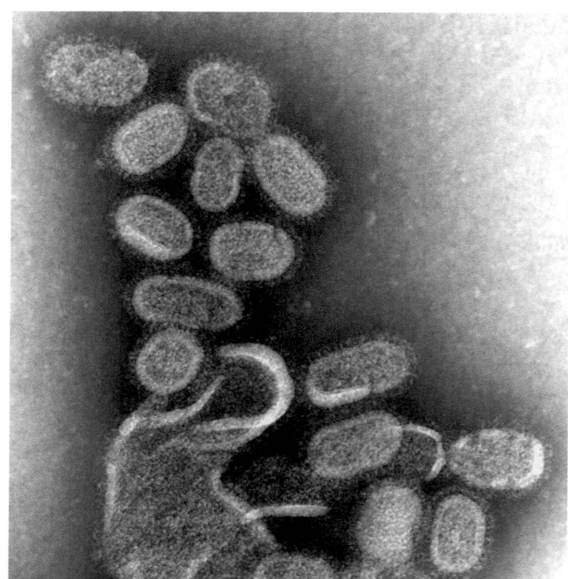

FIGURE 10.10 **1918 influenza A virus.** Reverse genetics was used to recreate the 1918 influenza A virus. This transmission electron micrograph shows virions of the 1918 H1N1 influenza A virus. *Image courtesy of the CDC/Dr. Terrence Tumpey and Cynthia Goldsmith.*

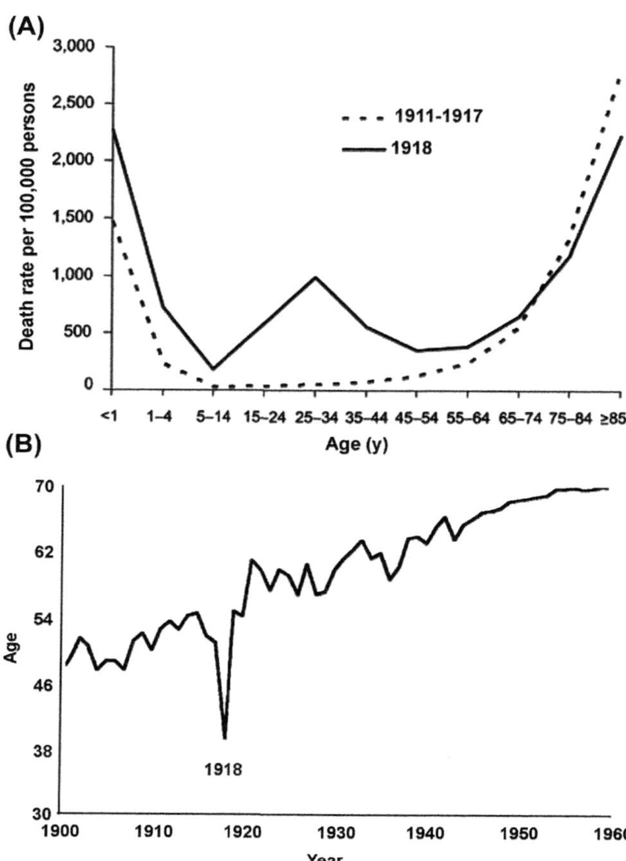

FIGURE 10.11 **Effects of the 1918 H1N1 influenza.** (A) Mortality from influenza and pneumonia by age in the United States during 1918 (*solid line*) and the years preceding the antigenic shift (*dashed line*). (B) The only time in the 20th century that life expectancy patterns dropped in the United States was in 1918, due to deaths from the 1918 H1N1 influenza. *Figure adapted from Taubenberger, J.K., 1999. ASM News 65, 473–478.*

This virus could have mutated, through genetic drift, as it was passed through the hundreds of thousands of troops in the area, resulting in the virulent strain that then spread throughout the world.

Despite the fact that the H1N1 virus was of avian origin, the virus had also been causing similar symptoms in pigs at the time. Another hypothesis is that reassortment between a human and avian virus could have occurred while the virus spread through the swine population, but alternatively, pigs could have contracted the virus from humans after the virus adapted to infect humans. As there were three distinct "waves" of infection (Spring 1918, September–November 1918, and early 1919), it is certainly a possibility that certain of these aspects contributed to the different waves. Without having influenza samples from infected humans, pigs, and birds from the years preceding and during the 1918 outbreak, we may never definitively know which of these scenarios is most accurate.

The first influenza virus was isolated 11 years after the outbreak was over, in 1930—a swine influenza virus isolated by Richard Shope and Paul Lewis—and so doctors and researchers at the time of the 1918 pandemic could only characterize the clinical symptoms and consider the claims at the time that the disease was caused by a bacterium. Considering that "filterable viruses" were being characterized during this period, however, several studies were carried out at the time with the 1918 agent (on both animals and humans) that seemed to clearly show that the infectious disease was also caused by a filterable virus, lending credence to the idea that influenza was a viral—not bacterial—disease.

As described previously in Section 10.2, children and the elderly are generally at greatest risk of dying from seasonal influenza. The mortality curve for seasonal influenza is generally "U-shaped," with the highest rates of deaths occurring in the youngest and oldest age groups. In the 1918 outbreak, children were still infected at higher rates than any other age group, but in contrast to seasonal influenzas where the elderly account for >90% of influenza-associated deaths, 99% of the people who died were under age 65. The mortality curve for the 1918 outbreak was instead "W-shaped," with an additional peak of deaths peaking in the 25–34 year age group (Fig. 10.11A). Individuals in this group generally have the strongest immune systems and likely fell victim to the excessive innate immune responses elicited by the 1918 virus that caused severe lung damage. It is also possible that the elderly had been exposed to a similar influenza virus in the past and had retained some immunity against the 1918 virus. Overall, the average life expectancy dropped by 12 years in 1918 because of this H1N1 avian-like influenza (Fig. 10.11B).

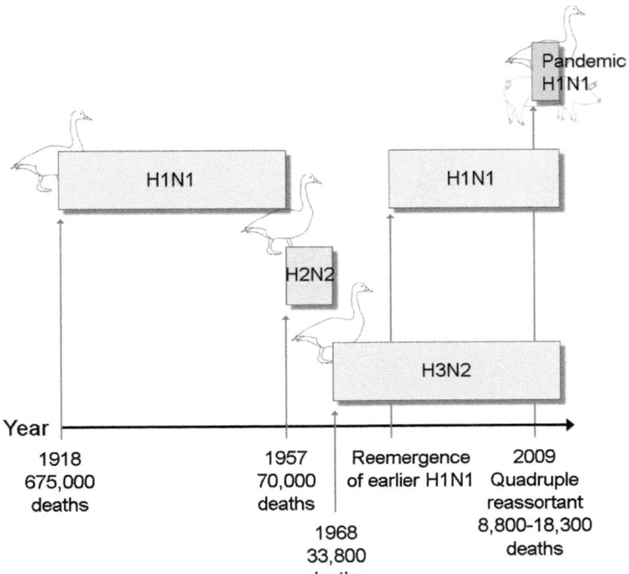

FIGURE 10.12 **Pandemics of the 20th century caused by antigenic shift.** The pandemic of 1918 began when an H1N1 virus began circulating in the human population, likely a zoonosis from the bird population, causing the "Spanish flu." In 1957, a reassortment event with an avian influenza created the H2N2 "Asian influenza." In 1968, the reassortant H3N2 began circulating, causing the "Hong Kong influenza." A similar virus to the earlier H1N1 reappeared in 1977 and began circulating alongside the H3N2. In 2009, an H1N1 quadruple-assortant of avian, swine, and human viruses began spreading among humans. The H3N2 and H1N1 viruses are the current seasonal human influenza A subtypes.

The interesting thing about pandemic viruses is that after humans have been exposed to the virus and generate immunity, the virus is no longer able to maintain the transmission rates and pandemic potential that it had in a nonimmune population. After initially infecting the population, pandemic viruses become the seasonal influenzas that continue undergoing genetic drift.

A handful of other antigenic shifts have occurred in the last 100 years (Fig. 10.12). In 1957, a shift from H1N1 to H2N2 originated in China and had spread throughout the world within half a year. This "Asian influenza" virus was found to be a reassortant between human and avian viruses, carrying avian HA, NA, and PB1 genes. Again, 30% of the population was infected, and possibly more in some areas, resulting in a mortality rate of 1 in 4000 individuals and an estimated 70,000 deaths in the United States. In 1968, the "Hong Kong influenza," which also originated in China, spread throughout Asia, the Middle East, and Australia, eventually reaching the United States through the return of troops from Vietnam. This H3N2 virus completely replaced the circulating H2N2 and was considered a moderate pandemic, possibly because antibodies against the NA protein moderated the infectivity of the virus even though the HA gene was novel to the human population.

In 1977, the H1N1 virus mysteriously reemerged in northern China and spread through Russia, and then the rest of the world. The H1N1 virus from 1918 had continued circulating and undergoing genetic drift until it was replaced in 1957 with the H2N2 subtype, and the H1N1 "Russian influenza" had high genetic similarity to an H1N1 strain that had been circulating in the 1950s. The pandemic was mild because of this, since most people alive at the time had created antibodies against the previous strain. Because the 1977 strain was so similar to the 1950s strain and had not undergone a typical amount of genetic drift, it has been suggested that the 1977 outbreak occurred through an accidental release of the virus by researchers. Interestingly, the 1977 H1N1 did not completely replace the circulating H3N2; for the first time in history, both strains circulated together.

After 1918, the H1N1 virus continued to undergo genetic drift within the swine population of the United States. The virus also underwent several reassortments with avian influenzas over the years, and by 1998, a triple-reassortant swine influenza virus was circulating in US pig populations that was composed of the avian PA and PB2 gene segments and a human PB1 segment. In 2008–09, this virus reassorted again with a different swine influenza circulating within the pig population to create a quadruple-reassortant virus that now contained the NA and M1 gene segments from a Eurasian H1N1 strain (Fig. 10.13). This quadruple-reassortant virus jumped into the human population in April of 2009, after cases of the seasonal H3N2 and H1N1 influenza were waning (Fig. 10.14). It was first identified in two boys 150 miles apart in California, the first on April 15 and the second on April 17, with no known exposure to pigs, indicating that the virus was able to undergo human-to-human transmission. Realizing that human-to-human transmission of swine influenzas had been rare historically, the CDC sprang into action in an effort to prevent a severe pandemic. They authorized the use of oseltamivir and zanamivir for pandemic H1N1 infections (the strain was resistant to M2 inhibitors amantadine and rimantadine) and pushed to educate the public on ways to prevent the spread of the virus. By April 21, the CDC had also begun efforts to include the pandemic H1N1 virus in the current flu vaccine formulation. On April 25, the Director General of the WHO declared the outbreak a "Public Health Emergency of International Concern," and within a week, the WHO had raised the pandemic alert from phase 4 to phase 5 (of 6 total phases), indicating "a pandemic is imminent." (The pandemic alert was raised to phase 6 on June 11.) On April 28, the diagnostic test for the pandemic H1N1 was approved, allowing laboratories to specifically identify the pandemic strain. By May, the CDC had determined that the 2009 H1N1 virus did not share similar virulence characteristics with the 1918 H1N1 virus, and 33% of people >60 years of age

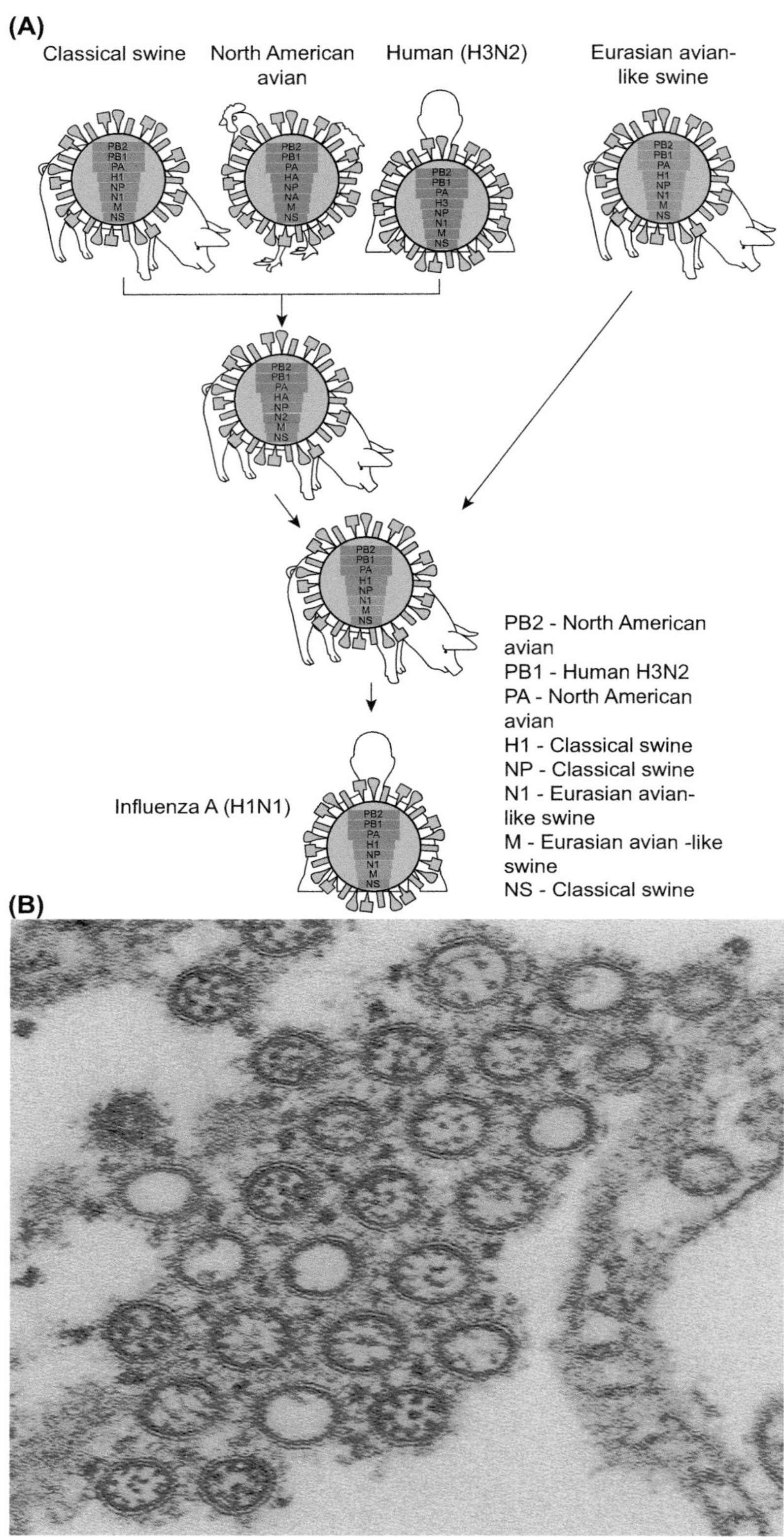

FIGURE 10.13 **Origin of the quadruple-reassortant 2009 H1N1 influenza A virus.** (A) In the late 1990s, reassortment between classical swine, avian, and human viruses resulted in a triple-reassortant virus circulating within North American pig populations. This virus reassorted once more with a Eurasian avian-like swine strain to create the H1N1 that appeared in 2009. The gene segments derived from each of these strains is indicated. *(Reprinted with permission from Neumann, G., Noda, T., Kawaoka, Y., 2009. Emergence and pandemic potential of swine-origin H1N1 influenza virus. Nature 459, 931–939, Macmillan Publisher Ltd, copyright 2009.)* (B) Transmission electron micrograph of the 2009 H1N1 influenza A virus. *(CDC/Cynthia Goldsmith.)*

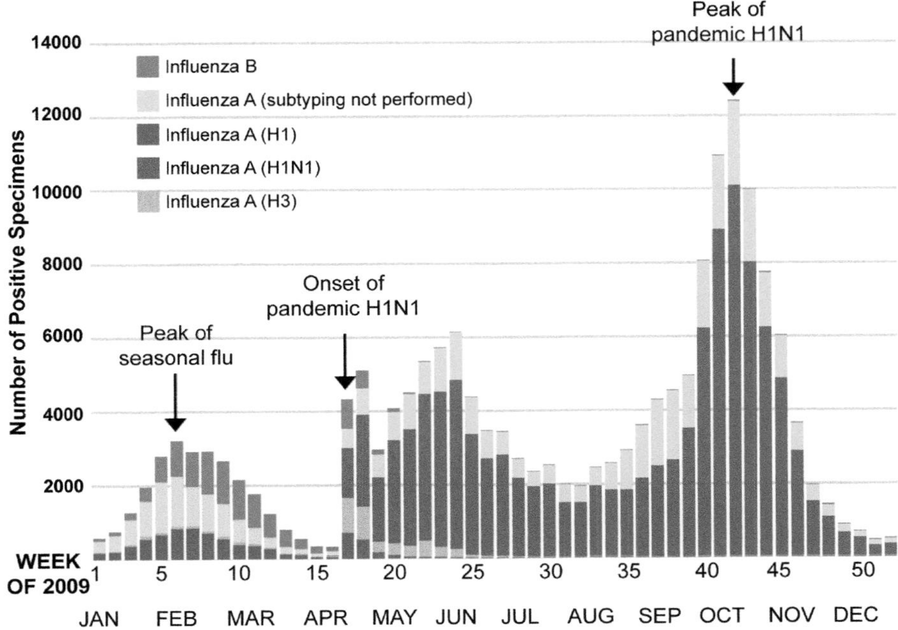

FIGURE 10.14 **Introduction of quadruple-reassortant H1N1 into humans in 2009.** Cases of seasonal influenza (in green, yellow, and blue) had peaked and were decreasing during the first few months of 2009. In mid-April, a new reassortant H1N1 strain (red) entered the US population, peaking in October of 2009. *Data obtained from FluView (internet), Atlanta, GA. Centers for Disease Control and Prevention. Available at: http://www.cdc.gov/flu/weekly.*

possessed antibodies against the 2009 H1N1, indicating that they had been exposed to a similar virus before, either through natural infection or through vaccination. At the end of June, the CDC estimated that 1 million cases of the pandemic H1N1 virus had occurred in the United States; in July, strains emerged that were resistant to oseltamivir. Cases of the virus peaked in the United States in October (Fig. 10.14), and the first doses of the vaccine were given on October 5 for those at highest risk. The vaccine became available for the public at large in December of 2009.

Overall, there were an estimated 43–89 million cases of the 2009 H1N1 in the United States, with influenza-associated hospitalizations numbering between 195,000–403,000. Interestingly, the number of deaths attributed to the pandemic H1N1 strain was estimated to be 8800–18,300, much less than a typical season with the seasonal H3N2 virus, and 90% of those who died were younger than 65 years of age. This reduction is attributed to better public health efforts and prior immunity in people over age 60, who are at the highest risk of dying from influenza-associated illnesses. Considering the mildness of the pandemic, critics questioned the necessity of the CDC's substantial efforts. Looking back, however, it took 6 months of accelerated development to release the 2009 H1N1 vaccine to the highest risk individuals, and if the 2009 antigenic shift were as virulent as the 1918 H1N1 virus, then hundreds of thousands of people could have died within this short time period even with modern medical advances.

Study Break

Create a timeline that shows when each antigenic shift occurred and which subtype jumped into the human population.

10.6 HIGHLY PATHOGENIC AVIAN INFLUENZA

Influenza A viruses of many different HA and NA subtypes circulate in wild bird populations, particularly waterfowl, where the disease is **epizootic**, meaning that it is epidemic in an animal population. Eighteen different HA subtypes (H1–H18) and eleven different NA subtypes (N1–N11) exist, but only three HA subtypes have circulated in humans: H1, H2, and H3. Therefore, the possibility remains that more human pandemics could occur through direct infection of humans with a mutated avian influenza, or through reassortment of an avian influenza with a human influenza in a host infected by both, such as swine. To infect humans, however, the avian HA protein needs to acquire the ability to infect human cells efficiently. The avian influenza, which normally is a gastrointestinal virus in birds, binds its HA protein to sialic acids linked to galactose with an alpha-2,3 linkage, meaning that the second carbon of sialic acid binds to the third carbon of the galactose sugar under it (see Fig. 10.5). In contrast, human influenzas bind to alpha-2,6-linked sialic acids. This slight change in receptor preference means that avian HA proteins do not efficiently bind to the human influenza virus receptor. For the H2 and H3 subtypes, the ability of the avian virus to bind human cells occurred through mutations in the HA protein's receptor binding site so that the avian HA protein acquired the ability to bind 2,6-linked sialic acids. For the H1 subtype, however, the HA location that directly binds the human receptor remains the same as in the avian HA protein, but an overall change in the structure of the HA protein allows it to bind the human alpha-2,6 sialic acid receptor.

Only over the past 20 years have we directly observed that **highly pathogenic avian influenza (HPAI)** viruses could be directly transmissible from birds to humans and cause severe respiratory syndromes, pneumonia, and death. The first known case of this occurred in Hong Kong in 1997, when a 3-year-old boy died of influenza-like symptoms. The virus was isolated and confirmed to be of the H5N1 strain, a strain that had not previously been known to infect humans, with an HA gene derived from geese (Fig. 10.15A). The same H5N1 strain was killing poultry

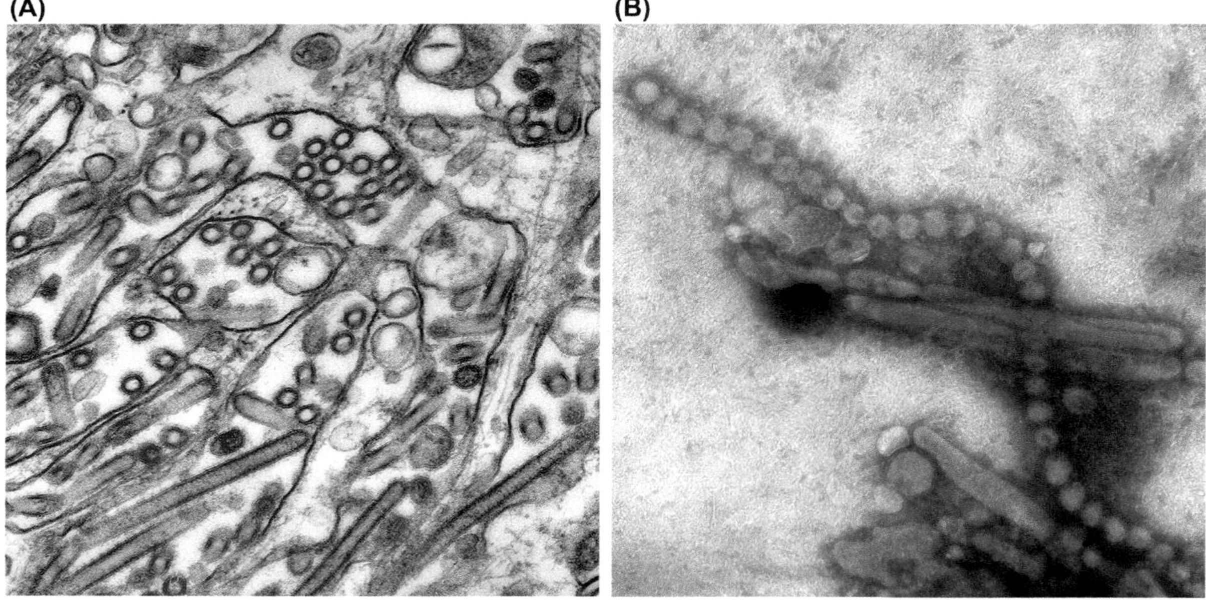

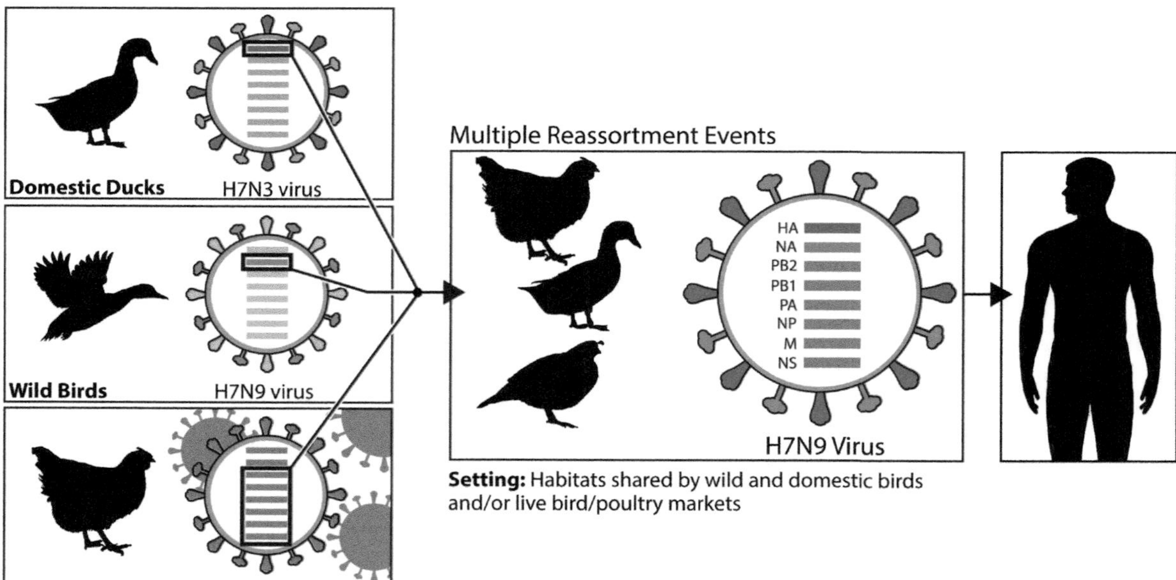

FIGURE 10.15 **Highly pathogenic avian influenza viruses.** Transmission electron micrographs of H5N1 (A) and H7N9 (B). Notice the pleomorphic sphere and filament morphology. (C) The HPAI H7N9 influenza A virus reported in 2013 in China likely obtained its HA gene from domestic ducks, NA gene from wild birds, and six remaining genes from multiple H9N2 viruses in domestic poultry. Around a third of humans infected with H7N9 have died of virus-induced pathology. *Images courtesy of the CDC and Cynthia Goldsmith, Jacqueline Katz, and Sherif R. Zaki (A); Thomas Rowe and Cynthia Goldsmith (B); and Dan Higgins and Douglas E. Jordan (C).*

Case Study: Cases of Influenza A (H5N1)—Thailand, 2004

Excerpted from Centers for Disease Control Morbidity and Mortality Weekly Report, February 13, 2004: 53(05), 100–103.

The human H5N1 viruses identified in Asia in 2004 are antigenically and genetically distinguishable from the 1997 and February 2003 viruses. To aid surveillance and clinical activities, this report provides a preliminary clinical description of the initial five confirmed cases in Thailand.

Of the five laboratory-confirmed cases in Thailand, four were in male children aged 6–7 years, and one was in a female aged 58 years; all patients were previously healthy (Table 10.4). Four patients reported deaths in poultry owned by the patient's family, and two patients reported touching an infected chicken. One patient had infected chickens in his neighborhood and was reported to have played near a chicken cage. None of the confirmed cases occurred among persons involved in the mass culling of chickens.

Patients reported to hospitals 2–6 days after onset of fever and cough (Table 10.4). Other early symptoms included sore throat (four), rhinorrhea (two), and myalgia (two). Shortness of breath was reported in all patients 1–5 days after symptom onset. On admission, clinically apparent pneumonia with chest radiograph changes was observed in all patients. Diarrhea and vomiting were not reported.

All patients had respiratory failure and required intubation a median of 7 days (range: 4–10 days) after onset of illness. Two patients had a collapsed lung (pneumothorax). Three patients required support for decreased cardiac function; two patients had renal (kidney) impairment as a later manifestation. None had documented evidence of secondary bacterial infection.

Late in the course of illness, three patients were treated with oseltamivir for 3–5 days. All received broad-spectrum antibiotics for community-acquired pneumonia while the cause of illness was under investigation. Four were treated with systemic steroids for increasing respiratory distress and clinically diagnosed acute respiratory distress syndrome with compatible chest radiograph changes.

Three children died 2–4 weeks after symptom onset, and one child and the adult died 8 days after symptom onset. All patients had laboratory evidence of influenza A (H5N1) by reverse transcriptase-polymerase chain reaction. In three cases, the virus was isolated in tissue culture, and in three cases, the viral antigens were identified by immunofluorescent assay.

Reported by T Chotpitayasunondh, S Lochindarat, P Srisan, Queen Sirikit National Institute of Child Health; K Chokepaibulkit, Faculty of Medicine, Siriraj Hospital, Mahidol Univ, Bangkok; J Weerakul, Buddhachinaraj Hospital, Phitsanulok; M Maneerattanaporn, 17th Somdejprasangkaraj Hospital, Suphanburi; P Sawanpanyalert, Dept of Medical Sciences, Ministry of Public Health, Thailand. World Health Organization, Thailand. CDC International Emerging Infections Program, Thailand.

TABLE 10.4 Clinical Features, Treatment, and Outcomes in Five Patients With Laboratory-Confirmed Influenza A (H5N1), by Sex and Age of Patient—Thailand, 2004

Sex	Age (years)	Signs and symptoms on admission	Subsequent clinical complications	Treatment and outcome
Male	7	Fever, cough, sore throat for 6 days. Dyspnea[a] on day 6; lung pathology on chest radiograph.	Respiratory failure on day 10; cardiac failure, ARDS,[b] gastrointestinal bleeding.	Oseltamivir on days 18–22. Died on day 29.
Male	6	Fever, cough, rhinorrhea for 5 days. Dyspnea[a] on day 6; lung pathology on chest radiograph.	Respiratory failure on day 8; hepatitis, ARDS.[b]	Oseltamivir on days 18–20. Died on day 20.
Male	6	Fevere, cough, rhinorrhea, sore throat for 4 days. Dyspnea[a] on day 5; lung pathology on chest radiograph.	Respiratory failure on day 6; pneumothorax,[c] ARDS.[b]	Died on day 18.
Female	58	Fever, cough, sore throat, myalgia for 2 days. Dyspnea[a] on day 2; lung pathology on chest radiograph.	Respiratory failure on day 4; cardiac failure, renal failure, ARDS.[b]	Died on day 8.
Male	6	Fever, cough, sore throat, myalgia for 4 days. Dyspnea[a] on day 5; lung pathology on chest radiograph.	Respiratory failure on day 5; cardiac failure, renal failure, ARDS.[b]	Oseltamivir on days 5–8. Died on day 8.

[a]*Labored breathing.*
[b]*Acute respiratory distress syndrome, a life-threatening lung condition.*
[c]*Collapsed lung.*

in the live bird markets of Hong Kong, and 17 additional human cases of H5N1 influenza occurred, all through direct contact with chickens. The government ordered the culling of all poultry within the region, which was effective in controlling the local transmission of the virus to humans. The virus replicated very efficiently in humans, and of the 18 people infected, 6 of them died (33%). No human-to-human transmission was reported, although it was feared that reassortment of the H5N1 strain with a circulating seasonal influenza within an infected person could result in a highly pathogenic virus able to be transmitted between humans.

Part of the reason that the H5N1 subtype is so virulent in the human host was because of the tropism of the virus. Avian influenza viruses more readily bind alpha-2,3-linked sialic acids (Fig. 10.5). In the human lung, these can be found deeper in the lung on the nonciliated cuboidal epithelial cells at the junction between the bronchioles and the alveoli (Fig. 10.1A). Researchers have shown that the H5N1 subtype exhibits tropism for these cuboidal epithelial cells, as well as alveolar macrophages and type II pneumocytes, both found in the alveoli of the lungs. The tropism of the H5N1 virus for the deepest part of the lower respiratory tract is a contributing factor to why the virus is so pathogenic in humans.

There have been over 600 cases of H5N1 transmission to humans since the WHO began recording them in 2003. The majority of cases have occurred in children and young adults, and 60% of those infected have died. Most of the infections occurred through direct bird-to-human transmission, although a handful of cases suggest that human-to-human transmission occurred between the sick individual and a close family member who was caring for the infected person.

The H5N1 virus continues to circulate throughout the world in both wild birds and poultry populations, currently endemic in China, Bangladesh, Indonesia, Vietnam, India, and Egypt. In addition to birds, the HPAI H5N1 has also been shown to infect pigs, dogs, ferrets, housecats, and wild cat species, among others.

In 2013, human infections with a novel HPAI H7N9 virus were reported in China. The H7N9 strain was also found within bird populations as a new reassortant infecting domestic ducks, wild birds, and domestic poultry (Fig. 10.15B and C). Like HPAI H5N1, the virus was transmitted to humans through direct contact with birds and caused severe respiratory infection. Of 163 human cases reported in China in 2013, 50 people (31%) died. H7N9 strains were identified that were able to efficiently bind both avian and human cell surface receptors. Interestingly, there have been no reports of H7N9 infection in swine populations.

SUMMARY OF KEY CONCEPTS

Section 10.1 Influenza Taxonomy and Types

- Influenza viruses are enveloped, helical −ssRNA viruses. There are three genera in the family that each contains a single virus type: Influenza A virus, Influenza B virus, and Influenza C virus. Subtypes of influenza A are determined by the HA and NA proteins possessed by the virus.
- Influenza A and B viruses are associated with seasonal epidemics. Currently, influenza A H1N1 and H3N2 subtypes are circulating in humans, as well as influenza B/Yamagata and B/Victoria strains. Influenza A viruses have caused several known pandemics during the last 100 years.

Section 10.2 Clinical Course of Infection

- Influenza virions are transmitted in respiratory droplets and bind ciliated epithelial cells of the sinuses, nasopharynx, larynx, trachea, bronchi, and bronchioles. These cells express the receptor for influenza, alpha-2,6-linked sialic acids.
- The average incubation period for influenza infection is 2 days. Symptoms of the disease last 3–7 days and include fever, myalgia, malaise, sore throat, rhinorrhea, and a nonproductive cough. The virus is communicable for 5–10 days after symptoms subside.
- Children, the elderly, and immunocompromised individuals are at highest risk of influenza-associated hospitalization. The most frequent complications arise from secondary bacterial pneumonia.
- Influenza cases generally peak between December and March. An average of 36,000 people die each year when H3N2 subtypes are circulating.

Section 10.3 Molecular Virology

- Influenza virions contain segmented negative-sense genomic RNA. Two viral proteins protrude from the envelope, the trimer HA and tetramer NA. The HA0 protein is cleaved into HA1 and HA2 subunits that remain bound to each other; HA1 is the portion responsible for binding sialic acid, which triggers clathrin-mediated endocytosis. In the endosome, HA2 initiates fusion of the viral envelope and endosomal membrane, assisted by the activities of the M2 ion channel protein. NA is an enzyme that cleaves sialic acid upon nascent virion release from the cell.
- Each vRNA segment is wound around the NP protein and associates with the RdRp, composed of PB1, PB2, and PA, to form an RNP complex. RNPs are imported into the nucleus, and the RdRp begins vmRNA transcription using a cap-snatching process. vRNA segments 7 and 8 undergo alternative splicing.
- The creation of full-length antigenome cRNA functions as a template for replication of the vRNA segments.
- Following transcription, HA and NA begin assembly at the plasma membrane while RNP complexes are exported from the nucleus by M1 and NS2/NEP to the site of assembly. The virion buds from the plasma membrane, acquiring its envelope. NA cleaves local sialic acids to prevent virions from aggregating at the cell surface.

Section 10.4 Genetic Changes in the Influenza Genome

- Antigenic drift occurs as a result of point mutations in the influenza genome, resulting in the formation of new strains that may not be recognized by previously generated host antibodies.
- Antigenic shift occurs as a result of vRNA segment reassortment between two virus subtypes, leading to the creation of a new subtype. Because only H1N1, H1N2, H2N2, and H3N2 subtypes have circulated in humans, the introduction of avian or swine subtypes into the human population could lead to pandemics, since no individual human would have any immunity against the new subtype and infection rates would be very high.

Section 10.5 Historical Influenza Antigenic Shifts

- Major antigenic shifts occurring in the last 100 years have resulted in significant morbidity and mortality. The 1918 "Spanish flu" was a particularly virulent strain that spread in waves throughout the world, leading to the deaths of an estimated 20–50 million people worldwide.
- Experiments that used reverse genetics to recreate the virus indicate that the 1918 H1N1 HA protein caused a cytokine storm from innate immune cells that caused major damage to the lungs, suggesting an explanation for the high mortality of individuals in the 25–34 age group and the "W-shaped" mortality curve.
- In 1957, antigenic shift occurred as the H1N1 virus was replaced with the H2N2 subtype, a reassortant with avian influenza viruses. Another antigenic shift, to H3N2, occurred in 1968.
- The H1N1 influenza reappeared in 1977 and began circulating alongside H3N2 in the human population.
- In 2009, a quadruple-reassortant H1N1 virus containing vRNA segments from avian, human, and swine influenza viruses began infecting the human population. The 2009 H1N1 "swine flu" was not as virulent as the 1918 H1N1 strain, and a third of people over age 60 possessed neutralizing antibodies against it, suggesting a similar strain had circulated previously in the human population.

Section 10.6 Highly Pathogenic Avian Influenza

- Influenza A viruses are epizootic in wild bird populations. Subtypes that infect birds have higher affinity for alpha-2,3 sialic acid linkages, which are located deeper within the human respiratory tract. As a result, influenzas that have been transmitted from birds to humans generally cause more severe respiratory syndromes. They are known as HPAI viruses.

- HPAI H5N1 and H7N9 subtypes continue to circulate in bird populations and occasionally are transmitted to humans in close contact with birds. Both strains exhibit high mortality in infected humans.

FLASH CARD VOCABULARY

Influenza type/subtype	Myocarditis
Hemagglutinin	Segmented genome
Neuraminidase	Nucleocapsid protein
Myalgia	Ribonucleoprotein complex
Malaise	Antigenome
Rhinorrhea	Cap-snatching
Pneumonia	Apical
Antigenic drift	Reverse genetics
Escape mutants	Epizootic
Antigenic shift	Highly pathogenic avian influenza virus
Reassortment	

CHAPTER REVIEW QUESTIONS

1. Write out the taxonomical classification of influenza viruses, noting the family, genera, species (types), and subtypes, if they exist for the virus.
2. A swine influenza A virus was discovered in Iowa in 2005 to be an H1N1 virus. The strain number was 15. How would you properly write the name of this strain?
3. What are the symptoms of influenza infection, and how is it differentiated from a cold?
4. Make a chart of the 11 proteins encoded by influenza A virus and indicate their functions.
5. What is an RNP and of what is it composed?
6. From nose to alveoli, explain the architecture of the respiratory tract and the type of epithelium that is found at each location. Which epithelia express alpha-2,6-linked sialic acids?
7. Describe how cap-snatching occurs.
8. Spell out what the following acryonyms stand for: vRNA, vmRNA, cRNA, RdRp, RNP, NP, HA, and NA. Provide a definition for each.
9. Since both are transcribed by the viral RdRp, what are the differences between a vmRNA transcript and the cRNA antigenome?
10. What would be the result of an influenza virion that lacks a functional NA protein?
11. Explain what antigenic drift and antigenic shift are. Why is one more likely to lead to pandemics?
12. How does reassortment occur?
13. Why was the 1918 H1N1 virus so virulent? Provide at least three reasons.
14. Why are avian influenzas particularly virulent in humans?

FURTHER READING

Center for Disease Control and Prevention (CDC), 2015. Influenza. In: Haborsky, J., Kroger, A., Wolfe, S. (Eds.), Epidemiology and Prevention of Vaccine-Preventable Diseases, thirteenth ed. Public Health Foundation, Washington, DC, pp. 187–207. Chapter 12.

Gamblin, S.J., Haire, L.F., Russell, R.J., et al., 2004. The structure and receptor binding properties of the 1918 influenza hemagglutinin. Science 303, 1838–1842.

Gibbs, M.J., Gibbs, A.J., 2006. Molecular virology: was the 1918 pandemic caused by a bird flu? Nature 440, E8 discussion E9–10.

Harrison, S.C., 2008. Viral membrane fusion. Nat. Struct. Mol. Biol. 15, 690–698.

Holmes, E.C., 2004. Virology. 1918 and all that. Science 303, 1787–1788.

Horby, P., Nguyen, N.Y., Dunstan, S.J., Kenneth Baillie, J., 2013. An updated systematic review of the role of host genetics in susceptibility to influenza. Influenza Other Respir. Viruses 7, 37–41.

Hutchinson, E.C., Fodor, E., 2013. Transport of the influenza virus genome from nucleus to nucleus. Viruses 5, 2424–2446.

Kash, J.C., Tumpey, T.M., Proll, S.C., et al., 2006. Genomic analysis of increased host immune and cell death responses induced by 1918 influenza virus. Nature 443, 578–581.

Kobasa, D., Jones, S.M., Shinya, K., et al., 2007. Aberrant innate immune response in lethal infection of macaques with the 1918 influenza virus. Nature 445, 319–323.

Kobasa, D., Takada, A., Shinya, K., et al., 2004. Enhanced virulence of influenza A viruses with the haemagglutinin of the 1918 pandemic virus. Nature 431, 703–707.

Neumann, G., Macken, C.A., Kawaoka, Y., 2014. Identification of amino acid changes that may have been critical for the genesis of A(H7N9) influenza viruses. J. Virol. 88, 4877–4896.

Neumann, G., Noda, T., Kawaoka, Y., 2009. Emergence and pandemic potential of swine-origin H1N1 influenza virus. Nature 459, 931–939.

Rumschlag-Booms, E., Guo, Y., Wang, J., Caffrey, M., Rong, L., 2009. Comparative analysis between a low pathogenic and a high pathogenic influenza H5 hemagglutinin in cell entry. Virol. J. 6, 76.

Rumschlag-Booms, E., Rong, L., 2013. Influenza A virus entry: implications in virulence and future therapeutics. Adv. Virol. 2013. http://dx.doi.org/10.1155/2013/121924.

Saito, T., Gale, M., 2007. Principles of intracellular viral recognition. Curr. Opin. Immunol. 19, 17–23.

Shaw, M.L., Palese, P., 2013. Orthomyxoviridae. In: Knipe, D.M., Howley, P.M. (Eds.), Fields Virology, sixth ed. Wolters Kluwer|Lippincott Williams and Wilkins, pp. 1151–1185.

Shi, Y., Zhang, W., Wang, F., et al., 2013. Structures and receptor binding of hemagglutinins from human-infecting H7N9 influenza viruses. Science 342, 243–247.

Steinhauer, D.A., Skehel, J.J., 2002. Genetics of influenza viruses. Annu. Rev. Genet. 36, 305–332.

Stevens, J., Corper, A.L., Basler, C.F., Taubenberger, J.K., Palese, P., Wilson, I.A., 2004. Structure of the uncleaved human H1 hemagglutinin from the extinct 1918 influenza virus. Science 303, 1866–1870.

Tamerius, J., Nelson, M.I., Zhou, S.Z., Viboud, C., Miller, M.A., Alonso, W.J., 2011. Global influenza seasonality: reconciling patterns across temperate and tropical regions. Environ. Health Perspect. 119, 439–445.

Taubenberger, J.K., 2006. The origin and virulence of the 1918 "Spanish" influenza virus. Proc. Am. Philos. Soc. 150, 86–112.

Taubenberger, J.K., Morens, D.M., 2006. 1918 Influenza: the mother of all pandemics. Emerg. Infect. Dis. 12, 15–22.

Tong, S., Zhu, X., Li, Y., et al., 2013. New world bats harbor diverse influenza a viruses. PLoS Pathog. 9. http://dx.doi.org/10.1371/journal.ppat.1003657.

World Health Organization, 2010. Evolution of a Pandemic: A(H1N1) 2009. World Health Organization.

Ye, Q., Krug, R.M., Tao, Y.J., 2006. The mechanism by which influenza A virus nucleoprotein forms oligomers and binds RNA. Nature 444, 1078–1082.

Zhao, H., Zhou, J., Jiang, S., Zheng, B.-J., 2013. Receptor binding and transmission studies of H5N1 influenza virus in mammals. Emerg. Microbes Infect. 2, e85.

Zimmer, S.M., Burke, D.S., 2009. Historical perspective — emergence of influenza A (H1N1) viruses. N. Engl. J. Med. 361, 279–285.

Chapter 11

Human Immunodeficiency Virus

The **human immunodeficiency virus (HIV)** is one of the most intriguing and challenging viruses to have existed. Evidence suggests that HIV first originated in Africa around 1920–30 as a result of cross-species infections of humans by **simian** (ape and monkey) viruses. The United States became aware of the disease that HIV causes, **acquired immune deficiency syndrome (AIDS)**, in 1981, and the virus was first identified 2 years later. HIV infects helper CD4 T cells of the immune system, causing their gradual decline in numbers. Scientifically, HIV is an enigmatic challenge that is being deciphered, molecule by molecule, in the search for a vaccine or cure. Sociologically, HIV began as a disease that caused fear and stigma but is now no longer a death sentence, manageable for years with antiviral medications. However, around 1.5 million people worldwide die each year of HIV/AIDS, making it the sixth most-common cause of death in the world.

11.1 HISTORY OF HIV INFECTION

The history of HIV/AIDS is a story without an ending that contains both victorious scientific occasions and incidents that were not the finest moments of humankind. The early 1980s were central in the characterization of AIDS and the virus that causes it. On June 5, 1981, the United States Centers for Disease Control and Prevention (CDC) published an article in its *Morbidity and Mortality Weekly Report* that documented five uncommon cases of pneumonia caused by the fungus *Pneumocystis carinii* (now known as *Pneumocystis jirovecii*) in young, previously healthy gay men. Within days, additional cases were being reported to the CDC. Also in 1981, cases of Kaposi's sarcoma—a rare skin cancer generally only seen in older men—were being reported in previously healthy, young gay men in New York and California. By the end of the year, 270 cases of this severe immunodeficiency had been reported, causing the death of 121 of those individuals that year. In 1982, the CDC proposed the name "acquired immune deficiency syndrome" to replace what had been referred to as "gay-related immune deficiency" in an effort to not misclassify the disease as one that only infected homosexuals. The CDC had also been receiving reports of AIDS in people who had received blood transfusions and in **hemophiliacs**, who lack clotting factor VIII and therefore had to receive it through intravenous injections of clotting factor concentrates derived from human plasma. (About half of all people with hemophilia, and 90% of those with severe hemophilia that required transfusions more often, became infected with HIV through contaminated factor VIII.)

By 1983, the CDC had identified all major routes of HIV transmission, which does not include through food, water, air, or personal contact. Regardless, fears of the virus grew while public health officials worked to educate the public (Fig. 11.1A). For years, discrimination and even violence occurred against people with HIV/AIDS; they were fired from their jobs, evicted from their homes, and denied access to medical care. Those that targeted homosexuals for their sexual preference had another reason to discriminate against them; Ryan White, a 13-year-old hemophiliac diagnosed in 1984, fought to attend public school while the parents of other children petitioned to ban him from school; and three hemophiliac brothers living in Florida—Ricky Ray, Robert Ray, and Randy Ray—were diagnosed with HIV in 1986 and were not allowed to attend public school because of their HIV status. Their parents sued the school board, and their house was burned down a week after the federal court ruled in their favor. Fear of stigma added to the fear of death in those presenting with AIDS symptoms.

The first major scientific breakthrough against the disease occurred in 1983. Pasteur Institute researchers Françoise Barré-Sinoussi and Luc Montagnier isolated a retrovirus, which they named lymphadenopathy-associated virus, from the lymph node of an individual with AIDS. (They shared the 2008 Nobel Prize in Physiology or Medicine for this discovery.) In the same issue of the scientific journal *Science*, National Cancer Institute researcher Robert Gallo described a retrovirus that he named the third of the human T-lymphotropic viruses, HTLV-III. The following year, Montagnier and Gallo held a joint press conference to announce the two viruses were in fact the same virus and likely the cause of AIDS. The first blood test for the virus was developed in 1985, allowing blood banks to begin testing for the virus. By this time, AIDS had been identified in all regions of the world, and 1.5 million people were estimated to have been infected.

In 1986, the International Committee on Taxonomy of Viruses officially named the retrovirus "human immunodeficiency virus." Many milestones occurred the following

(A)

You won't get AIDS from everyday contact.
You won't get AIDS from being a friend.
You won't get AIDS from a mosquito bite.
You won't get AIDS from a kiss.
You won't get AIDS by talking.
You won't get AIDS by listening.
You won't get AIDS from a public pool.
You won't get AIDS from a pimple.
You won't get AIDS from a toilet seat.
You won't get AIDS from a haircut.
You won't get AIDS by donating blood.
You won't get AIDS from an airplane.
You won't get AIDS from tears.
You won't get AIDS from food.
You won't get AIDS from a hug.
You won't get AIDS from a towel.
You won't get AIDS from a telephone.
You won't get AIDS from a crowded room.
You won't get AIDS from an elevator.
You won't get AIDS from a greasy spoon.
You won't get AIDS from a bump.
You won't get AIDS by laughing.
You won't get AIDS by watching a movie.
You won't get AIDS from a cat.
You won't get AIDS from a schoolyard.
You won't get AIDS from going to a party.
You won't get AIDS from taking a trip.
You won't get AIDS from a dog bite.
You won't get AIDS from visiting a city.
You won't get AIDS from a cab.
You won't get AIDS from a bus.
You won't get AIDS at a play.
You won't get AIDS by dancing.
You won't get AIDS because someone is different from you.
You won't get AIDS from a classroom.

Stop Worrying About How You Won't Get AIDS. And Worry About How You Can.

You *can* get AIDS from sexual intercourse with an infected partner.
You *can* get AIDS from sharing drug needles with an infected person.
You *can* get AIDS by being born to an infected mother.

AMERICA RESPONDS TO AIDS

(B)

(C)

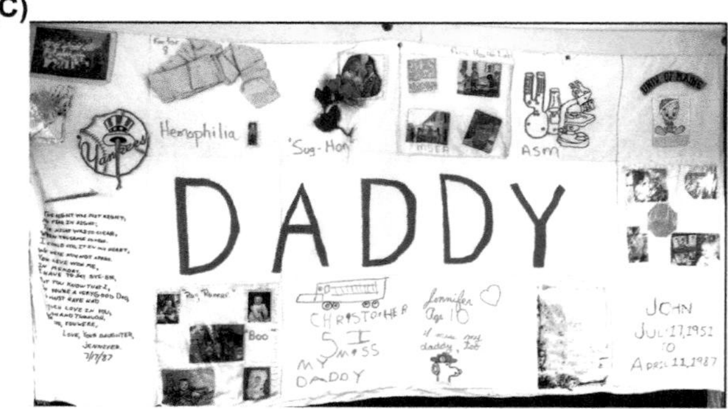

FIGURE 11.1 **The early years of HIV.** (A) In the 1980s and 1990s, public health officials distributed brochures and posters, such as this one from the Centers for Disease Control and Prevention, that described how HIV can and cannot be transmitted. (B) The AIDS quilt, shown here in an exhibit at the National Mall, first went on display in 1997. The quilt, which was nominated for the Nobel Peace Prize in 1989, is the largest community art project in the world, commemorating the lives of those who have died of HIV/AIDS. *(Photograph by Alan Kotok, used under a Creative Commons Attribution 2.0 Generic license.)* (C) One panel of over 48,000 found in the AIDS quilt. This panel honors a hemophiliac father who died of HIV in 1987, at the age of 35. *(Photo courtesy of Cindy Smith.)*

year, in 1987: the first antiretroviral drug, zidovudine (AZT), was approved; then President Ronald Reagan first discussed HIV/AIDS in a speech to the public; the AIDS quilt went on display, with 1920 panels (Fig. 11.1B and C); and the first HIV vaccine entered clinical trials.

Slow but steady progress, both scientifically and socially, occurred over the following 10 years. By 1994, AIDS was the leading cause of death for all Americans age 25–44, but **highly active antiretroviral therapy (HAART)**, a cocktail of *several* antiviral drugs rather than just one drug, began extending the lives of those with HIV in the mid-1990s. This led to a decline in the number of new AIDS cases in 1996. HAART is also referred to as **combination antiretroviral therapy (cART)**.

Since then, our scientific knowledge of HIV and its progression to AIDS has increased at an exponential rate. A new era of antiviral drugs has extended the life span of HIV+ individuals, but the epidemic is far from over, as evidenced by over 2 million people still being infected each year with HIV. Much remains to be learned about HIV, and a successful vaccine has yet to be created, over 30 years after the discovery of the virus.

11.2 TAXONOMY AND ORIGINS OF HIV

There are two species, or types, of HIV: HIV-1 and HIV-2 (Fig. 11.2). HIV-1 and HIV-2 are in the *Retroviridae* family of viruses that replicate through reverse transcription of their +ssRNA genome into cDNA, which becomes permanently integrated into the host genome. They are within the *Orthoretrovirinae* subfamily (the other retrovirus subfamily is *Spumaretrovirinae*, which contains foamy viruses whose replication and morphology are notably different than the *Orthoretrovirinae* members). Six retrovirus genera exist within the *Orthoretrovirinae* subfamily: *Alpharetrovirus, Betaretrovirus, Deltaretrovirus, Epsilonretrovirus, Gammaretrovirus,* and *Lentivirus*. HIV-1 and HIV-2 are the human members of the *Lentivirus* genus *(lenti* is Latin for "slow," reflecting the slow infections caused by these viruses). The two virus types cause clinically indistinguishable conditions, although viral loads tend to be lower during HIV-2 infections. HIV-2 is also less easily transmitted, possibly due to the reduced viral loads present, and progression to AIDS is much slower, if it occurs at all.

HIV-1 and HIV-2 are further broken down into genetically distinct **groups** (Fig. 11.2). HIV-1 contains group M (main/major), O (outlier), and N (non-M, non-O), as well as a new group named "group P." The HIV-1 group M viruses are the cause of the worldwide pandemic, responsible for more than 90% of HIV infections. (It is for this reason that when someone refers to "HIV," he/she is likely referring to HIV-1.) There are nine distinct **subtypes** (or **clades**) within group M: subtypes A, B, C, D, F, G, H, J, and K (subtypes E and I were prematurely classified as new subtypes before

Case Study—First Case Reports of AIDS

Excerpted from the Morbidity and Mortality Weekly Report, June 5, 1981 30 (21):1–3.

The following report began the investigation of the infectious disease that was causing opportunistic infections, including uncommon Pneumocystis pneumonia, in previously healthy individuals. The condition was termed AIDS the following year, and the cause of the immunosuppression was later determined to be HIV.

***Pneumocystis* Pneumonia—Los Angeles**

In the period October 1980–May 1981, five young men, all active homosexuals, were treated for biopsy-confirmed *P. carinii* pneumonia at three different hospitals in Los Angeles, California. Two of the patients died. All five patients had laboratory-confirmed previous or current cytomegalovirus (CMV) infection and candidal mucosal infection. Case reports of these patients follow.

Patient 1: A previously healthy 33-year-old man developed *P. carinii* pneumonia and oral mucosal candidiasis (*yeast infection*) in March 1981 after a 2-month history of fever associated with elevated liver enzymes, leukopenia (*low white blood cells*), and CMV viruria. The patient's condition deteriorated despite courses of treatment with trimethoprim–sulfamethoxazole (TMP/SMX), pentamidine, and acyclovir. He died May 3, and postmortem examination showed residual *P. carinii* and CMV pneumonia, but no evidence of neoplasia.

Patient 2: A previously healthy 30-year-old man developed *P. carinii* pneumonia in April 1981 after a 5-month history of fever each day and of elevated liver-function tests, CMV viruria, and documented seroconversion to CMV. Other features of his illness included leukopenia and mucosal candidiasis. His pneumonia responded to a course of intravenous TMP/SMX, but, as of the latest reports, he continues to have a fever each day.

Patient 3: A 30-year-old man was well until January 1981 when he developed esophageal and oral candidiasis that responded to amphotericin B treatment. He was hospitalized in February 1981 for *P. carinii* pneumonia that responded to TMP/SMX. His esophageal candidiasis recurred after the pneumonia was diagnosed, and he was again given amphotericin B. Material from an esophageal biopsy was positive for CMV.

Patient 4: A 29-year-old man developed *P. carinii* pneumonia in February 1981. He had had Hodgkins disease 3 years earlier, but had been successfully treated with radiation therapy alone. He did not improve after being given intravenous TMP/SMX and corticosteroids and died in March. Postmortem examination showed no evidence of Hodgkins disease, but *P. carinii* and CMV were found in lung tissue.

Patient 5: A previously healthy 36-year-old man with clinically diagnosed CMV infection in September 1980 was seen in April 1981 because of a 4-month history of fever, dyspnea, and cough. On admission he was found to have *P. carinii* pneumonia, oral candidiasis, and CMV retinitis. The patient has been treated with two short courses of TMP/SMX that have been limited because of a sulfa-induced neutropenia. He is being treated for candidiasis with topical nystatin.

The diagnosis of *Pneumocystis* pneumonia was confirmed for all five patients by closed or open lung biopsy. The patients did not know each other and had no known common contacts or knowledge of sexual partners who had had similar illnesses. Two of the five reported having frequent homosexual contacts with various partners. All five reported using inhalant drugs, and one reported parenteral drug abuse. Three patients had profoundly depressed in vitro proliferative responses to mitogens (*substances that cause lymphocyte proliferation*) and antigens. Lymphocyte studies were not performed on the other two patients.

Reported by MS Gottlieb, MD, HM Schanker, MD, PT Fan, MD, A Saxon, MD, JD Weisman, DO, Div of Clinical Immunology-Allergy; Dept of Medicine, UCLA School of Medicine; I Pozalski, MD, Cedars-Mt. Siani Hospital, Los Angeles; Field services Div, Epidemiology Program Office, CDC.

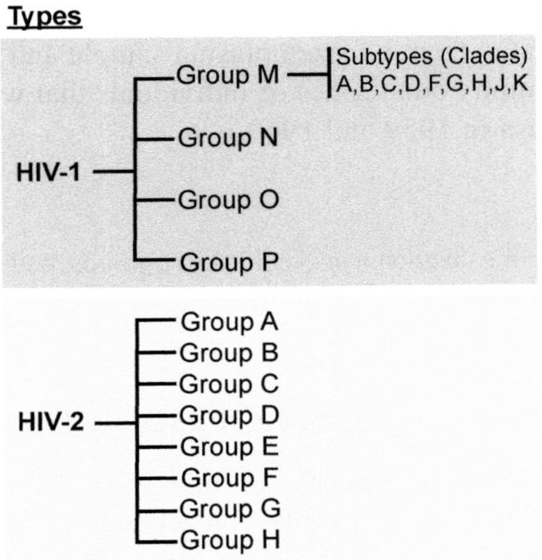

FIGURE 11.2 **HIV classification.** There are two types of HIV, HIV-1 and HIV-2. Strains within each of these types are further subdivided into HIV-1 Groups M, N, O, and P, and HIV-2 Groups A–H. HIV-1 Group M also has nine official subtypes/clades and a variety of circulating recombinant forms.

being reclassified into other subtypes). Numerous circulating recombinant forms also exist. The subtypes are geographically distributed throughout the world, with group C responsible for ~50% of total worldwide infections (Fig. 11.3). HIV-1 group O is responsible for <1% of HIV-1 infections and is generally restricted to Cameroon, Gabon, and neighboring countries. Group N is even more rare: only around 15 cases of HIV-1 group N have been documented, all of which were found in Cameroon. Group P viruses were identified in 2009, and only 2 cases have been documented, both from individuals living in Cameroon.

HIV-2 is composed of at least eight distinct groups, designated groups A through H. Found throughout Western Africa and Côte d'Ivoire, respectively, group A and B are the most prevalent HIV-2 groups, while groups C–H are rare. No HIV-2 subtypes currently exist.

More than 70% of emerging infectious diseases originate from wildlife, and HIV originated in this way as well. Each HIV-1 and HIV-2 group resulted from a separate

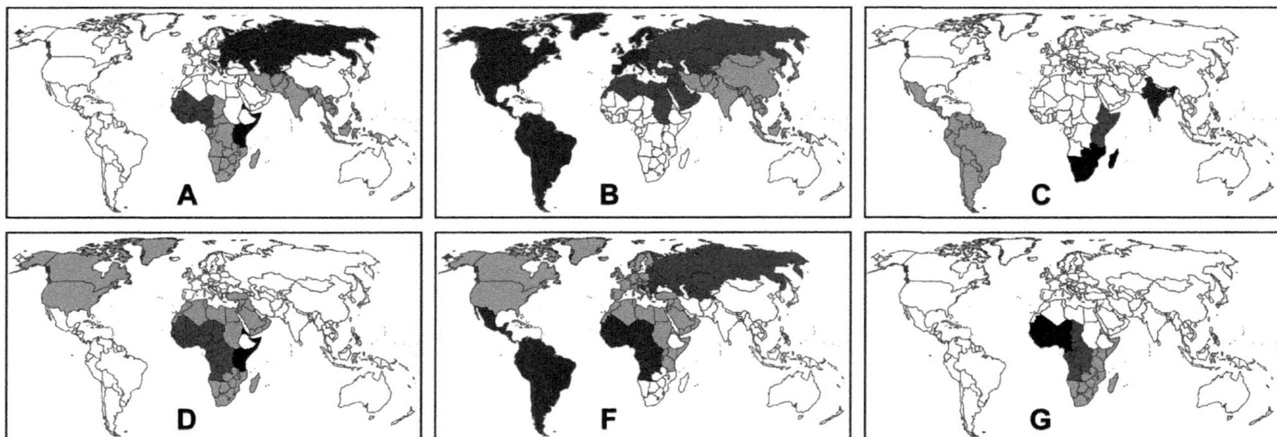

FIGURE 11.3 **Worldwide distribution of HIV-1 Group M subtypes.** Subtype C is responsible for ~50% of infections and is most prevalent in Eastern and Southern Africa and in India. Subtype B is the most prevalent subtype in Europe and the Americas. *Darker shading* indicates a greater proportion of infections of the subtype specified. *Map image reprinted with permission of Wolters Kluwer Health, Inc., from Hemelaar et al., 2011. Global trends in molecular epidemiology of HIV-1 during 2000–2007. AIDS 25 (5), 679–689.*

cross-species infection of humans from nonhuman primates, specifically from apes and Old World monkeys. **Simian immunodeficiency viruses (SIVs)** have circulated in these populations for at least 30,000 years and are thought to have been transmitted to humans through contact with the blood of an SIV-infected animal. The hunting of **bushmeat**, the meat of wild animals from Africa, involves several activities that could put the hunter in contact with the blood of an infected animal: butchering the meat to sell it, using body parts for traditional medicine, collecting carcasses, being injured during hunting, and even keeping the animal as a pet. SIVs are designated using lowercase letters that describe the species in which the virus was isolated. For instance, SIVcpz is an SIV from chimpanzees (such as *Pan troglodytes*), SIVgor is from gorillas (*Gorilla gorilla*), and SIVsmm was isolated from sooty mangabey monkeys (*Cercocebus atys*).

The four HIV-1 and eight HIV-2 groups appear to have each arisen from a separate cross-species infection of a human with an SIV (Fig. 11.4A). All eight HIV-2 groups appear to have originated from an SIV of sooty mangabey origin (SIVsmm), linked to strains originating in Côte d'Ivoire, Liberia, and Sierra Leone (Fig. 11.4B). HIV-1 groups M and N derive from a chimpanzee SIV (SIVcpz) from Cameroon, while HIV-1 groups O and P are most closely related to a SIVgor circulating in western lowland gorillas, also of Cameroon. A recent scientific report proposes that the spread of HIV-1 and the origin of the AIDS pandemic traces back to around 1920 in the former Belgian Congo city of Leopoldville, renamed Kinshasa in 1966 (currently the capital of the Democratic Republic of the Congo (DRC), formerly Zaire). The virus would have first jumped into the human population by infection of an individual in Cameroon that then traveled to Kinshasa, likely using the Sangha River system (Fig. 11.4C). In Kinshasa, the virus could have spread through the reuse of needles that were used to treat syphilis in the city's clinics. It is likely the virus spread from Kinshasa via the railway system that was constructed by the Belgian colony for trade and commerce purposes (Fig. 11.4C and D). The railways, used by over 1 million passengers in 1948, were the major mode of transportation in the country. Commercial sex workers also likely contributed to the sexual transmission of the virus. The railways connected Kinshasa to other populated cities within the DRC, such as Mbuji-Mayi and Lubumbashi, which would have given the virus ample numbers of hosts in which to begin its expansion across Africa. In support of this, the oldest known human samples to contain HIV-1 were isolated from a frozen plasma sample and from preserved lymph node tissues of individuals that were living in Kinshasa in 1959 and 1960.

Study Break
Describe the taxonomical classification, groups, and subtypes of HIV-1 and HIV-2.

11.3 EPIDEMIOLOGY OF HIV/AIDS

HIV is transmitted primarily in three ways. First, a person can be infected through sexual contact that exposes a mucosal epithelium to semen, vaginal secretions, rectal secretions, or blood containing the virus. This generally occurs through unprotected vaginal or anal intercourse. Condom use results in an 80% reduction in the transmission of HIV.

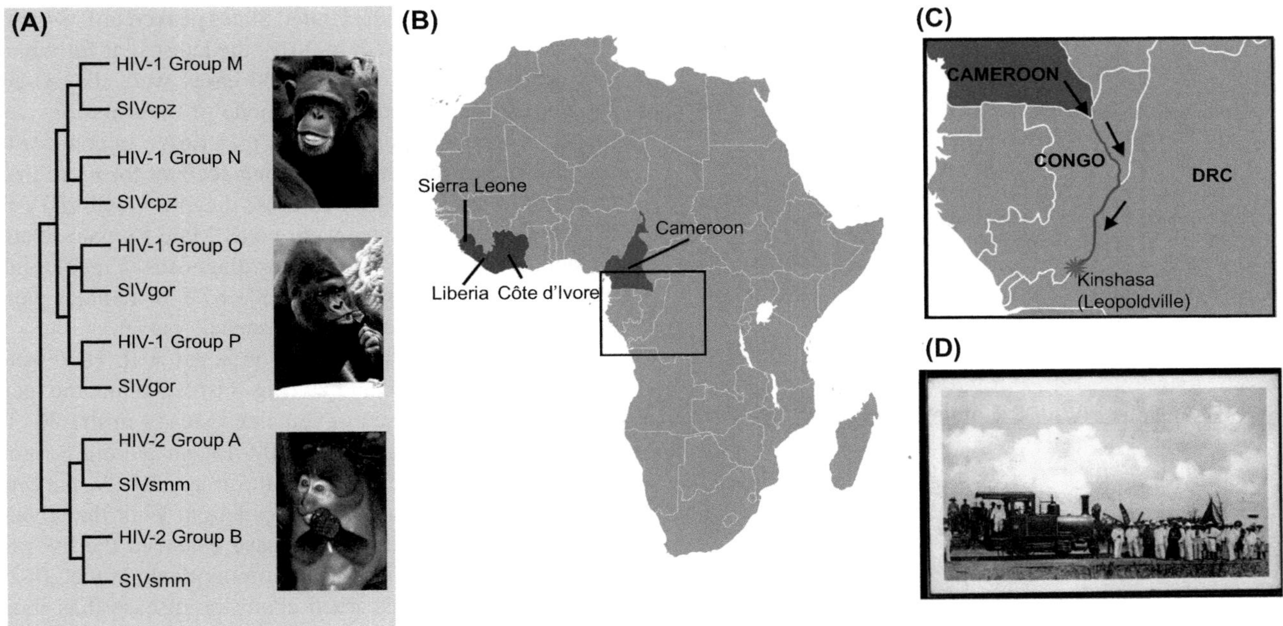

FIGURE 11.4 **Origin of HIV-1 and HIV-2 from simian immunodeficiency viruses (SIVs).** (A) The viruses from each HIV group are thought to have originated from a separate cross-species infection of a human with an SIV. This phylogenetic tree comparing virus similarity shows the probable origin of HIV Groups M (from SIVcpz), Group N (from SIVcpz), Group O (from SIVgor), and Group P (from SIVgor). Two of the eight HIV groups, which are thought to have originated from SIVsmm strains, are shown. *Note that the branches of the tree do not indicate evolutionary time in this phylogenetic tree. Sooty mangabey photo courtesy of Daegling et al., 2011. PLoS One 6(8), e23095.* (B) The first HIV-1 infections are thought to have occurred in Cameroon (red), while HIV-2 traces back to Côte d'Ivoire, Liberia, and Sierra Leone (blue). (C) This inset shows how the first HIV-1 infections likely spread from Cameroon using the Sangha River system to Leopoldville (currently Kinshasa). From here, people using the railways spread HIV to other populated areas within the current Democratic Republic of the Congo (DRC). (D) A postcard showing the first train to have arrived in Leopoldville in 1898.

The virus can also be transmitted from mother to child, which can occur transplacentally, during birth from exposure to the mother's genital secretions or blood, or from breast milk. The viral load at the time of delivery plays a role in the intrapartum transmission of HIV from mother to newborn. Up to 40% of infants become infected during birth, but the rate of HIV-1 transmission from mother to infant can be reduced by up to 70% through C-section delivery. The third major way that HIV is transmitted is through blood, either by injectable-drug use, improperly sterilized needles and equipment, or through accidental exposure to infectious materials. Infection rates vary depending upon the transmission route, with maternal transmission and anal sex having the highest probability of infection (Table 11.1). HIV is not transmitted through ingestion, inhalation, or touching. Insects, including mosquitos, have also never been shown to transmit HIV.

HIV infections began from cross-species events in Africa and have spread throughout the world since that time (Fig. 11.5A). In 2014, 36.9 million people were living with HIV globally, 2 million people became newly infected, and 1.2 million people died from AIDS-related illnesses. It is important not to lose sight of the magnitude of this infectious disease, although there is some good news: new HIV

TABLE 11.1 HIV Transmission Routes and Infection Rates

Exposure route	Transmission fluid	Risk of transmission per exposure
Vagina (vaginal-penile sex)	Semen	1 in 1250
Penis (vaginal-penile sex)	Vaginal, cervical secretions	1 in 2500
Receptive anal intercourse	Semen	1 in 72
Insertive anal intercourse	Blood, mucous secretions	1 in 900
Receptive oral sex	Semen	Low but possible
Maternal transmission	Genital secretions, blood, breast milk	1 in 2.5
Blood transfusions	Blood, blood products	1 in 1.08
Injectable-drug needle sharing	Blood	1 in 159

(A) 36.9 million people living with HIV (2 million new infections)

North America, Western/Central Europe 2.4 million (85,000)
Eastern Europe and Central Asia 1.5 million (140,000)
Caribbean 280,000 (13,000)
Middle East and North Africa 240,000 (22,000)
Asia and the Pacific 5 million (340,000)
Central and South America 1.7 million (87,000)
Sub-Saharan Africa 25.8 million (1.4 million)

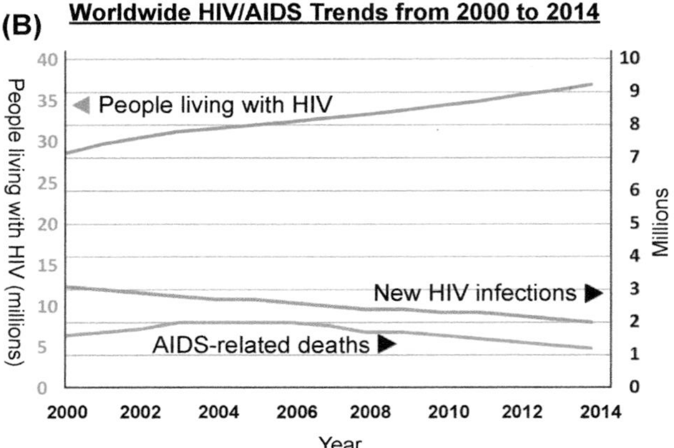

FIGURE 11.5 **Global HIV/AIDS trends.** (A) This map shows the number of people living with HIV/AIDS and the number of new infections (in parentheses) in 2014. In 2014, 36.9 million people were living with HIV worldwide, 70% of which were found in Sub-Saharan Africa. (B) This graph shows the total number of people living with HIV from 2000 to 2014 (*blue line*, left axis). Although the total number of people living with HIV/AIDS continues to increase, the number of new infections (*orange line*, right axis) and AIDS-related deaths (*gray line*, right axis) each year continues to drop. *Data from (A) and (B) derived from UNAIDS 2014 Global Statistics. Available at http://www.unaids.org/sites/default/files/media_asset/20150714_FS_MDG6_Report_en.pdf (accessed 28.07.15.).*

infections have fallen by 35% since 2000, and AIDS-related deaths continue to fall since the peak in 2004 (Fig. 11.5B).

Worldwide, heterosexual transmission is the most common route of infection. In Sub-Saharan Africa, where 70% of the HIV+ people in the world are located, over 50% of those infected are women. In this area of the world, the majority of new infections occur in stable couples who do not report outside partners or in people with multiple sexual partners. The Sub-Saharan country of Swaziland has the world's highest prevalence of HIV+ individuals: 26% of those age 15–49 are infected with HIV. Congenital transmission accounts for around 15% of new infections in Sub-Saharan Africa.

In North America and Western/Central Europe, 2.4 million people were living with HIV in 2014, about half of which were located in the United States. The incidence of new HIV infections in the United States peaked in 1984 with 130,000 cases diagnosed. In 2013, the latest year for which data are available, around 47,000 cases were diagnosed. The number diagnosed and the mode of transmission vary greatly depending upon gender (Fig. 11.6A). In contrast to international patterns where women account for more than half of those infected, 80% of those diagnosed with HIV in 2013 in the United States were male. Male-to-male sexual contact accounted for 81% of new diagnoses in men, totaling almost 38,000 new cases, while 87% of females were infected with HIV through heterosexual contact.

Over a third of those newly infected with HIV were between the ages of 20 and 29 (Fig. 11.6B). While the incidence of HIV decreased or remained steady in the 30–39 and 40–49 age groups from 2009–13, an alarming trend is that it has been increasing in those age 20–29, particularly in the male homosexual population. With the advent of effective antiviral drugs that have extended the lives of those with HIV, it is possible that individuals in the 20–29 age group are less concerned about the disease than individuals from previous generations, viewing it as a chronic condition rather than a death sentence.

In the 1980s, most cases of HIV in the United States occurred in Caucasians. Cases among Black/African Americans steadily increased from the 1980s until 1996, when the prevalence in this population became the largest of any race/ethnicity. In 2013, the largest percentage of new HIV infections was in the Black/African American population at 46%, followed by Caucasians at 28% and Hispanics at 21%. More Black/African American homosexual men, heterosexual men, and heterosexual women were diagnosed with HIV in 2013 than their Caucasian or Latino counterparts, which is striking when considering the overall proportion of these groups in the United States: Black/African Americans, Caucasian, and Latino groups accounted for 13.2%, 77.7%, and 17.1% of the US population in 2014, respectively, according to the United States Census Bureau.

States located in the South account for over 50% of new HIV diagnoses in the United States (Fig. 11.6C and D). The majority of HIV/AIDS cases in the United States are found in **metropolitan statistical areas (MSAs)**, areas that contain a core urban area with a population of 50,000 or more. Nationwide, the largest number of HIV diagnoses tend to correspond to MSAs with larger populations. In 2013, the five MSAs with the highest number of new HIV cases were New York City, Miami, Los Angeles, Washington, DC, and Atlanta, in that order. The highest *rates* of HIV diagnosis in 2013, however, are a different story: Southern cities Miami, New Orleans, Baltimore, Baton Rouge, and Atlanta top the list. The South is currently the region with the largest proportion of HIV/AIDS cases from suburban and nonurban areas as well.

Without drug treatment, it takes about 10 years for HIV to progress to AIDS, and people with AIDS typically

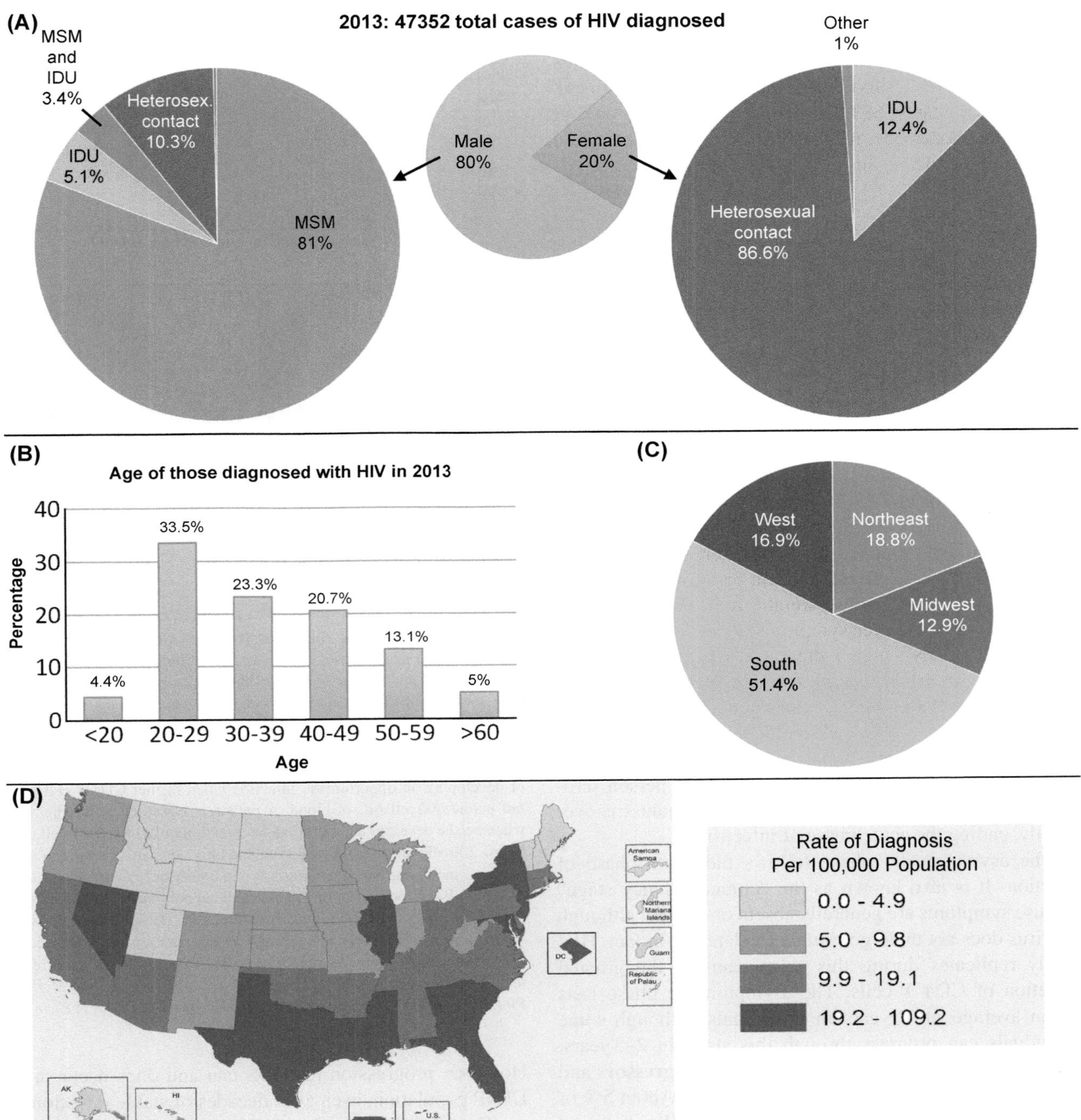

FIGURE 11.6 Epidemiology of HIV/AIDS in the United States. (A) Of the 47,352 cases of HIV diagnosed in the United States in 2013, 80% were male and 20% were female. Of the male cases, men who have sex with men (MSM) were 81% of the case patients, 10.3% contracted the disease through heterosexual contact, 5.1% were injectable-drug users (IDU), and 3.4% were MSM and IDU. Of the females diagnosed with HIV, 86.6% were through heterosexual contact. (B) The largest age group diagnosed with HIV in 2013 was people between ages 20–29. (C) Over 50% of the cases of HIV diagnosed in the United States in 2013 were located in the South. (D) This map shows the rate of HIV diagnoses in 2013 among adults and adolescents by area of residence. Southern states and the District of Columbia currently have the highest rates of diagnosis, although New York, Illinois, and Nevada also fall within the highest rate category. *Data from parts (A), (B), and (C) from the Centers for Disease Control and Prevention, 2013. HIV Surveillance Report, vol. 25. http://www.cdc.gov/hiv/library/reports/surveillance/. Published February 2015 (accessed 28.07.15.). Map in part (D) from Centers for Disease Control and Prevention, Maps Based on Data from 2013 HIV Surveillance Report. www.cdc.gov/hiv/pdf/library/slidesets/cdc-hiv-us-maps.pdf (accessed 28.07.15.).*

survive about 3 years without treatment. AIDS deaths in the United States peaked around 1995, although the prevalence of HIV+ individuals living with AIDS continues to rise due to HIV+ individuals surviving much longer with treatment. In the United States, 683,000 people in total have died of HIV/AIDS. Worldwide, nearly 39 million people have died since the epidemic began.

Study Break
Take a look at the graphs in Fig. 11.6 and summarize what each one shows. How can the prevalence of HIV be highest in the US' most-populated cities, but the incidence is higher in other cities?

11.4 CLINICAL PROGRESSION OF HIV/AIDS

The clinical course of HIV infection is divided into three stages: acute (or primary) infection, asymptomatic infection, and AIDS (Fig. 11.7A). Within 2–4 weeks after infection, >50% of HIV-infected individuals develop severe flu-like symptoms characterized by fever, swollen lymph nodes, a sore throat, **arthralgia** (joint pain), **myalgia** (muscle aches), headache, fatigue, weight loss, and sometimes a rash. Within a week after symptoms begin, HIV replicates to very high levels—higher than at any other time during infection—with 10^6–10^7 copies of viral RNA per milliliter of plasma. Replication is associated with an increase in CD8 cytotoxic T cells and a reduction in the number of circulating CD4 helper T cells found in the blood. Symptoms typically last 2–4 weeks, after which the person seroconverts against the virus and CD4 T cell counts recover slightly, ending the acute phase of infection.

The asymptomatic phase follows the acute phase of infection. It is also known as the "clinical latency stage" because symptoms are generally absent or limited, although the virus does not undergo *replication* latency. In fact, HIV slowly replicates during this stage, causing the gradual depletion of CD4 T cells. The asymptomatic phase lasts for an average of 10 years in individuals, although some individuals can progress through this stage in 2–3 years. These individuals are referred to as **rapid progressors** and account for 10–15% of the HIV+ population. About 5% of HIV+ individuals maintain normal CD4 T cell counts and low HIV titers in the absence of any **antiretroviral therapy (ART)**. These **long-term nonprogressors (LTNPs)** make up <5% of HIV+ individuals. A subset of LTNPs, known as **elite controllers**, even maintain plasma viral loads at levels that are below the limits of detection. The reasons behind control of infection in these individuals are still being elucidated, but there is evidence to implicate genetic factors, the type of immune response that occurs (activation of cytotoxic lymphocytes, rather than antibody production, correlates with protection), and differences in HIV-1 strains.

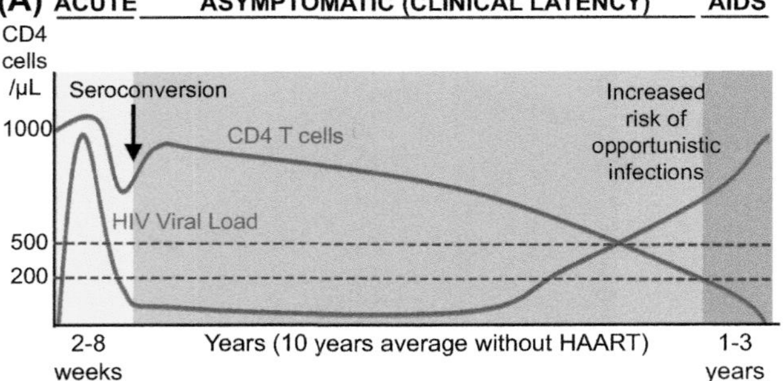

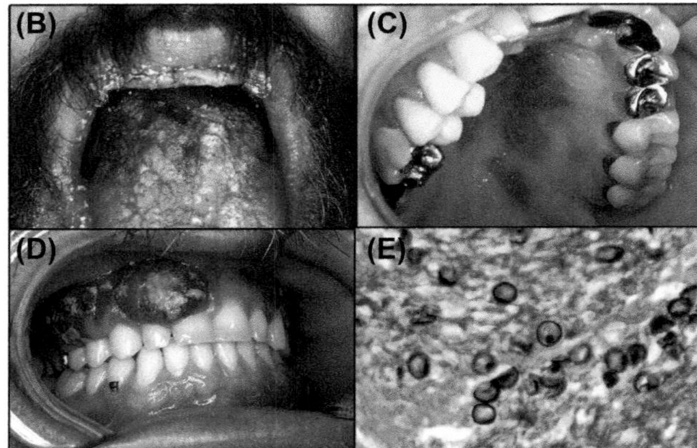

FIGURE 11.7 **Clinical course of HIV infection.** (A) HIV infection begins with the acute phase, when over half of individuals develop severe flu-like symptoms within 2–4 weeks of being infected. Viral loads reach high levels, and the individual seroconverts as symptoms resolve. The asymptomatic or clinical latency phase can last for years (and for decades in an individual on HAART) and is characterized by slow viral replication and gradual decline of CD4 T cells. A person has an increased likelihood of developing an opportunistic infection when his/her CD4 T cell levels fall below 500 cells/μL of blood. A person is classified as having AIDS when he/she develops an opportunistic infection or has a CD4 T cell count below 200 cells/μL. Several opportunistic infections are common in AIDS patients, including oral candidiasis (yeast infection) caused by *Candida albicans* (B), Kaposi's sarcoma, here of the upper palate of the mouth (C) and of the gums with a *C. albicans* coinfection (D), or *Pneumocystis jirovecii,* a fungus formerly known as *Pneumocystis carinii,* which causes pneumonia, shown here as round cysts (E). *Images (B)–(E) courtesy of the CDC, specifically (A) Sol Silverman, Jr., D.D.S., and John Molinari, Ph.D.; (B) and (C) Sol Silverman, Jr., D.D.S.; (D) Dr. Edwin P. Ewing, Jr.*

However, progression to AIDS can and does occur in the LTNP population, even after decades of stable infection.

The CD4 T cells of an HIV+ person are progressively depleted as the virus continues its replication, causing direct and indirect damage to CD4 T cells. An uninfected person has 500–1500 CD4 T cells/μL of blood, and when the CD4 T cell count falls below 500 cells/μL, the likelihood of contracting an opportunistic infection increases. **Opportunistic infections** are caused by pathogens that are usually combatted by the immune system but cause disease when an individual's immune system becomes impaired. They are referred to as "opportunistic" because the weakened immune system provides an opportunity for the pathogen

TABLE 11.2 Common Opportunistic Diseases That Occur in People With AIDS

Cause	Disease	Pathogen
Fungus		
	Candidiasis	*Candida albicans* (yeast) infections of bronchi, trachea, esophagus, lungs, oral cavity, vagina
	Pneumonia	*Pneumocystis jirovecii* (formerly *carinii*) pneumonia
	Cryptosporidiosis	*Cryptosporidium* infection of intestine
	Cryptococcosis	*Cryptococcus* lung infections and meningitis
	Coccidioidomycosis	*Coccidioides* infection of lung
	Histoplasmosis	*Histoplasma* infection of lung
Bacterium		
	Septicemia	*Salmonella* infection in bloodstream
	Mycobacterium avium complex	*Mycobacterium avium* lung/intestine infections
	Tuberculosis	*Mycobacterium tuberculosis* infection of lung
Virus		
	Retinitis, pneumonia, encephalitis, gastroenteritis	Cytomegalovirus of retina, lung, brain, colon
	Ulcers, bronchitis, pneumonitis, esophagitis	Herpes simplex virus infections of skin, bronchi, lungs, esophagus
	Progressive multifocal leukoencephalopathy	JC polyomavirus infection of brain and spinal cord
	Cervical/anal cancer	Human papillomavirus
	Kaposi's sarcoma	Kaposi's sarcoma-associated herpesvirus
	Lymphoma	Epstein–Barr virus
Parasite		
	Toxoplasmosis	*Toxoplasma* infection of brain
	Isosporiasis	*Isospora* infection of intestine

to replicate when normally it would be controlled. A person is classified as having progressed to the AIDS stage of infection when he/she develop one or more opportunistic infections or has a CD4 T cell count below 200 cells/μL, at which point the HIV+ person is at serious risk of developing a life-threatening opportunistic disease. A person diagnosed with AIDS lives around 3 years, or about 1 year if an opportunistic infection is acquired. Table 11.2 lists some common opportunistic infections that occur in people with AIDS. They are caused by fungi, bacteria, viruses, and parasites (Fig. 11.7B, D, and E). Herpesvirus reactivation from latency is common, and as described in Chapter 9, "Viruses and Cancer," the risk of developing cancer caused by the herpesviruses Epstein–Barr virus (EBV) or Kaposi's sarcoma-associated herpesvirus (KSHV) are much higher in HIV+ individuals than in the general population. Before ART, as many as 20% of patients with AIDS developed Kaposi's sarcoma from KSHV (Fig. 11.7C and D). Cervical and anal cancers associated with human papillomaviruses are also more frequent.

Other diseases manifest in AIDS patients due to the effects of the virus itself. Kidney disease is a common occurrence in HIV+ individuals, due to the side effects of antiviral drugs, infection of kidney cells by HIV, and the creation of antigen-antibody **immune complexes** that become lodged in the kidneys. More than 50% of HIV+ individuals develop **neurological** (nervous system) complications. Infection can cause **encephalitis**, the inflammation of the brain, or **meningitis**, the inflammation of the protective lining that covers the brain and spinal cord (meninges). Cognitive issues and HIV-associated dementia were also seen in 40–60% of HIV+ individuals before the advent of

ART. In this condition, HIV is thought to indirectly damage neurons by infecting other cells of the brain, particularly macrophage-like microglia, which produce substances that are directly neurotoxic or attract other damaging inflammatory cells into the brain. In support of this, HIV RNA can be found in the cerebral spinal fluid of individuals with HIV-associated dementia.

Ultimately, why does the immune system fail to keep HIV at bay? End-stage progression is characterized by a reduction in the number of HIV-specific cytotoxic T lymphocytes. With the increased diversity of the HIV population due to mutation, and the reduction of CD4 T cells and the help they provide to activate cytotoxic T lymphocytes and B cells, it is likely that the immune system becomes overwhelmed and unable to maintain effective responses against HIV and opportunistic pathogens.

The clinical progression of HIV-1 infection to AIDS can be predicted based upon the viral load and CD4 T cell counts in an individual. ART is currently the only way to reduce HIV replication and slow the decline of CD4 T cells. Forty percent of the nearly 37 million people living with HIV/AIDS are accessing HAART/cART, the simultaneous administration of a combination of at least three drugs together that target different stages of viral replication. As described in Chapter 8, "Vaccines, Antivirals, and the Beneficial Uses of Viruses," antivirals have been developed against HIV that target several stages of the virus's replication, including attachment, entry, reverse transcription, integration, and maturation (Table 11.3). If only one antiviral is administered instead of HAART, the virus quickly mutates, often within days or months, into a strain that is no longer susceptible to the drug. For example, a single amino acid change in the HIV reverse transcriptase—from methionine to isoleucine at position 184—is sufficient to render the virus insusceptible to certain reverse transcriptase inhibitors. The use of HAART has led to significant declines in the progression of HIV to AIDS and concurrent increases in life expectancy: in 2006–07, a 20-year-old HIV+ individual on HAART was expected to live an additional 51.4 years, an increase from 36 years in 2000–02. This is over 90% of the average life expectancy of an uninfected individual.

TABLE 11.3 Antiretroviral Drugs That Target Stages of HIV Replication

Stage	Drug type	Examples
Attachment	CCR5 antagonist	Maraviroc
Penetration	Fusion inhibitor	Enfuvirtide
Reverse transcription	Nucleoside/nucleotide reverse transcriptase inhibitors	Emtricitabine, didanosine, tenofovir, abacavir, zidovudine, stavudine, lamivudine, zalcitabine
	Nonnucleoside/ nucleotide reverse transcriptase inhibitors	Nevirapine, etravirine, efavirenz, rilpivirine, delavirdine
Integration	Integrase inhibitor	Raltegravir, elvitegravir, dolutegravir
Maturation	Protease inhibitor	Simeprevir, boceprevir, telaprevir, lopinavir, fosamprenavir, indinavir, darunavir, ritonavir, tipranavir, atazanavir, nelfinavir, amprenavir, saquinavir

Not an exhaustive list.

11.5 MOLECULAR VIROLOGY AND REPLICATION OF HIV-1

HIV is an enveloped retrovirus possessing a capsid with complex symmetry (Fig. 11.8A). Instead of conforming to strict helical or icosahedral architecture, the HIV capsid is built from a single protein, called the **capsid protein (CA)**, that forms an asymmetrical cone- or bullet-shaped core composed of pentamers and hexamers around the 9.7 kb genome (Fig. 11.8B). Because there are two complete copies of the +ssRNA genome, HIV is considered diploid, a feature unlike other more typical viruses, but is more accurately *pseudodiploid*, since only one of the two copies is used for reverse transcription. The +ssRNA genome possesses a 5′ cap and a 3′ poly(A) tail, but as a retrovirus, its genome is not directly translated upon entering the cell but is instead reverse transcribed into cDNA. Found within the core are the **nucleocapsid protein (NC)**, which coats the viral RNA, and the three viral enzymes required for replication: **reverse transcriptase (RT)**, **integrase (IN)**, and **protease (PR)**. An additional protein found within virions, **Vpr**, functions during the transport of viral cDNA into the nucleus.

An envelope surrounds the core to create spherical virions of ~120 nm in diameter. The **matrix protein (MA)** attaches to the inner surface of the envelope, maintaining the structure of the virion. Embedded within the envelope are trimers of the **envelope (Env) glycoproteins**, which consists of the heterodimer gp120 (the surface subunit, SU) and gp41 (the transmembrane subunit, TM). These are cleaved by cellular proteases during replication from the same viral polyprotein, gp160. gp120 and gp41 remain noncovalently bound to each other, with gp41 providing the transmembrane portion of the heterodimer and gp120 protruding from the envelope surface.

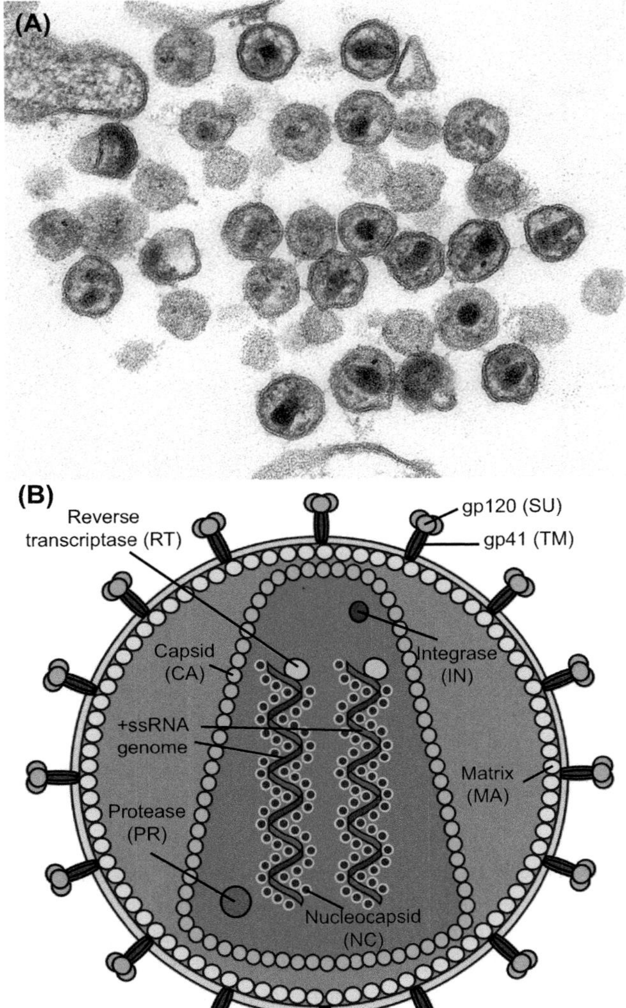

FIGURE 11.8 **HIV virion.** (A) Several enveloped HIV virions are seen in this transmission electron micrograph. Note the presence of a cone-shaped capsid. *(Courtesy of CDC/Maureen Metcalfe and Tom Hodge.)* (B) The HIV virion is composed of several proteins that are necessary for infection, including viral enzymes reverse transcriptase (RT), integrase (IN), and protease (PR). The nucleocapsid protein (NC) coats the viral RNA, which is enclosed within a cone formed by multiple capsid (CA) proteins. The matrix (MA) protein attaches to the inner surface of the plasma membrane envelope. gp120 (SU) and gp41 (TM) proteins facilitate attachment and fusion, respectively.

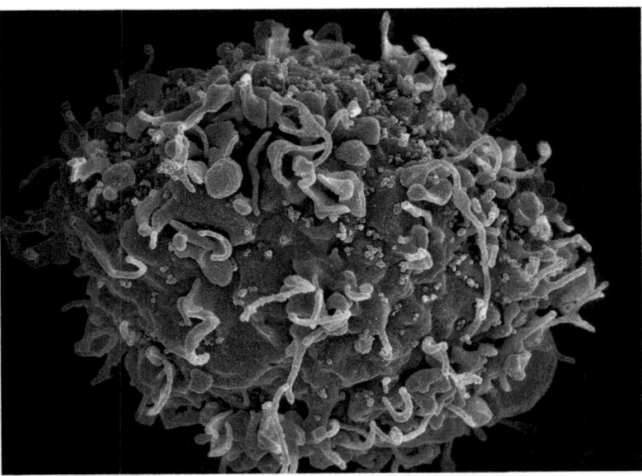

FIGURE 11.9 **HIV infection of a CD4 T cell.** This pseudocolored scanning electron micrograph shows multiple HIV virions (yellow) infecting a human T cell (blue). *Image courtesy of Seth Pincus, Elizabeth Fischer, and Austin Athman, National Institute of Allergy and Infectious Diseases.*

11.5.1 HIV Attachment, Penetration, Uncoating, and Reverse Transcription

The Env glycoprotein trimer mediates adsorption of the virion to the cell surface receptor, CD4, found on helper T cells (Fig. 11.9) and a population of macrophages and dendritic cells. First, however, the virus must come into contact with these cells. HIV gains entry into the body through blood or by exposure of the virus to genital or intestinal mucosal epithelial surfaces. At the mucosal surface, virions gain entry past the epithelium and into the tissue below by crossing through the tight junctions that attach cells together, or by penetrating through tears in the epithelium that occur during sexual activity or a result of concurrent infections with other pathogens. In the tissue, the virus comes into contact with the many dendritic cells and macrophages that reside there. Macrophages with CD4 can be directly infected by CCR5-tropic strains of HIV. As professional antigen-presenting cells, the dendritic cells increase their surface area by extending projections of the plasma membrane, resembling arms or dendrites, into the surrounding tissue to capture antigen. HIV comes into contact with the dendritic cells and is able to bind to a receptor known as DC-SIGN (Fig. 11.10A), which stands for "dendritic cell-specific ICAM3-grabbing nonintegrin." The binding of the virus to DC-SIGN triggers the dendritic cell to endocytose the virus into vesicles, some of which maintain infectious virions within them. As antigen-presenting cells, the dendritic cells leave the tissue and travel to the lymph node, where they present antigen to T cells. When the dendritic cell comes in contact with a CD4 T cell in the lymph node, a **virological synapse** is formed where the HIV virions become exocytosed and come into contact with the T cell (Fig. 11.10B). In this way, HIV uses the dendritic cell as a "Trojan horse" for transport from the tissue to the lymph nodes, even though HIV does not efficiently infect dendritic cells.

Upon coming in contact with the cell, gp120 binds the CD4 molecule and one of two **coreceptors**, either CCR5 or CXCR4. Variations in the gp120 molecule determine the tropism of the virus for CCR5 or CXCR4. CCR5 is the primary coreceptor used during initial infection, and CCR5 strains of HIV display tropism for macrophages and a subset of mucosal-associated memory CD4 T cells. Because a coreceptor is necessary for infection, individuals with

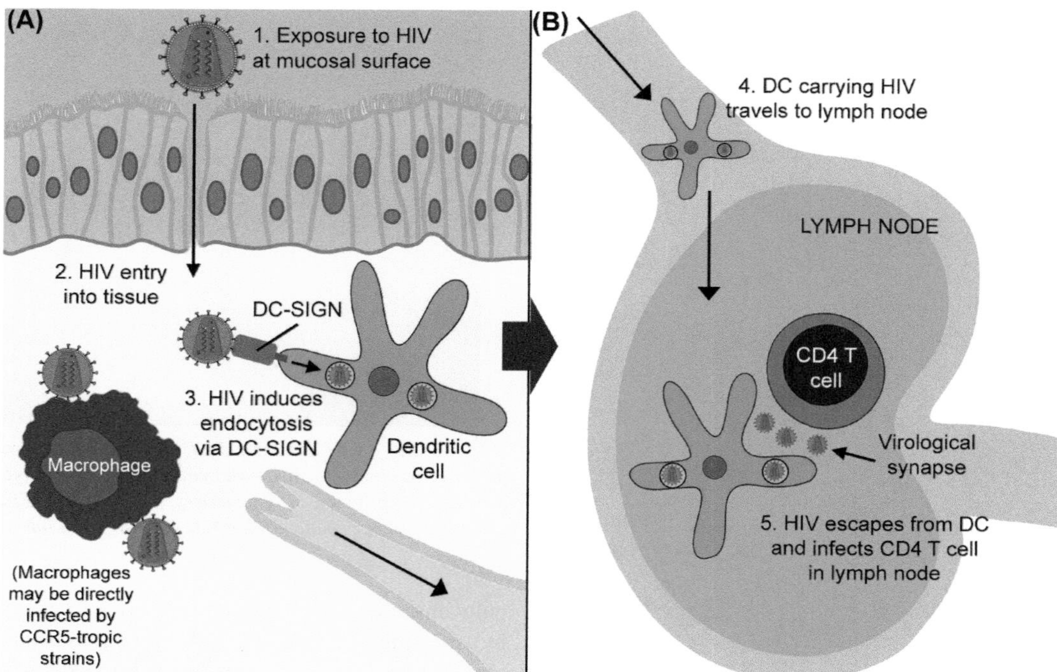

FIGURE 11.10 **Initial events upon HIV infection.** (A) At the mucosal surface, HIV gains entry into the tissue by crossing through the tight junctions that connect cells or by penetrating through tears in the epithelium. Within the tissue, HIV comes into contact with dendritic cells, which display DC-SIGN. HIV binding to DC-SIGN causes the internalization of the virions, some of which remain infectious within intracellular vesicles. The dendritic cell leaves the tissue and travels to the lymph node (B), where CD4 T cells come into contact with HIV when the dendritic cells make contact with the T cells. The location at which HIV virions escape the dendritic cell and make contact with the T cell is known as the virological synapse.

homozygous mutations in their CCR5 gene exhibit resistance to HIV infection (see *In-Depth Look*). These strains are often known as *R5* or *M-tropic* isolates, due to their ability to bind CCR5 and macrophages. On the other hand, CXCR4 strains of HIV-1, known as *X4* or *T-tropic* isolates, tend to arise later during infection as the virus changes its tropism. X4 strains preferentially infect naïve T cells, which express CXCR4. Coreceptor switching from CCR5 to CXCR4 is associated with accelerated disease progression to AIDS, likely due to the consequences of infecting a large pool of important immune system cells.

The binding of gp120 to CD4 induces a conformational change in gp120 that reveals a coreceptor-binding site (Fig. 11.11B). The binding of gp120 to the coreceptor further modifies the heterodimer, revealing a fusion peptide within gp41 that is inserted into the membrane of the cell (Fig. 11.11C). gp41 refolds to create a fusion pore in the membrane, likely after internalization of the virion into an endosome through clathrin-mediated endocytosis. The core (or nucleocapsid) is released through the pore into the cytoplasm of the cell (Fig. 11.11D).

The next step in the replication process is uncoating, the escape of the viral genome from the core. Much remains to be learned about HIV uncoating. It is known that the viral capsid influences reverse transcription of the +ssRNA genome into cDNA, and it is thought that the capsid must dissociate at least partially to allow viral enzymes access to cellular factors, such as nucleotides, during reverse transcription. The HIV reverse transcriptase possesses RNA-dependent DNA polymerase, DNA-dependent DNA polymerase, and RNase H activities. Although it is a DNA polymerase, it does not have proofreading ability and is as error-prone as the RNA-dependent RNA polymerases of traditional RNA viruses, introducing an incorrect nucleotide once every ~10^5 bases.

The retrovirus genome has several different domains that are of importance during viral replication. The two ends of the genome are flanked by redundant sequences, termed R (see Fig. 4.12). Located interior to the R domain is the U5 (unique to the 5′) and U3 (unique to the 3′) domains on the 5′ and 3′ ends of the RNA, respectively. These domains form **long terminal repeats** (LTRs) on each side of the cDNA that are important for **integration** of the HIV cDNA into the host's DNA. *Note that the repeats are only created during reverse transcription and so are only present within the cDNA, not the genomic RNA.* Only one of the two +ssRNA strands serves as a template for reverse transcription, which begins with the binding of a tRNA—usually one specific for lysine or proline—that the virus obtained from its previous cell. The entire process of reverse transcription is described in detail in Chapter 4, "Virus Replication." The resulting proviral DNA is double-stranded with repeated LTR ends that are composed of the U3, R, and U5 domains (Fig. 4.12).

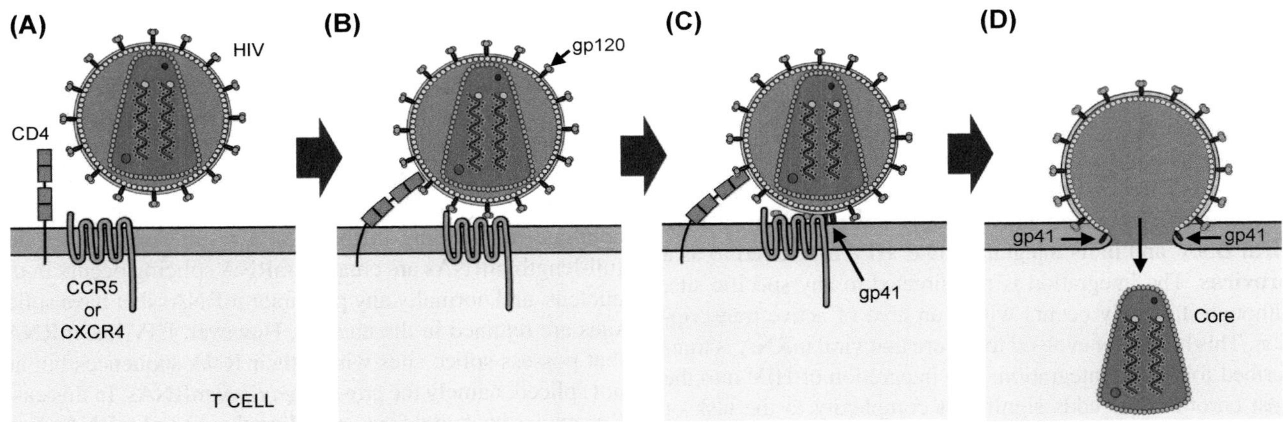

FIGURE 11.11 **HIV attachment and penetration.** (A) CD4 and CCR5 or CXCR4 are found on the surface of a CD4 T cell. (B) In the process of attachment, HIV gp120 binds to CD4. This produces a conformational change in the gp120 molecule that reveals the binding site for the coreceptor CCR5 or CXCR4. (C) The binding of gp120 to the coreceptor reveals the gp41 fusion peptide, which becomes inserted into the plasma membrane of the cell. (D) gp41 creates a pore in the membrane through which the nucleocapsid is released. Fusion likely occurs in an endosome following clathrin-mediated endocytosis.

In-Depth Look: Individuals With CCR5 Mutations

HIV uses one of two coreceptors to infect cells, either CCR5 or CXCR4. Within the Caucasian population, some individuals have a 32-base pair deletion in CCR5, termed CCR5Δ32, that causes a frame shift mutation in CCR5 that leads to a shortened, nonfunctional receptor that is unable to be utilized as a coreceptor for HIV entry. Only 1% of these individuals are homozygous—possessing two Δ32 alleles—while 10–15% of people with Δ32 mutations are heterozygous and carry one normal and one mutated allele. It has been found that homozygotes appear to be more resistant to infection with CCR5-tropic strains of HIV-1, although they are still susceptible to CXCR4 strains. Heterozygotes display delayed progression to AIDS, compared to individuals with two normal copies of the CCR5 gene. The Δ32 allele has not been found in people of African, East Asian, or Native American descent, suggesting that it may have evolved into the Caucasian population due to its selective advantage against another infectious disease, possibly smallpox. CCR5 is a receptor for **chemokines**, small proteins that cause the migration of cells. There are many chemokine receptors and several have redundant effects, which is why cells lacking CCR5 are still functional.

Discovery of the Δ32 mutation has left researchers wondering if it could be a key to treating, or even curing, HIV. In 2007, an HIV-infected man with leukemia, a cancer of white blood cells, underwent **myeloablative therapy**. This treatment uses radiation or chemotherapy to eliminate the leukemia cells, killing off all immune system cells and precursors in the process. In a process known as **hematopoietic stem cell transplantation**, he then received bone marrow stem cells—the precursors that develop into immune system cells—from an individual homozygous for the Δ32 mutation. HIV viral loads became undetectable in this individual, initially known as the "Berlin patient" and later identified as Timothy Ray Brown, who was able to discontinue HAART and is still HIV-free to this day. Other attempts to recreate this result have not been successful, unfortunately. One of the challenges has been eliminating the reservoir of T cells with integrated proviral DNA, because it appears that the virus can slowly reestablish infection even if small numbers of HIV-infected T cells remain after the myeloablative therapy.

In an attempt to block attachment of HIV to cells, a CCR5 **antagonist** (inhibitor) called maraviroc was developed and approved in 2008 as an antiviral treatment. The drug was effective in combination with other antivirals, but clinical trials showed that in some of those individuals that began the trial with only CCR5-tropic strains of HIV, it was not uncommon for individuals to switch to having CXCR4-tropic or mixed/dual infections, reemphasizing the plasticity of HIV and the difficulty in creating effective treatments against it. Researchers continue their efforts to harvest the potential of receptor and coreceptor antagonists in the treatment of HIV.

The core is thought to protect the reverse transcribed cDNA until it enters the nucleus, the site of integration. Most retroviruses must wait until mitosis, when the nuclear envelope dissolves, in order to gain entry into the nucleus. As a lentivirus, HIV cDNA is able to cross through the nuclear pores with the assistance of proteins, therefore allowing the virus to replicate in nondividing cells. The **preintegration complex (PIC)** is the term for the HIV cDNA and proteins, both of viral and cellular origins, that facilitate entry of the cDNA into the nucleus. Several HIV proteins found in the virion are involved in the process, including structural proteins (CA, NC, and MA), enzymes (RT and IN), and accessory proteins (Vpr). These are thought to interact with the **nucleoporin** proteins that constitute the nuclear pore complex, resulting in the import of the cDNA into the nucleus.

11.5.2 Integration and Replication

Within the nucleus, integrase removes two base pairs from each end of the cDNA, creates a nick in the host chromatin, and joins the cDNA to the host DNA (Fig. 4.13). DNA repair enzymes within the cell seal the break in the sugar-phosphate backbone to permanently join the HIV cDNA to the host chromosome. The integrated viral cDNA is now referred to as **proviral DNA**, and in its integrated state, HIV is referred to as a **provirus**. The integration is not directed to any specific site, although it usually occurs within an area of active transcription. This has likely evolved to ensure that viral mRNA is transcribed following integration. The integration of HIV into the host chromosome adds significant complexity to the task of "curing" someone of HIV that has already been infected.

All retroviruses possess the three major *gag, pol,* and *env* genes that are translated into polyproteins that are later cleaved to form the structural proteins and enzymes (Fig. 11.12A). HIV also encodes several essential and accessory regulatory proteins that are involved in virus replication (Table 11.4). The HIV proviral DNA contains at least 9 open reading frames that encode 15 individual proteins in total, and HIV relies upon extensive splicing of viral mRNAs to generate its mature transcripts (Fig. 11.12B).

As with all retroviruses, the HIV LTR contains a viral promoter. Host transcription factors bind to the U3 portion of the LTR and recruit RNA polymerase II to the viral DNA template. Initially, HIV produces doubly spliced mRNAs that encode regulatory proteins **Tat** (**t**rans**a**ctivator of **t**ranscription) and **Rev** (**r**egulator of **e**xpression of **v**iral proteins). Tat is an unusual protein. Instead of binding to DNA, it binds to mRNA strands as they are being transcribed. This stabilizes the transcribing RNA polymerase II complex, allowing it to continue transcribing mRNA. As a result, Tat ensures that full-length mRNAs are created. mRNA splicing occurs in the nucleus, and normally any precursor mRNAs that have splice sites are retained in the nucleus. However, HIV has mRNAs that possess splice sites within their RNA sequences but are not spliced, namely the *gag* and *gag-pol* mRNAs. In this case, Rev cycles back and forth, shuttling these viral mRNAs from the nucleus to the cytoplasm. Together, Tat and Rev promote

(A)

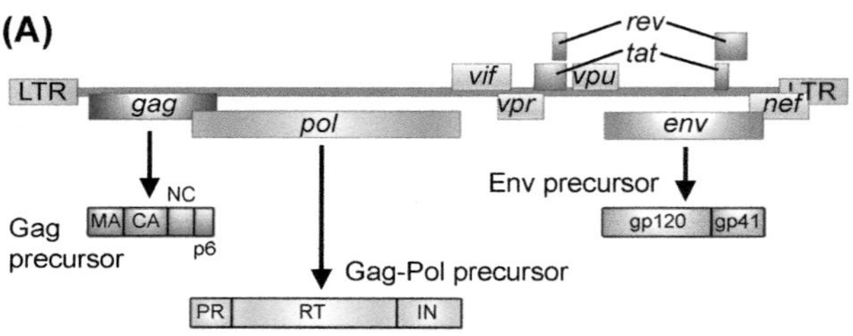

(B)

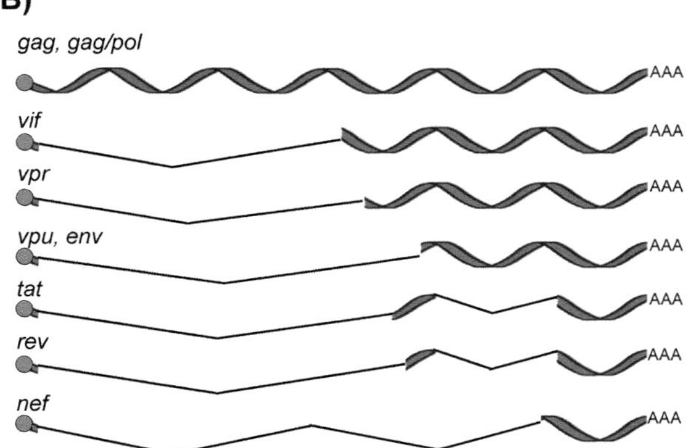

FIGURE 11.12 **HIV genome organization and gene expression.** (A) The proviral DNA contains nine open reading frames, indicated by colored boxes. Three genes (*gag, pol,* and *env*) encode polyprotein precursors that are cleaved by proteases into individual proteins. The Gag precursor encodes the MA, CA, NC, and p6 proteins; the Gag-Pol precursor encodes the PR, RT, and IN enzymes; and the Env precursor encodes envelope glycoproteins gp120 and gp41. (B) HIV mRNAs are extensively spliced. The mRNAs for the Gag and Gag-Pol precursors are unspliced, but mRNAs for *vif, vpr, vpu,* and *env* are singly spliced, while mRNAs for *tat, rev,* and *nef* are multiply spliced.

TABLE 11.4 Function of HIV Proteins

Protein	From gene	Function
Matrix (MA)	*gag*	Maintains structure of virion
Capsid (CA)	*gag*	Forms cone-shaped capsid
Nucleocapsid (NC)	*gag*	Coats viral genome
p6	*gag*	Interacts with viral and cellular proteins during replication
Protease (PR)	*pol*	Cleaves Gag precursor within virion
Reverse transcriptase (RT)	*pol*	Reverse transcribes viral ssRNA into double-stranded cDNA
Integrase (IN)	*pol*	Inserts viral cDNA into host chromosome
gp120 (surface, SU)	*env*	Facilitates attachment of virion to receptor and coreceptors
gp41 (transmembrane, TM)	*env*	Induces fusion of viral envelope with cell membranes
Viral infectivity factor (Vif)	*vif*	Inhibits intracellular antiviral defense mechanisms
Viral protein R (Vpr)	*vpr*	Part of preintegration complex; blocks cell division
Transactivator of transcription (Tat)	*tat*	Recruits cellular transcription factors to HIV promoter; interferes with immune surveillance; induces apoptosis
Regulator of expression of viral proteins (Rev)	*rev*	Exports unspliced mRNAs from nucleus
Viral protein unique (Vpu)	*vpu*	Degrades CD4; facilitates release of virion from cell surface
Negative factor (Nef)	*nef*	Downregulates CD4 and MHC Class I on T cells, manipulates intracellular signaling pathways to prevent apoptosis

the transcription and nuclear export of HIV mRNAs, which are translated in the cytoplasm by ribosomes.

The HIV structural, enzyme, and envelope proteins are derived from translated polyproteins (Fig. 11.12A). The **Gag precursor** is cleaved by the viral protease into the MA, CA, NC, and p6 proteins only after release from the cell. During translation of the Gag polyprotein, about 5% of the time the ribosome undergoes a frameshifting event that leads to continued translation of the protein. This causes the creation of the **Gag-Pol precursor**. Later, during maturation, autocatalysis cleaves off the Pol portion, which is further chopped into the PR protein, the IN protein, and the two RT subunits. The **Env precursor** p160 is highly glycosylated in the rough ER, and a cellular enzyme in the Golgi complex cleaves it into the two envelope subunits, gp120 and gp41. The remaining six HIV proteins are the product of spliced mRNAs and are translated individually.

Known collectively as **accessory proteins**, several HIV proteins assist in certain aspects of the replication process by modifying cellular processes and interacting with host proteins. Although not essential for replication, the accessory proteins Vif, Vpr, Vpu, and Nef increase the infectivity and replication of HIV virions.

The viral +ssRNA genome is translated by RNA pol II as a full-length, unspliced, capped, and polyadenylated mRNA. Transcription starts within the LTR, transcribing the 5′ R and U5 regions, coding sequences, and 3′ U3 and R regions to reconstitute the HIV genome.

11.5.3 Assembly, Maturation, and Release

Following its translation, the Gag precursor polyprotein is targeted to the plasma membrane of the cell, where the MA portion interacts with the inner side of the plasma membrane (Fig. 11.13A). The other end of the Gag polyprotein associates with the viral genome. The Env subunits gp120 and gp41, already cleaved in the Golgi complex into the two separate proteins, noncovalently associate with each other and are targeted to the site of virion assembly. The gp41 portion provides the transmembrane domain, while the gp120 portion associates with gp41 completely outside of the virion envelope. The assembled virion buds from the surface of the cell, obtaining its envelope in the process (Fig. 11.13B). During maturation, which occurs *after* budding from the plasma membrane, the viral protease cleaves several sites within the Gag and Pol precursors. This releases their individual proteins, dramatically altering the architecture of the capsid to create a fully infectious HIV virion (Fig. 11.13C). The entire process is summarized in Fig. 11.14.

In-Depth Look: Why Doesn't an HIV Vaccine Exist Yet?

Vaccine efforts began almost immediately following the discovery of HIV in 1983. At a press conference concerning the discovery of the virus in 1984, US Health and Human Services Secretary Margaret Heckler said, "we hope to have a vaccine ready for testing in about two years." And yet, over 30 years later, no successful vaccine has been created against the virus. It is not for lack of trying. The first clinical trial of an HIV vaccine occurred in 1986, and over 200 trials have taken place since that time.

There have been several difficulties encountered in the creation of an HIV vaccine. First, traditional vaccine methods that use live attenuated virus or inactivated virus cannot be used in an HIV vaccine due to safety concerns. Second, HIV does not infect typical animal models, such as mice or rats, and so testing candidate vaccines for their safety and efficacy is a challenge. Third, HIV mutates quickly due to the low fidelity of the reverse transcriptase, and the number of HIV subtypes makes it unreasonable and likely impossible to design a "universal vaccine" against them all. The greatest challenge in designing an HIV vaccine, however, has been deciphering how to stimulate the immune system to generate protective immunity against HIV. For viruses against which we have vaccines, such as influenza or measles, infected individuals generate an immune response against the virus that leads to the generation of immunity. By studying natural infection, researchers have learned the characteristics of a successful immune response and can create vaccines that stimulate the immune system in the same way. No individuals thus far are known to have been infected with HIV and successfully cleared the virus, so we do not know the natural course of immunity against HIV. Vaccines that generate neutralizing antibodies against virus attachment proteins are often very effective in preventing infection, but HIV vaccine candidates that generated antibody responses against gp120 were found to not be protective. Following its translation in a cell, the gp120 molecule is highly glycosylated in the rough ER. This provides a protective sugar coating that prevents antibodies from effectively interacting with the gp120 protein itself.

Following the discovery that antibodies and the humoral response are ineffective, vaccine design efforts shifted to generating cellular immunity—a cytotoxic T cell response that could kill off infected cells. A vaccine that used adenovirus to deliver genes for HIV Gag, Pol, and Nef proteins was highly immunogenic in monkeys, but in human trials, the vaccine was unable to prevent infection with HIV. In fact, the individuals that received the vaccine were infected at a higher rate!

Many novel vaccine ideas are being employed in the development of an HIV vaccine. The most promising candidate thus far used a "prime-boost" **strategy** where two different unsuccessful vaccine formulations were used together in the same individual in an attempt to provide humoral *and* cellular responses. The first vaccine (the "prime" vaccine) was a recombinant vector vaccine that used canarypox to deliver genes for HIV Gag, Pol, and Env proteins. This was administered six times. The last two injections also contained the second vaccine (the "boost" vaccine), which contained the HIV gp120 protein. Results were encouraging: the vaccine was 31.2% effective in preventing HIV. However, only one of three statistical tests that were performed showed statistical significance of this result. Taken together, slow but promising strides have been made in the engineering of a successful HIV vaccine, one of the greatest challenges that scientists have encountered.

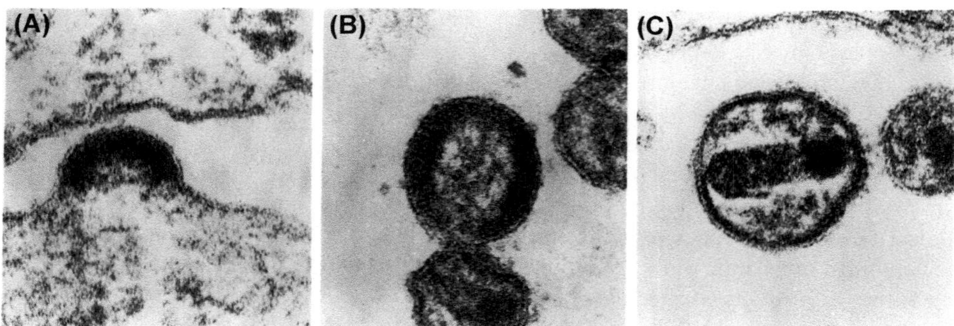

FIGURE 11.13 **Assembly, release, and maturation.** This transmission electron micrograph shows the assembly, release, and maturation of HIV as it replicates within a T cell. (A) HIV proteins assemble at the plasma membrane. (B) The virion buds from the surface. Following release, the protease cleaves the Gag precursor into individual MA, CA, NC, and p6 proteins, changing the architecture within the virion (C). This creates an infectious particle. *Image courtesy of Dr. Matthew Gonda/National Cancer Institute.*

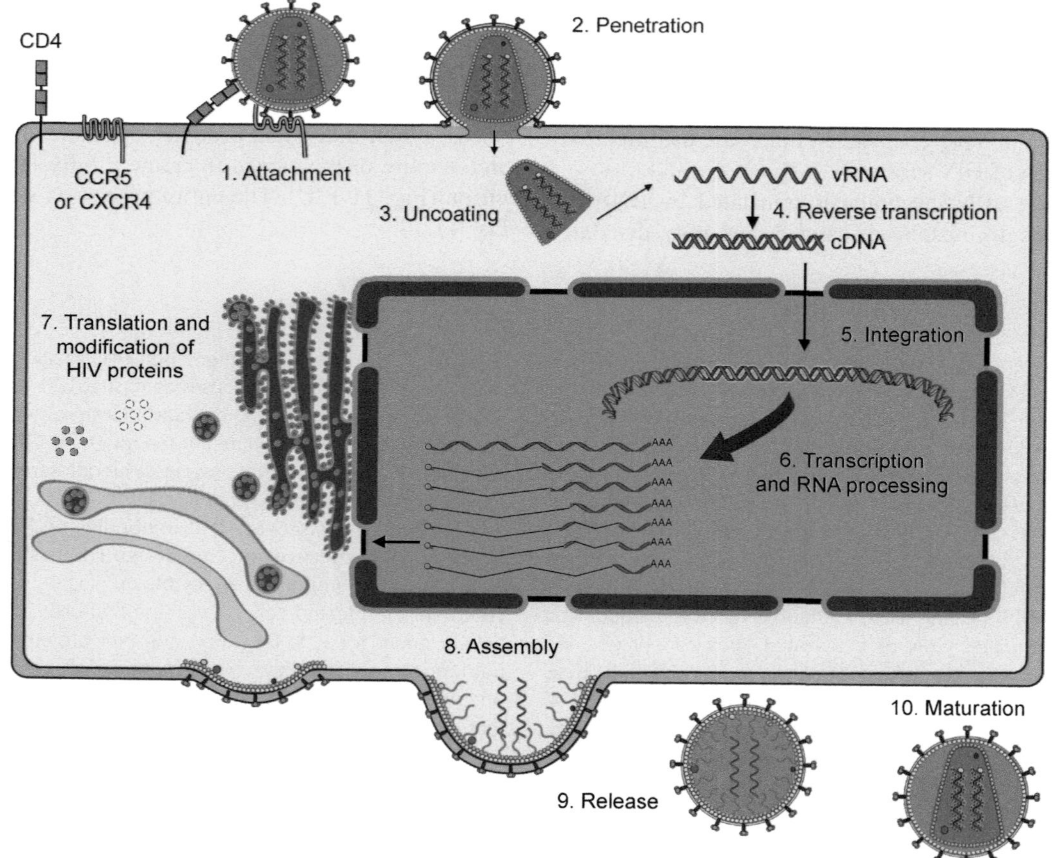

FIGURE 11.14 **The process of HIV replication.** 1. HIV replication begins when the gp120 of the virion binds to its cell surface receptor, CD4, and one of two coreceptors, CCR5 or CXCR4. 2. gp41 causes fusion of the virion envelope with the cellular membrane, which may occur at the cell surface or within an endosome following clathrin-mediated endocytosis. 3. The capsid is released into cytoplasm and partially uncoats while reverse transcription takes place within it. 4. Reverse transcriptase transcribes viral cDNA from one copy of the +ssRNA genome, creating LTRs on each end. The preintegration complex assists the cDNA in entering the nucleus. 5. Integrase joins the proviral DNA into the host's chromosomal DNA. 6. Transcription of viral genes by cellular RNA polymerase II generates viral mRNA. *vif, vpr, vpu, env, tat, rev, and nef* mRNAs are spliced before exiting the nucleus. Rev assists in the export of *gag* and *gag-pol* mRNAs. 7. Viral mRNAs are translated into viral proteins. gp160 is heavily glycosylated in the rough ER before being cleaved by cellular proteases within the Golgi complex into gp120 and gp41. 8. Viral proteins assemble at the plasma membrane, and the immature virion is released through budding (9). 10. In the process of maturation, the HIV protease cleaves the Gag precursor polyprotein to release the MA, CA, NC, and p6 proteins. These reorganize within the capsid to form the infectious virion.

SUMMARY OF KEY CONCEPTS

Section 11.1 History of HIV Infection

- The world became aware of AIDS in 1981 when cases of opportunistic infections began appearing in previously healthy individuals in the United States.
- HIV, the virus that causes AIDS, was discovered in 1983. A blood test for the virus was developed in 1985 and allowed the testing of blood and blood products.
- HIV/AIDS was the top cause of death for all Americans age 25–44 by 1994, but the invention of antiretroviral drugs that were used as part of HAART led to the decline of AIDS cases as infected people were living longer with HIV. HAART is still essential in the treatment of HIV+ individuals.

Section 11.2 Taxonomy and Origins of HIV

- There are two types of HIV, HIV-1 and HIV-2, both within the *Retroviridae* family, *Orthoretrovirinae* subfamily, and *Lentivirus* genus. HIV-1 contains groups M, N, O, and P. There are nine subtypes within group M, which cause >90% of HIV infections: subtypes A, B, C, D, F, G, H, J, and K. HIV-2 contains eight distinct groups, A–H, with no additional subtypes.
- Each HIV group originated from a separate cross-species infection event of humans with a SIV, possibly through the hunting and handling of bushmeat. HIV-1 groups M and N originated from a chimpanzee SIV from Cameroon, and HIV-1 groups O and P are most closely related to a western lowland gorilla SIV from Cameroon. All eight HIV-2 groups appear to have originated from an SIV of sooty mangabey origin.

Section 11.3 Epidemiology of HIV/AIDS

- HIV is most often transmitted through vaginal or anal sex, through blood, or perinatally. HIV is not transmitted through ingestion, inhalation, or touching.
- In 2014, 36.9 million people were living with HIV globally. Two million people were newly diagnosed, and 1.2 million died of HIV/AIDS. Worldwide, heterosexual transmission is the major mode of transmission of the virus.
- In the United States, 1.3 million people over age 13 are living with HIV/AIDS and ~47,000 new cases are diagnosed each year. Eighty one percent of cases in men are associated with homosexual contact, while 87% of cases in women are associated with heterosexual activity.
- The incidence of HIV has been increasing in younger age groups and in Black/African American populations. States located in the Southern United States account for >50% of new HIV infections. The majority of cases are found in MSAs, although the largest cities in the country do not necessarily have the highest rates of infection, which are found in Miami, New Orleans, Baltimore, Baton Rouge, and Atlanta, in that order.
- Worldwide, nearly 39 million people are estimated to have died of HIV/AIDS. Over 683,000 people have died in the United States of HIV/AIDS.

Section 11.4 Clinical Progression of HIV/AIDS

- HIV infection is divided into three major stages: acute infection, asymptomatic infection (or clinical latency), and AIDS. Acute infection is characterized by severe flu-like symptoms, in addition to weight loss and a rash. Symptoms last 2–4 weeks.
- The asymptomatic phase of HIV infection is characterized by the slow replication of the virus and gradual depletion of CD4 T cells. The clinical progression of the disease can be predicted by the viral load and the CD4 T cell counts in an infected person. An uninfected individual generally has between 500 and 1500 CD4 T cells/μL of blood. The likelihood of contracting an opportunistic infection increases when CD4 T cells fall below 500 cells/μL, and a person is characterized as having AIDS when they fall below 200 CD4 T cells/μL. In addition to opportunistic infections, HIV+ individuals have higher rates of cancers and HIV-associated kidney and neurological diseases.
- Use of HAART has led to significant declines in the rate of HIV progression and has dramatically increased the life expectancy of a person infected with HIV.

Section 11.5 Molecular Virology and Replication of HIV-1

- HIV is an enveloped retrovirus of ~120 nm in diameter with a 9.7 kb +ssRNA genome.
- HIV is transported to lymph nodes by tissue dendritic cells, which endocytose the virus after it binds to DC-SIGN. HIV gp120 attaches to CD4 and one of two coreceptors, CCR5 or CXCR4, on the surface of helper T cells. gp41 fuses the viral envelope with the cell membrane, likely during the process of clathrin-mediated endocytosis, releasing the core into the cytoplasm.
- Reverse transcription takes place in a partially uncoated capsid. Reverse transcriptase possesses RNA-dependent and DNA-dependent DNA polymerase functions, and it uses its RNase H activity to degrade RNA from an RNA:DNA duplex. Reverse transcription results in viral cDNA that is flanked by LTR sequences.
- Following the transport of the PIC into the nucleus, integrase removes two base pairs from each end of the cDNA, creates a nick in the host chromatin, and joins the cDNA to the host DNA in a location of active transcription. The integrated viral DNA is known as proviral DNA.
- Cellular RNA polymerase II transcribes HIV mRNAs, several of which are spliced. Gag, Gag-Pol, and Env precursor polyproteins are each cleaved into multiple viral proteins. The HIV genome is also transcribed as a full-length, capped, and polyadenylated mRNA.

- Assembly of nascent virions occurs at the plasma membrane, to which gp120 and gp41 localize. On the interior of the membrane, the Gag and Gag-Pol precursors assemble alongside the viral genome. The virion buds from the surface of the cell to obtain its envelope. Following release, the viral protease cleaves the polyprotein precursors to create an infectious virion.

FLASH CARD VOCABULARY

Human immunodeficiency virus (HIV)	Antiretroviral therapy (ART)
Acquired immune deficiency syndrome (AIDS)	Long-term nonprogressors (LTNPs)
	Elite controllers
Simian	Opportunistic infections
Hemophiliacs	Immune complexes
HAART/cART	Neurological
HIV type/group/subtype	Encephalitis
Simian immunodeficiency virus	Meningitis
Bushmeat	Capsid protein
Metropolitan statistical areas	Nucleocapsid protein
Arthralgia	Reverse transcriptase
Myalgia	Integrase
Rapid progressors	Protease
Vpr	Antagonist
Matrix protein	Long terminal repeat (LTR)
Envelope glycoprotein	Integration
Virological synapse	Proviral DNA
Coreceptor	Provirus
Preintegration complex (PIC)	Tat/Rev
Nucleoporins	Gag precursor
Chemokines	Gag-Pol precursor
Myeloablative therapy	Env precursor
Hematopoietic stem cell transplantation (HSCT)	Accessory proteins
	Prime-boost strategy

CHAPTER REVIEW QUESTIONS

1. In the early 1980s, how did doctors first become aware that a novel disease was appearing?
2. Why is HAART/cART used for HIV+ individuals, instead of a single drug?
3. Examine Fig. 11.3 and identify the geographical location of the HIV-1 subtypes within Group M.
4. From what biological source are HIV-1 and HIV-2 thought to have originated? How did the virus spread from its initial site of infection?
5. What are the major ways in which HIV is transmitted?
6. Where in the world is HIV most prevalent?
7. For the United States, summarize in which regions, age groups, and ethnicities HIV incidence is highest. Is there a difference in these categories between males and females?
8. Describe the three clinical stages of HIV infection.
9. Why do opportunistic infections cause disease in HIV+ individuals but not in uninfected people?
10. What do CD4 T cells have to do with the progression of HIV to AIDS?
11. From what you learned about CD4 or helper T cells in Chapter 6, "The Immune Response to Viruses," why do you think creating a CD4 antagonist as an antiviral drug would not be a prudent decision?
12. Create a chart of the viral proteins that were discussed in Section 11.5 and define the function of each one.
13. Make a list of the seven stages of viral replication and describe how HIV accomplishes each step.
14. How are LTRs generated and why are they important for HIV replication?
15. HIV creates three polyproteins: Gag, Gag-Pol, and Env. What is a polyprotein and how does each one end up forming several individual viral proteins?
16. What are the reasons that a successful HIV vaccine has not yet been developed?

FURTHER READING

Allers, K., Schneider, T., 2015. CCR5Δ32 mutation and HIV infection: basis for curative HIV therapy. Curr. Opin. Virol. 14, 24–29.

Barré-Sinoussi, F., Chermann, J.C., Rey, F., et al., 1983. Isolation of a T-lymphotropic retrovirus from a patient at risk for acquired immune deficiency syndrome (AIDS). Science 220, 868–871.

Benjelloun, F., Lawrence, P., Verrier, B., Genin, C., Paul, S., 2012. Role of HIV-1 envelope structure in the induction of broadly neutralizing antibodies. J. Virol. 86, 13152–13163.

Buonaguro, L., Tornesello, M.L., Buonaguro, F.M., 2007. Human immunodeficiency virus type 1 subtype distribution in the worldwide epidemic: pathogenetic and therapeutic implications. J. Virol. 81, 10209–10219.

Bush, D.L., Vogt, V.M., 2014. Vitro assembly of retroviruses. Annu. Rev. Virol. 1, 561–580.

Centers for Disease Control and Prevention, 2015. HIV/AIDS. http://www.cdc.gov/HIV.

Centers for Disease Control and Prevention, 2015. HIV Surveillance Report, 2013, vol. 25. http://www.cdc.gov/hiv/library/reports/surveillance/.

D'arc, M., Ayouba, A., Esteban, A., et al., 2015. Origin of the HIV-1 group O epidemic in western lowland gorillas. Proc. Natl. Acad. Sci. 112, E1343–E1352.

Denton, P.W., Long, J.M., Wietgrefe, S.W., et al., 2014. Targeted cytotoxic therapy kills persisting HIV infected cells during ART. PLoS Pathog. 10. http://dx.doi.org/10.1371/journal.ppat.1003872.

Di Nunzio, F., 2013. New insights in the role of nucleoporins: a bridge leading to concerted steps from HIV-1 nuclear entry until integration. Virus Res. 178, 187–196.

Faria, N.R., Rambaut, A., Suchard, M.A., et al., 2014. The early spread and epidemic ignition of HIV-1 in human populations. Science 346, 56–61.

Friant, S., Paige, S.B., Goldberg, T.L., 2015. Drivers of bushmeat hunting and perceptions of zoonoses in Nigerian hunting communities. PLoS Negl. Trop. Dis. 9, e0003792.

Gallo, R.C., Sarin, P.S., Gelmann, E.P., et al., 1983. Isolation of human T-cell leukemia virus in acquired immune deficiency syndrome (AIDS). Science 220, 865–867.

Goepfert, P., Bansal, A., 2014. Human immunodeficiency virus vaccines. Infect. Dis. Clin. North Am. 28, 615–631.

Gummuluru, S., Ramirez, N.-G.P., Akiyama, H., 2014. CD169-Dependent cell-associated HIV-1 transmission: a driver of virus dissemination. J. Infect. Dis. 210, S641–S647.

Joint United Nations Programme on HIV/AIDS. UNAIDS. http://www.unaids.org.

Karn, J., Stoltzfus, C.M., 2012. Transcriptional and posttranscriptional regulation of HIV-1 gene expression. Cold Spring Harb. Perspect. Med. 2, a006916.

Lewis, G.K., DeVico, A.L., Gallo, R.C., 2014. Antibody persistence and T-cell balance: two key factors confronting HIV vaccine development. Proc. Natl. Acad. Sci. 111, 15614–15621.

Li, Y., Yang, D., Wang, J.-Y., et al., 2014. Critical amino acids within the human immunodeficiency virus type 1 envelope glycoprotein V4 N- and C-terminals contribute to virus entry. PLoS One 9, e86083.

Loutfy, M.R., Wu, W., Letchumanan, M., et al., 2013. Systematic review of HIV transmission between heterosexual serodiscordant couples where the HIV-positive partner is fully suppressed on antiretroviral therapy. PLoS One 8. http://dx.doi.org/10.1371/journal.pone.0055747.

Melikyan, G.B., 2014. HIV entry: a game of hide-and-fuse? Curr. Opin. Virol. 4, 1–7.

Menéndez-Arias, L., 2013. Molecular basis of human immunodeficiency virus type 1 drug resistance: overview and recent developments. Antivir. Res. 98, 93–120.

Samji, H., Cescon, A., Hogg, R.S., et al., 2013. Closing the gap: increases in life expectancy among treated HIV-positive individuals in the United States and Canada. PLoS One 8, 6–13.

Sharp, P.M., Hahn, B.H., 2011. Origins of HIV and the AIDS epidemic. Cold Spring Harb. Perspect. Med. 1, 1–22.

Shaw, G.M., Hunter, E., 2012. HIV transmission. Cold Spring Harb. Perspect. Med. 2, 1–22.

Smiley, S.T., Singh, A., Read, S.W., et al., 2014. Progress toward curing HIV infections with hematopoietic stem cell transplantation. Clin. Infect. Dis. 60, 292–297.

United States Department of Health and Human Services. AIDS.gov. http://www.aids.gov.

Zhu, T., Korber, B.T., Nahmias, A.J., Hooper, E., Sharp, P.M., Ho, D.D., 1998. An African HIV-1 sequence from 1959 and implications for the origin of the epidemic. Nature 391, 594–597.

Chapter 12

Hepatitis Viruses

Hepatitis refers to inflammation of the liver. Excessive alcohol consumption, drugs, toxins, and metabolic disorders can cause hepatitis, but several viruses are a major cause of the condition. The disparate collection of viruses that cause hepatitis are known as **hepatitis viruses**. Worldwide, viral hepatitis is one of the top 10 infectious disease causes of death. Hepatitis B and hepatitis C cause 80% of global liver cancer cases, and almost 400 million people around the world are living with chronic hepatitis. This chapter discusses the clinical aspects, epidemiology, and molecular virology of the major hepatitis viruses.

12.1 CLINICAL COURSE OF HEPATITIS VIRUS INFECTIONS

There are five major hepatitis viruses: hepatitis A, B, C, D, and E virus (HAV, HBV, HCV, HDV, and HEV). These viruses are all found in different viral families and possess diverse molecular properties, although they all cause similar clinical symptoms. Acute infection begins with an incubation period of 1–3 months, depending upon the virus (Table 12.1), followed by a prodromal period when nonspecific symptoms occur. These include malaise, loss of appetite, nausea, and vomiting, which often accompany low-grade fever. Occasionally, **myalgia** (muscle pain) and **arthralgia** (joint pain) are experienced. The illness period begins 3–10 days later when specific symptoms of hepatitis appear: abdominal pain in the general location of the liver, which is located in the upper-right portion of the abdomen, just below the right rib cage; dark but clear urine; and **jaundice,** the yellowing of the skin and the whites of the eyes (Fig. 12.1). Normally, macrophages in the spleen or liver remove old red blood cells and break down the hemoglobin within them into heme and globin portions. Globin is a protein that becomes broken down into amino acids for reuse, but the heme portion is converted into **bilirubin**. Liver enzymes modify the bilirubin to make

TABLE 12.1 Transmission and Epidemiology of the Major Hepatitis Viruses

	HAV	HBV	HCV	HDV	HEV
Mode of transmission	Fecal-oral	Blood, sexual contact, perinatal	Predominantly blood, also sexual contact, perinatal	Blood	Fecal-oral
Average incubation period (range)	28 days (15–50 days)	120 days (45–160 days)	45 days (14–180 days)	Coinfection: 90 days (45–60 days) Superinfection: 14–56 days	40 days (15–60 days)
Chronic infection?	No	Yes (occurs 5–10% overall, very likely in infants)	Yes (occurs 75–85%)	Yes (<5% with coinfection, up to 80% with superinfection)	No (chronic only in immunosuppressed)
Worldwide incidence of acute infection	1.4 million/year	4 million/year	4 million/year	200,000–400,000	20 million
Worldwide prevalence of chronic infection	No chronic infection	350 million	170 million	15–20 million	Very rare
US cases/year	3500	5000–8000	30,000	Unknown	Uncommon
Vaccine available?	Yes	Yes	No	Yes (HBV vaccine prevents)	Only in China

it water soluble so it can be more easily excreted into the urine. Liver dysfunction in patients with hepatitis leads to an excess of the yellow bilirubin in the blood and urine, resulting in jaundice and darker urine, respectively. Because the brown color of stool also results from the breakdown of bilirubin, patients with hepatitis may have gray-colored stools due to defective bilirubin metabolism. Tests for serum aspartate aminotransferase (AST) and alanine aminotransferase (ALT), two liver enzymes involved in amino acid metabolism that spill over into the bloodstream upon liver damage, are also useful in diagnosing acute hepatitis. Symptoms generally peak within 1–2 weeks. The jaundice fades during the convalescence period.

HAV and HEV do not generally cause chronic infections, and lifelong immunity against the virus occurs following primary infection. The likelihood of developing a persistent infection with HBV varies greatly depending upon age of infection: chronic infection develops in >90% of infants, 25–50% of children between 1 and 5 years of age, and 6–10% of older children and adults. In contrast, 75–85% of those infected with HCV will develop a persistent infection. Only 15–25% of those with acute HCV clear the virus.

12.2 TRANSMISSION AND EPIDEMIOLOGY OF HEPATITIS VIRUSES

Because the hepatitis viruses are all distinct viruses, they do not all exhibit the same modes of transmission. As such, their portals of entry are not all the same, either. They all cause a similar clinical condition, however, and so laboratory tests that test for antiviral antibodies, viral proteins, or virus nucleic acid are essential in differentiating infection between the viruses (see *In-Depth Look*).

In-Depth Look: Hepatitis Case Definitions

Because hepatitis viruses cause similar symptoms, it is necessary to examine laboratory test results, epidemiologic linkages, and incubation periods to determine the cause of clinical hepatitis. The following are the CDC case definitions used in 2016 to distinguish the viral causes of acute, perinatal, and chronic hepatitis:

Acute Hepatitis A

Clinical criteria: An acute illness with discreet onset of symptoms, and jaundice or elevated serum aminotransferase levels.

Laboratory criteria: IgM antibody to hepatitis A virus (anti-HAV positive).

Confirmed case: A case that meets the clinical case definition and is laboratory confirmed or a case that meets the clinical case definition and occurs in a person who has an epidemiologic link with a person who has laboratory-confirmed hepatitis A (ie, household or sexual contact with an infected person during the 15–50 days before the onset of symptoms).

Acute Hepatitis B

Clinical criteria: An acute illness with discreet onset of symptoms, and jaundice or elevated ALT levels >100 IU/L.

Laboratory criteria: HBsAg positive, and IgM antibody against the hepatitis B core antigen (anti-HBc positive).

Confirmed case: A case that meets the clinical criteria and is laboratory confirmed.

Perinatal Hepatitis B

Clinical description: Perinatal HBV infection in the newborn can range from asymptomatic to fulminant hepatitis.

Laboratory criteria: Positive for hepatitis B surface antigen.

Confirmed case: HBsAg positive in any infant >1–24 months old who was born to an HBsAg-positive mother.

Chronic Hepatitis B

Clinical description: Persons with chronic hepatitis B virus (HBV) infection may be asymptomatic. They may have no evidence of liver disease or may have a spectrum of disease ranging from chronic hepatitis to cirrhosis or liver cancer.

Laboratory criteria: Negative for anti-HBc IgM antibodies and positive for HBsAg, HBeAg, or HBV DNA; OR HBsAg positive or HBV DNA positive or HBeAg positive two times at least 6 months apart (any combination of these tests performed 6 months apart is acceptable).

Confirmed case: A case that is laboratory confirmed.

Acute Hepatitis C

Clinical criteria: An acute illness with discreet onset of symptoms of acute viral hepatitis, and jaundice or elevated serum ALT levels >200 IU/L.

Laboratory criteria: Anti-HCV positive and positive for HCV RNA.

Confirmed case: A case that meets clinical criteria and has a positive hepatitis C virus antigen or RNA, or a negative HCV antibody, antigen, or RNA test result followed by a positive result within 12 months.

Chronic Hepatitis C

Clinical criteria: Most HCV+ chronic infections are asymptomatic, although many have chronic liver disease ranging from mild to severe.

Laboratory criteria: Anti-HCV antibodies and HCV RNA positive.

Confirmed case: A case that does not meet clinical criteria for acute HCV, does not have conversion from HCV– to HCV+ within the previous 12 months (or no report of test conversion), and is positive for HCV RNA.

Acute/Chronic Hepatitis D

Hepatitis D virus is a defective virus, unable to replicate without HBV.

Acute Hepatitis E

There are no approved serological tests to diagnose HEV infection, although several are available for research or non-US purposes that test for anti-HEV antibodies or HEV nucleic acid.

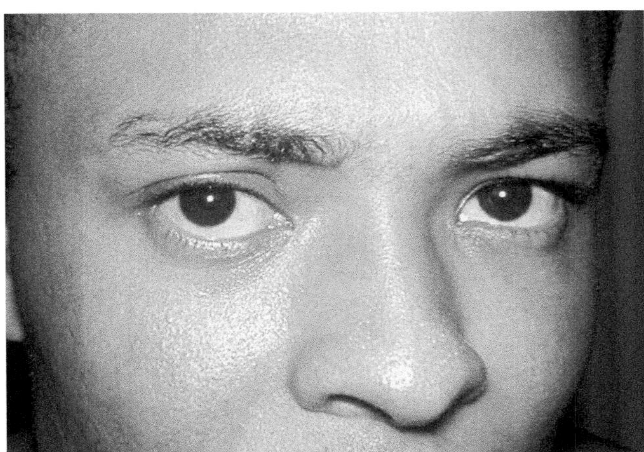

FIGURE 12.1 **Jaundice, a visible symptom of hepatitis.** Jaundice, or yellowing of the skin and whites of the eyes, is visible in this man infected with hepatitis A virus. Jaundice is caused by the accumulation of bilirubin in the blood. *Photo courtesy of the CDC/Dr. Thomas F. Sellers.*

12.2.1 Hepatitis A Virus

HAV is shed in the feces of infected people and transmitted through the fecal-oral route. The virus can be transmitted by drinking contaminated water, eating shellfish harvested from sewage-contaminated water, or consuming raw or undercooked foods that have come into contact with virions shed from an infected individual. The virus is nonenveloped and can survive for prolonged periods at low pH (even at the very acidic pH of 1), in fresh water or seawater, and in freezing to moderate temperatures, so contaminated frozen foods or ice made from contaminated water can still harbor infectious virus. Humans are the only natural reservoir for the virus.

After ingestion, the virus is absorbed by the gastrointestinal tract and travels via the hepatic portal vein to the liver, where it infects and replicates within hepatocytes. After 10–12 days—usually about 2 weeks before symptoms appear—virus becomes present in the bloodstream and is found in high concentration in the feces. It is at this point that the virus is most transmissible, although it is shed for up to 3 weeks after symptoms appear, and children can shed virus for up to 6 months following infection. The incubation period of the virus is approximately 28 days (range of 15–50 days). Clinical infection, which usually lasts less than 2 months, is indistinguishable from infection with the other hepatitis viruses and therefore requires laboratory confirmation (see *In-Depth Look*). Whereas jaundice occurs 70% of the time in older children and adults, 70% of infections are asymptomatic in children <6 years of age, who serve as an inconspicuous source of infection.

The virus does not cause persistent infections, and complications are rare—the majority of people fully recover from infection. Complications include prolonged jaundice and viral relapses over the course of months, which occur in 10–15% of people, and the bile duct from the liver to the intestine can also become blocked. Death can occur through **fulminant hepatitis,** also known as fulminant hepatic failure, whereby liver function ceases and the organ fails. Worldwide, the estimated mortality rate is 1 death per 1000 children <15 years old, 3 per 1000 adults ages 15–39, and 18–21 per 1000 adults age 40 and older.

HAV infections occur throughout the world, but the virus is highly endemic in Central America, South America, Africa, Asia, the Middle East, and the Western Pacific (Fig. 12.2A). Worldwide, over 1.4 million cases of HAV infection occur each year.

Until 2004, HAV was the most frequently reported type of hepatitis in the United States, occurring in large, nationwide epidemics. Hepatitis A vaccines were first licensed in 1995 and 1996, and the vaccine was added to the recommended childhood vaccine schedule in 2006. Since the introduction of the vaccine, HAV cases have been at historic lows, decreasing 93.7% from 1990 to 2009 (Fig. 12.2B). In 2013, an estimated 3500 cases of HAV occurred.

12.2.2 Hepatitis B Virus

After HAV, HBV is the second most common cause of acute viral hepatitis. Unlike HAV, HBV is transmitted through blood, semen, vaginal fluids, saliva, and other bodily fluids. The highest concentrations of virus are found in blood, and **percutaneous** transmission—through needle-puncture of the skin—can occur through sharing of contaminated needles during injection-drug use, unsafe reuse of needles or medical devices in health-care settings, or accidental needle pricks with infected human blood. As the virus is found in semen and vaginal fluid, transmission can also occur through sexual contact with an infected person or perinatally from an infected mother to her newborn. In the United States, perinatal and sexual transmission are the most common routes of transmission; in endemic areas of the world, perinatal transmission—primarily during the birthing process—is the most important factor involved in maintaining high prevalence rates. Although enveloped, HBV virions can remain infectious for more than 7 days at room temperature. Humans are the only known natural host of HBV.

Following infection with HBV, the incubation period averages 120 days (range of 45–160 days), the longest for any of the hepatitis viruses. The prodromal phase usually lasts 3–10 days and includes the symptoms described in Section 12.1. This is followed by jaundice and abdominal pain that last 1–3 weeks, on average. Although 95% of HBV infections are cleared by the immune system, malaise and fatigue may persist for months following infection, and fulminant hepatitis—with a fatality rate of 63–93%—occurs in 1–2% of acutely infected individuals. The overall case fatality rate for acute HBV infection is approximately 1%.

As mentioned above, the propensity to develop persistent HBV infection varies greatly with age: >90% of

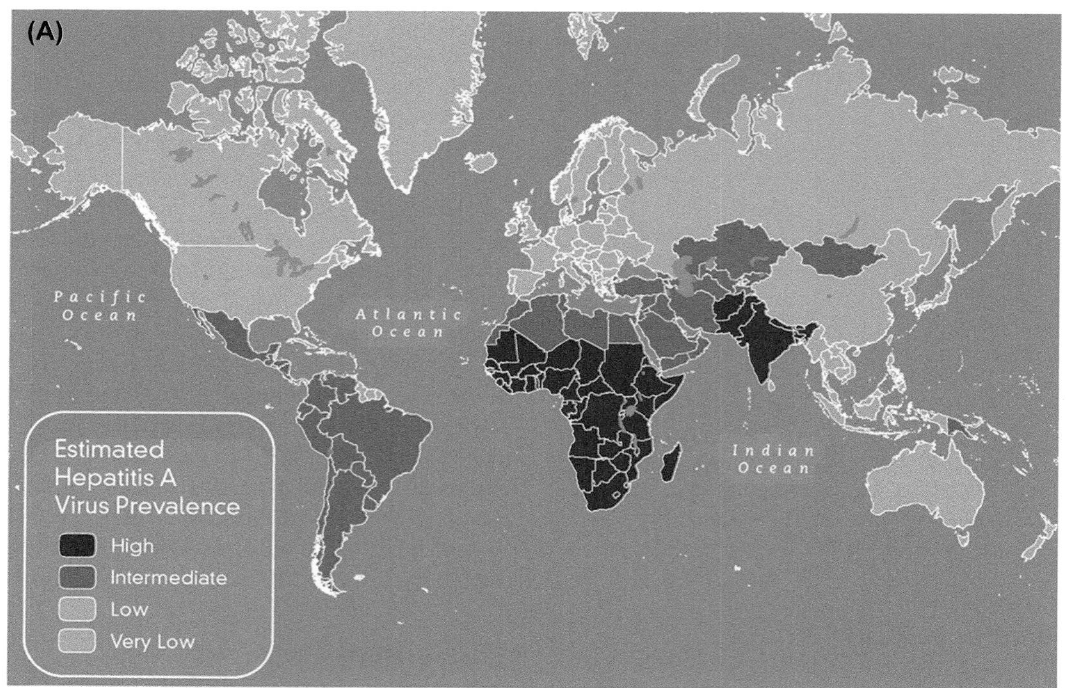

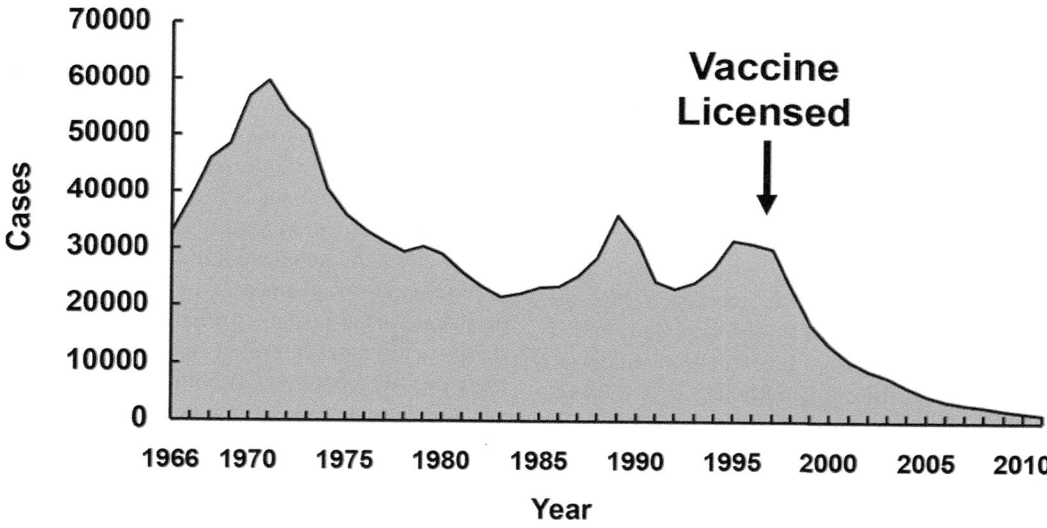

FIGURE 12.2 **Incidence of hepatitis A.** (A) Worldwide, HAV is highly endemic in Central America, South America, Africa, Asia, the Middle East, and the Western Pacific. (B) In the United States, HAV has been at historic lows since the introduction of the hepatitis A vaccine. The actual number of cases is estimated to be twice the number of cases that are reported in any year. *Images courtesy of the Centers for Disease Control and Prevention: (A) Screening for Hepatitis during the Domestic Medical Examination, http://www.cdc.gov/immigrantrefugeehealth/guidelines/domestic/hepatitis-screening-guidelines.html (accessed 15.06.15.) and (B) Centers for Disease Control and Prevention, 2015. Hepatitis A. In: Hamborsky, J., Kroger, A., Wolfe, C., (Eds.), Epidemiology and Prevention of Vaccine-Preventable Diseases, thirteenth ed. Public Health Foundation, Washington, DC, 175–186.*

infants, 25–50% of children between 1 and 5 years of age, and 6–10% of older children and adults develop chronic infection. Chronic infection can lead to chronic liver disease, cirrhosis, or liver failure, and up to 50% of the cases of hepatocellular carcinoma are attributable to HBV (see Chapter 9, "Viruses and Cancer," and Fig. 9.8).

HBV prevalence varies in different parts of the world. Forty-five percent of the global population lives in an area with high prevalence of chronic HBV infection. The prevalence of HBV is highest in sub-Saharan Africa, where up to 12% of children and adolescents have chronic HBV infection (Fig. 12.3A). High-intermediate prevalence (5–7%) occurs in

(A)

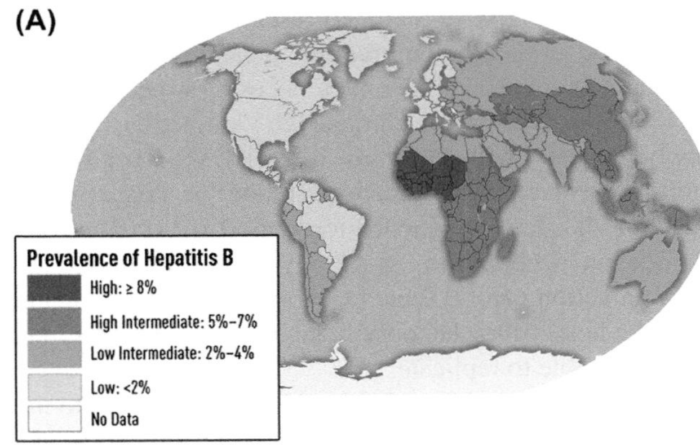

(B)

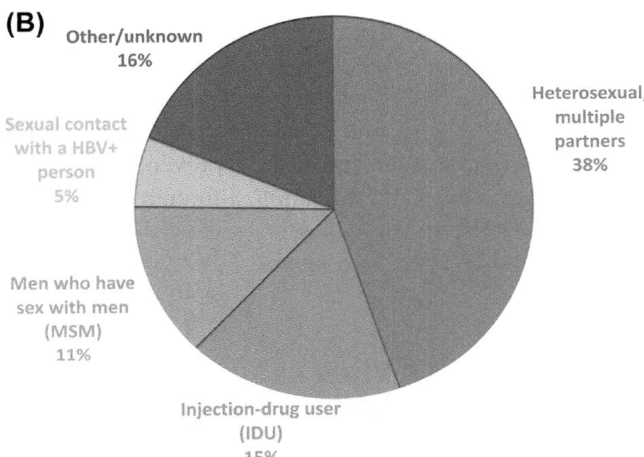

(C)

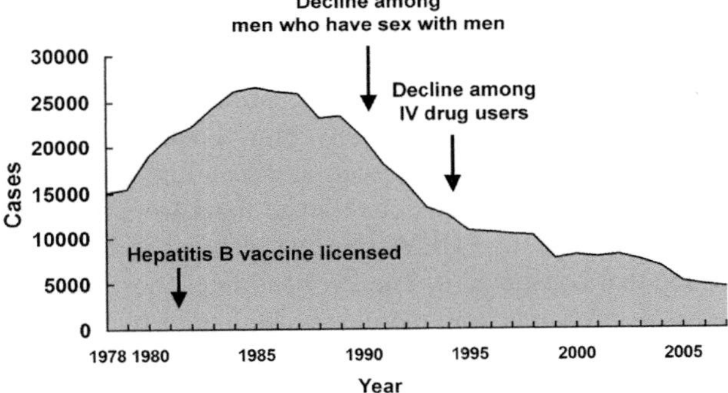

FIGURE 12.3 **Hepatitis B prevalence and risk factors.** (A) HBV prevalence is highest in sub-Saharan Africa and in areas of Asia and South America. *(Figure courtesy of Centers for Disease Control and Prevention, 2016. CDC Health Information for International Travel 2016. Oxford University Press, New York.)* (B) HBV infection is commonly associated with several high-risk behaviors, including heterosexuals with multiple sex partners, men who have sex with men, and injection-drug users. *(Data from Lyn Finelli, DrPH, and Beth P. Bell, MD, MPH. Hepatitis B. In: Roush, S.W., Baldy, L.M., (Eds.), 2011. Manual for the Surveillance of Vaccine-Preventable Diseases. Centers for Disease Control and Prevention Atlanta, GA, (Chapter 4).)* (C) Reported cases of HBV peaked in the 1980s and declined due to HIV prevention efforts in groups of injection-drug users and men who have sex with men that also were effective in reducing HBV transmission. Further declines have been attributed to increased HBV vaccination coverage. The number of actual cases is estimated to be 6.5 times the number of reported cases in any year. *(Figure from Hepatitis B. 2015. In: Hamborsky, J., Kroger, A., Wolfe, C., (Eds.), Epidemiology and Prevention of Vaccine-Preventable Diseases, thirteenth ed., Public Health Foundation, Washington, DC, 175–186.)*

East Asia, Southeast Asia, Papua New Guinea and Oceanic islands, and the South American countries of Bolivia, Peru, and Ecuador. Altogether, over 400 million people in the world have chronic, lifelong infections. 780,000 people die each year from hepatitis B infection: 130,000 from acute hepatitis B and 650,000 from HBV-associated cirrhosis and liver cancer.

HBV prevalence is low in the United States, Western Europe, and Australia, although 1.4 million people in the United States are chronically infected, with 5000–8000 new acute cases reported each year (although the actual cases of acute HBV infection are estimated to be 6.5 times the number of reported cases). Infection is commonly associated with certain high-risk behaviors (Fig. 12.3B), specifically heterosexuals with multiple sexual partners, men who have sex with men, and injection-drug users. In the United States, reported cases of HBV peaked in the mid-1980s (Fig. 12.3C). The drastic decline over the following 10 years was attributed to HIV prevention efforts in injection-drug users and men who have sex with men that also resulted in reduced HBV transmission.

Subsequent declines in HBV have been attributed to an increase in HBV vaccine coverage. The first HBV vaccine was licensed in the United States in 1981, although it was not well-accepted because it used purified HBsAg particles from the serum of chronically infected individuals (*HBsAg* refers to the **HB**V **s**urface **a**nti**g**en, or protein). Recombinant subunit vaccines composed of yeast-produced HBsAg (see Fig. 8.5) were created against the virus in 1986 and 1989, and the HBV vaccine became part of the recommended US childhood vaccine schedule in 1991. The World Health Organization began recommending global childhood HBV vaccination in 1992, and as of 2013, 183 member states vaccinate infants against HBV as part of their vaccination schedules. The three-shot series, the first of which is given at birth, provides 95% immunity against the virus. Infants born to HBsAg-positive mothers receive at birth the first dose of the HBV vaccine, as well as hepatitis B immunoglobulin to provide passive immunity, a regimen that is 85–95% effective in preventing chronic HBV infection. Worldwide, increased vaccination will result in lower prevalence and transmission of the virus, especially to newborns who are more likely to develop chronic infection. In developed countries, continued efforts to vaccinate high-risk individuals will further reduce the burden of HBV infection.

Treatment for HBV is supportive and attempts to prevent the progression of liver disease. Interferon-α and nucleoside analog reverse transcriptase inhibitors, such as tenofovir or entecavir, are often used, depending upon the stage of liver disease and levels of HBV virus.

12.2.3 Hepatitis C Virus

HCV is also a blood-borne pathogen, although unlike HBV, it is almost exclusively associated with exposure to contaminated blood, most frequently through sharing of contaminated needles or paraphernalia by injection-drug users. People who received blood transfusions or organ transplants before the blood supply started being screened for HCV in 1992 are also at risk for HCV. Less frequent modes of transmission include sexual contact or perinatal transmission from mother to child. As with HAV and HBV, HCV naturally infects only humans.

The incubation period for HCV averages 45 days, with a range of 14–180 days, although 80% of people are asymptomatic. When symptoms do occur, they resemble those of the other hepatitis viruses. In contrast to HBV, which most individuals clear, 75–85% of people infected with HCV develop persistent infections. The most common symptom of chronic infection is fatigue, but the long-term effects of persistent HCV infection are severe: of those that are acutely infected with HCV, 60–70% will develop chronic liver disease, 5–20% will develop cirrhosis over the following 20–30 years, and 1–5% will die from cirrhosis or hepatocellular carcinoma (Fig. 12.4A).

Worldwide, HCV is a significant burden. 170 million people are living with chronic HCV infection, and half a million die of HCV-related liver disease each year. Prevalence is highest (10%) in Egypt, with >3% prevalence in the Middle East, Central Asia, China, South Asia, and in North African countries bordering the Mediterranean Sea (Fig. 12.4B). In these areas, improper sterilization of needles and medical equipment is thought to be the primary mode of HCV transmission. In the United States, it is estimated that up to 3.9 million people have chronic HCV infection, with almost 30,000 new cases of HCV each year (Fig. 12.4C). Over 12,000 people die each year of HCV-related liver diseases, and HCV is the leading cause of liver transplants in the United States. Around half of the individuals who have chronic HCV do not realize it until years later, when liver problems appear.

No vaccine currently exists for HCV. The first treatment for HCV infection was the antiviral cytokine IFN-α in 1991, and the addition of ribavirin in 1998 increased the effectiveness of treatment. Since that time, newer antivirals have been added to the treatment regimen, including sofosbuvir, a nucleoside analog that inhibits HCV NS5A polymerase activity, and simeprevir or ledipasvir, two HCV protease inhibitors. Current drug combinations result in an impressive >90% **sustained viral response** rate, meaning that the viral load is reduced to levels that remain undetectable after treatment is stopped.

Study Break

Both HBV and HCV can cause chronic liver disease and hepatocellular carcinoma. How are both viruses transmitted, and for each virus, what are the odds of an acute infection becoming chronic?

12.2.4 Hepatitis D Virus

HDV is a **defective virus**, meaning that it is unable to replicate on its own. Also known as "hepatitis delta virus," HDV requires infection of the same cell with HBV, which contributes the HBsAg that HDV uses for virion assembly. This can occur through **coinfection,** meaning that HDV and HBV are contracted at the same time, or by **superinfection**, in which a person with chronic HBV infection becomes newly infected with HDV as well. HDV is transmitted through contact with infectious blood or contaminated injection devices. The incubation period is 90 days when coinfection occurs, and 14–56 days after superinfection. HDV infection can be acute or chronic; if HBV is cleared from the host, which happens in the majority of cases, then HDV is unable to replicate.

The geographical prevalence of HDV is identical to that of HBV, since it requires HBV to create infectious virions. Worldwide, 15–20 million people are estimated to be infected with HDV, less than 10% of those with HBV. However, several studies have indicated that chronic coinfection with HBV and HDV leads to more severe liver disease than chronic HBV infection alone. The nucleoside analog reverse transcriptase inhibitors used to prevent HBV infection do not prevent HDV replication, although interferon-α treatment has been shown to have some efficacy in reducing HDV viral loads. Vaccination against HBV protects against HDV infection.

12.2.5 Hepatitis E Virus

Like HAV, HEV is transmitted via the fecal-oral route. It is the number one cause of waterborne jaundice outbreaks worldwide, primarily transmitted by fecal contamination of drinking water. This is largely due to poor sanitation or inadequate sewage treatment. Shellfish, such as clams and mussels, can also concentrate the virus within their tissues if harvested from waters that have been polluted with sewage. HEV is unique among the hepatitis viruses in that it infects animals as well, namely pigs, boar, and deer. In certain parts of the world, such as Japan and Europe, the consumption of inadequately cooked meat from these animals has transmitted HEV from the infected animal to the person that consumed it. The liver and intestines, including sausages prepared from them, are a source of high titers of virus. HEV is relatively resistant to heat so foods must be cooked completely to avoid risk.

The incubation period of acute hepatitis E infection averages 40 days (range of 15–60 days). When symptoms are present, acute HEV results in the same symptoms as the other hepatitis viruses, and virions are shed into the stool from 1 week before to 30 days after the onset of jaundice. Chronic infection is very rare and only reported in those undergoing immunosuppressive treatment for organ transplantation. The overall fatality rate is around 1%, although HEV is a very serious disease in pregnant women. Transmission of the virus to a fetus can result in miscarriage or premature delivery of the child, and the pregnant woman is at much higher risk of developing fulminant hepatitis,

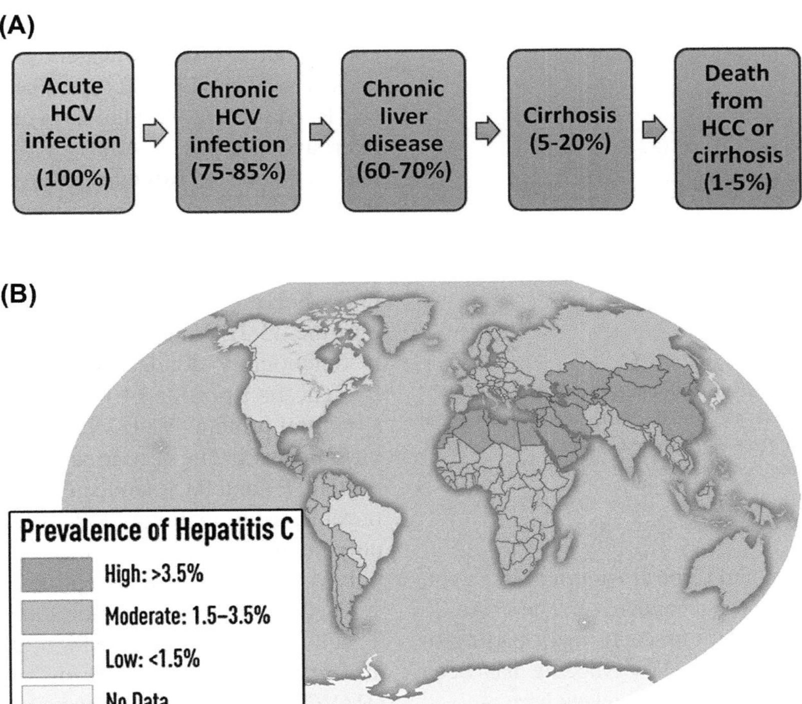

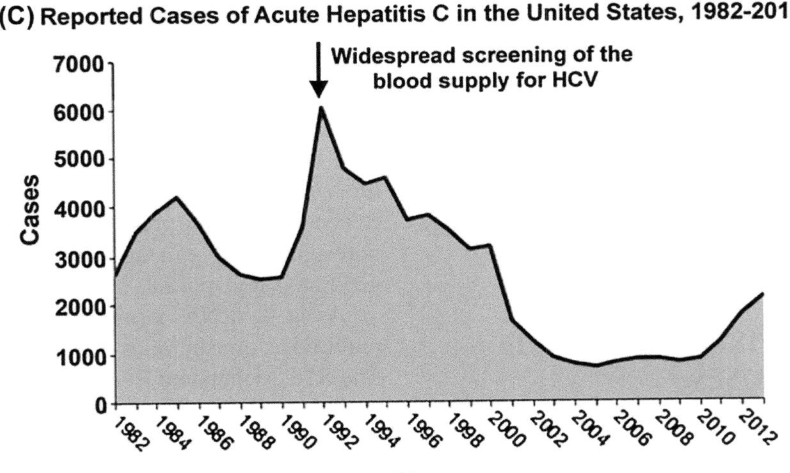

FIGURE 12.4 **Hepatitis C course of infection and prevalence.** (A) Of those that are acutely infected with HCV, 75–80% will develop chronic infection, 60–70% will develop chronic liver disease, 5–20% will develop cirrhosis over the following 20–30 years, and 1–5% will die from cirrhosis or hepatocellular carcinoma. (B) HCV prevalence is highest in China, the Middle East, Central Asia, South Asia, and North African countries bordering the Mediterranean Sea. *(Figure courtesy of Centers for Disease Control and Prevention, 2016. CDC Health Information for International Travel 2016. Oxford University Press, New York.)* (C) Reported cases of acute hepatitis C infection in the United States. Cases dropped precipitously when a blood test for HCV became widely available in 1992. The actual number of cases of HCV is estimated to be 14 times the number of reported cases in any year. *(Figure from Centers for Disease Control and Prevention. Viral Hepatitis—Hepatitis C Information, Statistics and Surveillance. Available at http://www.cdc.gov/hepatitis/hcv/statisticshcv.htm (accessed 15.06.15.).)*

which causes death in 10–30% of women infected during their third trimester.

Because of its route of transmission, HEV is most prevalent in developing countries that are lacking effective means of providing sanitized drinking water. Sixty percent of all HEV infections occur in East and South Asia, where 25% of individuals in some age groups are seropositive for the virus (Fig. 12.5). The virus is also highly endemic

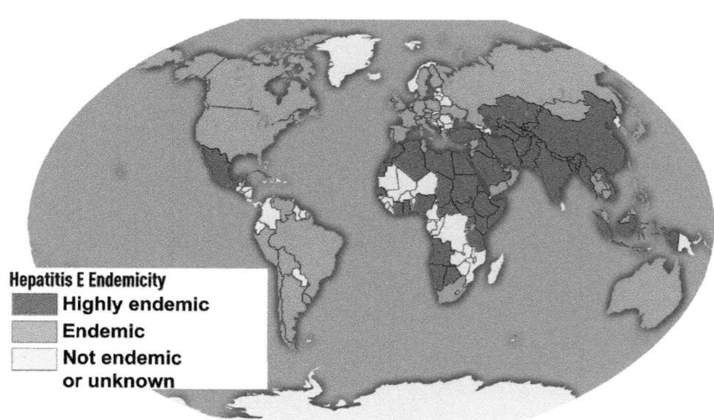

FIGURE 12.5 **Worldwide hepatitis E incidence.** The incidence of HEV is highest in East and South Asia, the Middle East, Africa, and Central America. *Figure courtesy of Centers for Disease Control and Prevention, 2016. CDC Health Information for International Travel 2016. Oxford University Press, New York.*

in the Middle East, Africa, and Central America; in Egypt, 50% of people over age 5 are seropositive. There are 20 million worldwide HEV infections each year, leading to an estimated 56,000 deaths. An HEV vaccine has been created in China but is not yet available elsewhere.

12.2.6 Other Hepatitis Viruses

Additional cases of viral hepatitis exist that are not caused by HAV, HBV, HCV, HDV, or HEV. In the 1990s, two viruses were identified that were prematurely named hepatitis F virus and hepatitis G virus, although scientists have been unable to substantiate that these viruses cause human hepatitis. Newly identified viruses are also being investigated for their possible role in inducing human liver disease.

12.3 MOLECULAR VIROLOGY OF THE HEPATITIS VIRUSES

Despite causing similar clinical conditions, each of the hepatitis viruses belongs to different viral families, possesses distinct virus properties, and uses different replication strategies (Table 12.2).

12.3.1 Hepatitis A Virus

Discovered in 1979, HAV is a small nonenveloped icosahedral virus of 27–32 nm in the family *Picornaviridae,* which includes rhinovirus and poliovirus (Fig. 12.6A and B). Officially, the virus' species name was changed from "hepatitis A virus" to "hepatovirus A" in 2014 by the International Committee for Taxonomy of Viruses. HAV possesses a +ssRNA genome of 7.5 kb; the 5′ end is not capped but is instead covalently linked to a viral protein known as VPg (Fig. 12.6C). The 5′ nontranslated region (NTR, equivalent with the term untranslated region, UTR) is around 740 bases—nearly 10% of the HAV genome—and contains an **internal ribosome entry site (IRES)** that initiates cap-independent translation (see Chapter 4, "Features of Host Cells: Cellular and Molecular Biology Review"). A single open reading frame is translated into a polyprotein that is cleaved by the viral 3C protease into the 11 HAV proteins (Fig. 12.6C). The polyprotein is functionally divided into three regions: P1, P2, and P3. The first portion of the polyprotein, P1, encodes the structural proteins VP1, VP2, VP3, and VP4 that form the virion capsid. The second and third portions of the polyprotein, P2 and P3, encode nonstructural proteins involved in replication, including VPg, the protease, and the RNA polymerase. The genome also possesses a 3′ NTR with a poly(A) tail of 40–80 nucleotides.

HAV binds to a receptor found on T cells known as T cell immunoglobulin and mucin domain 1 (TIM-1), also referred to as the hepatitis A virus cellular receptor 1 (HAVCR1), which is not expressed on hepatocytes. Instead, HAV binds the asialoglycoprotein receptor (ASGPR) of hepatocytes (Fig. 12.7). Both TIM-1 and ASGPR appear to bind IgA, the antibody isotype that is produced at mucosal surfaces, and it has been proposed that the binding of IgA antibody to HAV in the gastrointestinal tract might support the binding of HAV to either TIM-1 or ASGPR. However, can occur in the absence of IgA. Not much is known about how HAV penetrates the cell, although evidence suggests this occurs through nonclathrin-mediated endocytosis. Recently, the VP4 protein has been implicated in creating pores, presumably in the endosome, through which the viral genome is transported.

As a +ssRNA virus, entry into the nucleus is not required as translation is the first cellular event that occurs. The RNA-dependent RNA polymerase (RdRp) transcribes vmRNA and creates an antigenome to replicate the +ssRNA genome, which is packaged into capsids composed of VP1, VP2, VP3, and possibly VP4. HAV does not visibly affect host cells and is not lytic, implicating exocytosis as the method of capsid release.

A single **serotype** of HAV exists, meaning that antibodies against the virus equally recognize all strains of HAV because they are antigenically similar to each other, although four human **genotypes** with differing genomic sequence have been identified. The antigen recognized by host antibodies is composed of parts of both VP1 and VP3 proteins. Either protein subunit individually is not immunogenic, explaining the choice of inactivated virus in the HAV vaccine.

In-Depth Look: Family Names

Each hepatitis virus is found in a different family (or genus for HDV, which is not categorized into a family). Fortunately, we can decipher the origins of these family names to make it easier to remember which virus is found in each. Hepatitis A virus is in the *Picornaviridae* family. Breaking down this name, *pico* means "small" and is followed by "*rna*"—these are small RNA viruses. (Of the hepatitis viruses, four have RNA genomes, but HAV possesses the smallest capsid.) Hepatitis B virus is in the *Hepadnaviridae* family. Following the same process as with *Picornaviridae*, you can see that HBV is a "*Hepa*" *(liver)* "*dna*" virus, and HBV is the only hepatitis virus with a DNA genome. HCV is one of several viruses in the *Flaviviridae* family. "*Flavi*" is Latin for "yellow," because the family was named after yellow fever virus. (Of course, HCV causes jaundice, too!) The HDV genus is *Deltavirus*, because HDV is often called hepatitis *delta virus*. Finally, hepatitis E virus is in the *Hepeviridae* family—it is the "*Hep-e*" virus.

TABLE 12.2 Molecular Properties of Hepatitis Viruses

	HAV	HBV	HCV	HDV	HEV
Family	*Picornaviridae*	*Hepadnaviridae*	*Flaviviridae*	Genus: *Deltavirus*	*Hepeviridae*
Genome type	+ssRNA	Partially ss/dsDNA	+ssRNA	−ssRNA	+ssRNA
Genome size	7.5 kb	3.2 kb	9.6 kb	1.7 kb	7.2 kb
Virion diameter	27–32 nm	42 nm	55–60 nm	42 nm	32–34 nm
Infection target	Humans	Humans	Humans	Humans	Humans, pigs, wild boar, deer
Genotypes	4	8	11	3	4

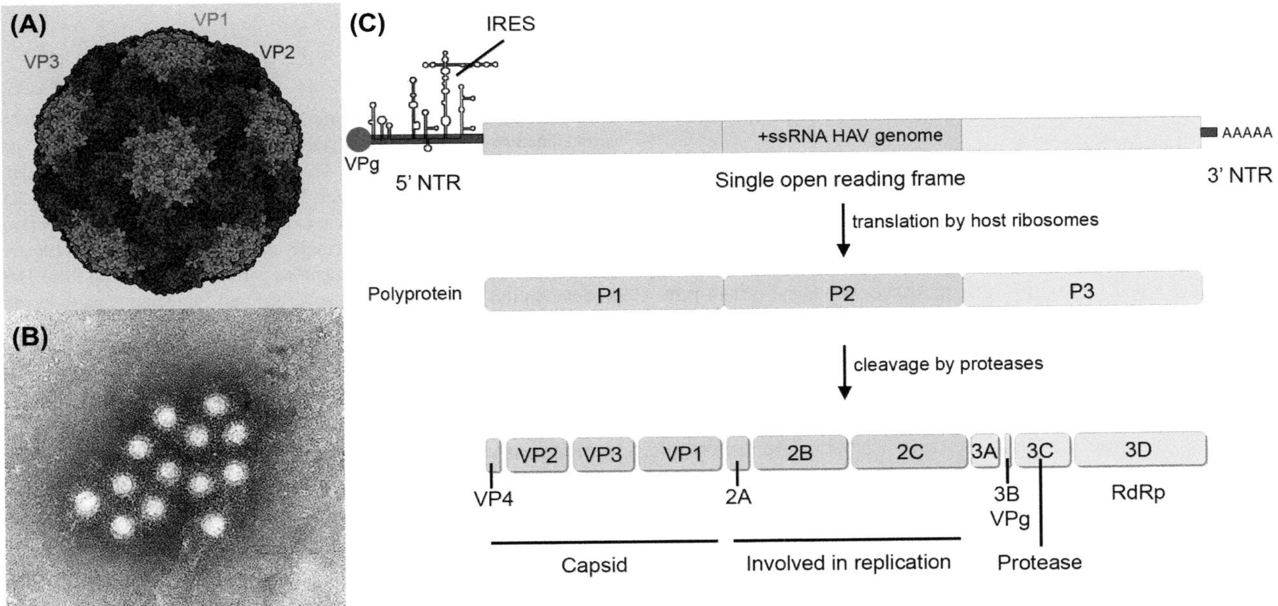

FIGURE 12.6 **Hepatitis A virus.** (A) HAV is a small, non-enveloped icosahedral virus of 27–32 nm. VP1 (tan), VP2 (brown), and VP3 (green) form the capsid. VP4 is required for capsid formation but is not found in the final capsid. This 3D representation is visualized looking at the 5′ axis of rotation, through a vertex. *(Image generated using QuteMol (IEEE Trans Vis Comput Graph 2006. 12, 1237–1244) using the 4QPI PDB assembly (Nature 2015. 517, 85–88.)* (B) Electron micrograph of HAV virions. *(Image courtesy of CDC/Betty Partin.)* (C) The genome structure of HAV. A single +ssRNA of 7.5 kb contains an IRES for ribosome translation initiation and a single open reading frame that is translated into a polyprotein, which is cleaved into the 11 HAV proteins.

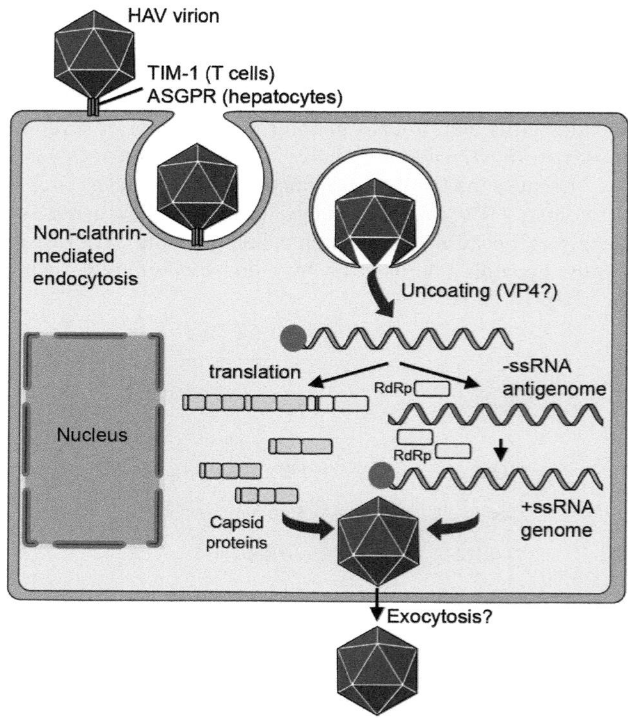

FIGURE 12.7 **Hepatitis A replication.** HAV is thought to bind to TIM-1 on T cells or ASGPR on hepatocytes. Following nonclathrin-mediated endocytosis, the +ssRNA genome is released into the cell and translated by host ribosomes into a polyprotein that is cleaved into the HAV proteins. The RdRp creates an antigenome that is used to create additional vmRNAs, as well as the +ssRNA genome. The genome associates with VPg and is encapsidated by VP1, VP2, and VP3 to form the virion, which is possibly released via exocytosis.

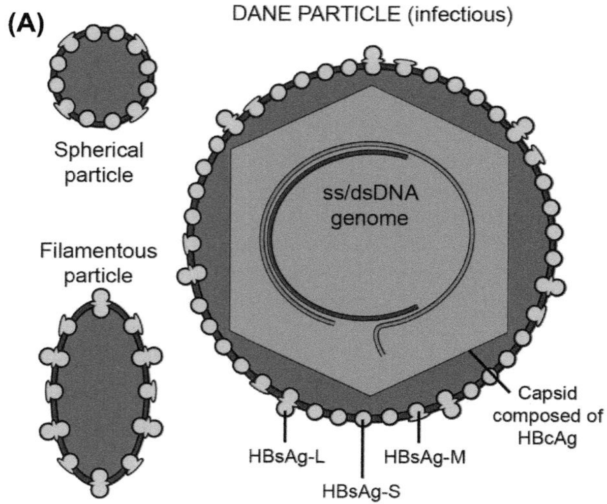

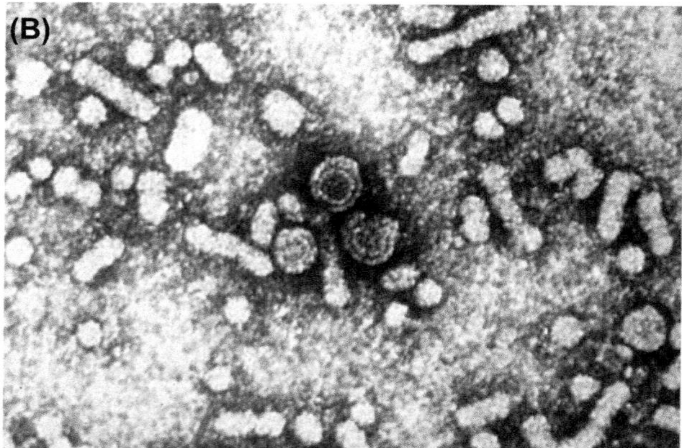

FIGURE 12.8 **Particles generated during HBV infection.** (A) HBsAg is translated from three different start codons to yield three different sizes of the same protein—HBsAg L, M, and S. These incorporate into membranes to create spherical particles, filamentous particles, and Dane particles, which are the only ones that enclose the capsid and are infectious. The genome of HBV is partially single- and double-stranded. (B) Electron micrograph of spherical and filamentous particles surrounding three Dane particles in the center. Note that spherical and filamentous particles outnumber Dane particles. *Image courtesy of the CDC.*

12.3.2 Hepatitis B Virus

HBV is a small enveloped icosahedral virus of the *Hepadnaviridae* family that possesses a 3.2-kb circular DNA genome that is partially single- and partially double-stranded (Fig. 12.8A and B). During infection, three types of viral particles are generated: a complete, infection particle of 42 nm known as the Dane particle, after the scientist who first visualized it in 1970 using electron microscopy; a 22 nm spherical particle; and a filamentous particle that is an elongated version of the spherical particle (Fig. 12.8B). The spherical and filamentous forms outnumber the infectious Dane particles 100- to 10,000-fold and are composed of membrane containing only the HBsAg glycoprotein, which also comes in three forms—small (S), medium (M), and large (L)—based upon their molecular weight. L-HBsAg is enriched in Dane particles and is responsible for virion attachment to the cell surface receptor.

HBV begins infection when the L-HBsAg embedded within the envelope of the infectious Dane particle attaches to its cell surface receptor on the hepatocyte (Fig. 12.9). This has recently been identified as the sodium taurocholate cotransporting polypeptide (NTCP), a hepatocyte transporter that is involved in the uptake of bile salts. The virus is taken in via clathrin-mediated endocytosis. The nucleocapsid is thought to escape from the endocytic vesicle and disassemble at the nuclear pore or within the nucleus.

The circular HBV genome is composed of a complete negative-sense strand and incomplete positive-sense strand (Fig. 12.9), known as the **relaxed circular DNA (rcDNA).** In the nucleus, the gapped segment is repaired into double-stranded DNA, creating the **covalently closed circular DNA (cccDNA)** that remains in the nucleus as a separate episome. Host RNA polymerase II transcribes three viral mRNAs from the cccDNA that are polyadenylated and transported to the cytoplasm for translation. The first transcript encodes the HBeAg, also known as the precore protein, which is an nonstructural protein that enters

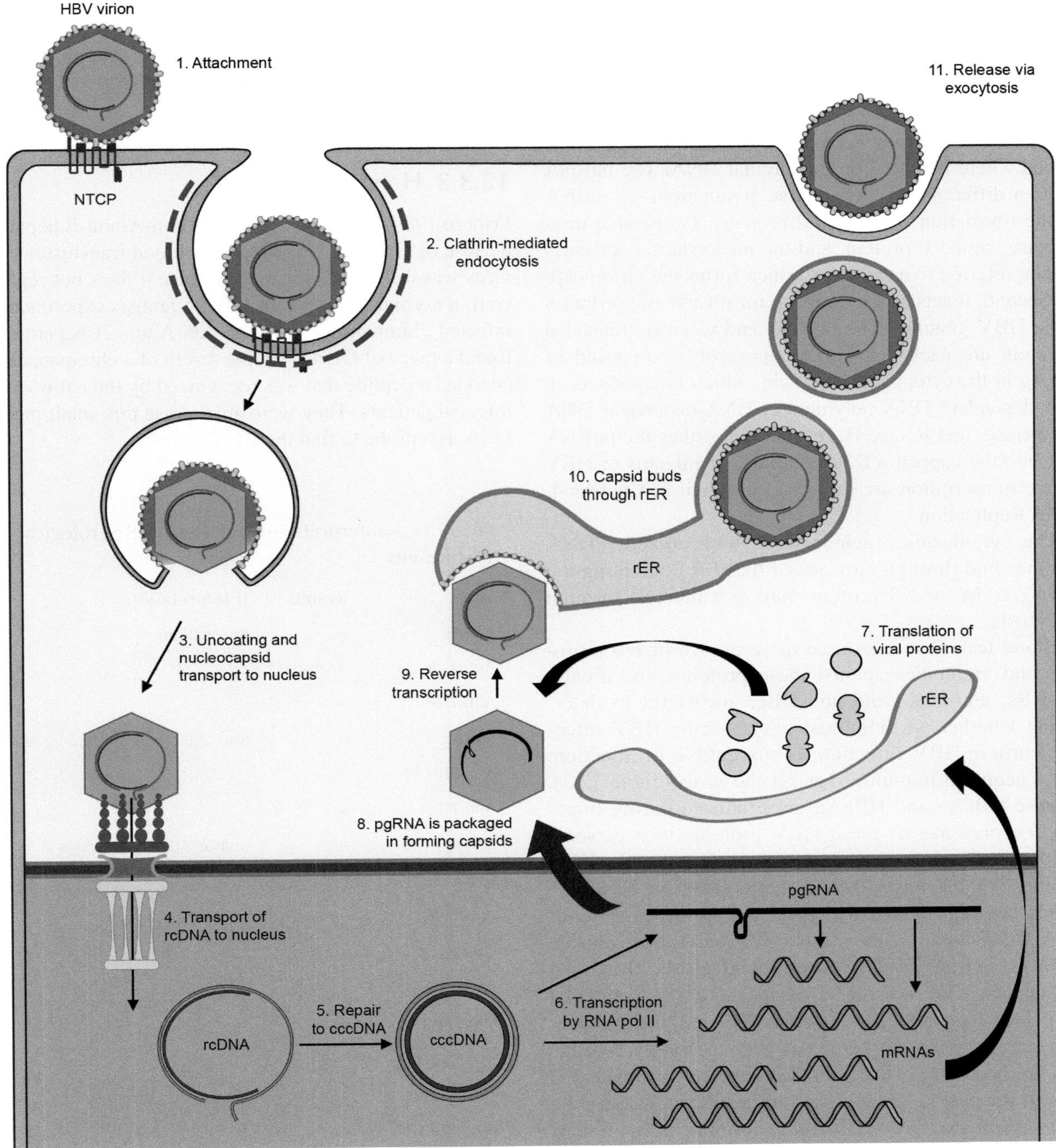

FIGURE 12.9 **Replication of HBV.** (1) Attachment of HBV to the hepatocyte occurs via binding to NTCP. (2) The virion is taken in by clathrin-mediated endocytosis. (3) Uncoating occurs and the nucleocapsid is transported to the nucleus. (4) The rcDNA is transported into the nucleus and repaired to cccDNA (5). (6) RNA polymerase II transcribes three mRNAs, which are translated to form viral HBsAg, HBeAg, and HBx proteins (7). RNA pol II also transcribes an RNA pregenome (pgRNA) that encodes P protein (the reverse transcriptase) and HBcAg, which are also translated by host ribosomes in the cytoplasm (7). (8) The P protein and copies of the pgRNA are packaged into assembling capsids. (9) Reverse transcription is carried out by P protein within the capsid, using the pgRNA as a template to form the gapped rcDNA genome. (10) The nucleocapsid buds through the rER to obtain its envelope containing HBsAg L, M, and S. (11) The virion is released through exocytosis.

the rough endoplasmic reticulum (rER) and ends up being secreted from the cell. The second transcript encodes the three membrane HBsAg glycoproteins—L, M, and S—that are differentially sized due to initiation of translation at different in-frame start codons. The third mRNA encodes the X protein (HBx), which transactivates both viral and cellular genes. As described in Chapter 9, "Viruses and Cancer," HBx plays a role in oncogenesis by stimulating cell proliferation pathways, interfering with DNA repair mechanisms, and inhibiting the activity of p53.

At the same time it transcribes these mRNAs, RNA polymerase II also creates an RNA **pregenome (pgRNA)**. This piece of positive-sense RNA is longer than the genome because RNA polymerase II ends transcription at a site beyond where it started on the circular DNA. The pgRNA has two different functions. First, it functions as mRNA for the translation of two viral proteins: the reverse transcriptase, called P protein, and the nucleocapsid (or core) protein, referred to as HBcAg, which forms the virion capsid. Second, it acts as the template for reverse transcription of the HBV genome. The pgRNA and recently translated P protein are packaged into nucleocapsids composed of HBcAg in the cytoplasm. P protein, which functions as an RNA-dependent DNA polymerase, DNA-dependent DNA polymerase, and RNase H, reverse transcribes the pgRNA back into the gapped rcDNA genome. The details of HBV reverse transcription are described in detail in Chapter 4, "Virus Replication."

The cytoplasmic nucleocapsids with copied rcDNA genomes bud through portions of the rER containing the HBsAg L, M, and S proteins and exit the cell through exocytosis.

Blood tests that determine the presence of HBV proteins and antibodies against these proteins, specifically anti-HBs, anti-HBc, and anti-HBe, are useful in determining whether an individual has an acute HBV infection, chronic HBV infection, is susceptible to infection, or has acquired immunity against the virus (Table 12.3). Because HBsAg and HBeAg are produced during infection, the presence of these HBV proteins in a person's blood indicates the person is infectious. Because the HBV vaccine contains only HBsAg, people who have been vaccinated but not exposed will only have anti-HBs antibodies. Anti-HBc, anti-HBe, and anti-HBs antibodies indicate that a person has been exposed to infectious virus, and IgM against HBc indicates a recent infection. Generally, if a person has antibodies against HBV but no viral proteins present, then the person is immune to HBV. Anti-HBs antibodies are sufficient and necessary to clear the virus. If the person has any viral antigens present, then the person is infected with HBV, and examining the antibodies that have been generated help to determine if it is an acute infection that has a chance of clearance, or whether the person is chronically infected.

Genome sequencing has identified eight genotypes of HBV, named A through H, the prevalence of which varies depending upon geographical location. All eight genotypes have been found in the United States, although genotypes A and C each account for around 30% of cases. Certain genotypes are associated with faster progression of disease; for instance, genotype C has been associated with a higher risk of developing hepatocellular carcinoma than the other genotypes.

Study Break
When considering blood tests for hepatitis viruses, what is the difference between a viral protein antigen and an antiviral antibody?

12.3.3 Hepatitis C Virus

Prior to 1989, HCV was known as "non-A/non-B hepatitis," a cause of viral hepatitis seen after blood transfusions. The virus was difficult to identify because it does not replicate well in normal cell lines. In 1989, scientists experimentally infected chimpanzees with the non-A/non-B hepatitis and found a piece of DNA from the serum of a chimpanzee that encoded a peptide that was recognized by the antibodies of infected patients. They were able to use this small piece of DNA as a probe to find the HCV genome.

TABLE 12.3 Interpretation of Hepatitis B Serological Test Results

Tests	Results	Interpretation
HBsAg	+	Susceptible
Anti-HBc	–	
Anti-HBs	–	
HBsAg	–	Immune due to natural infection
Anti-HBc	+	
Anti-HBs	+	
HBsAg	–	Immune due to hepatitis B vaccine
Anti-HBc	–	
Anti-HBs	+	
HBsAg	+	Acutely infected
Anti-HBc	+	
IgM anti-HBc	+	
Anti-HBs	–	
HBsAg	+	Chronically infected
Anti-HBc	+	
IgM anti-HBc	–	
Anti-HBs	–	
HBsAg	–	Interpretation unclear[a]
Anti-HBc	+	
Anti-HBs	–	

[a]*Most commonly indicates a resolved infection. Can also indicate a false-positive anti-HBc test, low level chronic infection, or a resolving acute infection.*
Table from the Centers for Disease Control and Prevention, Viral Hepatitis—Hepatitis B Information. http://www.cdc.gov/hepatitis/hbv/hbvfaq.htm (accessed 15.06.15.)

HCV is a small enveloped icosahedral +ssRNA virus in the *Flaviviridae* family and *Hepacivirus* genus (Fig. 12.10A). Unlike other flaviviruses, including yellow fever virus, dengue virus, and West Nile virus, HCV is not spread by arthropods. The 9.6-kb HCV genome is organized much in the same way as HAV: HCV has 5′ and 3′ NTRs, a 5′ IRES for cap-independent translation, and is composed of a single open reading frame. This is translated to produce a polyprotein of ~3000 amino acids in length that is cleaved into the 10 viral proteins by cellular and viral proteases (Fig. 12.10C). The first segment of the polyprotein encodes the core (capsid) protein, envelope proteins E1 and E2, and ion channel p7. The core protein makes up the icosahedral capsid, which is wrapped with an envelope into which E1 and E2 are embedded (Fig. 12.10B). The remainder of the polyprotein encodes the nonstructural (NS) proteins: two proteases that facilitate cleavage of the polyprotein (NS2 and NS3), the RdRp that copies the genome and generates vmRNA (NS5B), and other nonstructural proteins required for infection (NS4A, NS4B, NS5A).

Elucidating the molecular mechanisms of HCV replication has been extremely difficult because the virus has historically not replicated well in cell culture or in mouse models, only in humans or experimentally infected chimpanzees. Only in the last decade have scientists been able to engineer cell culture systems and transgenic mice that support the replication of HCV. As such, there are still steps within the HCV replication cycle that are not completely understood, although much has been elucidated.

An interesting facet of HCV replication is that the virus associates with molecules involved in the distribution of lipids within the body. The liver is responsible for producing proteins that shuttle lipids to body cells and back to the liver. **Very low-density lipoprotein (VLDL)** is a protein that contains triglycerides—oils and fats—that are shuttled to body cells that use them as energy or to fat cells that store them. **Low-density lipoprotein (LDL)** transports cholesterol to cells. (Both VLDL and LDL are associated with the "bad" type of cholesterol that has negative effects upon arteries.) HCV virions associate with VLDL and LDL, which contribute to infection by possibly transporting the virion to the surface of the liver cell.

At least four cellular proteins are required for infection: CD81, scavenger receptor class B type 1 (SRB1), claudin-1 (CLDN1), and occludin (OCLN). The viral E1 and E2 proteins associate with these receptors, inducing clathrin-mediated endocytosis (Fig. 12.11). The HCV +ssRNA genome escapes from the endosome and is translated on rER-associated ribosomes into the polyprotein that is cleaved by host proteases and viral proteases NS2 and NS3. Viral proteins NS3, NS4A, NS4B, NS5A, and NS5B remodel the rER membrane, producing a "membranous web" of double-membrane vesicles (DMVs) and multimembrane vesicles (MMVs) (Fig. 12.12). This is the site of genome replication: the RdRp NS5B creates a −ssRNA antigenome that functions as a template to create additional mRNAs or multiple copies of the +ssRNA genome, which are encapsidated by the core protein within newly forming virions. The nucleocapsid buds through the rER, which contains the E1 and E2 envelope proteins. The formation of nascent virions follows the VLDL synthesis pathway within the hepatocyte, and as a result, virions readily incorporate lipids into them and even contain the protein components of the VLDL molecule. These enveloped progeny virions, around 55–60 nm in diameter, are released through exocytosis.

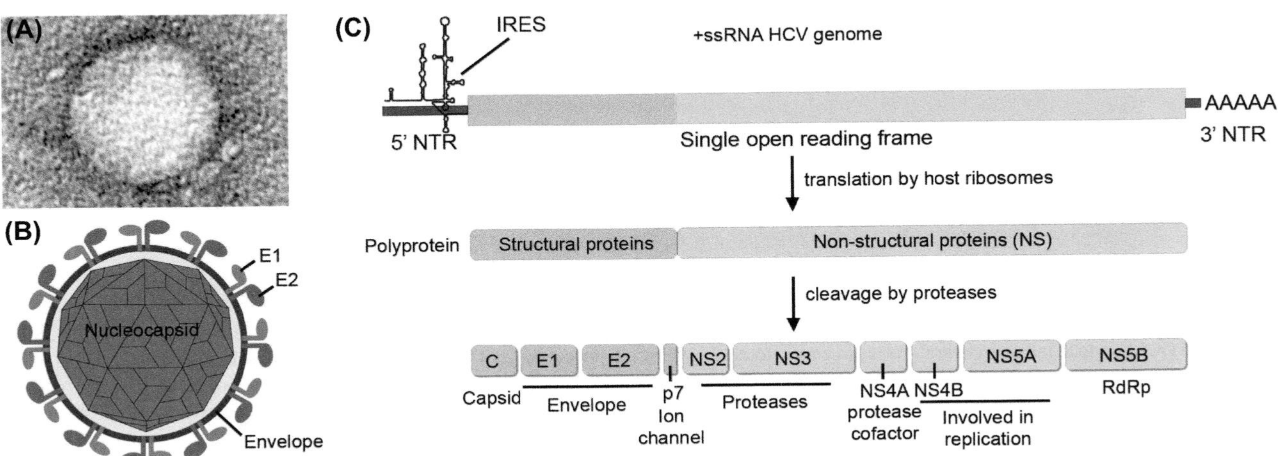

FIGURE 12.10 **Hepatitis C virus virion and +ssRNA genome.** (A) HCV is a small enveloped icosahedral virus. *Electron micrograph courtesy of Maria Theresa Catanese, Martina Kopp, Kunihiro Uryu, and Charles Rice.* (B) The HCV +ssRNA genome of 9.6 kb has a 5′ NTR that contains an IRES for ribosome initiation of translation, a single open reading from that encodes a polyprotein, and a 3′ UTR with a poly(A) tail. The polyprotein is cleaved by cellular and viral proteases into 10 HCV proteins. (C) HCV is an enveloped, icosahedral virus. The capsid is composed of the core protein, and E1 and E2 proteins are embedded into the viral envelope.

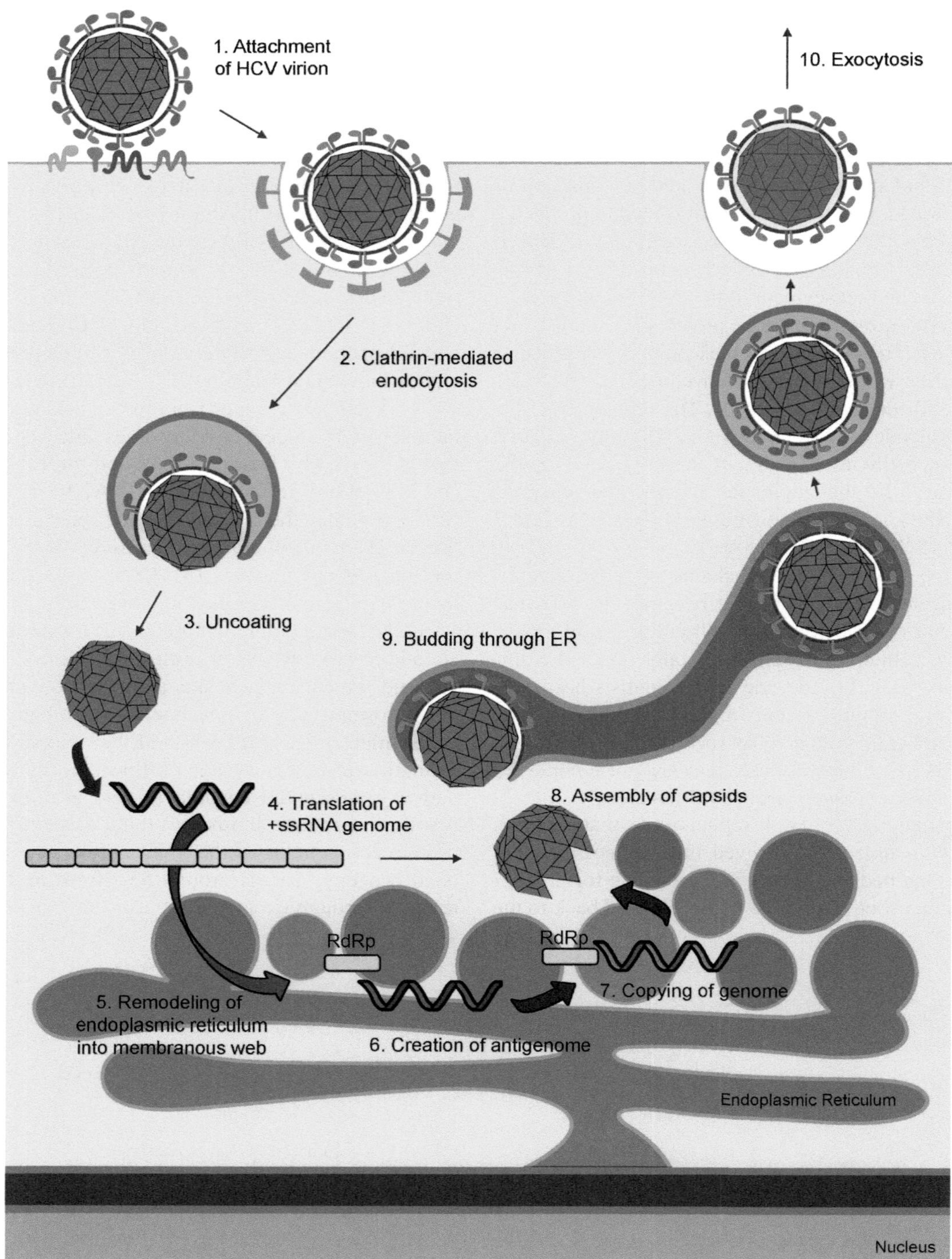

FIGURE 12.11 **HCV replication.** In the process of attachment (1), HCV E1 and E2 proteins associate with at least four cellular proteins: CD81, SRB1, CLDN1, and OCLN, which initiates clathrin-mediated endocytosis (2). Following uncoating (3), the +ssRNA genome is translated into a polyprotein (4) that is cleaved by host proteases and viral proteases NS2 and NS3. Viral proteins remodel the rER membrane (5), creating a membranous web of double- and multimembrane vesicles that become the site of genome replication. The RdRp NS5B creates the –ssRNA antigenome (6), that is used to create additional mRNAs or copies of the +ssRNA genome (7). The nucleocapsid is assembled (8), and buds through the rER to obtain its envelope (9), before being exocytosed (10).

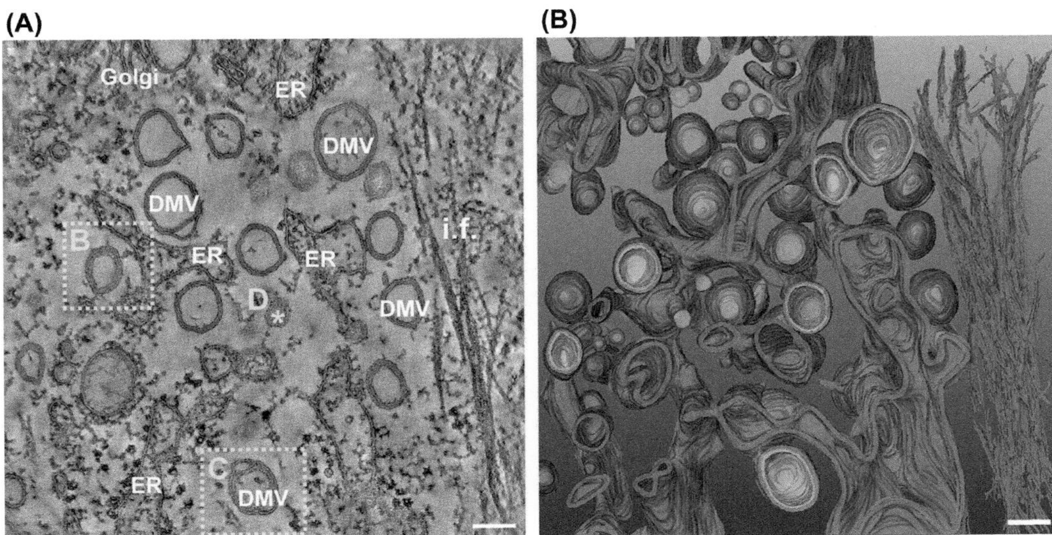

FIGURE 12.12 **The membranous web remodeled by HCV viral proteins.** HCV viral proteins NS3, NS4A, NS4B, NS5A, and NS5B remodel the rough ER membrane, creating double- and multi-membraned vesicles where the genome is replicated. (A) This image was generated using 3D electron tomography. It shows several double-membrane vesicles (DMVs), including those highlighted with yellow letters, derived from the rough ER. (B) shows the 3D reconstruction of the image in A. DMVs are in yellowish/light brown, the ER is in dark brown, the Golgi is in green, intermediate filaments are in blue, and single-membrane vesicles are in purple. *Image courtesy of Romero-Brey, I., Merz, A., Chiramel, A., Lee, J.-Y., Chlanda, P., Haselman, U. et al., 2012. Three-Dimensional Architecture and Biogenesis of Membrane Structures Associated with Hepatitis C Virus Replication. PLoS Pathog 8(12), e1003056. http://dx.doi.org/10.1371/journal.ppat.1003056.*

Mathematical modeling suggests that around 10^{12} HCV virions are produced each day in an infected person, 100-fold more than are produced upon HIV infection.

There are 11 major HCV genotypes that vary more than 30% in their nucleotide sequences. Genotypes 1–3 are prevalent throughout the world, and genotypes 1a and 1b are the most common, accounting for ~70% of infections within the United States. Genotype 4 is found in the Middle East, Egypt, and central Africa; genotype 5 is found in South Africa; and genotypes 6–11 are distributed throughout Asia. Although the particular genotype does not significantly influence the virulence of HCV, it does predict the response to HCV treatments. For instance, individuals with genotype 1 do not respond well to IFN-α alone, whereas genotypes 2 and 3 show better response rates against the cytokine.

12.3.4 Hepatitis D Virus

HDV is a defective virus, also known as a **satellite virus**, because it does not encode all the genes it requires to create nascent infectious virions. In this case, HBV is the **helper virus** that provides the L, M, and S HBsAg that become embedded within the HDV envelope (Fig. 12.13A).

HDV is an enveloped virus of ~42 nm in diameter with no distinct nucleocapsid structure (possibly icosahedral) that contains a circular −ssRNA of around 1.7 kb. It is the only known animal virus with a circular RNA genome. It does not belong to an order or family but is the sole species within the *Deltavirus* genus. Its species is officially *Hepatitis delta virus*. It encodes two forms of a single protein, known as the delta antigen (HDAg), that are 195 (HDAg-S) and 214 (HDAg-L) amino acids in length. Within the nucleocapsid, the −ssRNA genome associates with 60–70 molecules of HDAg-S and HDAg-L.

Because the envelope of HDV virions contains the HBV L, M, and S HBsAg, it uses the same receptor for attachment that HBV does, namely NTCP. It also therefore enters via clathrin-mediated endocytosis and is released from endosomes. Replication of the circular −ssRNA genome takes place in the nucleus, and transport of the genome into the nucleus likely occurs with the assistance of the HDAg. Although the virus possesses an RNA genome, it appears that cellular RNA polymerase II, a *DNA-dependent* RNA polymerase, is involved in the replication of HDV RNA. The one HDV protein possesses no polymerase activity, although it binds to the HDV genomic RNA and may function as a transcriptional activator. The HDV genome is circular, and a circular antigenome along with a linear 5′ capped and 3′ polyadenylated mRNA is transcribed from it (Fig. 12.13B). The mRNA encodes the HDAg-S. The antigenome becomes posttranscriptionally modified by cellular enzymes, leading to the modification of a stop codon and translation of a protein 19 amino acids longer, the HDAg-L. HDAg-S is necessary in the process of genome replication, and

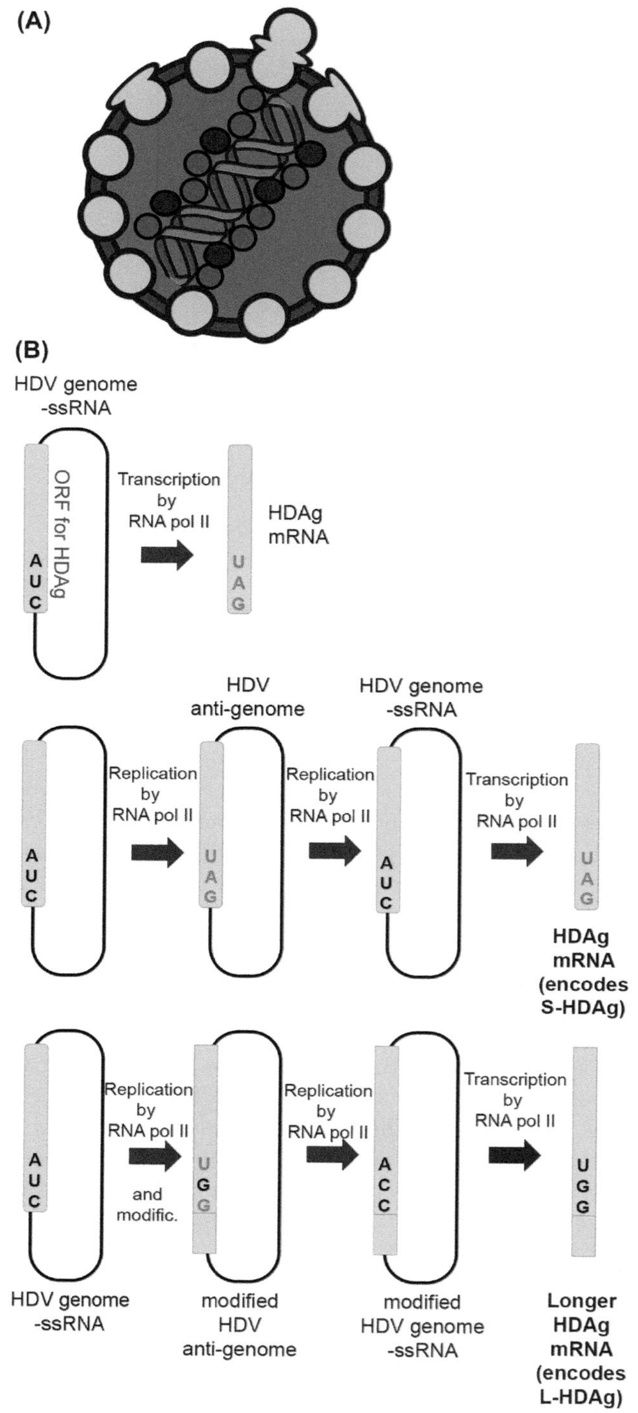

FIGURE 12.13 **Hepatitis D virus.** (A) The structure of the HDV virion. The HBsAg envelope proteins are of HBV origin. The –ssRNA genome associates with the S-HDAg and L-HDAg proteins, two proteins of slightly different sizes that are derived from the same HDV gene. (B) Cellular RNA polymerase II transcribes mRNA from the –ssRNA HDV genome that is used to translate the S-HDAg (top panel). An antigenome is also transcribed by RNA polymerase II, used as a template for additional copies of the HDV genome (which can also be used for transcription of mRNAs, middle panel). During HDV replication, the antigenome undergoes posttranscriptional modification, which introduces a point mutation that eliminates a stop codon. This results in the transcription of the L-HDAg protein, 19 amino acids longer than L-HDAg.

HDAg-L induces the assembly of the HDV virion, which is facilitated by the HBsAg L, M, and S proteins found in the cell as a result of HBV replication. In contrast to the envelope of HBV virions that contain primarily L-HBsAg, HDV virions contain S, M, and L HBsAg in a ratio of 95:5:1. Although it is not prevalent, L-HBsAg is still required for attachment. Because HBV proteins facilitate assembly, it is possible that the HDV virions are also released through exocytosis.

There are at least three HDV genotypes. Genotype I is found worldwide and is the only genotype present in North America and Europe, genotype II is found in Asia, and genotype III has been found only in South America.

12.3.5 Hepatitis E Virus

HEV shares many similarities with HAV: HEV is also a nonenveloped, icosahedral +ssRNA virus transmitted through the fecal-oral route (Fig. 12.14A and B). HEV is in the *Hepeviridae* family, however, and is genetically distinct from HAV—and most other viruses, for that matter. Instead of encoding a single polypeptide from its +ssRNA genome, the 7.2-kb HEV genome encodes three open reading frames (ORFs) (Fig. 12.14C). It does not contain an IRES; instead, the 5′ end of the genome is capped and followed by a small NTR. The nonstructural proteins, including the RdRp, are encoded by the largest ORF, found on the 5′ end of the +ssRNA. The capsid protein is encoded by the smaller ORF2 that begins 37 nucleotides downstream of ORF1, and a third ORF encoding a single protein overlaps with the 5′ end of ORF2. It is likely that ORF2 and ORF3 proteins are translated from a single mRNA that is derived not from the +ssRNA genome directly but by transcription of negative-sense **subgenomic RNA.** At the other end of the genome is found a 3′ NTR followed by a poly(A) tail.

Elucidating the replication strategy of HEV has been difficult due to the lack of a simple cell culture system or a rodent model of infection. HEV replicates within hepatocytes, although replication within the gastrointestinal tract and kidney has also been detected. The cell surface receptor for the HEV capsid protein is unknown, although it is believed the virus enters through receptor-dependent clathrin-mediated endocytosis. In the cytoplasm, the capped +ssRNA genome ORF1 is translated by ribosomes to produce the nonstructural proteins. These are still being characterized but include a methyltransferase that provides the 5′ caps for nascent +ssRNA genomes, a protease that may cleave the polyprotein produced by ORF1, an RNA helicase, and the RdRp. The RdRp is able to transcribe complementary RNA starting at the 3′ poly(A) tail of the genome to create full-length negative-sense subgenomic RNA from which new +ssRNA genome copies are made. The subgenomic RNA

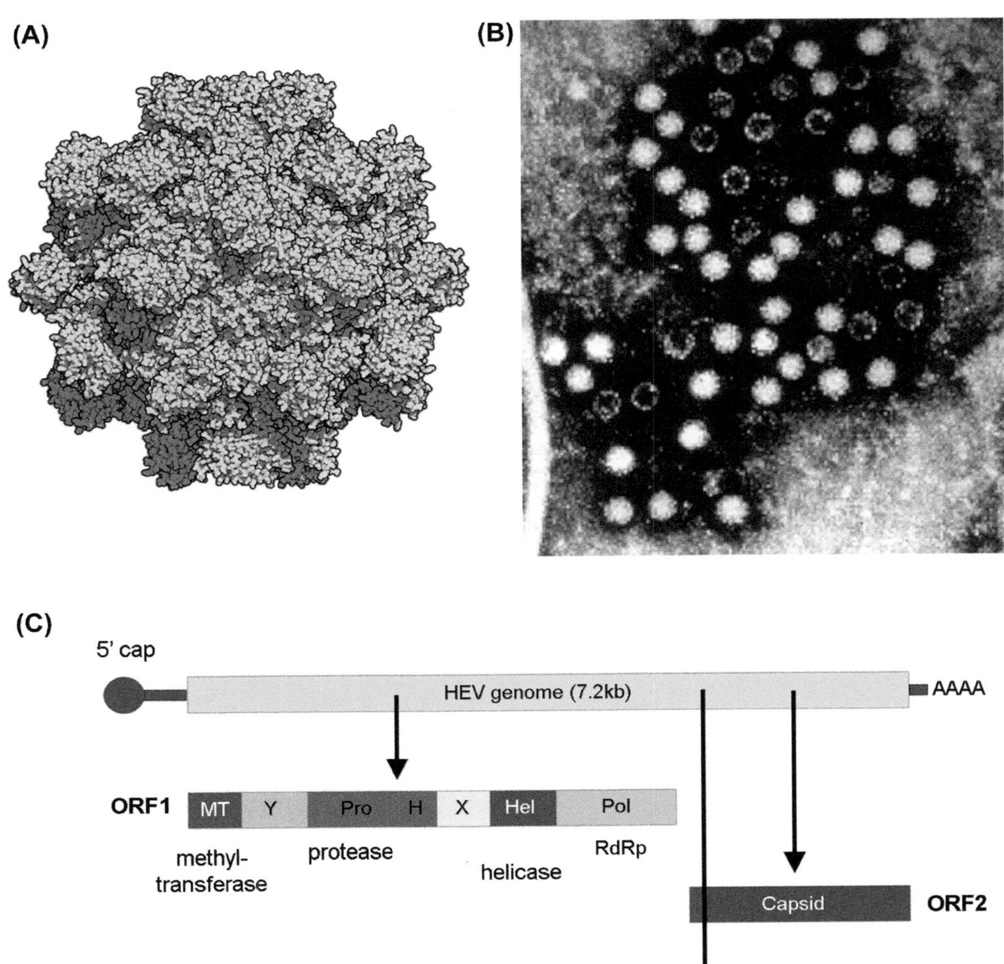

FIGURE 12.14 **Hepatitis E virus.** (A) HEV is composed of a nonenveloped icosahedral capsid. *(Image generated using QuteMol (IEEE Trans Vis Comput Graph 2006. 12, 1237–1244) using the 3HAG PDB assembly (Proceedings of the National Academy of Sciences USA 2009. 106, 12, 992–88).)* (B) Electron micrograph of hepatitis E virions. *(Courtesy of the CDC.)* (C) The 7.2-kb HEV genome is +ssRNA with a 5′ cap and 3′ poly(A) tail. It contains three open reading frames that encode the capsid protein and proteins with methyltransferase, protease, helicase, and RdRp functions.

is also transcribed by the RdRp to form the vmRNA for ORF2 and ORF3 (Fig. 12.14C). ORF2 encodes the capsid protein, which is thought to facilitate assembly around the copied +ssRNA genome in the cytoplasm. The details of assembly and release are unknown, although the non-enveloped nucleocapsids produced are 32–34 nm in diameter.

There are four genotypes of HEV. Humans are infected by all four, but genotypes 3 and 4 are able to infect pigs, wild boar, and deer. Genotype 1 is endemic in Asia and Africa, genotype 2 in Mexico and West Africa, genotype 3 in developed countries, and genotype 4 in Taiwan, China, and Japan. Genotypes 1 and 2 are transmitted through poor sanitation in developing countries, whereas genotypes 3 and 4 are associated with zoonotic transmission from infected pork or deer meat. Genotype 3 is able to cause chronic infection in immunosuppressed organ transplant recipients.

Case Study: Delta Hepatitis—Massachusetts

Excerpted from the Morbidity and Mortality Weekly Report, September 7, 1984, 33(35);493–494.

An outbreak of hepatitis B (HB) began in September 1983 in Worcester, Massachusetts, primarily involving parenteral drug abusers (PDAs) and their sexual contacts. As of August 1, 1984, 75 cases of acute HB have been identified, 50 of which are considered outbreak-related. Fulminant hepatitis has been a prominent feature of this outbreak. Six deaths have occurred, for an outbreak-related case fatality ratio of 12%.

Patients meeting all the following criteria were considered outbreak-related HB cases: (1) an acute clinical illness compatible with HB; (2) elevated serum glutamic-oxaloacetic transaminase (SGOT) or serum glutamic-pyruvic transaminase (SGPT) two or more times greater than the upper limit of normal (when such results were available); (3) positive serology for hepatitis B surface antigen (HBsAg); (4) residence and/or primary diagnosis and treatment within the city of Worcester; and (5) a PDA or a direct contact of a PDA.

Patients with acute HB who could be located were interviewed regarding their drug and alcohol use, as well as risk factors for HB. Serum samples were obtained to test for markers of hepatitis B virus (HBV) infection and delta virus infection.

Of the 50 outbreak-related case patients, 35 were male. Twenty-nine were white, non-Hispanic; 17 were Hispanic; two were black; and two were of unknown race. Ages ranged from 15 years to 43 years (median 25 years). Forty-three patients used needles; six were sexual contacts of PDAs; and one had direct contact with open wounds of a person with hepatitis. Of the six patients who died, three were male; five were white, non-Hispanic, and one was Hispanic. Ages ranged from 19 years to 34 years of age (median 27 years). Five were PDAs, and one was a sexual contact of a known PDA.

Testing for HB markers confirmed HB in all cases. Serum specimens were available from four patients who died; three had immunoglobulin M (IgM) anti-delta virus antibodies. IgM anti-delta virus antibodies were also present in four of 22 PDAs with nonfulminant acute HB, one of seven PDA contacts with nonfulminant acute HB, and none of 11 nonoutbreak-related patients with acute HB. In addition, two of 13 non-ill HBsAg-positive PDAs had serologic markers of delta virus infection (one with IgG antibodies and one with IgM).

Reported by: T. Ukena, MD, Worcester Hahnemann Hospital, L.J. Morse, MD, A. Gurwitz, MD, W.G. Irvine, J.G. McCarthy, E.M. Macewicz, M. Smith, Worcester Dept of Public Health, R. Bessette, MD, C. Pelletier, St. Vincent Hospital, A. Decelles, Worcester City Hospital, M. Bemis, R. Glew, MD, Memorial Hospital, S. Weinstein, H. Kotilainen, University of Massachusetts Hospital, G.F. Grady, MD, Acting Director, Communicable Diseases and Venereal Diseases, Massachusetts Dept of Public Health; Hepatitis Br, Div of Viral Diseases, Center for Infectious Diseases, CDC.

SUMMARY OF KEY CONCEPTS

Section 12.1 Clinical Course of Hepatitis Virus Infections

- Hepatitis refers to inflammation of the liver. HAV, HBV, HCV, HDV, and HEV are unrelated viruses that cause hepatitis.
- Regardless of the viral origin, hepatitis is characterized by a prodromal period of malaise, loss of appetite, nausea, vomiting, low-grade fever, myalgia, and/or arthralgia. Specific symptoms of hepatitis include jaundice, dark urine, and abdominal pain in the general location of the liver. Tests for AST and ALT are useful in verifying hepatitis. Symptoms peak within 1–2 weeks; viral titers are highest before specific symptoms appear.
- HBV, HCV, and HDV cause acute infections that can become chronic infections, while HAV and HEV usually cause only acute infections that are successfully cleared by the host immune response. Fulminant hepatitis is a serious complication of acute infection with high mortality rates.

Section 12.2 Transmission and Epidemiology of Hepatitis Viruses

- Blood tests that measure specific hepatitis antigens or antibodies against the hepatitis viruses are required to determine the cause of clinical hepatitis.
- HAV is transmitted through the fecal-oral route, most often through sewage-contaminated water, eating shellfish from contaminated water, or consuming raw/undercooked food that has come in contact with virions from an infected person. Humans are the only known reservoir of the virus.
- The incubation period for HAV averages 28 days. Adults are more likely than children to display symptoms. Infections occur worldwide but are highly endemic in areas with poor sanitation and untreated drinking water. A vaccine is available against HAV.
- HBV is transmitted through blood, semen, vaginal fluids, and other bodily fluids. The greatest risks of HBV transmission occur in heterosexuals with multiple sexual partners, injection-drug users, men who have sex with men, and people who have sexual contact with an HBV+ individual. HBV is also transmitted perinatally, the most important risk factor in areas of the world where HBV is endemic.
- The incubation period for HBV averages 120 days. Fulminant hepatitis occurs in 1–2% of acutely infected individuals, and the overall case fatality rate is approximately 1%.
- The propensity to develop persistent HBV infections varies greatly with age: >90% of infants, 25–50% of children between 1 and 5 years of age, and 6–10% of older children and adults develop chronic infection. Chronic infection can lead to chronic liver disease, cirrhosis, or liver failure, and up to 50% of the cases of hepatocellular carcinoma are attributable to HBV.
- The prevalence of HBV is highest in sub-Saharan Africa, East Asia, Southeast Asia, Papua New Guinea and Oceanic islands, Bolivia, Peru, and Ecuador. 400 million people in the world have chronic infections, and 780,000 people die each year from acute and chronic HBV. An HBV vaccine is available.
- HCV is almost exclusively transmitted through blood, although it can also be transmitted perinatally or through sexual contact. The incubation period averages 45 days. 75–85% of people infected with HCV develop persistent infections, and 1–5% of those acutely infected will die from cirrhosis or hepatocellular carcinoma.
- Worldwide, 170 million people are living with chronic HCV infection, and half a million die of HCV-related liver disease each year. Prevalence is highest in North African countries, the Middle East, and Central and South Asia. No vaccine exists for HCV, although recent combinations of antivirals can achieve >90% sustained viral response rates.
- HDV is a defective virus that requires HBV to replicate. This can occur through coinfection or HDV superinfection. The average incubation period is 90 days for coinfection or 14–56 days after superinfection of an HBV+ individual. 15–20 million people worldwide are infected with HDV, which can lead to more severe liver disease than infection with HBV alone.
- HEV is transmitted through the fecal-oral route due to contaminated water or eating contaminated shellfish. HEV also infects pigs, boar, and deer. Inadequately cooking and ingesting the meat or entrails of an infected animal can also transmit the virus. In areas where HEV is endemic, which include East and South Asia, the Middle East, Africa, and Central America, the virus is most often transmitted through untreated drinking water.
- The HEV incubation period averages 40 days. HEV is a very serious disease in pregnant women, 10–30% of which die if infected during their third trimester. Infection can also cause miscarriage or premature delivery. 15–20 million people are infected with HEV each year.

Section 12.3 Molecular Virology of the Hepatitis Viruses

- HAV is a nonenveloped icosahedral virus of 27–32 nm in the *Picornaviridae* family with a 7.5-kb +ssRNA genome, which contains an IRES and is translated into a polyprotein that is cleaved into the 11 HAV proteins. HAV binds to TIM-1 on T cells and ASGPR on hepatocytes. It is endocytosed through nonclathrin-mediated endocytosis and creates an antigenome to replicate its +ssRNA genome. It is released through exocytosis.
- HBV is an enveloped icosahedral virus of 42 nm in the *Hepadnaviridae* family with a 3.2-kb partially single-/double-stranded DNA genome. The S, M, and L HBsAg

proteins form three types of particles, infectious Dane particles, spherical particles, and filamentous particles. The L-HBsAg is required to bind the virus's receptor, NTCP. The virus is internalized through clathrin-mediated endocytosis. The circular rcDNA is repaired into cccDNA in the nucleus. RNA pol II transcribes three viral mRNAs and a pregenome RNA, which is used for transcription of two additional mRNAs and as the template for reverse transcription of the genome, which occurs by P protein in the nucleocapsid after packaging. Budding through the rER occurs, and the virion exits through exocytosis.

- HCV is an enveloped icosahedral +ssRNA virus of 55–60 nm in the *Flaviviridae* family with a 9.6-kb genome, which contains an IRES and is translated into a polyprotein that is cleaved into the 10 HCV proteins. At least four cellular proteins—CD81, SRB1, CLDN1, and OCLN—are required for infection of a cell, which induces clathrin-mediated endocytosis. Nonstructural proteins remodel the rER membrane into a membranous web of double- and multimembrane vesicles, where genome replication occurs. Nucleocapsids bud through the rER and are released through exocytosis.
- HDV requires the HBV HBsAg to replicate. It is an enveloped virus of 42 nm with a circular −ssRNA genome of 1.7 kb. It encodes two forms of the HDAg protein due to posttranscriptional modification of its antigenome. RNA pol II transcribes the HDAg vmRNA and replicates the genome.
- HEV is a nonenveloped icosahedral virus of 32–34 nm of the *Hepeviridae* family that encodes three ORFs. Revealing the replication strategy of HEV has been difficult due to the lack of a simple cell culture system or a rodent model of infection. HEV encodes a capsid protein and proteins that have methyltransferase, protease, helicase, and RdRp activity, among others.

FLASH CARD VOCABULARY

Hepatitis	Superinfection
Myalgia	Serotype
Arthralgia	Genotype
Jaundice	rcDNA
Bilirubin	cccDNA
Fulminant hepatitis	pgRNA
Percutaneous	Satellite virus
Sustained viral response	Helper virus
Defective virus	Subgenomic RNA
Coinfection	

CHAPTER REVIEW QUESTIONS

1. What are the nonspecific and specific symptoms of acute hepatitis?
2. Why does jaundice occur during hepatitis?
3. Which hepatitis viruses can result in chronic infections?
4. You are writing a movie plot where an evil character forces his victims to become infected with one of the hepatitis viruses. If you were one of the victims, which virus would you choose (assuming you want to live a long, healthy life)? Which virus would be the worst to choose?
5. Which of the hepatitis viruses are primarily transmitted through the fecal-oral route?
6. Which species are infected by each of the hepatitis viruses?
7. What is the order of progression from acute HCV infection to hepatocellular carcinoma?
8. What is the difference between HDV coinfection and superinfection?
9. For which hepatitis viruses are vaccines available?
10. Create a chart that lists the five hepatitis viruses, their types of nucleic acids, and the size of their genomes. List one interesting fact concerning the molecular biology of each virus.
11. What is an IRES? Which hepatitis viruses possess one in their genomes?
12. For each of the hepatitis viruses, draw out a summary of how the virus replicates in cells.
13. You are an emergency room doctor. You see a 45-year old woman with a 5-day history of malaise, fatigue, low-grade fever, and nausea. She noted her urine became very dark a day before, and you note her liver is enlarged and she is jaundiced. She has had three sexual partners in the past few months, has not used illegal drugs, and is generally is excellent health otherwise. You order serological tests, and the results are that she is positive for HBsAg and IgM anti-HBc, but negative for anti-HAV and anti-HCV. Use the *In-Depth Look* and Table 12.3 to determine what type of hepatitis she may have. Why do you think this?
14. You are a doctor, and a 25-year old man appears in your emergency room with nausea, vomiting, and abdominal pain. He has not had any sexual partners for a year, and he does not drink alcohol or use illegal drugs. He mentioned he had been on a trip to Mexico about a month earlier and had a roast of wild boar. You note that the whites of his eyes are yellow and his upper-right abdomen is painful when palpated. You order blood tests, and the results are negative for HBsAg, IgM anti-HBc, IgM anti-HAV, and anti-HCV. His ALT levels are 40 times the normal upper limit, however. Which viral hepatitis do you think caused his condition? Why?
15. Compare the genomes of the three hepatitis viruses that possess +ssRNA genomes. Can you **ACE** this question?

FURTHER READING

Akkina, R., 2013. New generation humanized mice for virus research: comparative aspects and future prospects. Virology 435, 14–28.

Alves, C., Branco, C., Cunha, C., 2013. Hepatitis delta virus: a peculiar virus. Adv. Virol. 2013. http://dx.doi.org/10.1155/2013/560105.

Bartenschlager, R., Lohmann, V., Penin, F., 2013. The molecular and structural basis of advanced antiviral therapy for hepatitis C virus infection. Nat. Rev. Microbiol. 11, 482–496.

Cao, D., Meng, X.-J., 2012. Molecular biology and replication of hepatitis E virus. Emerg. Microbes Infect. 1, e17.

Centers for Disease Control and Prevention, 2015a. Hepatitis A. In: Hamborsky, J., Kroger, A., Wolfe, C. (Eds.), Epidemiology and Prevention of Vaccine-Preventable Diseases, thirteenth ed. Public Health Foundation, Washington, DC, pp. 135–148.

Centers for Disease Control and Prevention, 2015b. Hepatitis B. In: Hamborsky, J., Kroger, A., Wolfe, C. (Eds.), Epidemiology and Prevention of Vaccine-preventable Diseases, thirteenth ed. Public Health Foundation, Washington, DC, pp. 149–174.

Centers for Disease Control and Prevention, 2013. Viral Hepatitis Surveillance, United States, 2013. Atlanta, GA. http://www.cdc.gov/hepatitis/statistics/2013surveillance/index.htm.

Centers for Disease Control and Prevention, 2016. CDC Health Information for International Travel 2016. Oxford University Press, New York.

Gerlich, W.H., 2013. Medical virology of hepatitis B: how it began and where we are now. Virol. J. 10, 239.

Jacka, B., Lamoury, F., Simmonds, P., Dore, G.J., Grebely, J., Applegate, T., 2013. Sequencing of the hepatitis C virus: a systematic review. PLoS One 8. http://dx.doi.org/10.1371/journal.pone.0067073.

Jones, S.A., Hu, J., 2013. Hepatitis B virus reverse transcriptase: diverse functions as classical and emerging targets for antiviral intervention. Emerg. Microbes Infect. 2, e56.

Kim, C.W., Chang, K.-M., 2013. Hepatitis C virus: virology and life cycle. Clin. Mol. Hepatol. 19, 17–25.

Lindenbach, B.D., Rice, C.M., 2005. Unravelling hepatitis C virus replication from genome to function. Nature 436, 933–938.

Lohmann, V., Bartenschlager, R., 2013. On the history of hepatitis C virus cell culture systems. J. Med. Chem. http://dx.doi.org/10.1021/jm401401n.

Macovei, A., Radulescu, C., Lazar, C., et al., 2010. Hepatitis B virus requires intact caveolin-1 function for productive infection in HepaRG cells. J. Virol. 84, 243–253.

Mahy, B., Van Regenmortel, M.H.V., 2010. Desk Encyclopedia of Human and Medical Virology. Academic Press/Elsevier, Oxford, United Kingdom.

Mohd Hanafiah, K., Groeger, J., Flaxman, A.D., Wiersma, S.T., 2013. Global epidemiology of hepatitis C virus infection: new estimates of age-specific antibody to HCV seroprevalence. Hepatology 57, 1333–1342.

Morgan, R.L., Baack, B., Smith, B.D., Yartel, A., Pitasi, M., Falck-Ytter, Y., 2013. Eradication of hepatitis C virus infection and the development of hepatocellular carcinoma. Ann. Intern Med. 158, 329–337.

Nassal, M., 2008. Hepatitis B viruses: reverse transcription a different way. Virus Res. 134, 235–249.

Ott, J.J., Stevens, G.A., Groeger, J., Wiersma, S.T., 2012. Global epidemiology of hepatitis B virus infection: new estimates of age-specific HBsAg seroprevalence and endemicity. Vaccine 30, 2212–2219.

Panda, S.K., Varma, S.P.K., 2013. Hepatitis E: molecular virology and pathogenesis. J. Clin. Exp. Hepatol. 3, 114–124.

Pascarella, S., Negro, F., 2011. Hepatitis D virus: an update. Liver Int. 31, 7–21.

Romero-Brey, I., Merz, A., Chiramel, A., et al., 2012. Three-dimensional architecture and biogenesis of membrane structures associated with hepatitis C virus replication. PLoS Pathog. 8, e1003056.

Schädler, S., Hildt, E., 2009. HBV life cycle: entry and morphogenesis. Viruses 1, 185–209.

Shukla, A., Padhi, A.K., Gomes, J., Banerjee, M., 2014. The VP4 peptide of hepatitis A virus (HAV) ruptures membranes through formation of discrete pores. J. Virol. 88, 12409–12421.

Tami, C., Silberstein, E., Manangeeswaran, M., et al., 2007. Immunoglobulin A (IgA) is a natural ligand of hepatitis A virus cellular receptor 1 (HAVCR1), and the association of IgA with HAVCR1 enhances virus-receptor interactions. J. Virol. 81, 3437–3446.

Tseng, C.-H., Lai, M.M.C., 2009. Hepatitis Delta virus RNA replication. Viruses 1, 818–831.

Watashi, K., Urban, S., Li, W., Wakita, T., 2014. NTCP and beyond: opening the door to unveil hepatitis B virus entry. Int. J. Mol. Sci. 15, 2892–2905.

Wedemeyer, H., Manns, M.P., 2010. Epidemiology, pathogenesis and management of hepatitis D: update and challenges ahead. Nat. Rev. Gastroenterol. Hepatol. 7, 31–40.

World Health Organization, 2015. Hepatitis. http://who.int/topics/hepatitis/en/ (accessed 15.06.15.).

Yan, H., Zhong, G., Xu, G., et al., 2012. Sodium taurocholate cotransporting polypeptide is a functional receptor for human hepatitis B and D virus. Elife 2012. http://dx.doi.org/10.7554/eLife.00049.

Chapter 13

Herpesviruses

Herpesviruses are ubiquitous in nature and infect a range of animals, from oysters to humans. Over 100 herpesvirus species have so far been identified, but considering that at least one herpesvirus has been discovered in each mammalian species investigated, it is very likely that hundreds other herpesviruses will eventually be revealed. In humans, these large dsDNA viruses cause some well-known conditions, including chickenpox, cold sores, and genital herpes, and also include viruses that can lead to cancer, as discussed in Chapter 9, "Viruses and Cancer." The name *herpes* derives from a Greek word that means "to creep," referring to the creeping rash that is characteristic of some herpesviruses.

13.1 HERPESVIRUS CLASSIFICATION

Herpesviruses infect a range of invertebrate and vertebrate animals. The viruses are classified into their respective taxons based upon similar characteristics, including nucleic acid and protein sequences. The *Herpesvirales* order contains three families: *Alloherpesviridae* (which infect amphibians and fish), *Malacoherpesviridae* (which infect mollusks such as snails and oysters), and *Herpesviridae* (which infect reptiles, birds, and mammals, including humans). In this chapter, we will focus on the viruses within this last family.

A hallmark of all herpesviruses is that they establish **latency**, meaning that the virus is never completely cleared from the host but remains within cells of the body in a dormant state. The *Herpesviridae* family is divided into three subfamilies, depending upon the type of cell in which the virus becomes latent: the *Alphaherpesvirinae* establish latency in neurons, *Betaherpesvirinae* establish latency in immune cells, specifically monocytes or T lymphocytes, and *Gammaherpesvirinae* establish latency in B lymphocytes. As could be reasonably hypothesized based upon their targets of latency, *Betaherpesvirinae* and *Gammaherpesvirinae* viruses share more protein similarities, while the *Alphaherpesvirinae* are genetically less similar to the other two subfamilies. It is estimated that the three subfamilies branched from one viral ancestor around 400 million years ago.

The nine human herpesviruses (HHVs) are distributed within these three subfamilies. The majority of the HHVs were given names as they were discovered (for example, Epstein–Barr Virus (EBV), named after its discoverers Anthony Epstein and Yvonne Barr), but in 1976, the International Committee for Taxonomy of Viruses (ICTV) began numbering and assigning proper names to the herpesviruses based upon the natural host species they infect. Starting at this time, any herpesvirus that infects humans was termed a "human herpesvirus" (HHV), and each different virus was assigned a number in order of discovery. Currently, HHV-1, HHV-2, HHV-3, HHV-4, HHV-5, HHV-6A, HHV-6B, HHV-7, and HHV-8 have been identified. This system also applies to other herpesviruses, such as murid herpesviruses (of mice), bovine herpesviruses (of cattle), or equid herpesviruses (of horses and related animals), for instance. The numbering does not necessarily correspond to similarities between the viruses but was done to establish a naming convention that could be used to simplify the large number of herpesviruses of different species. The historic names of the viruses are still often used, in association with their assigned taxonomical designations.

As shown in Table 13.1, the nine HHV species are spread between the three subfamilies of the *Herpesviridae* family. Herpes simplex virus type 1 (HSV-1, also HHV-1), herpes simplex virus type 2 (HSV-2, also HHV-2), and varicella zoster virus (VZV, also HHV-3) are all within the *Alphaherpesvirinae* subfamily. These viruses generally infect epithelial cells in vivo, cause vesicular lesions, and establish latency in neurons. Viruses within the *Betaherpesvirinae* subfamily include cytomegalovirus (CMV, also HHV-5), and HHV-6A, HHV-6B, and HHV-7, which do not have common names. These viruses have a slow growth cycle, establish latency in monocytes, T lymphocytes, or bone marrow progenitor cells, and often cause inapparent (subclinical) infection. The *Gammaherpesvirinae* subfamily includes EBV (also known as HHV-4) and Kaposi's sarcoma-associated herpesvirus (KSHV, also HHV-8), two viruses that productively infect and become latent within lymphocytes or lymphoid tissues. Both of these viruses can induce transformation of infected cells, leading to cancer.

13.2 CLINICAL CONDITIONS CAUSED BY HERPESVIRUSES

Members of the *Herpesviridae* family are large, enveloped viruses possessing an icosahedral capsid surrounding a dsDNA core (Fig. 13.1). With the exception of VZV, which is a respiratory virus, all other HHVs generally require close

TABLE 13.1 Human Herpesviruses

	Abbreviation	Official Name	Seroprevalence[a]	Initial Site of Infection	Site of Latency	Notable Disease(s) Caused
Alphaherpesvirinae						
Herpes simplex virus-1	HSV-1	HHV-1	53.9%[b]	Epithelial cells	Trigeminal ganglia	Oral herpes, herpes keratitis, herpes gladiatorum, congenital defects, genital herpes
Herpes simplex virus-2	HSV-2	HHV-2	16%[b]	Epithelial cells	Sacral nerve root ganglia (S2–S5)	Genital herpes, congenital defects
Varicella zoster virus	VZV	HHV-3	>98%[c]	Epithelial cells	Dorsal root ganglia	Chickenpox, shingles, herpes zoster ophthalmicus, postherpetic neuralgia
Betaherpesvirinae						
Human cytomegalovirus	HCMV	HHV-5	50–80%[b]	Epithelial cells, monocytes, endothelial cells	Monocytes, CD34+ monocyte progenitors	Congenital defects, mononucleosis
Human herpesvirus-6A	–	HHV-6A	~100%[b]	Lymphocytes, monocytes, neural cells	Hematopoietic bone marrow stem cells	None (orphan virus)
Human herpesvirus-6B	–	HHV-6B	88–100%[b]	Lymphocytes, monocytes, neural cells	Hematopoietic bone marrow stem cells	Roseola infantum
Human herpesvirus-7	–	HHV-7	73–100%[b]	Lymphocytes	Lymphocytes	Roseola infantum
Gammaherpesvirinae						
Epstein–Barr virus	EBV	HHV-4	89%[c]	B cells, epithelial cells	Memory B cells	Mononucleosis, Burkitt's lymphoma, Hodgkin's lymphoma, nasopharyngeal carcinomas, some gastric carcinomas
Kaposi's sarcoma-associated herpesvirus	HSHV	HHV-8	<5%[b]	B cells, endothelial cells	Memory B cells	Kaposi's sarcoma, primary effusion lymphoma, multicentric Castleman's disease

[a]*In the United States.*
[b]*Among 14–49 year olds.*
[c]*Among adults.*

contact with infected tissues or secretions for transmission to occur because of the fragility of the virion envelope. Herpesviruses undergo a *lytic* infection and then establish latency within host cells. In this case, "lytic" refers to a productive infection that generates nascent virions, because although infected cells are often lysed by natural killer (NK) cells or cytotoxic T cells, the viruses generally exit the cell through exocytosis, rather than virally induced lysis.

Once latency occurs, infection can be reactivated in response to cellular stresses, such as ultraviolet light or fever, or because of immunosuppression, making these viruses of concern to those with genetic or acquired immunodeficiency. Several herpesviruses also cause congenital infections (present at birth) or perinatal infections (acquired around the time of birth) of the newborn, particularly when the mother is first infected or has an active infection during pregnancy.

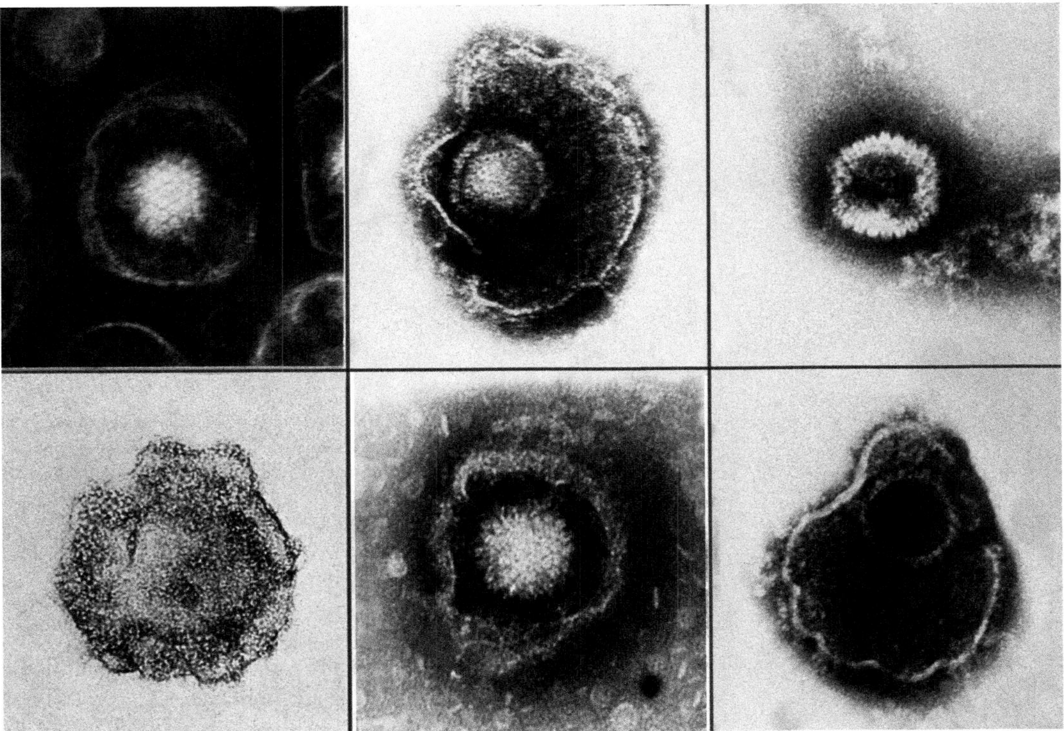

FIGURE 13.1 **Herpesvirus virion morphologies.** This figure shows several virions from different herpesviruses, illustrating the "fried egg" appearance of the icosahedral capsid surrounded by the envelope. *Courtesy of CDC / Dr. Erskine Palmer.*

13.2.1 Herpes Simplex Virus Type 1 and Type 2 (HSV-1/HHV-1 and HSV-2/HHV-2)

HSV-1 and HSV-2 are two closely related viruses, sharing approximately 85% homology in protein-coding regions of their viral genomes, although their envelope proteins generate different antibody profiles within a host. Both viruses can infect at either site, but HSV-1 causes 80–90% of oral herpes (herpes labialis), while historically most cases of genital herpes have been caused by HSV-2. Both viruses cause similar herpetic lesions, which are **papules** (raised bumps) that progress into **vesicles** (fluid-filled blisters) and **ulcers** (open sores) that crust over or rupture, leaving raw skin that takes weeks to heal. For HSV-1 infections, these cold sore-like lesions are found in and around the mouth, while HSV-2 infects the genital, anal, or perianal regions (Fig. 13.2).

Primary infection refers to the first exposure of a person to the virus. With HSV-1 or HSV-2, primary infection is often asymptomatic. However, when symptoms are present, they are generally severe, as it will take weeks for an antibody response to be mounted against the virus. **Recurrent infections**, when the virus reactivates after being latent, are generally milder and of shorter duration than the primary infection. Interestingly, because of the similarity between the two viruses, a person with antibodies against one of the HSV viruses generally has less severe manifestations of disease when infected with the other HSV virus. This is known as a **nonprimary first episode infection**.

People generally contract oral HSV-1 as children less than 5 years of age, and infections are asymptomatic over 90% of the time. In cases where symptoms are present, the incubation period is 2–12 days, with an average of 4 days. A sore throat and a fever of 101–104°F can be involved, and visible symptoms begin as clusters of papules that develop into infectious fluid-filled vesicles. These can progress into ulcers, particularly on sites inside the mouth, that take 2–3 weeks to resolve. Shedding of the virus into saliva can occur for weeks following infection. Asymptomatic children can also shed virus into saliva, and healthy children and adults do so occasionally, as well. The virus is spread through direct contact or through saliva, and so touching, sharing utensils, or kissing can transmit HSV-1 from an infected individual.

HSV-1 also causes infections at body sites other than the oral cavity. The virus is capable of infecting nearly any epithelium, and close contact with herpetic lesions or shed virions can transfer the infection to susceptible epithelial cells. **HSV keratitis** results from the virus infecting the corneal epithelium. Severe infections can cause scarring or permanent damage to the cornea, and HSV keratitis is the leading cause of corneal blindness in the United States. **Herpes gladiatorum**, or "mat herpes," is a cutaneous infection with

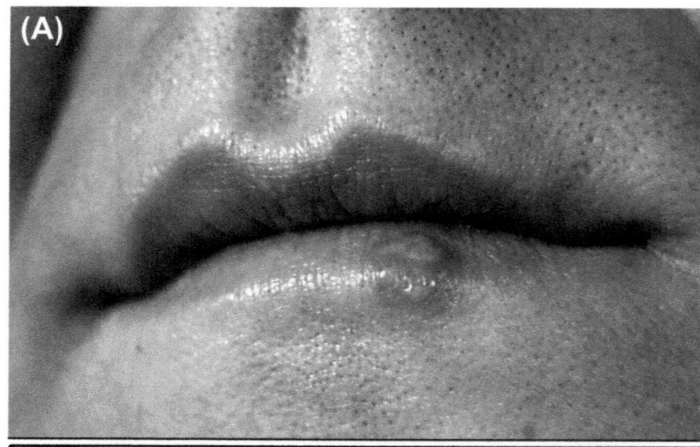

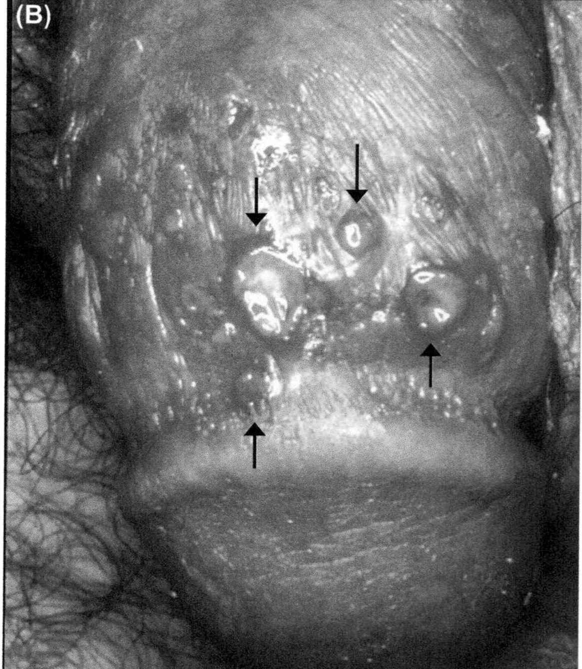

FIGURE 13.2 Oral and genital herpes. HSV-1 and HSV-2 cause herpetic lesions at the site of infection. Cold sores form in and around the mouth during oral herpes (A), and vesicles (*arrows*) and ulcers appear on genital, anal, or perianal regions during genital herpes (B). *Courtesy of CDC / Dr. Hermann, Dr. N.J. Flumara, and Dr. Gavin Hart.*

HSV-1 that is transmitted by wrestlers through close skin contact during practice or competition. The herpetic lesions are usually located on the head and neck, areas of close contact during wrestling. This condition is also commonly observed in rugby players, where it is termed *herpes rugbiorum*. Recurrent infections at these sites are also common.

In industrialized countries such as the United States, the age of HSV-1 **seroconversion**—the initial presence of antibodies against the virus in a person—has been increasing, indicating that people are being exposed to the virus at a later age (*sero* derives from the word *serum*, the acellular fraction of the blood where antibodies are located). **Seroprevalence** refers to the percentage of people in a population that have antibodies against a particular pathogen. Because a person's immune system generates a long-lived antibody response, measuring the presence of antibodies against a virus is often easier than measuring the amount of virus present in a person, especially during latent infections when virus may not be found in the blood.

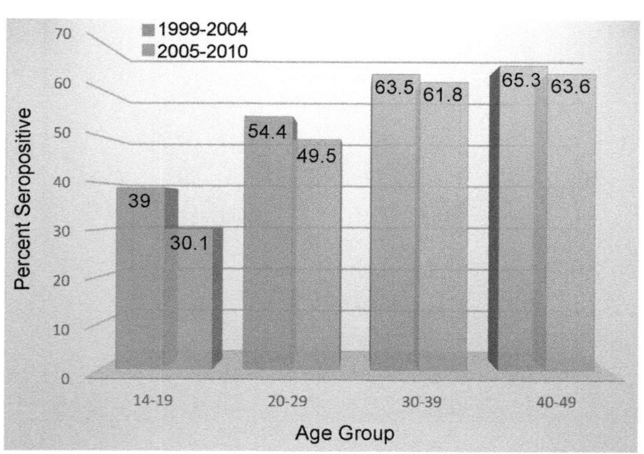

FIGURE 13.3 Seroprevalence of HSV-1. The age at which individuals are first being exposed to HSV-1 is increasing, as shown by comparing HSV-1 seroprevalence in different age groups during 1999–2004 (*blue bars*) and 2005–2010 (*orange bars*). *Data from Bradley, H., Markowitz, L.E., Gibson, T., McQuillan, G.M., 2014. Seroprevalence of herpes simplex virus types 1 and 2-United States, 1999–2010. J. Infect. Dis. 209, 325–333.*

Researchers at the US Centers for Disease Control and Prevention carried out a study to examine HSV-1 and HSV-2 seroprevalence in different age groups. They found that the general seroprevalence of HSV-1 in the United States was 53.9% among 14–49 year olds in 2005–10. Only 30.1% of people in the 14–19 age group had antibodies against HSV-1, a reduction from the previous 5 years, indicating that children and adolescents are being exposed to HSV later in life than previous generations were (Fig. 13.3). This decline has been attributed to better living conditions, less crowding, and better hygiene habits, all factors that reduce the transmission of the virus from one person to the next.

While the rate of HSV-1 seroprevalence has been decreasing in this population, the rate of *genital* HSV-1 infection is increasing. The results of an HSV vaccine trial published in 2012 and 2013 indicated that 60% of genital herpes cases are now attributable to HSV-1, and previous studies have suggested that in college students, upward of 78% of genital herpes cases are caused by HSV-1, rather than HSV-2. The reason for this is still under investigation, but it is thought that the decreasing seroprevalence of HSV-1 in children leads to a situation where a person is first exposed to HSV-1 through sexual contact, and without antibodies from a prior oral exposure to HSV-1, the virus is able to infect cells in the anogenital region and establish latency from this area. In addition, people who have been exposed to HSV-1 tend to have less severe HSV-2 infections due to some cross-reactivity between HSV-1-generated antibodies and the HSV-2 virus.

HSV-2 causes a similar condition as HSV-1, except that HSV-2 is more often associated with genital herpes, rather than oral herpes. HSV-2 more commonly infects the genital, anal, and perianal epithelium than oral epithelia, although the incubation period and herpetic lesions are the same as with HSV-1. Being a sexually transmitted disease, the rates of HSV-2 seroconversion increase with age. An estimated 50 million individuals in the United States, and 16% of those between ages 14 and 49, are infected with HSV-2. Ninety percent of the people who harbor the virus have not been diagnosed because it causes asymptomatic infection in the majority of those infected. Asymptomatic individuals with HSV-2 shed virus intermittently; the result is that the majority of HSV-2 is transmitted by people who do not realize they are infected.

HSV-2 and HSV-1 can cause severe problems if transmitted during birth from an infected mother to her child. **Neonatal herpes simplex virus infection** is most often acquired through intrapartum transmission by the shedding of virus from the mother's genital tract. The infection can be localized to the skin of the newborn, causing characteristic herpetic lesions, but the virus can also disseminate to other organs or infect the central nervous system, where it can cause encephalitis (inflammation and swelling of the brain). Neonatal HSV infection occurs in about 1 in 3000 births in the United States, with a significant mortality rate of 60%. The highest transmission rates are associated with primary infections of the mother with HSV-1 or HSV-2: 50% of these cases result in transmission to the newborn, whereas about a third of nonprimary first episode infections and less than 4% of recurrent or asymptomatic infections result in transmission. Mothers with active genital herpes lesions near the time of delivery will be encouraged to deliver the baby by cesarean section to prevent exposing the baby to the virus. Overall, transmission occurs about 5% of the time when the mother is shedding virus.

Anogenital HSV infections are also associated with increased transmission of the human immunodeficiency virus (HIV). A person infected with HSV-2 is two to three times more likely to acquire HIV, and four times more likely to transmit it.

As *Alphaherpesvirinae* subfamily members, both HSV-1 and HSV-2 establish latency in sensory neurons. The virus establishes a productive infection in skin epithelial cells and spreads within the tissue, making contact with the termini of local sensory neurons. The virion enters the neuron axon and undergoes **retrograde transport** to the nucleus, which can be quite a distance from the initial site of entry into the cell (Fig. 13.4). The virus remains latent in the nucleus until reactivation, which can be induced by immunosuppression or in response to physiological stresses such as fatigue, menstruation, or malnutrition, and environmental stresses such as UV light, heat, or cold. The virus then undergoes **anterograde transport**, traveling back down the neuron, where it initiates a recurrent infection at the initial site of infection. In oral herpes, HSV becomes latent in the trigeminal ganglia, while the site of latency in genital herpes is the sacral nerve root ganglia (S2–S5). The molecular process of latency will be discussed in more detail later in this chapter.

Serological assays (blood tests) are able to differentiate between infection with HSV-1 and HSV-2 by testing for antibodies against viral surface glycoprotein G1, found in HSV-1, or glycoprotein G2, found in HSV-2. A limitation of serology, however, is that the site of infection cannot be determined from the presence of antibodies in the blood. Newer FDA-approved tests are based upon amplifying specific HSV-1 or HSV-2 DNA sequences directly from swabs of lesion fluid.

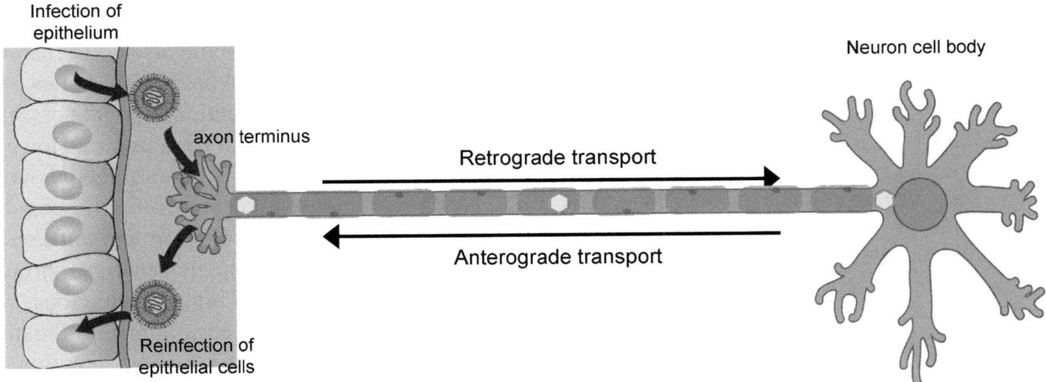

FIGURE 13.4 **Retrograde and anterograde transport of herpesvirus virions.** The Alphaherpesviruses establish latency in neurons, which become infected during the primary infection. Virions undergo retrograde transport by traveling along axon microtubules to reach the nucleus in the neuron cell body. They uncoat at the nuclear envelope, and virus DNA is transported into the nucleus, where it remains in a latent state. Upon reactivation, the virus activates the lytic cycle, assembles the nucleocapsid in the nucleus, and travels down the axon back to the site of initial infection. Here it reinfects epithelial cells and establishes lytic infection.

Case Study: Herpes Gladiatorum at a High School Wrestling Camp—Minnesota

Excerpted from MMWR Weekly, February 09, 1990/39(5); 69–71.

In July 1989, the Minnesota Department of Health investigated an outbreak of HSV-1 dermatitis (herpes gladiatorum) in participants at a Minnesota wrestling camp. The camp was held July 2 through July 28 and attended by 175 male high school wrestlers from throughout the United States. The participants were divided into three wrestling groups according to weight (group 1, lightest; group 3, heaviest). During most practice sessions, wrestlers had contact only with others in the same group. The outbreak was detected during the final week of camp, and wrestling contact was subsequently discontinued for the final 2 days.

A case was defined as isolation of HSV-1 from involved skin or eye or the presence of cutaneous vesicles. To identify cases, a clinic was held at the camp to obtain viral cultures and examine skin lesions. Additional clinical data were obtained from review of emergency department records at the facility where all affected wrestlers were referred for medical care. A questionnaire was administered to wrestlers by telephone following the conclusion of camp.

Clinical and questionnaire data were available for 171 (98%) persons. The mean age of these participants was 16 years (range: 14–18 years); 153 (89%) were white; 137 (80%) were high school juniors or seniors. The median length of time in competitive wrestling was 4 years.

Sixty (35%) persons met the case definition, including 21 (12%) who had HSV-1 isolated from the skin or eye. All affected wrestlers had onset during the camp session or within 1 week after leaving camp. Two wrestlers had a probable recurrence of HSV, one oral and one cutaneous, during the first week of camp. Lesions were located on the head or neck in 44 (73%) persons, the extremities in 25 (42%), and the trunk in 17 (28%). Herpetic conjunctivitis occurred in five persons; none developed keratitis. Associated signs and symptoms included lymphadenopathy (60%), fever and/or chills (25%), sore throat (40%), and headache (22%). Forty-four (73%) persons were treated with acyclovir.

Attack rates increased by weight group: of 55 wrestlers in group 1, 12 (22%) were affected; of 57 in group 2, 17 (30%); and of 59 in group 3, 31 (53%). Thirty-eight (22%) wrestlers interviewed reported a past history of oral HSV-1 infection. Twenty-three percent of affected wrestlers continued to wrestle for at least 2 days after rash onset. Athletes who reported wrestling with a participant with a rash were more likely to have confirmed or probable HSV-1 infection.

Reported by: JL Goodman, MD, EJ Holland, MD, CW Andres, MD, SR Homann, MD, RL Mahanti, MD, MW Mizener, MD, A Erice, MD, Univ of Minnesota Hospital and Clinic, Minneapolis; MT Osterholm, PhD, State Epidemiologist, Minnesota Dept of Health. Div of Field Svcs, Epidemiology Program Office, CDC.

Nucleoside analogs are the standard treatment for people infected with HSV-1 or HSV-2 (Fig. 8.11). As described in Chapter 8, "Vaccines, Antivirals, and the Beneficial Uses of Viruses," a **nucleoside** is a DNA or RNA nucleotide without the phosphate group. As such, it is comprised of a nitrogenous base (adenine, guanine, cytosine, thymine, or uracil) attached to a sugar (deoxyribose for DNA or ribose for RNA). During DNA or RNA synthesis, nucleosides are converted into nucleotides that are incorporated into the replicating DNA or RNA. Nucleoside analogs are drugs that compete with normal nucleotides during DNA or RNA replication. When incorporated into a growing strand, they are unable to bond to subsequent nucleotides, thereby terminating the growing strand and halting the replication of the genome. Acyclovir, valacyclovir, and famciclovir are common nucleoside analogs used to treat primary and recurrent infections with HSV-1 and HSV-2. Antiviral therapy diminishes the duration and severity of symptoms during primary or recurrent infection and can reduce the frequency of recurrences.

Considering that the majority of HSV infections are asymptomatic, transmission often occurs without any knowledge that a person is infected. Control measures to prevent transmission include the use of antiviral drugs to reduce viral titers and the frequency of active lesions, and male condoms can reduce the spread of genital herpes during sexual intercourse.

13.2.2 Varicella Zoster Virus (VZV/HHV-3)

VZV, officially classified as HHV-3, is the third *Alphaherpesvirinae* subfamily member. Better known as the virus that causes chickenpox (which is clinically termed varicella, from the Latin *varius*, meaning "spotted"), VZV induces an acute, productive infection and then establishes latent infection in sensory neurons, specifically dorsal root ganglia.

Reactivation of the virus results in herpes zoster, also known as shingles. This generally occurs after age 60 or in immunocompromised individuals and is characterized by a very painful rash on the skin. Since shingles only occurs as a reactivation of VZV infection, it is not possible to contract shingles from a person that has shingles. However, it is possible to contract chickenpox from someone with shingles.

VZV is the only herpesvirus that is spread through the respiratory route, generally through infected respiratory secretions or airborne droplets generated through coughing or sneezing. The virus is highly contagious, with an incubation period around 2 weeks after initial exposure (range of 10–21 days). The virus begins infection by replicating locally within the nasopharynx and in the tonsils and regional lymph nodes. A primary viremia develops within a week, spreading the virus to organs such as the spleen and liver. Sensory neurons that will later harbor latent virus are also infected at this time. A secondary viremia occurs, and it is thought that infected memory T lymphocytes, which circulate in tissue capillaries, spread the virus to the skin, where visible symptoms begin developing.

Fever and tingling in the skin are the first prodromal symptoms to occur, most often in adults. Unlike HSV-1 and HSV-2, where the majority of cases are asymptomatic, it is rare for a person infected with VZV to not present with symptoms. One to two days after the prodrome, if one is present, red macules form and quickly develop into raised papules and fluid-filled vesicles that crust over, creating the characteristic chickenpox rash (Fig. 13.5A and B). These itchy, 1–4 mm lesions appear asynchronously, starting on the head, moving to the trunk, and finally spreading to the appendages. The **centripetal rash**, concentrated on the trunk of the body and lesser toward the extremities (Fig. 13.6), occurs in two to four overlapping phases over the course of 3–5 days, so it is common to observe macules, papules, vesicles, and dry scabs all within the same area of the body. In children, 200–500 lesions appear during the course of infection, in total.

Before the vaccine was available, 85% of cases occur red in children under 15 years of age; 39% of cases were in children between 1 and 4 years old. Varicella is usually mild and self-limiting in children, but complications can occur in the immunocompromised and in adults, both of whom can have more severe disease. Most complications from varicella relate to viral/bacterial pneumonia, secondary infections of skin lesions, and neurological problems,

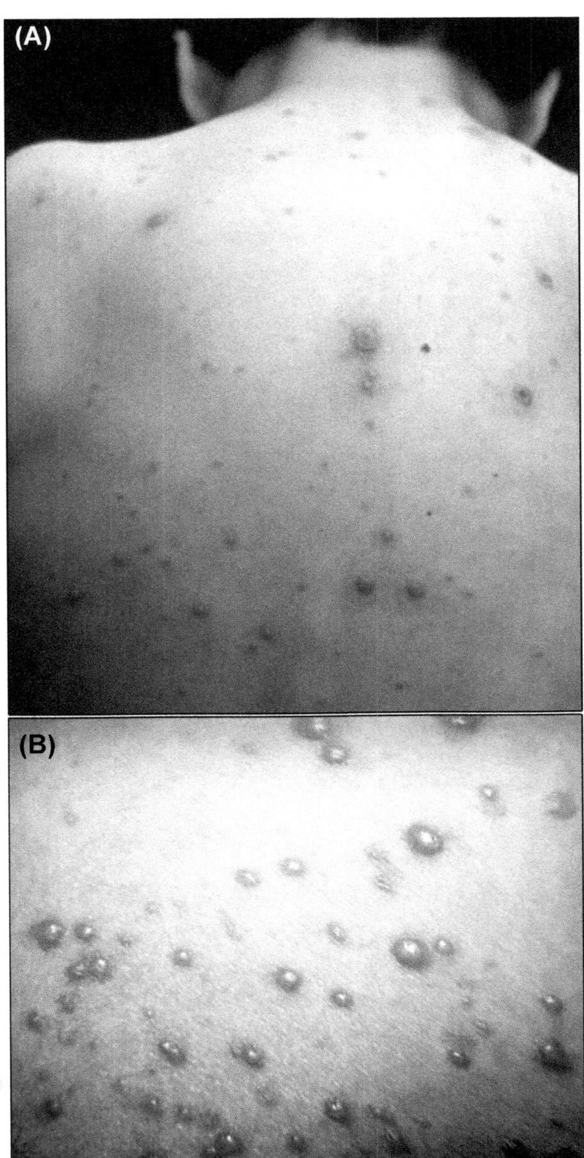

FIGURE 13.5 **Varicella zoster virus (VZV) (chickenpox) rash.** VZV causes chickenpox and its characteristic rash (A). The lesions of the rash appear and mature over the course of several days. Consequently, macules, papules, and vesicles are found together in one location (B). *Courtesy of CDC / Joe Miller.*

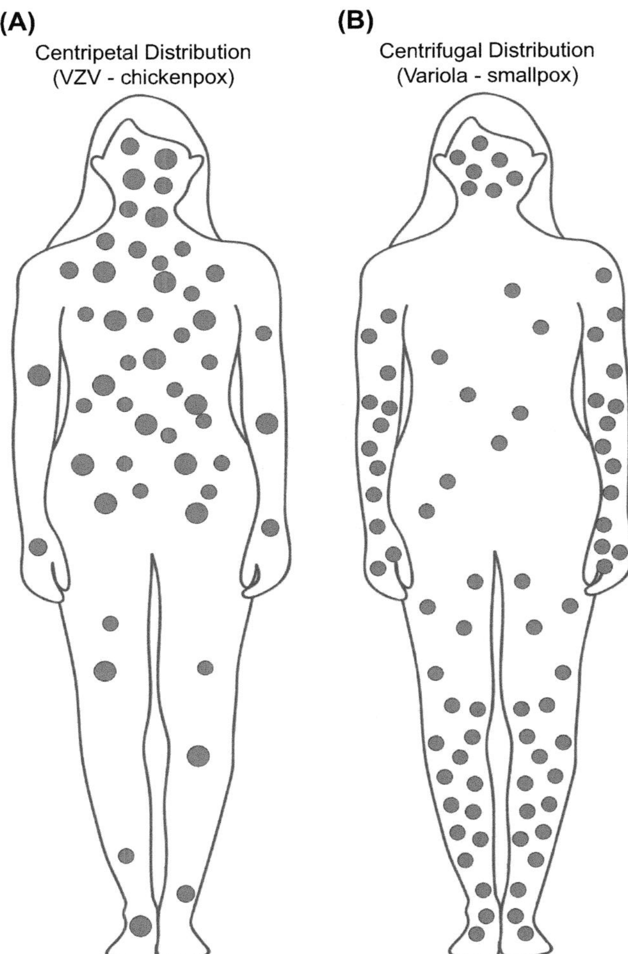

FIGURE 13.6 **Centripetal distribution of chickenpox rash.** (A) VZV causes a centripetal rash, meaning that majority of the lesions appear on the trunk of the body, with fewer on the appendages. (B) Other viruses, including variola virus (smallpox), cause the opposite, a centrifugal rash with the greatest concentration of lesions on the face and distal extremities.

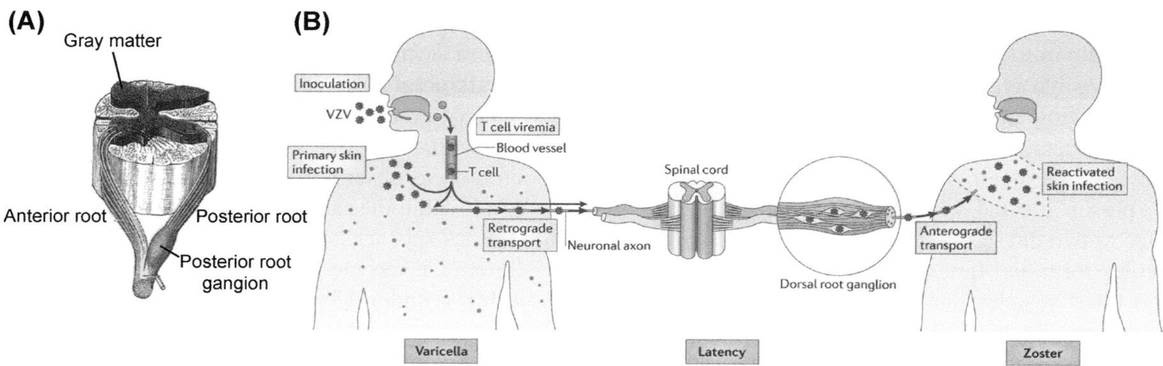

FIGURE 13.7 **VZV latency in and reactivation from the posterior (dorsal) root ganglia.** (A) Illustration showing the site of VZV latency, the posterior root ganglion portion of a sensory neuron entering the spinal column. *(Illustrations from Henry Gray, Anatomy of the Human Body. Philadelphia and New York, Lea and Febinger, 1918.)* (B) During the primary infection and associated viremias, VZV makes contact with sensory neurons in the skin. The virus undergoes retrograde transport and establishes latency in the neuron nucleus in the dorsal root ganglion. Following reactivation, anterograde transport of the virus enables reinfection of the original site, causing herpes zoster (shingles). *Reprinted by permission of Macmillan Publishers Ltd: Zerboni, L., Sen, N., Oliver, S.L., Arvin, A.M., 2014. Molecular mechanisms of varicella zoster virus pathogenesis. Nat. Rev. Microbiol. 12 (3), 197–210, copyright 2014.*

such as encephalitis and meningitis. Before the approval of the chickenpox vaccine, the virus caused an estimated 11,000 hospitalizations each year in the United States. Interestingly, more deaths occurred in previously healthy individuals, rather than the immunocompromised. With the exception of newborns, the fatality rate increases with the age of the individual. From 1970 to 1994, 41% of the 2262 people who died of varicella were ≥20 years of age.

VZV establishes latency in posterior root ganglia (also called dorsal root ganglia) (Fig. 13.7). A posterior root ganglion is a cluster of nerve cell bodies found immediately outside of the spinal column whose axons are normally responsible for transmitting sensory signals to the spinal cord. The virus makes contact with the neurons during the viremias or from local replication in skin lesions. Like HSV-1 and HSV-2, VZV undergoes retrograde transport to establish latency in the nucleus of the neuron, where it remains indefinitely. The precise stimuli that cause reactivation of the virus are still under investigation, but immunosuppression and increasing age correlate with an increased risk of reactivation. A third of individuals will reactivate VZV during their lifetimes, and half of those that live until age 85 will experience a recurrence.

When VZV reactivates, it causes shingles. The clinical condition is officially known as **herpes zoster**, from the Greek word *zoster/zooster*, referring to a thick ancient metal belt that encircled half the trunk of the body. When VZV reactivates, it travels down the sensory neuron to the initial site of infection, usually on the trunk of the body. Here, a preeruptive prodrome lasting 1–10 days (average of 2 days) is characterized by pain and tingling at the site of reactivation. The acute eruptive phase begins as an area of redness before characteristic vesicles develop on the skin, accompanied by severe pain. These herpetic lesions begin in one area and tend to spread in a wide patch on one side of the body (Fig. 13.8), resembling the wide *zoster* belt of antiquity. Reactivation can occur within several neurons simultaneously; up to seven at once have been reported. Symptoms tend to resolve by 2 weeks, although skin lesions take longer to heal.

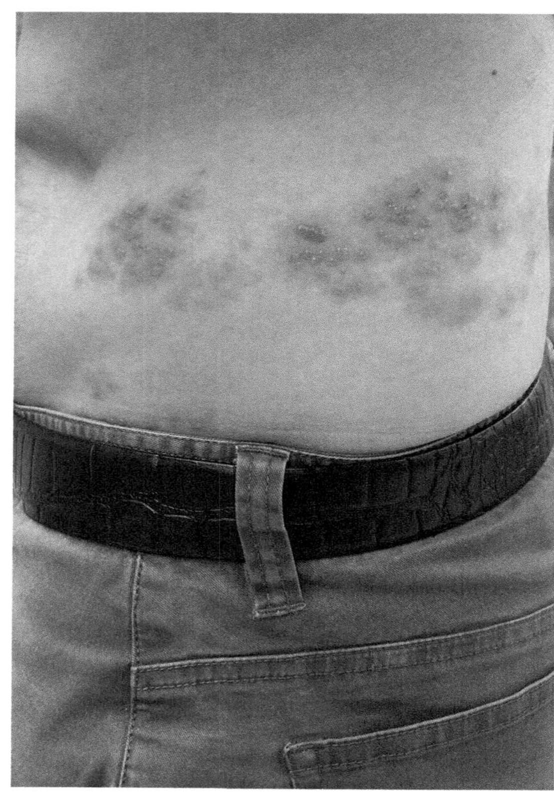

FIGURE 13.8 **Herpes zoster.** Image showing the reactivation of VZV on the left side and back of an affected individual. Note the spreading of the rash in one region, rather than a diffuse distribution over the entire body. *Image courtesy of Dr. Michael Beach.*

Herpes zoster ophthalmicus occurs in 10–25% of recurrent infections. In this condition, VZV reactivates from the trigeminal neuron found in the head and reestablishes infection in the eye area, often accompanied by a rash and pain in accompanying areas. This can result in damage to the cornea (keratitis) or inflammation of the eye tissue (uveitis). These can be severe and result in scarring, with a chance of chronic eye issues or vision loss.

Postherpetic neuralgia (PHN) is a possible complication of herpes zoster. *Neuralgia* refers to severe pain that occurs along a nerve, and 5% (for those <60 years old) to 20% (for those ≥80 years old) of people who have herpes zoster will have sensory nerve damage that results in PHN. PHN is characterized by severe persistent or intermittent pain that can last for weeks, months, or even years.

Acyclovir, valacyclovir, and famciclovir are effective in reducing the replication of VZV and are prescribed for high-risk individuals that contract the virus and develop varicella, including adults and pregnant women. These nucleoside analogs reduce the symptoms and duration of herpes zoster and reduce the frequency of PHN.

Considering that varicella can have high mortality rates in certain populations, and herpes zoster is a painful disease that occurs in a third of the US population, vaccine efforts were initiated to develop a vaccine against VZV. Three different VZV vaccines, all delivered subcutaneously, have been approved by the US Food and Drug Administration:

1. In 1995, a live attenuated virus vaccine was approved for people 12 months of age and older. The vaccine contains a minimum VZV titer of 1350 PFU.
2. In 2005, a combined Measles-Mumps-Rubella-Varicella (MMRV) vaccine was approved for people 12 months of age through 12 years of age. The vaccine contains the same attenuated viruses and in the same concentration as the stand-along MMR vaccine, along with 9772 PFU of the attenuated VZV to ensure adequate antibody responses.
3. In 2006, a herpes zoster vaccine was approved for people age 60 and older, which was extended to people age 50 and older in 2011. This vaccine contains the same attenuated VZV as the varicella vaccine but with a minimum of 19,400 PFU of VZV per vaccine. The higher dose is to account for waning immune responses in older individuals.

The varicella vaccine is very effective and thought to provide long-lived immunity. Ninety-seven percent of children between 12 months and 12 years of age generate detectable antibody against VZV after a single dose of the vaccine. The vaccine is estimated to protect 70–90% of children from infection and 90–100% against moderate or severe disease. In adults, 99% of people aged 13 years and older develop antibodies after two doses. The herpes zoster vaccine has also been effective in reducing the incidence of herpes zoster by 51%. It has also reduced the severity of herpes zoster in those that developed the disease, reducing PHN rates by 66%.

Because all three vaccines involve an attenuated virus, the vaccines are generally not recommended in immunocompromised individuals. For children, two doses of the varicella vaccine are recommended, the first between 15 and 18 months and the second between 4 and 6 years of age. Either or both of these doses can be in the combined MMRV vaccine, although the first dose is recommended to be the stand-alone varicella vaccine due to a slight increase in febrile seizures using the combined MMRV vaccine in infants (but not older children). Any person that did not receive the varicella vaccine series or contract chickenpox naturally should also receive the vaccine series, especially since the virus is more virulent in adults. It is recommended that adults receive a single dose of the herpes zoster vaccine at age 60 or older as a booster for natural or vaccine-induced immunity.

13.2.3 Epstein–Barr Virus (EBV/HHV-4)

EBV, officially known as HHV-4, is one of the most common human viruses in the world. EBV was named after the first human lymphocyte cell line, named "EB" after the last names of Anthony Epstein and Yvonne Barr, a researcher and his lab technician who performed the cell culture. These two individuals, along with technician Bert Achong, published a seminal paper in 1964 that described the novel herpesvirus found in the EB cells. This virus was eventually referred to as "Epstein–Barr virus" after the cell line from which it was isolated.

In the United States, around 89% of people seroconvert against EBV before the age of 20. In children, EBV presents as one of many common childhood illnesses with typical coldlike symptoms. In teenagers or adults, however, EBV causes 90% of the cases of **mononucleosis**, also known as "mono" or the "kissing disease," which is characterized by a fever of 101–104°F, a severe sore throat, headaches, and significant fatigue. **Lymphadenopathy** (swollen lymph nodes) and **hepatosplenomegaly** (enlargement of the liver and spleen; "hepato" refers to the liver, "spleno" refers to the spleen, and "megaly" is Greek for "large") are common due to the increased proliferation of lymphocytes within these organs during mononucleosis. **Splenomegaly** (enlargement of the spleen) can be problematic because the organ can rupture with mild trauma, or even spontaneously. Rupture occurs in 1–2 per 1000 people with infectious mononucleosis.

EBV is primarily spread through saliva and can be transmitted by kissing or sharing items such as utensils, drinking glasses, or toothbrushes. In children, EBV can be spread by the sharing of toys that young children may have put in their mouths. EBV can also be transmitted through blood, semen, and organ transplants. An infected person can transmit the virus for weeks while infected, shedding the virus into saliva from infected cells in the **oropharynx**, the area in the back of the throat that includes the tonsils (see Fig. 10.1). The majority of infected cells within tonsils are B lymphocytes, although other cells in the oropharynx, such as epithelial cells, may produce virus during productive infection.

Symptoms typically resolve in 2–6 weeks but can last for months, in some cases. Complications are generally rare but can involve the hematological system (blood and bone marrow), nervous system, or the heart.

As a member of the *Gammaherpesvirinae* subfamily, EBV establishes latency in lymphocytes, specifically memory B cells. As discussed in Chapter 9, "Viruses and Cancer," reactivation of EBV in immunocompromised individuals can lead to the development of Burkitt's lymphoma, Hodgkin's lymphoma, gastric carcinomas, nasopharyngeal carcinoma, and posttransplant lymphoproliferative disease, a B cell tumor where B cells proliferate profusely in people who are intentionally immunosuppressed due to transplant rejection medications.

Several studies have suggested a link between EBV and systemic autoimmune diseases. Patients with systemic lupus erythematosus, rheumatoid arthritis, and Sjögren's syndrome have been shown to have higher titers of EBV in peripheral blood mononuclear cells and increased amounts of antibody against the virus. EBV-infected B cells have been found in their tissues, producing antibody that contributes to the autoimmune disease. So far a direct causal relationship has not been definitively shown, but it is very likely that EBV reactivation at least contributes to the ongoing pathology in these autoimmune diseases. This could occur by the reactivation of EBV-infected B cells to produce disease-causing antibody, or through the immune responses against EBV causing an inflammatory environment that enhances the disease and causes flare-ups. It is also a possibility that the autoimmune patient's immune system is altered in a way, either genetically or because of the disease, that allows more EBV reactivations.

13.2.4 Human Cytomegalovirus (CMV/HHV-5)

Human cytomegalovirus (HCMV), officially known as HHV-5, is a member of the *Betaherpesvirinae* subfamily of viruses characterized by slow replication and clinically asymptomatic infections in healthy individuals. HCMV has the largest genome of any of the HHVs—230 kb with an estimated 160–200 protein-encoding genes—and so possesses many mechanisms to manipulate the infected cell and interfere with host immune responses while ensuring its own replication. The name cytomegalovirus comes from the Greek roots *cyto* and *megalo,* meaning "big cell," because cytomegaloviruses generate large **cytomegalic inclusion bodies**, also known as "owl eyes," that function as a histological sign of disease (Fig. 13.9). These form in the nucleus of infected cells as a result of nuclear remodeling during replication.

Although HCMV infects only humans, CMV-like viruses have been isolated from a variety of mammalian hosts, including other primates and rodents. HCMV is spread through close contact with infectious bodily fluids, such as saliva, urine, blood, semen, or breast milk. The virus typically initiates primary infection in the mucosal epithelium at the site of entry, and viremia occurs as the virus infects and replicates within monocytes and CD34+ immature leukocytes (blood cells) of the monocyte lineage. Infection can last for months within these blood cells, which spread the virus to other organs and to the areas where shedding occurs, such as the salivary glands, kidney, or mammary glands. Interestingly, most cells of the body are susceptible but not permissive to infection, meaning that although the virus can bind and enter, the cell lacks the intracellular proteins and molecules that allow replication to occur. HCMV latency occurs in monocytes and their progenitors in the bone marrow, and the virus reactivates when the monocytes differentiate into macrophages and dendritic cells within the tissues. As with the other herpesviruses, low levels of persistent infection can lead to intermittent shedding in people with latent infections as well, but a robust immune response prevents recurrent infection.

Healthy children and adults generally do not display symptoms when infected with HCMV. When symptoms are present, they are mild and resemble the many other colds that children contract: sore throat, fever, swollen glands, and malaise. However, 10–20% of infectious mononucleosis

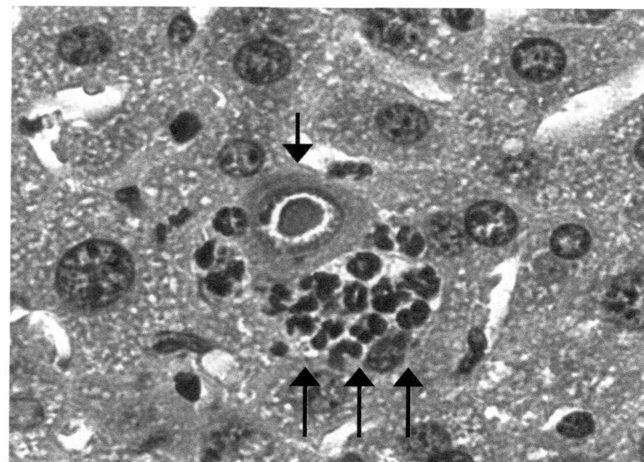

FIGURE 13.9 **Cytomegalic inclusion body.** Cytomegalovirus replication causes "owl eye" inclusion bodies within the nuclei of infected cells (*single arrow*), such as in this hematoxylin and eosin-stained liver section from an infected mouse. Noninfected hepatocyte nuclei are visible as large round purple circles. The multiple *arrows* highlight a focus of immune cells, which are much smaller, attacking the infected cell.

cases are due to HCMV. As discussed previously, the majority of infectious mononucleosis is caused by EBV, but infectious mononucleosis from HCMV is clinically indistinguishable, although lymphadenopathy and splenomegaly are less common with HCMV than with EBV.

In immunocompromised individuals, such as in AIDS patients or those undergoing organ or bone marrow transplants, HCMV is an opportunistic infection that can cause life-threatening disease. Reactivation of the virus is more problematic than primary infection and can lead to **retinitis** (inflammation of the retina), alveolar inflammation, inflammation of the gastrointestinal tract, and hepatitis, among other conditions.

HCMV primarily undergoes horizontal transmission in immunocompetent individuals, usually transmitted within the first decades of life. In certain locations of the world, such as in parts of Africa, the seroprevalence of HCMV is nearly 100%. In the United States, 50–80% of individuals are infected with HCMV before age 40. Not having seroconverted against HCMV can become problematic, however, in women that are not infected before childbearing age and contract the virus during pregnancy. HCMV is the only herpesvirus that undergoes **transplacental transmission**, from mother to fetus through the placenta, and so vertical transmission of the virus can also occur. In fact, HCMV is the most frequent congenital viral infection in the United States. The chance of transplacental transmission is highest when the mother has a primary infection during pregnancy, and around 40% of women of childbearing age are seronegative for HCMV. One to four percent of women that are seronegative for HCMV will contract the virus during pregnancy, and a third of these will pass the virus to their fetus. The main modes of transmission to seronegative pregnant women are through sexual intercourse or from other children in the household that become infected.

Of pregnant women who were already seropositive for HCMV before pregnancy, about 1–2% transmit the virus to their fetus. Although the transmission rate is lower for seropositive than seronegative women, this accounts for a significant number of HCMV-infected newborns.

Ninety percent of infected children show no symptoms and are not affected by the virus, but 1–2% of newborns who show symptoms will have permanent problems. These include deafness, blindness, physical disabilities, and varying degrees of mental retardation and cognitive delays. According to the CDC, about 0.13% of overall live births have permanent effects of HCMV infection, for a total of around 4000–5000 births each year in the United States.

Considering the morbidity and mortality associated with the virus, several different vaccine efforts are underway to prevent primary or recurrent infection in those of highest risk, particularly women of childbearing age and patients awaiting organ transplants.

13.2.5 HHV-6A, HHV-6B, and HHV-7

In addition to HCMV, the *Betaherpesvirinae* subfamily includes HHV-6 and HHV-7 (Fig. 13.10A). These viruses have smaller genomes than HCMV, around 162 and 153 kb, respectively, with about 85 gene products. HHV-6 is divided into two separate species, HHV-6A and HHV-6B. Although the two viruses are ~95% similar, they are distinct viruses that possess different cellular tropism, antigenicity, and pathogenesis. It is difficult to use serology to differentiate the two viruses due to the generation of cross-reactive antibodies.

All three viruses infect T lymphocytes, particularly mature CD4 T cells. In these cells, the viruses cause "balloonlike" cytopathic effects (see Fig. 7.6A). HHV-6A and HHV-6B can also infect a variety of other human immune cells, including dendritic cells, monocytes/macrophages, and NK cells. Both viruses are neurotropic: in a study examining brain samples taken from deceased individuals, 85% of the people had detectable HHV-6 DNA in their brain samples: 10% had HHV-6A DNA only, 57.5% had

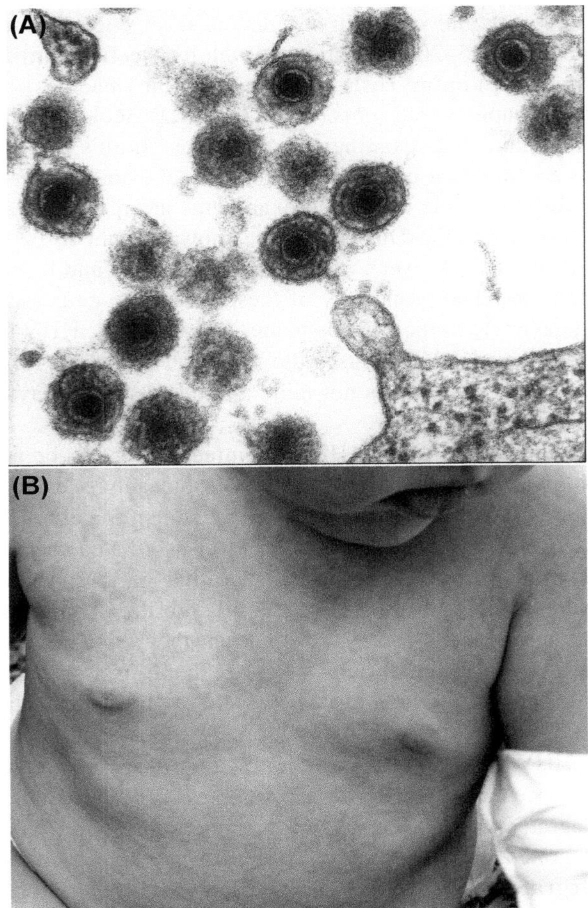

FIGURE 13.10 **HHV-6 and HHV-7.** (A) HHV-6 virions. *(Photo courtesy of NCI/Bernard Kramarsky.)* (B) Maculopapular rash of roseola infantum on the chest of an infant.

HHV-6B DNA only, and 17.5% had both HHV-6A and HHV-6B DNA. HHV-6 establishes latency in hematopoietic bone marrow stem cells, and HHV-7 DNA has been found in circulating lymphocytes not undergoing lytic infection.

HHV-6 and HHV-7 are ubiquitous viruses. Like HCMV, the salivary glands are major reservoirs of virus during infection due to the replication of the viruses there. Consequently, high titers exist in the saliva of seropositive individuals. All newborns receive passive antibodies against HHV-6 from their mother through the placenta, further emphasizing the prevalence of this virus. Infants begin seroconverting as soon as maternal antibodies wear off, and nearly all children are seropositive by 2 years of age, thought to contract the virus from parents or siblings. HHV-7, on the other hand, tends to be contracted slightly later in life, beginning at age 2, with most children seropositive by the teenage years.

HHV-6A is an orphan virus, meaning that a clinical condition has not definitively been associated with the virus. On the other hand, HHV-6B and HHV-7 cause childhood illness. Infection begins abruptly with a high fever, usually over 102°F, that lasts 3–5 days and is accompanied by a runny nose and irritability. After the fever begins to subside, about 15–20% of patients develop **roseola infantum**, a **maculopapular rash** composed of flat (macules) and raised (papules) red spots (Fig. 13.10B). Roseola infantum is also known as exanthema subitum or "sixth's disease," in reference to it being the sixth classical exanthem (rash) of children. (The numbered exanthems are measles, scarlet fever, rubella, Dukes' disease (now thought to be the same as scarlet fever), erythema infectiosum, and roseola infantum.) The propensity of these viruses to cause roseola gave rise to their genus name, *Roseolovirus*. HHV-6B, rather than HHV-7, causes the majority of roseola infantum cases. The rash is not itchy and takes several days to resolve.

Complications from HHV-6B infection tend to be neurological, due to the virus's neurotropic nature. Primary infection with HHV-6B or HHV-7 accounts for a quarter to a third of the **febrile seizures**, or fever-associated seizures, of children less than 2 years of age that go to a hospital emergency room for treatment. While febrile seizures have no complications in 95% of children, they are a common cause of **status epilepticus**, a seizure that lasts more than 5 min. Studies investigating whether these viruses can cause lasting brain damage in this setting are underway.

HHV-6A and HHV-6B have mammalian telomere sequences at the ends of their genomes that allow them to integrate into the telomeres of human chromosomes. When integrated, the viruses are referred to as **chromosomally integrated** HHV-6A (ciHHV-6A) and ciHHV-6B. Because the viruses have a broad tropism, integration can occur in several different cell types during infection. ciHHV-6A and ciHHV-6B are found in germline cells in about 1% of the population, meaning that the viruses could theoretically be transmitted vertically to offspring through a sperm or egg with DNA containing the integrated virus. During the process of fertilization, the DNA from one sperm and one egg combine to form a complete set of chromosomes, and if ciHHV-6A or ciHHV-6B was integrated into one of those chromosomes, the result would be that every cell of the resulting offspring would also have the integrated virus present within its DNA.

While most cases of the virus resolve in children, an association between HHV-6B and certain conditions has been noted. The virus reactivates in half of patients that receive bone marrow or solid organ transplants, many times from chromosomally integrated virus present in the transplant. This can result in rejection of the organ or in central nervous system diseases, including encephalitis and cognitive decline. There is also evidence that the virus correlates with multiple sclerosis, although a causative link has not yet been determined. Similarly, studies examining an association between HHV-6B and the progression of HIV are also ongoing.

13.2.6 Kaposi's Sarcoma-Associated Herpesvirus (KSHV/HHV-8)

KSHV (officially known as HHV-8) is the second human virus in the *Gammaherpesvirinae* subfamily and the most recently discovered HHV, in 1994. Like EBV, the virus infects and becomes latent in B lymphocytes and is associated with malignancies, as described in Chapter 9, "Viruses and Cancer." Discovered in 1994, KSHV is the most recent herpesvirus to be identified, named after the type of tumor from which the virus was isolated, Kaposi's sarcoma.

Unlike other herpesviruses, KSHV seroprevalence varies considerably depending upon the region of the world. The virus is most prevalent in sub-Saharan Africa, where over 50% of the population is seropositive. In this locale, the virus is most often seen in children, who likely spread it through saliva, although the virus can be transmitted sexually as well. KSHV exhibits intermediate prevalence in the Mediterranean, Middle East, Caribbean, and parts of South America, where 5–20% of people are infected. In North America, northern Europe, and Asia, the seroprevalence is low (1–7%), and transmission correlates with sexual activity. The virus is most commonly spread via homosexual activity, although heterosexual activity can also transmit the virus. Much higher amounts of virus are found in saliva than genital secretions or semen, suggesting that deep kissing may be involved in the transmission of the virus during sex. The virus appears to be transmitted inefficiently in blood, if at all.

Like several herpesviruses, KSHV infection is largely inapparent in immunocompetent individuals. Apparent infection of children is associated with a high fever followed by a maculopapular rash. Infections of adults result in mild symptoms, swollen lymph nodes, fatigue, and a localized rash. The greatest risk of viral infection is to immunocompromised individuals, who more frequently develop Kaposi's sarcoma. Patients undergoing organ transplantation have a 500–1000 times greater risk of developing Kaposi's sarcoma than the general population, and HIV+ individuals have a 50% risk of developing Kaposi's sarcoma.

13.3 MOLECULAR VIROLOGY

As described previously, the nine human viruses of the *Herpesviridae* family are subdivided into three subfamilies, the *Alphaherpesvirinae, Betaherpesvirinae,* and *Gammaherpesvirinae*, based upon their growth properties and targets of latency. Regardless of the subfamily in which it resides, all herpesviruses share some common characteristics. They are all large, intricate T = 16 viruses with a ~100 nm icosahedral capsid composed of 150 hexons and 12 pentons, for a total of 162 capsomers (Fig. 13.11). The capsid, which is composed of four major proteins and several minor proteins, is assembled in the nucleus and surrounded by an amorphous mass of 15–40 different proteins termed the **tegument** (Fig. 13.12A). The tegument proteins are generally found in two layers, an inner capsid-associated layer and an outer envelope-associated layer, and account for 40% of the mass of the virion. The virion derives its envelope from one of the membranes within the endomembrane system, and at least 11 different glycoproteins are integrated into the envelope. The entire virion is around 220 nm in size.

Each herpesvirus has a linear dsDNA core, although the genome size varies between approximately 125 and 230 kb, with VZV having the smallest genome and HCMV the largest. The number of translatable genes (open reading frames, or ORFs) is proportional to the size of the genome; as the largest, HCMV has ~200 genes. The gene-encoding regions of the herpesvirus genomes are referred to as "unique long" (U_L) or "unique short" (U_S) regions (Fig. 13.12B). There are repeated nucleotide sequences within the herpesvirus genomes, either terminal (at the end of the genome) or internal, and these can be direct repeats or inverted repeats. Found within these repeated sequences are promoters for UL or US genes, origins of DNA replication, a few protein-encoding genes, and DNA encoding latency-associated transcripts (LATs), which will be discussed later.

Like eukaryotic genes, most herpesvirus genes each have an individual promoter containing regulatory sequences and a TATA box, a transcription initiation site, and a polyadenylation signal sequence (see Chapter 4, "Virus Replication," for a review of these terms and the process of transcription). Host RNA polymerase II transcribes the majority of herpesvirus genes, although a few genes from EBV are transcribed by RNA polymerase III. Unlike eukaryotes, it is common for herpesviruses to have multiple genes within one section of DNA (**polycistronic** DNA), and often they end at the same polyadenylation sequence. About 10–20% of herpesvirus genes are spliced, more commonly occurring in *Betaherpesvirinae* and *Gammaherpesvirinae* than in *Alphaherpesvirinae* subfamily members.

The HHV all share homology among 40 essential proteins that are required throughout the process of replication. These include five envelope proteins, and several enzymes and proteins involved in DNA replication.

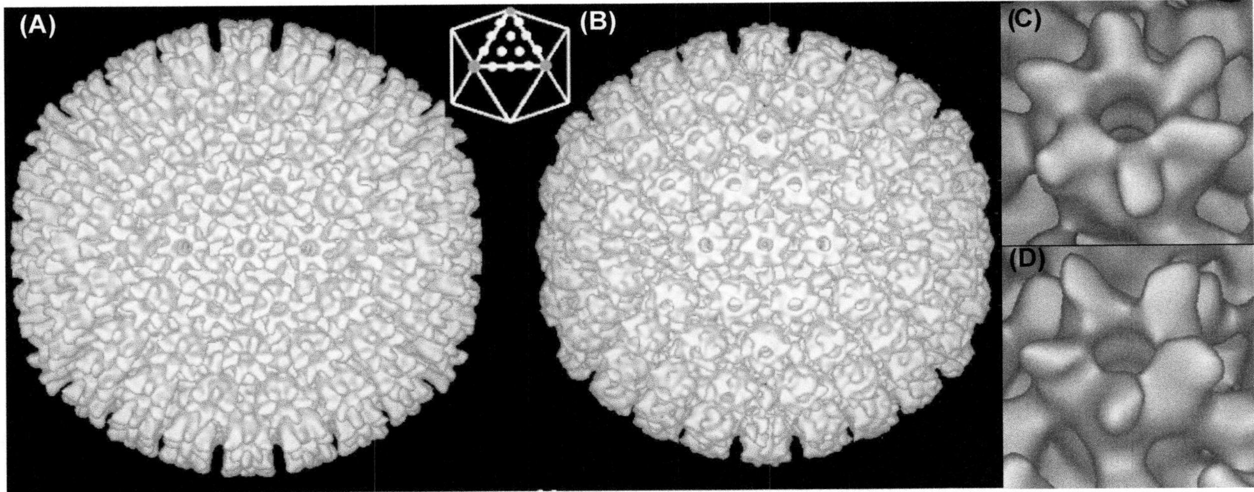

FIGURE 13.11 **Herpesvirus capsid structure.** Icosahedral herpesvirus capsids, including those for HCMV (A) and HSV-1 (B), are composed of pentons at the fivefold axes of symmetry and hexons on the faces of the icosahedron. As a reminder, the icosahedron figure in the middle has green circles at the fivefold axes. (C) and (D) show an individual hexon and penton capsomer from HCMV. *Reprinted from Butcher, S.J., et al., 1998. Structure of the human cytomegalovirus B capsid by electron cryomicroscopy and image reconstruction. J. Struct. Biol. 124 (1), 70–76, copyright 1998, with permission from Elsevier.*

(A)

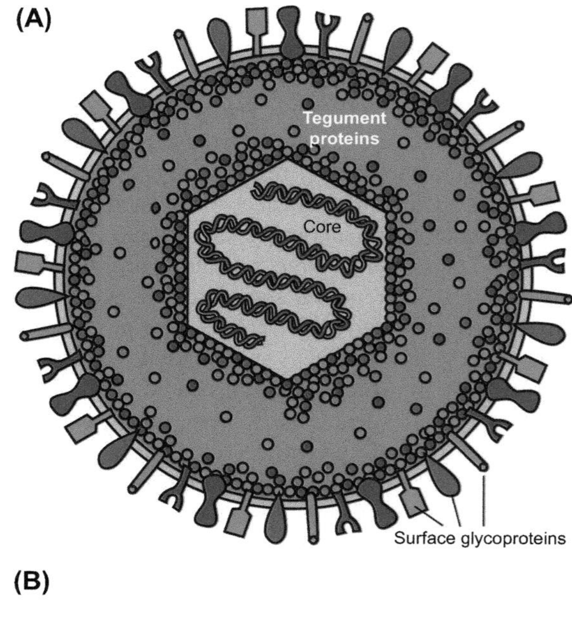

(B)

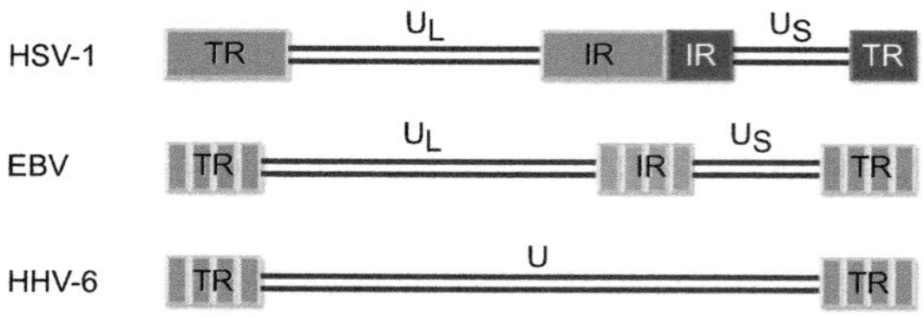

FIGURE 13.12 Herpesvirus virion structure. (A) Herpesviruses are large, enveloped icosahedral viruses with dsDNA genome cores surrounded by two layers of tegument proteins, an inner layer located around the capsid and an outer layer directly beneath the envelope. The viral envelope contains several glycoproteins projecting from the surface. (B) The gene-encoding regions of the herpesvirus genomes are known as "unique long" (U_L) and "unique short" (U_S) regions, referring to the length of DNA in between terminal or internal repeated sequences of DNA, known as TR and IR, respectively. Shown here are three examples of the architecture of herpesvirus genomes.

Their genomes also encode a range of proteins that mimic host proteins and intricately interfere with the immune response against the viruses.

The entire process of herpesvirus replication takes approximately 18–20 h to complete for alphaherpesviruses and 48–72 h for betaherpesviruses, which replicate more slowly. Replication involves several envelope glycoproteins that mediate attachment and fusion of the envelope with the plasma membrane. The capsid and surrounding tegument proteins are transported to the nuclear envelope of the cell, where uncoating occurs. Transcription and DNA replication occur in the nucleus, as does assembly of new capsids following protein translation. The capsids bud through the nuclear envelope and acquire a final envelope from one of the components of the endomembrane system, being released from the cell through exocytosis. Replication has been more thoroughly investigated in the herpesviruses that were first discovered, so the following subsections cover the detailed replication steps using HSV-1 as a model and noting major differences in other herpesvirus family members. As with all viruses, certain stages have been better characterized than others.

13.3.1 Attachment, Penetration, and Uncoating

Herpesviruses have many different glycoproteins embedded into their envelopes, over 11 for HSV-1 and HSV-2. Approximately 659 protein spikes in total extend from the surface of each virion. Unlike other viruses, such as influenza, that have a single glycoprotein that mediates both attachment and fusion, a subset of herpesvirus glycoproteins is involved in the process. First, glycoprotein B (gB) and gC bind reversibly to glycosaminoglycans (GAGs) extending from the cell surface, including heparan sulfate and chondroitin sulfate (Fig. 13.13). Because GAGs are found ubiquitously on the surface of cells, this helps to explain why some herpesviruses can attach to, but not enter, many different types of cells. Binding of gB and gC brings gD into close contact with one of the specific cell surface receptors to which it binds (Table 13.2).

HSVs are able to interact with several receptors. The first is a protein known as the herpesvirus entry mediator (HVEM). This protein is expressed in a variety of organs, although not brain or muscle, and shows high expression in T cells, B cells, and monocytes. Both HSV-1 and HSV-2 are able to bind HVEM. In addition, HSVs can bind members of the nectin family. Nectins are *inter*cellular adhesion molecules that form dimers on the surface of the cell and interact with other nectin dimers on the surface of adjacent cells. They play a role in the adherens junctions and tight junctions that anchor cells to each other. Nectins are broadly expressed in the body; they are found on epithelial cells and neurons, two cell types that are targeted by herpesviruses. Both HSV-1 and HSV-2 bind nectin-1, while HSV-2 alone is able to bind nectin-2. A final receptor that is bound by HSV-1 only is a modified GAG known as 3-O-sulfated heparan sulfate (3-OS HS). In this case, HSV-1 uses 3-OS HS, which is expressed on a variety of different cell types, for both initial binding and entry. Interestingly, although the gD of HSV-2 does not use 3-OS HS for entry, blocking this molecule inhibits attachment and fusion. The use of 3-OS HS as a receptor for HSV-1 and HSV-2 infection is currently being further characterized.

The binding of gD to its receptor allows the association of gD with gB, gH, and gL. All four proteins are required for fusion. Of the four, gH has portions that resemble a fusion peptide and likely mediates the fusion of the viral envelope with the plasma membrane.

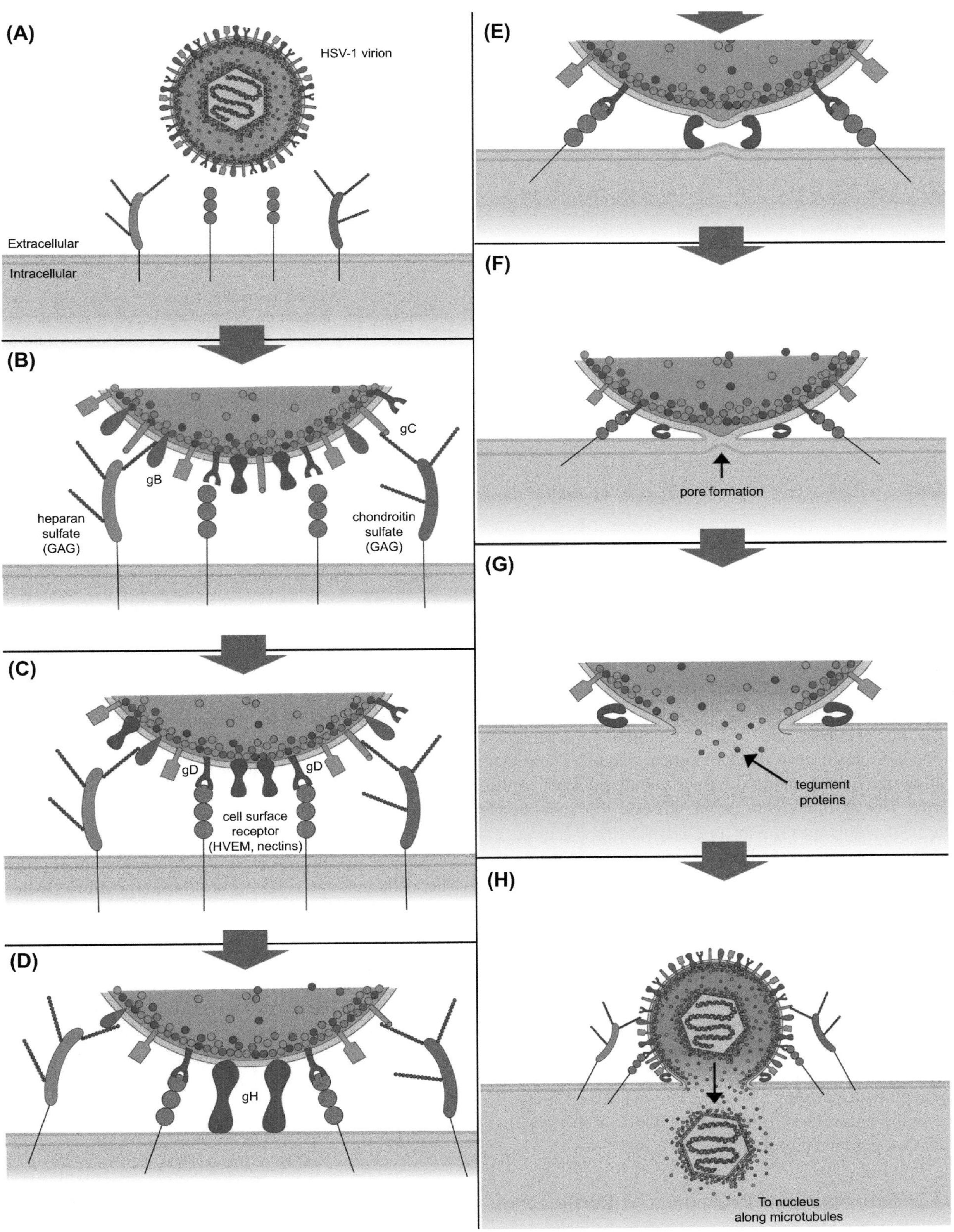

FIGURE 13.13 **Herpesvirus entry.** (A) Herpesviruses possess several envelope glycoproteins that facilitate the process of entry into the cell. (B) For HSV-1, gB and gC bind to glycosaminoglycans (GAGs), including heparan sulfate and chondroitin sulfate, on the cell surface. This brings gD into close contact with a specific cell surface receptor, such as HVEM, nectin-1, or 3-OS heparan sulfate (C). Binding of gD to the receptor allows the association of gD with gB, gH, and gL. All four proteins are required for fusion, but gH possesses a possible fusion peptide (D). The gH protein undergoes a conformational change, bringing the virion envelope and plasma membrane closer together (E, F) until they fuse together, creating a pore (G). The capsid is released into the cytoplasm (H), escorted by tegument proteins that assist in the binding of the capsid to the microtubule transport machinery that transports the capsid to the nuclear envelope.

TABLE 13.2 Human Herpesviruses

Name	Genome Size (kb)	Receptors
HHV-1 (HSV-1)	152	HVEM, Nectin-1, 3-OS HS
HHV-2 (HSV-2)	154	HVEM, Nectin-1, Nectin-2
HHV-3 (VZV)	125	Insulin degrading enzyme (IDE), mannose 6-phosphate receptor (CI-M6PR)
HHV-4 (EBV)	172	CD21, MHC Class II, CD35, β1 integrins (epithelial cells)
HHV-5 (HCMV)	230	Epidermal growth factor receptor (EGFR), platelet-derived growth factor receptor (PDGFR) α, BST2/tetherin, others
HHV-6A	156	CD46
HHV-6B	162	CD134, CD46
HHV-7	153	CD4, heparan sulfate
HHV-8 (KSHV)	138	αVβ3 integrin, DC-SIGN

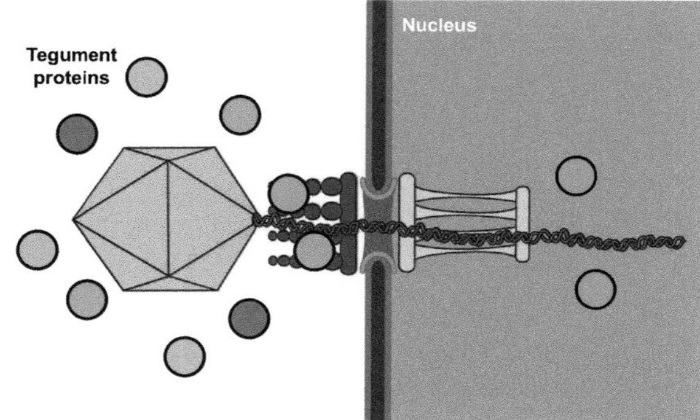

FIGURE 13.14 **Capsid uncoating.** Tegument proteins attach to cellular microtubule motor proteins dynein and dynactin, allowing for the transport of the capsid to the nucleus. The capsid is destabilized upon binding to the filaments extending from the nuclear pore complex, and tegument proteins assist in the transport of the dsDNA genome into the nucleus, where transcription and DNA replication occur.

Herpesviruses can also enter the cell via endocytosis. Within the endocytic vesicle, the four glycoproteins mentioned above are thought to function in the same manner to induce fusion of the viral envelope with the membrane of the endocytic vesicle.

The nucleocapsid and tegument proteins are released into the cytoplasm once deenvelopment occurs. The intact capsid is transported along the microtubule network to the nucleus. This is thought to occur through the attachment of an inner tegument protein (possibly UL36 and/or UL37 in HSV) to the microtubule motor protein dynein and its accessory protein dynactin. These motor proteins normally function by "walking" along microtubules, carrying intracellular cargo within vesicles. Herpesvirus capsids are thought to be transported to the nuclear pores by hijacking this cellular process. Once at the nucleus, the capsid binds to the cytoplasmic filaments extending from the nuclear pores. This destabilizes the capsid, and uncoating occurs as the viral DNA is transported into the nucleus through the pore, assisted by additional tegument proteins (Fig. 13.14). Some tegument proteins also enter the cytoplasm and will assist in the initiation of transcription. Once in the nucleus, the dsDNA genome circularizes.

13.3.2 Expression of Proteins and Replication of Nucleic Acid

Transcription of viral mRNA occurs in three stages: immediate early (IE) or α, early (E) or β, and late (L) or γ gene transcription. The IE genes become transcribed 2–4 h after infection takes place. The tegument protein VP16 recruits host general transcription factors and RNA polymerase II to the promoters of the six IE genes. Other host transcription factors, such as Sp1, are also recruited to viral promoter sequences and enhance transcription by RNA polymerase II. Some IE mRNAs are spliced following transcription, and all transcripts are transported to the cytoplasm for translation by host ribosomes. The translated IE proteins return to the nucleus, where they initiate transcription of E genes around 4–8 h following infection. E transcripts are translated by ribosomes, after which a number of these proteins return to the nucleus to become involved with the replication of the viral DNA. These include the viral DNA polymerase, the helicase-primase enzyme, and several DNA-binding proteins that together initiate DNA synthesis. Other notable E genes support the synthesis of nucleotides to ensure an adequate supply for replication. The DNA genome is replicated through **rolling circle replication** (Fig. 13.15). This process uses one of the circularized DNA strands as a template to replicate the other strand repeatedly, much in the same way a roller stamp uses one rolling template to create a long pattern. The new strand is used to create the second strand of the DNA duplex, and the **concatemers**, or genome copies in tandem, are then separated by enzymes before being packaged into the capsid. (The word "concatenate" means "to link together in a chain," the same way the genome copies are linked together during replication into concatemers.)

13.3.3 Assembly, Maturation, and Release

E proteins induce the expression of the L genes, which are transcribed from the replicated DNA. L gene products are all involved in assembly of the virion and include capsid proteins, membrane glycoproteins, and scaffold proteins

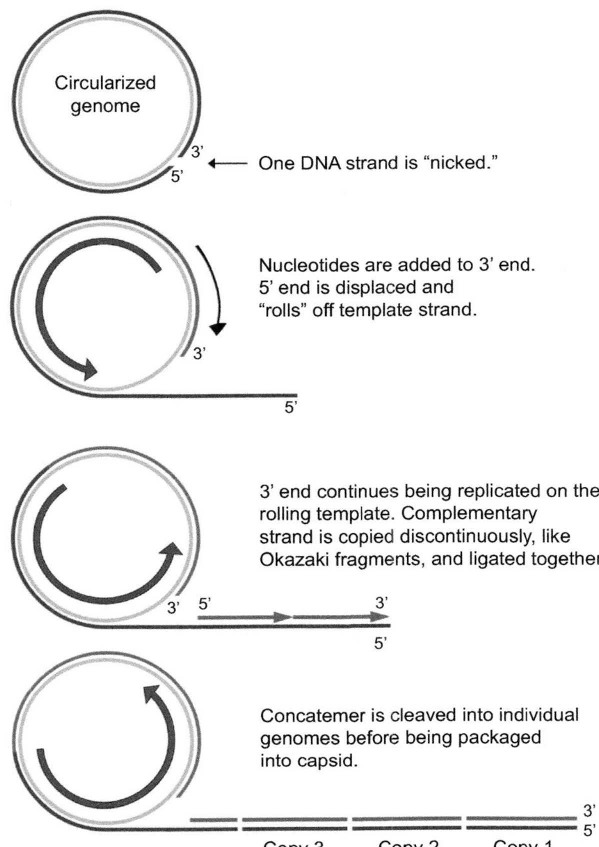
Circularized genome
3'
5'
One DNA strand is "nicked."
Nucleotides are added to 3' end.
5' end is displaced and
"rolls" off template strand.
3'
5'
3' end continues being replicated on the
rolling template. Complementary
strand is copied discontinuously, like
Okazaki fragments, and ligated together.
3' 5'
3'
5'
Concatemer is cleaved into individual
genomes before being packaged
into capsid.
3'
5'
Copy 3
Copy 2
Copy 1

FIGURE 13.15 **Rolling circle replication.** In rolling circle replication, one strand of the circularized genome is used as a template to create multiple copies of the other strand, which is then replicated. The double-stranded concatemer is cleaved into individual genomes before being packaged into capsids.

that assist with assembly. Over 30 L gene products end up as part of the final, mature virion, at least 11 of which end up as glycoproteins embedded in the envelope.

L proteins involved in the formation of the nucleocapsid, including scaffold proteins that assist in the assembly of the capsid, are transported back into the nucleus, where the capsid is assembled. In the process of maturation, the viral protease cleaves the scaffold proteins from the capsid, a step that is required before the replicated dsDNA can be packaged into the capsid.

The final nucleocapsid is too large to fit through the nuclear pores, so it undergoes an envelopment and de-envelopment process to escape from the nucleus (Fig. 13.16). Escorted by tegument proteins, it buds through the inner nuclear membrane into the intermembrane space, acquiring a membrane envelope in the process. The nuclear envelope is a two-membrane layer. Having passed through the first membrane, the viral envelope fuses with the outer nuclear membrane. This envelope remains part of the outer nuclear membrane as the naked nucleocapsid is released into the cytosol.

The nucleocapsid then travels to a location within the endomembrane system of the cell. Unlike some other enveloped viruses, herpesviruses do not obtain their envelope from the plasma membrane. The location varies depending upon the specific herpesvirus but includes the ER, ER-Golgi intermediate compartment, Golgi complex, or vesicles derived from the Golgi, including endosomes. The many envelope proteins assemble on the interior of this membrane, and budding of the nucleocapsid through the membrane forms the final envelope of the virion. The mature virion is released through exocytosis.

13.4 LATENCY

A hallmark of herpesviruses is that they undergo latency in host cells following productive infection. During latency, no new virions are produced by the infected cell. Herpesviruses express IE, E, and L gene products during lytic infection, but these genes are either not expressed or minimally expressed during latency, depending upon the virus.

During latency, HSV-1 and HSV-2 express high levels of latency associated transcripts (LATs), RNAs that are thought to be involved in maintaining latency. An unstable 8.3 kb primary transcript gives rise to two stable introns of 1.5 kb and 2 kb. LATs prevent activation of the IE genes that initiate the cascade that drives the cell into a lytic program.

Intriguingly, no proteins are translated from LATs, which begs the question as to how simple RNA transcripts could prevent viral genes from being transcribed. A possible answer to this question was revealed with the recent discovery of **microRNAs (miRNAs)**. miRNAs are small pieces of ssRNA, usually around 22 nucleotides in length, that bind to mRNAs in a complementary antisense fashion to prevent their transcription. miRNAs were first discovered in cells, but it has been determined that a handful of viruses also express miRNAs, including herpesviruses.

Several miRNAs have been discovered to be encoded within HSV-1 and HSV-2 LATs. The miRNAs form secondary hairpin structures that are cut out of the LATs by cellular enzymes in the nucleus (Fig. 13.17). The miRNAs are trimmed in the cytoplasm and associate with a protein complex, called the **RNA-induced silencing complex (RISC)**. The miRNA pieces target the RISC to its complementary sequence within the viral mRNA, and one of the RISC proteins cleaves the vmRNA. In this way, the vmRNA is degraded and does not become translated.

Several herpesvirus miRNAs have been found to degrade lytic gene transcripts, which results in the continuation of latency. For example, HSV-1 miRNA H2 is derived from the large LAT, and it binds vmRNA from the *ICP0* gene. The ICP0 protein is an IE protein that induces transcription of IE, E, and L genes. By blocking ICP0 mRNA, no ICP0 protein is produced, and therefore IE, E, and L genes

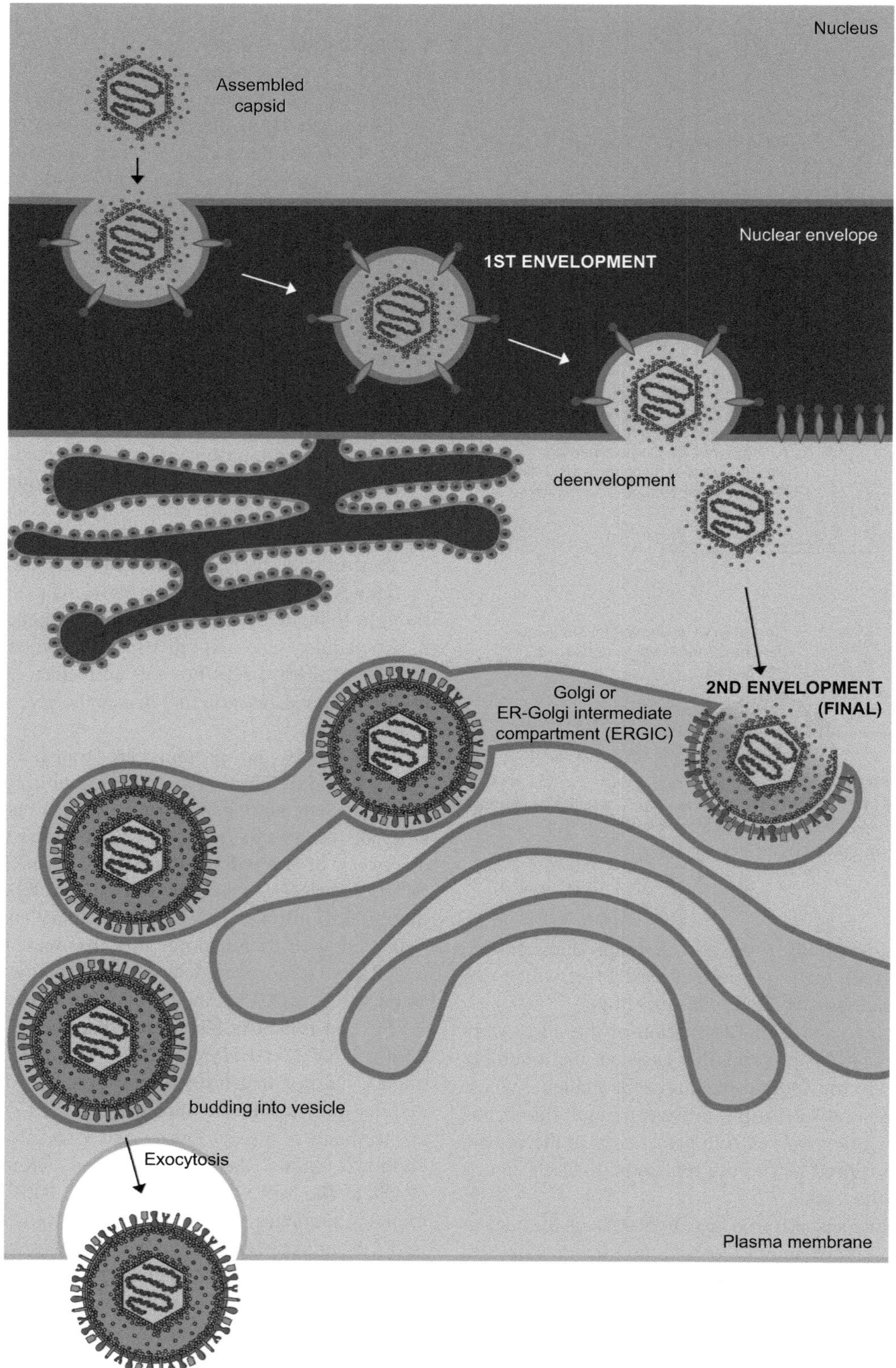

FIGURE 13.16 **Double envelopment, assembly, and release.** To pass from the nucleus into the cytoplasm, the nucleocapsid must traverse the double membrane of the nuclear envelope. To do so, it buds through the inner nuclear membrane, acquiring an envelope from it. Now in the intermembrane space, the temporary envelope fuses with the second nuclear membrane to release the capsid into the cytosol. Viral proteins assemble on one of the membrane organelles within the cell, from where the nucleocapsid acquires its final envelope. The virion is released via exocytosis.

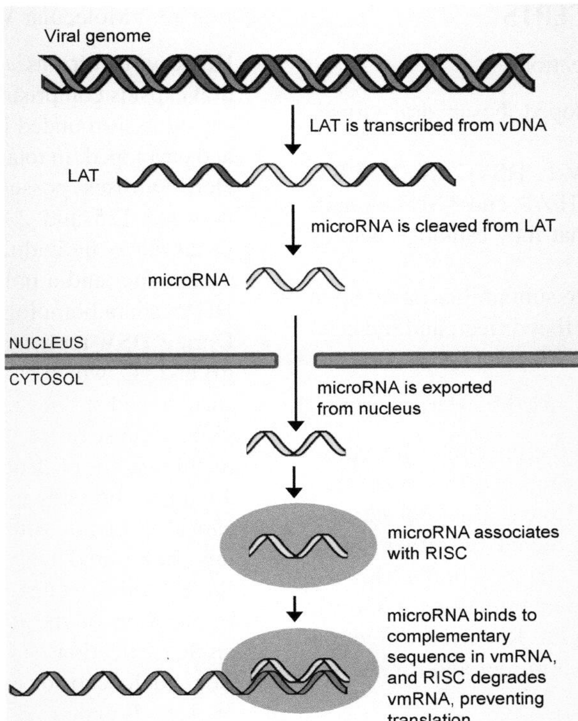

FIGURE 13.17 **microRNAs produced by herpesviruses.** Latency-associated transcripts (LATs) are encoded within the HSV-1 and HSV-2 genomes. Within the LATs are small sequences of RNA known as microRNAs (miRNAs) that are excised and leave the nucleus. In the cytosol, they are further processed and associate with the multiprotein RISC. The miRNA sequence targets the RISC to a specific location within vmRNA transcripts. Upon binding the vmRNA, the RISC cleaves the vmRNA, preventing translation. The RISC is also able to prevent the binding of ribosomes to the vmRNA, which also blocks translation. The result is that the virus-encoded miRNAs regulate the expression of viral proteins, and viral miRNAs also prevent cellular proteins from being translated.

are not turned on. The result is that the virus remains in latency. HSV-2 has a homologous miRNA that also inhibits ICP0 mRNA. The discovery of miRNAs within the LATs that retain the virus in a latent state by blocking the transcription of lytic genes may help to explain how latency is maintained within host cells.

The first viral miRNA ever discovered was found in EBV, and so far, miRNAs have been discovered in all HHVs except VZV and HHV-7. While HSV-1 and HSV-2 use miRNAs to maintain latency, other herpesviruses generate miRNAs that downregulate the translation of cellular proteins involved in the immune response against the virus, particularly NK cell responses and the production of Type 1 IFN. Current research continues to investigate how lytic and latent programs are regulated by the herpesviruses and how virally encoded miRNAs are involved in successful infection.

SUMMARY OF KEY CONCEPTS

Section 13.1 Herpesvirus Classification

- Herpesviruses are large, enveloped icosahedral viruses with large dsDNA genomes.
- There exist nine HHVs: HSV-1, HSV-2, VZV, EBV, HCMV, HHV-6A, HHV-6B, HHV-7, and KSHV. A hallmark of all herpesviruses is that they establish latency within cells of the body.
- Herpesviruses are classified into subfamilies based upon virus properties, the cell types they infect, and the cells in which they establish latency.

Section 13.2 Clinical Conditions Caused by Herpesviruses

- HSV-1 and HSV-2 infect epithelial cells and cause vesicles and ulcers. HSV-1 is more often associated with oral herpes while HSV-2 causes genital herpes, although both sites can be infected by either virus. Oral herpes infections result in latency in the trigeminal ganglia, and genital herpes infections establish latency in the sacral root ganglia. Recurrent infection results in similar symptoms, but milder and for shorter duration.
- VZV infects epithelial cells and causes chickenpox during primary infection, characterized by a fever and an itchy, centripetal rash. The virus becomes latent in dorsal root ganglia. Recurrent infection results in herpes zoster, better known as shingles.
- EBV causes 90% of the cases of mononucleosis, which can last for weeks to months. Reactivation of the virus is associated with a variety of different cancers, and several studies have suggested a link between EBV and systemic autoimmune diseases.
- HCMV is the leading cause of congenital infections and is the only herpesvirus able to undergo transplacental transmission. Infected newborns can present with deafness, blindness, physical disabilities, and cognitive delays. Other herpesviruses, including HSV-1, HSV-2, and VZV, also cause congenital or perinatal infections. HCMV also causes a minority of mononucleosis cases.
- HHV-6A is an orphan virus. HHV-6B and HHV-7 are associated with childhood illness and cause a rash known as roseola infantum. The HHV-6A and HHV-6B genomes are flanked by mammalian telomere sequences that allow integration into human chromosomes.
- KSHV causes subclinical infection in immunocompetent individuals. Immunosuppressed people have a high risk of developing Kaposi's sarcoma from KSHV reactivation.
- Herpesviruses frequently reactivate in immunocompromised individuals. This is a major concern and can lead to severe disease.

Section 13.3 Molecular Virology

- Herpesvirus virions are composed of ~100 nm icosahedral capsids comprising 150 hexons and 12 pentons. The capsid is surrounded by two layers of tegument proteins and enveloped. In total, the virion is ~220 nm in diameter.
- Herpesviruses possess large linear dsDNA genomes between 125 and 230 kb in length. Most herpesvirus genes have individual promoters, a transcription initiation site, and a polyadenylation signal sequence. All HHVs share homology between 40 essential proteins.
- During HSV replication, GAGs are bound by envelope gB and gC, which brings the virion gD into contact with specific cell surface receptors for entry. gD, gB, gH, and gL then cause fusion of the viral envelope with the plasma membrane for nucleocapsid release into the cytoplasm.
- Tegument proteins facilitate the transport of the nucleocapsid along microtubules to the nucleus, where uncoating occurs and DNA is transported into the nucleus. IE, E, and L genes are transcribed in distinct stages. Replication of the dsDNA genome occurs via rolling circle replication.
- The nucleocapsid is assembled and matures in the nucleus. It undergoes envelopment and de-envelopment to pass through the nuclear envelope, obtains an envelope from one of the membrane-bound organelles, and is released through exocytosis.

Section 13.4 Viral Latency

- No new virions are created during latency, and viral proteins are minimally translated, if at all.
- Most herpesviruses express miRNAs, small antisense pieces of ssRNA that bind to mRNA, to regulate latency or interfere with host antiviral responses. The miRNA associate with RISC, a cellular complex of proteins, to cleave mRNA or prevent ribosome binding.

FLASH CARD VOCABULARY

Latency	Vesicles
Macules	Ulcers
Papules	Primary infection
Recurrent infection	Splenomegaly
Nonprimary first episode infection	Oropharynx
HSV keratitis	Cytomegalic inclusion bodies
Herpes gladiatorum	Retinitis
Seroconversion	Transplacental transmission
Seroprevalence	Roseola infantum
Neonatal HSV infection	Maculopapular rash
Retrograde transport	Febrile seizures

Anterograde transport	Status epilepticus
Nucleoside/nucleoside analogs	Chromosomally integrated virus
Centripetal rash	Tegument
Herpes zoster	Rolling circle replication
Herpes zoster ophthalmicus	Concatemer
Postherpetic neuralgia	Double envelopment process
Mononucleosis	Latency-associated transcripts
Lymphadenopathy	microRNA
Hepatosplenomegaly	

CHAPTER REVIEW QUESTIONS

1. Create a chart that lists the official name of each virus, the commonly used name of the virus, and the conditions caused by each.
2. In what types of cells do viruses in each subfamily become latent?
3. HSV-1 seroprevalence rates are dropping in adolescents. What is a potential disadvantage of a person contracting HSV-1 later in life?
4. Make a list of the herpesviruses that cause congenital or perinatal infection. How is each specifically transmitted to the newborn?
5. Why can't a person contract herpes zoster from someone with herpes zoster?
6. Describe the difference between macules, papules, vesicles, and ulcers.
7. What is the difference between the varicella vaccine and the herpes zoster vaccine? Why are they different?
8. Which herpesviruses cause mononucleosis, and what are some major effects of the virus upon the body?
9. How could you distinguish the rash caused by varicella zoster virus and HHV-6B/HHV-7?
10. Why are herpesviruses of major concern to immunocompromised individuals?
11. What is meant by "lytic" versus "latent" infection?
12. Make a list of the seven stages of viral replication (attachment, penetration…), and indicate how herpesviruses accomplish each stage.
13. What are microRNAs? Describe how they block protein expression.

FURTHER READING

Ablashi, D., Chatlynne, L., Cooper, H., et al., 1999. Seroprevalence of human herpesvirus-8 (HHV-8) in countries of Southeast Asia compared to the USA, the Caribbean and Africa. Br. J. Cancer 81, 893–897.

Ali, M.M., Karasneh, G.A., Jarding, M.J., Tiwari, V., Shukla, D., 2012. A 3-O-Sulfated heparan sulfate binding peptide preferentially targets herpes simplex virus 2-Infected cells. J. Virol. 86, 6434–6443.

Arvin, A., Campadelli-Fiume, G., Mocarski, E. (Eds.), 2007. Human herpesviruses: Biology, therapy, and immunoprophylaxis. Cambridge University Press, New York.

Balfour, H.H., Sifakis, F., Sliman, J.A., Knight, J.A., Schmeling, D.O., Thomas, W., 2013. Age-Specific prevalence of Epstein–Barr virus infection among individuals aged 6-19 years in the United States and factors affecting its acquisition. J. Infect. Dis. 208, 1286–1293.

Belshe, R.B., Leone, P.A., Bernstein, D.I., et al., 2012. Efficacy results of a trial of a herpes simplex vaccine. N. Engl. J. Med. 366, 34–43.

Bernstein, D.I., Bellamy, A.R., Hook, E.W., et al., 2013. Epidemiology, clinical presentation, and antibody response to primary infection with herpes simplex virus type 1 and type 2 in young women. Clin. Infect. Dis. 56, 344–351.

Bradley, H., Markowitz, L.E., Gibson, T., McQuillan, G.M., 2014. Seroprevalence of herpes simplex virus types 1 and 2-United States, 1999–2010. J. Infect. Dis. 209, 325–333.

Chan, P.K., Ng, H.K., Hui, M., Cheng, A.F., 2001. Prevalence and distribution of human herpesvirus 6 variants A and B in adult human brain. J. Med. Virol. 64, 42–46.

Cousins, E., Nicholas, J., 2014. Molecular biology of human herpesvirus 8: novel functions and virus-host interactions implicated in viral pathogenesis and replication. Recent Results Cancer Res. http://dx.doi.org/10.1007/978-3-642-38965-8.

Crough, T., Khanna, R., 2009. Immunobiology of human cytomegalovirus: from bench to bedside. Clin. Microbiol. Rev. 22, 76–98.

Dodding, M.P., Way, M., 2011. Coupling viruses to dynein and kinesin-1. EMBO J. 30, 3527–3539.

Dowd, J.B., Palermo, T., Brite, J., McDade, T.W., Aiello, A., 2013. Seroprevalence of Epstein–Barr virus infection in U.S. children ages 6–19, 2003–2010. PLoS One 8, 1–7.

Eisenberg, R.J., Atanasiu, D., Cairns, T.M., Gallagher, J.R., Krummenacher, C., Cohen, G.H., 2012. Herpes virus fusion and entry: a story with many characters. Viruses 4, 800–832.

Guo, H., Shen, S., Wang, L., Deng, H., 2010. Role of tegument proteins in herpesvirus assembly and egress. Protein Cell 1, 987–998.

Chapter 22: Varicella. In: Hamborsky, J., Kroger, A., Wolfe, C. (Eds.), 2015. Epidemiology and prevention of vaccine-preventable diseases, thirteenth ed. Public Health Foundation, Washington, DC, pp. 353–376.

Kaufer, B.B., Flamand, L., 2014. Chromosomally integrated HHV-6: impact on virus, cell and organismal biology. Curr. Opin. Virol. 9, 111–118.

Knipe, D.M., Howley, P.M. (Eds.), 2013. Fields virology, sixth ed. Wolters Kluwer | Lippincott Williams and Wilkins.

Kotton, C.N., 2013. CMV: prevention, diagnosis and therapy. Am. J. Transpl. 13, 24–40.

Krug, L.T., Pellett, P.E., 2014. Roseolovirus molecular biology: recent advances. Curr. Opin. Virol. 9, 170–177.

Lopez, A., Schmid, S., Bialek, S., 2011. Chapter 17: Varicella. In: Roush, S.W., Baldy, L.M. (Eds.), Manual for the surveillance of vaccine-preventable diseases, fifth ed. Centers for Disease Control and Prevention, Atlanta, GA, pp. 1–16.

Lyman, M.G., Enquist, L.W., 2009. Herpesvirus interactions with the host cytoskeleton. J. Virol. 83, 2058–2066.

Maeki, T., Mori, Y., 2012. Features of human herpesvirus-6A and -6B entry. Adv. Virol. 2012. http://dx.doi.org/10.1155/2012/384069.

Osterrieder, N., Wallaschek, N., Kaufer, B.B., 2014. Herpesvirus genome integration into telomeric repeats of host cell chromosomes. Annu. Rev. Virol. 1, 215–235.

Razonable, R.R., 2013. Human herpesviruses 6, 7 and 8 in solid organ transplant recipients. Am. J. Transpl. 13 (Suppl. 3), 67–77 quiz 77–8.

Reynolds, M.A., Kruszon-Moran, D., Jumaan, A., Schmid, D.S., McQuillan, G.M., 2010. Varicella seroprevalence in the U.S.: data from the national health and nutrition examination survey, 1999–2004. Public Health Rep. 125, 860–869.

Shenk, T., Alwine, J.C., 2014. Human cytomegalovirus: coordinating cellular stress, signaling, and metabolic pathways. Annu. Rev. Virol. 1, 355–374.

Tesini, B.L., Epstein, L.G., Caserta, M.T., 2014. Clinical impact of primary infection with roseoloviruses. Curr. Opin. Virol. 9, 91–96.

Vanarsdall, A.L., Johnson, D.C., 2012. Human cytomegalovirus entry into cells. Curr. Opin. Virol. 2. http://dx.doi.org/10.1016/coviro.2012.01.001.

Young, L.S., Rickinson, A.B., 2004. Epstein–Barr virus: 40 on. Nat. Rev. Cancer 4, 757–768.

Zerboni, L., Sen, N., Oliver, S.L., Arvin, A.M., 2014. Molecular mechanisms of varicella zoster virus pathogenesis. Nat. Rev. Microbiol. 12, 197–210.

Chapter 14

Poliovirus

Poliomyelitis, often called "**polio**," is a disease caused by **poliovirus**, which is transmitted through the fecal–oral route. Poliovirus replicates within and causes damage to the central nervous system, predominantly to the brain stem and anterior horn cells, the motor neurons located in the spinal cord (see Fig. 14.9). The word "poliomyelitis" stems from the Greek words *polios* and *myelos*, meaning "gray marrow," in reference to the gray matter of the spinal cord that becomes inflamed and can cause paralysis (referred to as **paralytic poliomyelitis**) (Fig. 14.1). Because the disease afflicted mainly children, poliomyelitis was also known as "infantile paralysis." Interestingly, the permanent paralysis that is most commonly associated with poliovirus infection only occurs in less than 1% of total infections. In fact, ~72% of poliovirus infections in children are entirely asymptomatic, highlighting that those with visible symptoms are only a minority of the total number of infected individuals. The number of symptomatic cases of polio has historically paled in comparison to other common childhood diseases, such as measles, mumps, or rubella, but the visible paralysis that resulted from polio was a constant reminder of the permanent disability that the virus was capable of inflicting. Although many have already died of old age, 10–20 million people worldwide are still living with polio-induced paralysis.

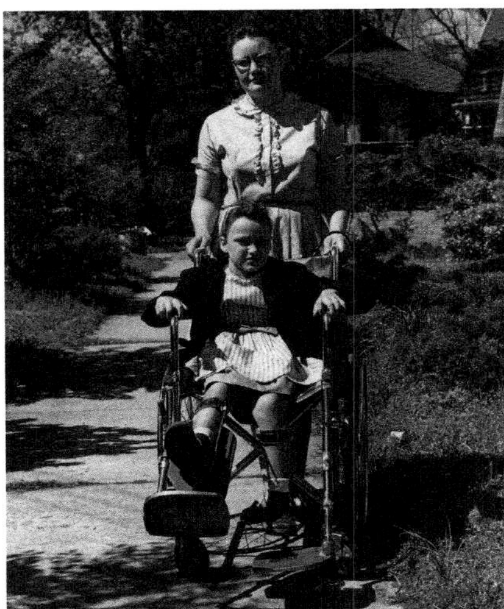

FIGURE 14.1 **Paralytic poliomyelitis.** Poliomyelitis is the condition caused by poliovirus infection of central nervous system tissues. Infection damages motor neurons and can cause paralysis. *Image courtesy of the CDC.*

14.1 THE EARLY YEARS OF POLIOVIRUS

Poliovirus has afflicted humans for thousands of years. The first documented depiction of a polio victim is an Egyptian stela that dates back to 1580–1350 BC (see Fig. 1.1A), a carving of a priest with a walking cane and **foot drop deformity**, a condition that prevents **dorsiflexion** of the foot (lifting of the foot at the ankle), a common visible sign of poliomyelitis. In modern history, the first person to document poliomyelitis was English physician Michael Underwood in 1789. In the second edition of his *Treatise on the Diseases of Children*, he describes a condition that "usually attacks children previously reduced by fever, seldom those under one or more than four or five years old…a debility of the lower extremities which gradually become more infirm, and after a few weeks are unable to support the body."

Although Underwood indicated that cases of poliomyelitis were infrequent, epidemics became common in Scandinavian countries in the late 19th century, and the first report of poliomyelitis in the United States occurred with a cluster of 8–10 cases of "infantile paralysis" in West Feliciana, Louisiana. The first major outbreak in the United States occurred in 1894, in Vermont, when 18 deaths and 132 cases of permanent paralysis were reported. Outbreaks became more common as the years passed. In 1907, 2500 cases of poliomyelitis were reported in New York City, causing 125 deaths. Several hypotheses have been proposed to explain why once sporadic cases of polio were now causing epidemics. As an RNA virus without proofreading ability, it has been proposed that polio may have mutated into a more virulent strain. The growth and urbanization of cities with increasing levels of hygiene may also have played a role: poliovirus is transmitted through the fecal–oral route, and improvements in sanitation and water quality would have reduced the exposure of people to the virus, delaying the age at which they contracted poliovirus. The reduced prevalence of poliovirus likely translated to less natural boosting of immunity against the virus, including in mothers who would consequently have lower levels of circulating antibody passed transplacentally or through breast milk to infants. Taken together, the reduced prevalence of the

Essential Human Virology.

FIGURE 14.2 **Notable individuals in the progress against polio.** The Polio Hall of Fame consists of bronze-sculpted busts of 15 scientists and two laymen that made important contributions to our understanding and eradication of polio. Notable individuals include Ivar Wickman (3rd from left), Karl Landsteiner (4th), Albert Sabin (8th), Isabel Morgan (11th), David Bodian (13th), John Enders (14th), Jonas Salk (15th), and Franklin D. Roosevelt (16th), 32nd President of the United States. President Roosevelt developed poliomyelitis in 1921, resulting in permanent paralysis below the waist. In 1926, he purchased a resort with natural warm springs, thought at the time to be effective in the physical therapy of polio patients, in Warm Springs, Georgia. He turned the resort into a polio rehabilitation center in 1927. The Polio Hall of Fame seen here is housed at the *Roosevelt Warm Springs Institute for Rehabilitation. Photograph courtesy of Henry Lytton Cobbold.*

virus would have resulted in a larger population of susceptible individuals at one time, resulting in an epidemic. The increased age at infection led to more severe disease and, as a result, an increase in the number of deaths.

Scientifically, the first decade of the 1900s provided several breakthroughs on the etiology of poliomyelitis. It was first suggested in 1905 by Swedish physician Ivar Wickman that polio was an infectious disease that could be transmitted from person to person, notably by those that show no symptoms of infection (Fig. 14.2). In 1909, physicians Karl Landsteiner and Erwin Popper identified that poliomyelitis was of viral origin. Others had been unsuccessful in recapitulating the symptoms of poliomyelitis in small animals, such as rabbits or guinea pigs, because the virus does not infect them. Landsteiner and Popper were successful in inducing spinal cord lesions similar to those observed in humans by injecting a baboon and rhesus monkey with a spinal cord suspension from a 9-year-old boy who had died of poliomyelitis. Although poliovirus does not normally infect apes and monkeys, this showed that they could be infected experimentally. They suggested that polio was of viral origin, and in the same year, Landsteiner and Constantin Levaditi of the Pasteur Institute successfully propagated the virus after filtering it through an ultrafilter, verifying polio was of viral origin.

Over the next 30 years, efforts toward understanding the virus, determining its mode of transmission, and finding a vaccine against it were plagued by technical difficulties, inaccurate scientific assumptions, and ineffective vaccine formulations. In 1910, physician Simon Flexner, the first director of the Rockefeller Institute for Medical Research, discovered protective substances—later determined to be neutralizing antibodies—in the blood of monkeys immune to poliovirus. However, Flexner erroneously came to the conclusion

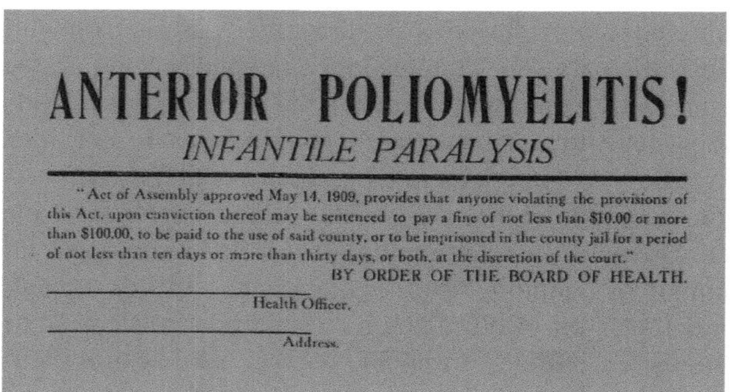

FIGURE 14.3 **Polio placard.** By law, individuals with polio and those who were in contact with them were quarantined in their houses in an attempt to reduce transmission to others. *Image courtesy of the National Library of Medicine.*

that because poliovirus could be transmitted between monkeys through mucosal secretions, it must travel to the spinal cord by interacting with nerve endings in the nasal passages following droplet transmission to mucosal surfaces. This assumption accompanied other difficulties that set back progress against polio. Research was limited to infection of monkeys, since the virus did not infect small mammals such as mice or rabbits. An ineffective vaccine was created from monkey spinal cords, and a dangerous vaccine occurred from the improper attenuation of live poliovirus.

Meanwhile, a major epidemic in New York City in 1916 resulted in over 9000 cases of paralytic polio and 2343 deaths in the city alone. More than 6000 deaths occurred nationwide in 1916. Epidemics were reported annually, generally peaking in the summer months, leading to the closure of swimming pools, parks, movie theaters, and public gathering areas in an attempt to reduce spread (Fig. 14.3).

FIGURE 14.4 **Inventors of the poliovirus vaccines currently used today.** (A) Jonas Salk, pictured at his laboratory at the University of Pittsburgh, was the inventor of the inactivated polio vaccine. (B) Albert Sabin (right) was the inventor of the oral polio vaccine. He is pictured here with Robert Gallo, best known for his work on HIV/AIDS. *Courtesy of National Cancer Institute.*

Although scientists knew that polio was caused by a virus, they did not yet understand how it was being transmitted from person to person, especially with the majority of cases not showing any symptoms at all.

This stagnant scientific period was spurred into advancement in the 1940s. In 1941, Albert Sabin and Robert Ward analyzed tissues obtained from fatal cases of human poliomyelitis and determined that primary infection appeared to occur in the gastrointestinal tract and not the respiratory tract, as had been previously proposed by Flexner. In fact, they were unable to find any virus in the nasal mucosa or associated nerves. These results strongly suggested that the virus was transmitted orally, not through the respiratory system. In 1949, John Enders, the director of the Boston Children's Hospital Research Laboratory, and colleagues Thomas Weller and Frederick Robbins worked out a method for growing poliovirus in a variety of human embryonic tissues using tissue culture techniques. Others had propagated poliovirus in culture but only in human embryonic brain tissue, creating an insurmountable challenge for the mass propagation of virus. Enders' team was able to propagate the virus in skin, muscle, and intestinal cells. The three researchers received the Nobel Prize in Physiology or Medicine in 1954 for this discovery, which was the breakthrough needed for the future production of a polio vaccine.

At the University of Pittsburgh in 1951, Jonas Salk used Enders' techniques to propagate poliovirus in monkey kidney cells, a technique that would later allow for the mass propagation of virus. It had been determined in 1949 by David Bodian and Isabel Morgan at Johns Hopkins University that there existed three serotypes of poliovirus, and so Salk's laboratory propagated one strain of each serotype for the development of a vaccine (Fig. 14.4A). Using attenuated, live virus vaccine formulations had been well established and commonly used at the time, but Salk decided to instead inactivate the viruses using formalin. The choice of an inactivated vaccine drew criticism, particularly from Albert Sabin, a renowned poliovirus researcher who was developing an attenuated vaccine preparation against the virus (Fig. 14.4B). Sabin expressed concerns about the immunogenicity of an inactivated vaccine and the prudence of using a virulent strain, even if it were to be inactivated.

Salk, backed by the National Foundation for Infantile Paralysis (founded by President Franklin D. Roosevelt in 1938 and now known as the March of Dimes), conducted successful small-scale trials with the inactivated vaccine. This provided the foundation for a massive vaccine trial that began in the US in April of 1954 and included over 1.5 million children in 211 counties in 44 states, the largest trial that had ever taken place. A year later, the results were announced: the vaccine was 80–90% effective in preventing paralytic polio. Parents breathed a sigh of relief, anxious to prevent their children from the virus that had caused over 21,000 paralytic cases in 1952. The incidence of paralytic cases dropped rapidly following the approval of Salk's **inactivated polio vaccine (IPV)** in 1955.

In 1963, the US government approved Albert Sabin's **oral polio vaccine (OPV)**, composed of three attenuated poliovirus strains (trivalent), and it became part of the US childhood immunization schedule in 1965. In comparison to the IPV that is administered intramuscularly as an injection, the oral polio vaccine is administered as a liquid dose that is swallowed (Fig. 14.5A) or given on a sugar cube

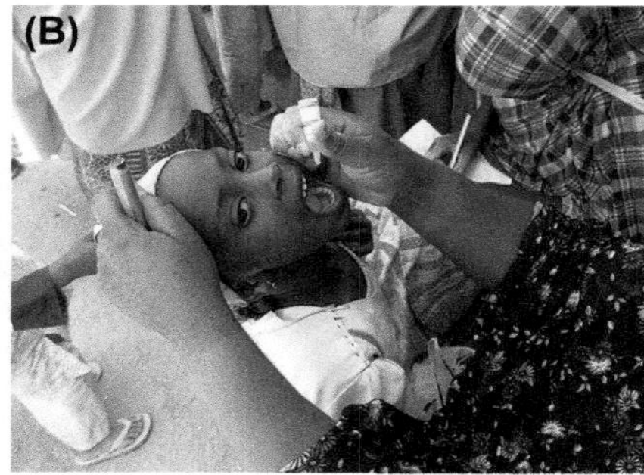

FIGURE 14.5 **The oral polio vaccine.** The oral polio vaccine uses three live attenuated poliovirus strains. It was administered by dripping a dose of the vaccine onto a sugar cube (A) or by simply dripping a dose into the mouth for swallowing (B), as shown in this 2014 image of a Nigerian girl receiving the vaccine. *Photo courtesy of the CDC / Molly Kurnit, MPH, and Alford Williams.*

(Fig. 14.5B). Both vaccines are highly effective in producing immunity against poliovirus after 3 or 4 doses.

Due to its ease of use, the OPV soon replaced the IPV and was paramount in eliminating wild poliovirus from the United States by 1979 and the Western Hemisphere by 1991. However, because the OPV contains three attenuated poliovirus strains and initiates infection via the gastrointestinal tract, the normal route of transmission, it is possible that a live attenuated virus could revert at low frequency back into a neurovirulent strain, particularly in immunocompromised individuals or those lacking certain aspects of the immune response. **Vaccine-derived poliovirus (VDPV)** infection occurs very infrequently but can cause **vaccine-associated paralytic polio (VAPP)**, equivalent to the paralytic poliomyelitis caused by wild poliovirus. When it happens, VAPP generally occurs 4–30 days after receiving the OPV. VAPP is a rare event, occurring in 1 individual for every 2.4 million doses of vaccine administered. In the United States, the OPV caused about 8–10 cases of VAPP each year (Fig. 14.6). Because the risk of VAPP from the OPV was greater than the risk of wild poliovirus infection after its elimination from the country, the United States discontinued the use of OPV in 2000. Doing so eliminated any risk of VAPP, since the IPV is inactivated and therefore unable to revert to a VDPV strain. Worldwide vaccination efforts that have relied upon the OPV are also encouraging the switch to the IPV due to the risk of VAPP. VDPVs can hinder vaccine efforts because they can be the source of virulent outbreaks capable of causing poliomyelitis. These efforts will be discussed in Section 14.4.

Study Break

Review the advantages and disadvantages of live attenuated virus vaccines versus inactivated virus vaccines. See Chapter 8, "Vaccines, Antivirals, and the Beneficial Uses of Viruses," if you need a refresher.

In-Depth Look: Simian Virus 40 and the Poliovirus Vaccine

In 1957, a polyomavirus named simian virus 40 (SV40) was isolated from the monkey kidney cells that were used to propagate the poliovirus strains found in the Salk and Sabin vaccines. It was later determined that an estimated 100 million individuals between 1955 and 1963 received vaccines containing SV40, found in both the OPV and in the IPV (it was found that SV40 was incompletely inactivated by the methods that inactivated poliovirus for the IPV). The presence of SV40 has been of concern because, not long after its discovery, SV40 was found to be a highly oncogenic virus, being able to transform cells and form tumors in hamsters.

Initially it was thought that SV40 was unable to infect humans, but neutralizing antibodies against the virus have been found in humans—both before and after the use of contaminated vaccines—suggesting that SV40 has been circulating within the human population. SV40 DNA has been found in a variety of human tumors, particularly brain tumors, bone cancers, non-Hodgkin's lymphomas, and mesotheliomas, a tumor of the lung that has been linked to asbestos exposure. Although SV40 has been found in these human tumors, studies have been unable to show that SV40 is the *cause* of these tumors. For example, SV40 has been found in tumors of people that could not have received a contaminated vaccine, and the frequency of cancer is the same in people that received a contaminated vaccine versus those that did not. In addition, SV40 has been detected in noncancerous tissue as well, making it difficult to determine if SV40 is the cause of the tumors or just happened to be present in the cells that later became cancerous.

New molecular biology techniques that allow the rapid and sensitive sequencing of viral genomes are assisting scientists in the creation of effective vaccines. These techniques are also being used to verify that unknown viruses, such as SV40, are not present within current vaccine formulations.

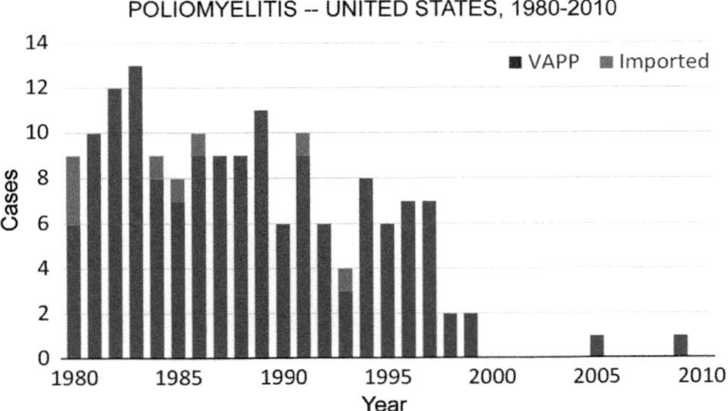

FIGURE 14.6 **Vaccine-associated paralytic poliomyelitis.** Following the elimination of wild poliovirus from the United States in 1979, an average of eight annual cases of paralytic poliomyelitis occurred through importation from other countries (blue) or as vaccine-associated paralytic polio (VAPP, gray), a low-frequency side effect of infection with the oral vaccine. Because of this, the United States switched to the IPV in 2000, eliminating the potential for VAPP. A case of VAPP in 2005 was due to the infection of an unvaccinated individual while out of the country, and a case in 2009 in an immunocompromised individual is thought to be a result of a long-term infection with a vaccine strain. *Data obtained from Centers for Disease Control and Prevention, 2015. Poliomyelitis. In: Hamborsky, J., Kroger, A., Wolfe, C., (Eds.), Epidemiology and Prevention of Vaccine-Preventable Diseases, thirteenth ed. Public Health Foundation, Washington DC, pp. 175–86.*

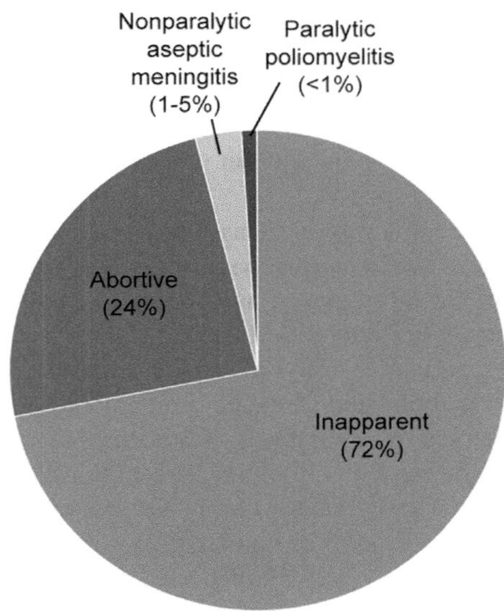

FIGURE 14.7 **Results of infection with poliovirus.** Of those infected with poliovirus, about 72% do not present with any symptoms. 24% of individuals have a mild, nonspecific illness known as abortive poliomyelitis. Between 1% and 5% of people infected develop nonparalytic aseptic meningitis characterized by stiffness of the neck, back, and/or legs. Less than 1% of infections result in acute flaccid paralysis, which may be temporary or result in varying degrees of permanent disability.

14.2 CLINICAL COURSE OF INFECTION

Poliovirus is in the *Picornaviridae* family, within the *Picornavirales* order that was created in 2008. It is in the *Enterovirus* genus that also includes rhinovirus, and its species is officially known as *Enterovirus C*. There exist three serotypes of poliovirus, known as poliovirus type 1, type 2, and type 3 (PV1, PV2, and PV3), and immunity against one serotype does not protect against infection with the other serotypes. All three serotypes are capable of causing paralytic poliomyelitis, although type 1 causes the most epidemics of paralytic poliomyelitis. Humans are the only reservoir for poliovirus, an important aspect in the potential for eradication of the virus.

Poliovirus is nonenveloped and acid resistant. Its capsid proteins can withstand the low pH of the stomach, a requirement of viruses that initiate infection via the gastrointestinal tract. The virus is also resistant to detergents and 70% alcohol, although 5% bleach or high heat (above 55°C/131°F) is effective in interrupting virion structure. Virions are stable for 1–2 months in fresh or salt water at 16°C (60.8°F). Greater than 10^8 poliovirus particles absorb to a gram of wastewater sludge, which extends the infectivity of virions in the environment, even at higher temperatures.

Poliovirus is transmitted primarily through the fecal–oral route. The virus can be recovered from saliva and throat swabs, and there is evidence to suggest that it can also be transmitted via the oral–oral route by droplets or sharing of **fomites**, inanimate items such as utensils, cups, or toys. Infection is inapparent in 72% of infections. An infected individual is most infectious from 7 to 10 days before infection until 7–10 days after symptoms appear, although virus is shed into stool for 3–6 weeks and is communicable even if symptoms are absent.

Following infection, poliovirus replicates in the oropharynx and gastrointestinal tract mucosa. From here, it gains entry at these respective locations into the tonsils and the Peyer's patches, superficial lymph nodes immediately under the epithelium of the small intestine (see Fig. 5.3D). From these sites, poliovirus spreads to the regional lymph nodes and a **primary** or **minor viremia** occurs. At this stage, about 3–6 days after initial infection, ~24% of individuals exhibit nonspecific symptoms that last for 2–3 days, including headache, fever, rash, and sore throat (Fig. 14.7). This is known as **abortive poliomyelitis**, which resolves completely.

During the minor viremia, if distal sites are infected, such as the spleen, bone marrow, muscle, and other lymph nodes, then a **secondary** or **major viremia** can occur as a result of increased replication of the virus. It is at this stage that the virus infects the central nervous system, either directly from the bloodstream or by infection of a peripheral nerve in which the virus undergoes retrograde transport to the central nervous system. The central nervous system is composed of the brain and the spinal cord. Poliovirus infects anterior horn cells, the motor neurons of the spinal cord, or the part of the brain stem known as the medulla oblongata. It should be noted that infection of nervous tissue is not required for continued transmission of the virus but may be due to an increase in virulence or a lack of adequate immune responses.

There are two forms of disease that can occur upon infection of the central nervous system: **nonparalytic aseptic meningitis** (also referred to as nonparalytic poliomyelitis) and **paralytic poliomyelitis**. About 1–5% of symptomatic cases progress to nonparalytic aseptic meningitis, an inflammation of the brain and spinal cord membranes (meninges) that is characterized by fever, headache, back pain, and stiffness in the neck, back, and/or legs. Abnormal or heightened sensations may also occur as a result of nerve damage. Nonparalytic aseptic meningitis resolves completely in most individuals. On the other hand, paralytic poliomyelitis is the most severe manifestation of poliovirus infection, occurring in <1% of infected individuals, and may have lasting effects. Paralytic symptoms appear 1–18 days following the appearance of prodromal symptoms. They may appear when it looks as though the person has begun recovering from an abortive infection, and they may also occur in the absence of typical prodromal symptoms. Fever and headache accompany muscle pain and spasms in the limbs or back. This progresses quickly, within 2–4 days, into **acute flaccid paralysis**, the weakness of muscles with reduced or absent reflexes that results from the destruction of motor neurons. When the neurons are damaged, control of muscle responses supplied by those nerves is reduced or lost. Paralysis is usually **asymmetrical**—occurring only on one side of the body—and more common in the legs than in the arms (Fig. 14.8). Paralysis usually occurs 7–21 days following infection. Without the ability to move, muscle **atrophy** (shrinkage) accompanies prolonged paralysis. Sensory neurons and cognition are unaffected.

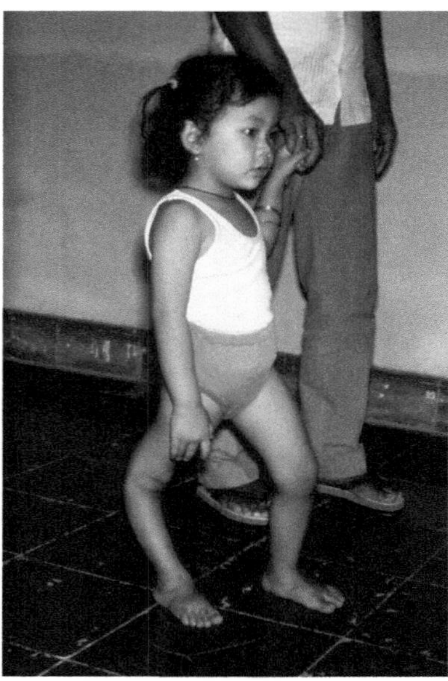

FIGURE 14.8 **Paralytic poliomyelitis.** Paralysis caused by poliovirus is usually asymmetrical and appears 7–21 days following infection. Without normal stimulation, the affected muscles atrophy, as shown in this image of a child left with permanent paralysis due to polio infection. *Image courtesy of CDC.*

Paralytic poliomyelitis is classified into three types: spinal, bulbar, and bulbospinal poliomyelitis. **Spinal poliomyelitis** refers to the condition that affects the motor neurons of the body (Fig. 14.9A), most commonly the legs but also including the back and neck. **Bulbar poliomyelitis** occurs as a result of destruction of the bulbar region of the lower brain stem, more commonly referred to as the medulla oblongata (Fig. 14.9B). The nerves originating from this site control muscles involved in autonomic functions, such as swallowing and breathing. **Bulbospinal poliomyelitis** occurs when both the spinal and bulbar neurons are affected. Generally, the amount of paralysis plateaus after an individual's fever departs, and strength begins to return within days or weeks. About 10% of people with paralytic poliomyelitis recover completely, while ~80% of these individuals have some degree of permanent paralysis. Heat and physical therapy can be used for rehabilitation to stimulate muscles and improve mobility, although weakness or paralysis remaining after 12–18 months is generally irreversible. The fatality rate is higher in adults than in children: 15–30% of adults presenting with paralytic poliomyelitis succumb to the condition, compared to 2–5% of children.

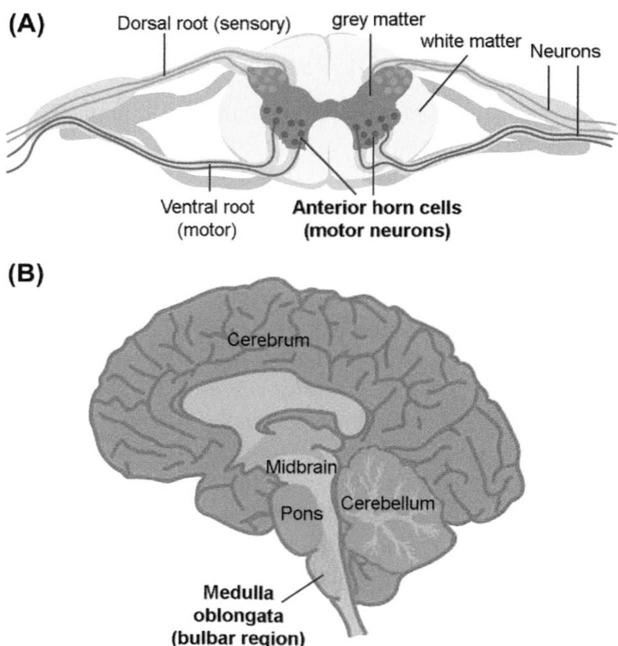

FIGURE 14.9 **Poliovirus damage to the central nervous system.** (A) Spinal poliomyelitis occurs when poliovirus affects anterior horn cells, the motor neurons that control movement of muscles. (B) Bulbar poliomyelitis results from the destruction of the bulbar region of the brain stem, more commonly referred to as the medulla oblongata. Bulbospinal poliomyelitis indicates both sites have sustained damage from poliovirus replication.

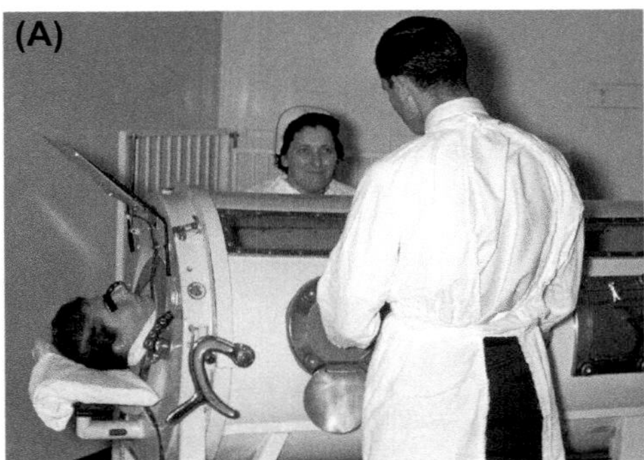

FIGURE 14.10 **The iron lung.** Poliovirus infection of the medulla oblongata can affect autonomic functions, including swallowing and breathing. The iron lung (A) was an artificial respirator that was invented to maintain respiration in individuals with bulbar involvement. A diaphragm at the bottom of the machine (B, *arrow*) was moved to change the pressure within the tube, which would cause air to enter or leave the lungs. Portable respirators became available in the 1950s and 1960s, but some people with permanent damage chose to continue using their iron lungs, even until their deaths many years later. *Photos courtesy of the CDC and Mary Hilpertshauser.*

Spinal, bulbar, and bulbospinal polio accounted for around 79%, 2%, and 17% of cases, respectively, of paralytic poliomyelitis in the United States from 1969 to 1979. Overall, the fatality rate is 5–10% for those with paralytic poliomyelitis, but the rate is higher (25–75%) for cases involving bulbar involvement, primarily as a consequence of suffocation due to an effect upon respiratory processes. In an attempt to treat this condition, artificial respirators known as the "iron lung" were invented, a sealed tube in which an individual was placed with only his or her head sticking outside the machine (Fig. 14.10A). The device would move a diaphragm at the bottom of the tube (Fig. 14.10B), causing the pressure inside the tube to decrease and air to be forced into the person's lungs. The machine would then push in the diaphragm, increasing the air pressure within the tube and causing the air to be forced out of the person's lungs. The machine was used until temporary damage declined or, in some cases, indefinitely due to permanent respiratory paralysis.

Around 25–40% of people that have recovered from paralytic polio experience **post-polio syndrome (PPS)**, 15–40 years after the original infection. Although the virus is no longer present, muscle weakness, fatigue, and atrophy slowly begin to develop in the muscles that were initially affected during the acute stage of the disease. The cause of PPS is unknown, although it is thought to be due to the eventual decline of neurons that have been overcompensating for the neurons that were damaged by infection. PPS is rarely life-threatening, but if bulbar nerves were initially involved then breathing or swallowing reflexes may be negatively affected.

Study Break

Describe the process of how poliovirus initiates infection, spreads throughout the body, and eventually infects motor neurons.

14.3 POLIOVIRUS REPLICATION

Poliovirus is a small, nonenveloped icosahedral virus of about 30 nm in diameter (Fig. 14.11A and B). It possesses a 7.5 kb +ssRNA genome that is infectious, translated immediately upon entry into the cell. The genome contains a 5′-nontranslated region (NTR) that contains an internal ribosome entry site (IRES), followed by a single open reading frame encoding a polyprotein that is cleaved sequentially into intermediate precursors and 11 mature poliovirus proteins (Fig. 14.12). The genome is also polyadenylated at the 3′-NTR. A viral protein known as **VPg (virion protein, genome-linked)** associates with the 5′-end of the genome and is thought to act as a protein primer for replication of the genome.

14.3.1 Attachment, Penetration, and Uncoating

The poliovirus capsid is composed of 60 copies each of 4 repeating proteins—**VP1, VP2, VP3**, and **VP4**. VP2 and VP4 are derived from autocatalysis of VP0 during maturation of the virion. VP1, VP2, and VP3 form the surface of the capsid, with VP4 associating on the inner capsid wall. VP1 forms a star-shaped plateau or "mesa" on the fivefold axis, which is surrounded by a deep canyon into which the cell surface receptor binds (Fig. 14.11C).

The poliovirus virion is acid resistant, which allows it to survive the low pH of the stomach to initiate infection within the small intestine. The cell surface receptor for all three poliovirus serotypes is CD155, a glycoprotein that functions as an adhesion molecule in adherens junctions. In addition, CD155 is also recognized by NK cells to induce their cytotoxicity. CD155 is also commonly referred to as the "poliovirus receptor," or PVR. It is expressed on the surface of intestinal epithelial cells and on M cells of Peyer's patches, which may facilitate their entry into the Peyer's patches following infection of the intestinal epithelium.

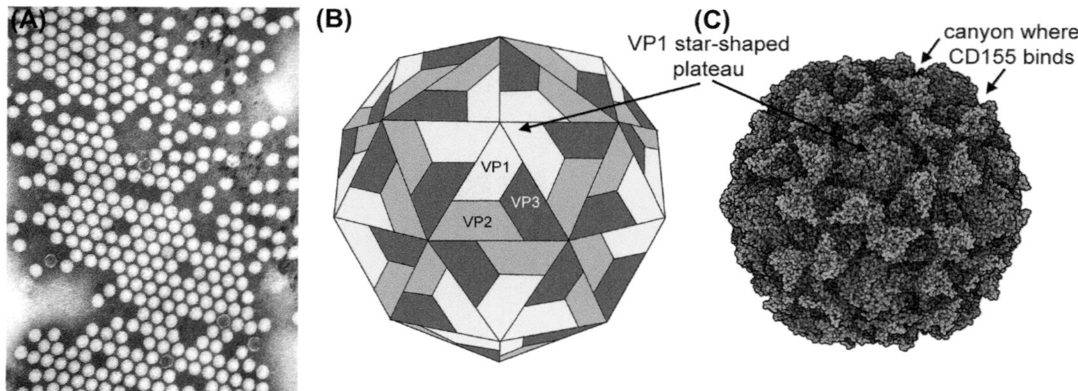

FIGURE 14.11 **Poliovirus virion.** (A) Icosahedral poliovirus virions. *Figure courtesy of the CDC, Dr. Fred Murphy, and Sylvia Whitfield.* (B) The poliovirus capsid, composed of structural units containing VP1, VP2, and VP3. VP4 associates with these proteins on the underside of the capsid. (C) The 3D structure of the poliovirus capsid, showing the star-shaped plateau formed by VP1 and the canyon in which CD155 binds. Rendering was performed using QuteMol *(IEEE Trans Vis Comput Graph 2006. 12, pp. 1237–44.)* with a 2PLV PDB assembly *(EMBO J. 1989. 8, pp. 1567–1579.).*

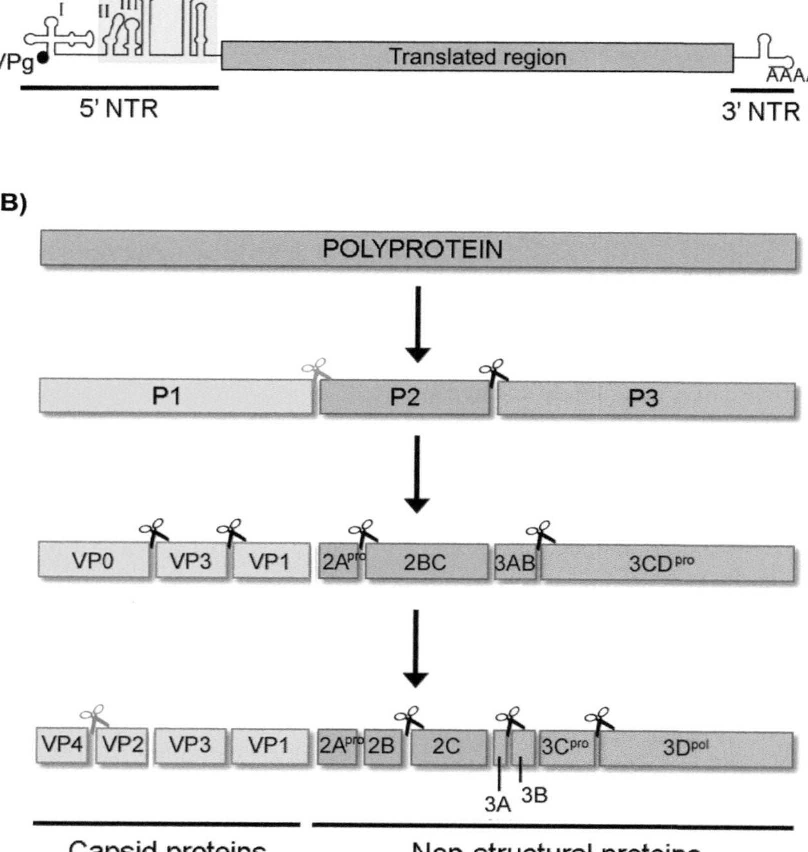

FIGURE 14.12 **Poliovirus genome and translation products.** (A) The +ssRNA poliovirus genome possesses a 5′-NTR that contains an IRES, followed by a single open reading frame that encodes the polyprotein. The 3′-NTR is polyadenylated. *(Reprinted with permission from De Jesus, N.H., 2007. Epidemics to eradication: the modern history of poliomyelitis. Virol. J. 4, 70.)* (B) The translated polyprotein is cleaved several times by both viral and host proteinases into 11 distinct proteins. Intermediate cleavage products also have important functions. The colored scissors indicate sites of cleavage: orange by $2A^{pro}$, black by $3C^{pro}$ or $3CD^{pro}$, and green by an unidentified cellular protease.

CD155 is not expressed in rodents and small mammals, which explains why experiments that attempted to infect these animals have historically been unsuccessful. In 1990, a transgenic mouse strain was engineered to express the human CD155 molecule. These mice were susceptible to infection, whereas the normal nontransgenic mice were not (see Fig. 4.3).

Upon interaction with CD155, the poliovirus capsid undergoes a conformation change whereby VP1 inserts into the cell membrane, forming a pore through which the viral genome is released. This was initially thought to occur at the plasma membrane, although now evidence suggests that this occurs after internalization of the virion by clathrin- and caveolin-independent endocytosis.

14.3.2 Translation and Replication

Within the cytoplasm, VPg is removed by a host DNA repair enzyme, TDP2, often referred to as "**VPg unlinkase**." The IRES found at the 5′-end of the +ssRNA genome recruits cellular proteins that direct the assembly of translation initiation complexes, and ribosomes translate the single polyprotein of ~3000 amino acids in length (~250 kD). The polyprotein undergoes several cleavages to produce hybrid and individual proteins of various functions (Table 14.1). It is first divided into three precursors: P1, P2, and P3. P1 comprises the capsid proteins, and P2 and P3 contain the nonstructural proteins, including the RNA-dependent RNA polymerase (RdRp) **3D^{pol}**, viral proteinases **2A^{pro}**, **3CD^{pro}**, and **3C^{pro}**, VPg (3B), and other proteins necessary for replication. Interestingly, several of the intermediate polypeptides have very important functions. Notably, 3CD^{pro} is one of the major viral proteinases.

The importance of the IRES in poliovirus virulence is apparent when examining the differences between wild poliovirus and the attenuated OPV strains. Sequencing of these genomes has shown that point mutations in the IRES of each attenuated strain cause defects during translation within neuronal cells, possibly resulting in a strain with reduced neurovirulence.

As a +ssRNA virus, translation of viral proteins precedes replication of the genome in order to produce the RdRp, called 3D^{pol}. Like all picornaviruses, poliovirus induces the formation of **replication complexes (RCs)**, membrane vesicles derived from the endoplasmic reticulum on which arrays of RdRp form for genome replication. The poliovirus VPg functions as a primer for synthesis of both negative- and positive-strand RNAs. For synthesis of the negative-sense antigenomic RNA, the RdRp adds two uracil-containing nucleotides to VPg, using the poly(A) tail at the 3′-end of the +ssRNA genome as a template. It uses VPg-pU-pU as a primer to copy the genomic RNA into −ssRNA, which functions as a template for genome replication and transcription of additional mRNAs.

For synthesis of the +ssRNA genome, VPg is again used as a primer, except that it adds the two uridine triphosphates using a ***cis*-acting replication element (CRE)** as a template. CREs are sequences that are found within the coding regions of the poliovirus genome. They form secondary stem-loop structures that bind viral proteins that assist the RdRp in identifying its viral RNAs from among the polyadenylated cellular mRNAs.

TABLE 14.1 The Function of Poliovirus Proteins

Precursor	Protein	Function
P1	VP1	Capsid protein, forms pore in endosome membrane
	VP2	Capsid protein
	VP3	Capsid protein
	VP4	Interacts with RNA genome on interior of capsid
P2	2A^{pro}	Protease that cleaves P1 from P2
	2BC	Forms replication complexes
	2B	Rearranges intracellular membranes
	2C	Forms replication complexes
P3	3AB	Remodels RNA, stimulates the RNA-dependent RNA polymerase (RdRp)
	3CD^{pro}	Major protease, binds CREs during replication
	3A	Inhibits secretion, interacts with replication membranes
	3B	VPg, functions as primer for replication
	3C^{pro}	Protease, binds CREs during replication
	3D^{pol}	RdRp

14.3.3 Assembly, Maturation, and Release

The 2A^{pro} cleaves the single poliovirus polyprotein between P1 and P2 to release the P1 precursor, which contains all components of the capsid. The P1 precursor is cleaved by 3CD^{pro} to release VP1 and VP3 proteins, along with VP0, an immature protein that will later be cleaved into VP2 and VP4 as part of the maturation process. VP1, VP3, and VP0 associate with each other to form the structural units, known as **protomers**. Five protomers spontaneously aggregate to form a pentamer, and twelve pentamers (for a total of 60 structural units) form the **procapsid**.

The +ssRNA genome and covalently associated VPg are packaged into the procapsid, or alternatively, the procapsid assembles around the genome. Virions undergo maturation into an infectious virion upon the cleavage of VP0 into VP2 and VP4 by what is thought to be a host protease. Poliovirus has been considered a lytic virus, although evidence has emerged recently that in certain conditions release can also occur in a nonlytic fashion through the release of virus within vesicles that have been hijacked from those involved in **autophagy**, a process that degrades and recycles damaged intracellular components, including organelles.

The entire poliovirus replication cycle is depicted in Fig. 14.13.

14.4 EPIDEMIOLOGY AND WORLDWIDE ERADICATION EFFORTS

The success of the smallpox vaccination campaign convinced scientists, governments, and public health organizations worldwide that serious infectious diseases could be eliminated permanently. Following on the success of polio vaccination campaigns in the Americans, the World Health Organization (WHO) declared in 1988 that polio would be their next target, aiming for its eradication by the year 2000: a "gift from the 20th to the 21st century." Poliovirus has several biological characteristics that make it a good target for eradication. Like variola virus, the virus that causes smallpox, poliovirus also exclusively infects humans. There are a limited number of serotypes of poliovirus—three—and the virus does not usually cause persistent infections that act as indefinite reservoirs. Immunity is lifelong. Unlike smallpox, however, the vast majority of poliovirus infections are asymptomatic, which means that *all* children in endemic areas must be vaccinated, not just those that were in contact with other infected individuals. For full immunity, four vaccine doses must also be administered. This has

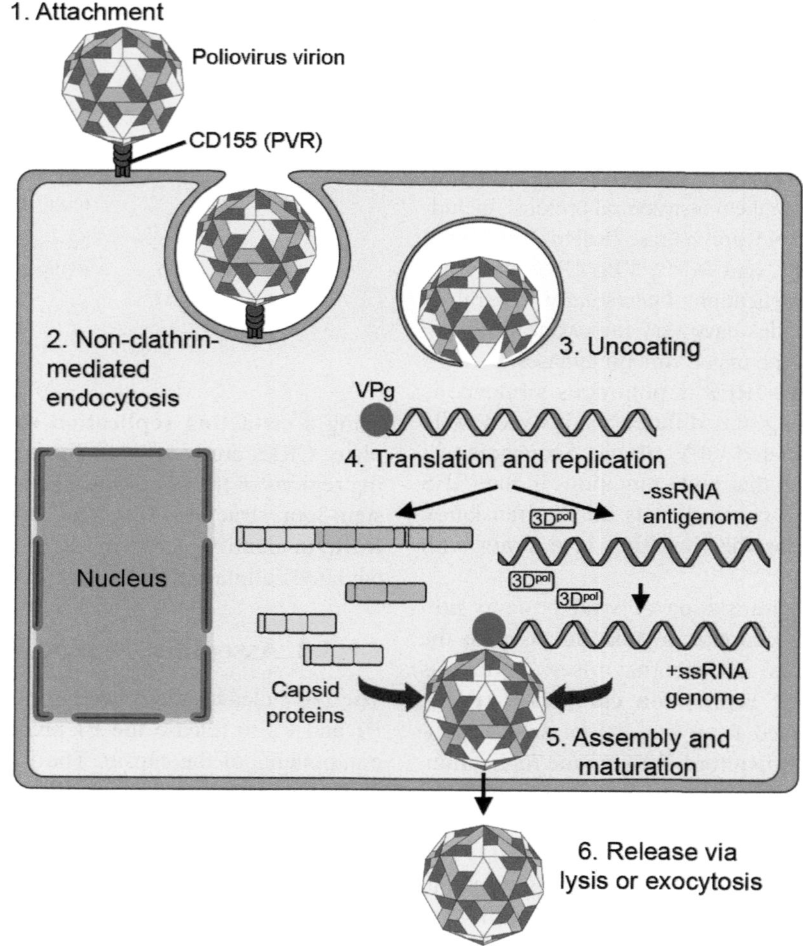

FIGURE 14.13 **Poliovirus replication.** Poliovirus binds to CD155, which induces endocytosis. A conformational change in the capsid creates a pore through which the genome is transported. Host ribosomes translate the genome into a polyprotein, which is cleaved into separate proteins by viral proteinases. The viral RdRp $3D^{pol}$ replicates the antigenome template and genomic RNA on replication complexes *(not shown due to space constraints)*. The genome is packaged into capsids composed of VP1, VP3, and VP0. In the process of maturation, VP0 is cleaved into VP2 and VP4 by an unidentified cellular protease. Release then occurs via lysis or exocytosis involving autophagosome vesicles.

made attempts at eradicating poliovirus significantly more difficult than may have first been anticipated.

The Sabin OPV vaccine was chosen for vaccination efforts because it does not require skilled labor to administer and it costs a tenth of the IPV. The **Global Polio Eradication Initiative** focused on high routine immunization coverage; supplementary immunization through "**National Immunization Days**" (Fig. 14.14), which had previously proved very effective in reducing polio in South American countries; effective surveillance; and, during the final stages in an area, door-to-door immunization campaigns to immunize the last of the unvaccinated. The polio global eradication initiative has been a concerted global public-private effort by national governments, WHO, the US Centers for Disease Control and Prevention (CDC), Rotary International (who has funded the massive polio immunization campaigns), the United Nations Children's Fund (UNICEF, who supplied the oral polio vaccine), and the Bill and Melinda Gates Foundation (which alone is funding a third of the total eradication budget between 2013 and 2018, nearly $2 billion).

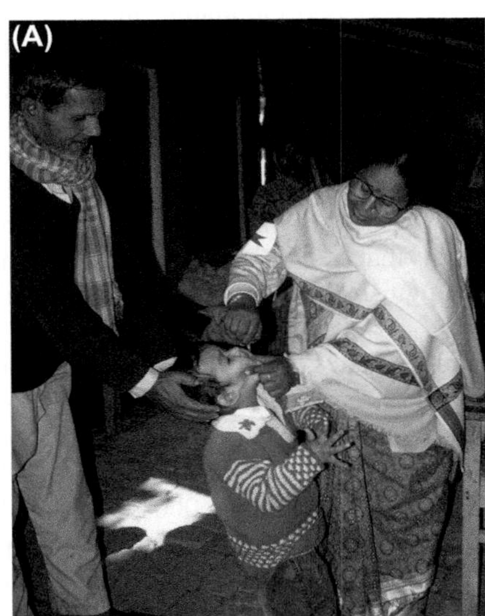

FIGURE 14.14 **National Immunization Days.** All children within a country were immunized on National Immunization Days. (A) A clinic worker administers an oral polio vaccine to a child in Uttar Pradesh, India. (B) Scouts wearing their "no Polio" headbands as part of the India's celebration of National Immunization Day in 2000. *Photographs courtesy of the CDC and Christine Zahniser, BSN, MPH.*

In 1988, the year the global initiative was announced, it is estimated that 350,000 annual polio cases occurred globally, paralyzing 1000 children each day. Polio was endemic is 125 countries at the time. Significant and even life-threatening efforts have been made by millions of health workers and volunteers; negotiations and "humanitarian pauses" have been drawn up to immunize children in warring countries; millions and millions of homes have been visited in door-to-door immunization campaigns, which even took place by boat, bicycle, horse, or on foot (Fig. 14.15). In 1995, over 93 million children under 3 years of age were immunized in one day during India's first national immunization day. In 1996, almost two thirds of the children in the world under 5 years of age—420 million children—were immunized during national immunization days. Efforts were successful in substantially reducing the number of reported cases: by 1996, only 3995 cases of poliovirus were reported. Serotype 2 ceased circulating in 1999.

The goal of polio eradication by the year 2000 was not achieved, although endemic poliovirus infections had been reduced to 20 countries. This number had dropped to four countries by 2006—India, Nigeria, Pakistan, and Afghanistan. India has since been removed from the list after its last case was reported in 2011, and so has Nigeria, which logged its last case (of type 3) in November 2012 (Fig. 14.16). The effort is close: a total of 72 cases of wild poliovirus were reported in 2015, 54 from Pakistan and 20 from Afghanistan.

Immunization efforts have been plagued by several consistent challenges. A major issue has been the importation

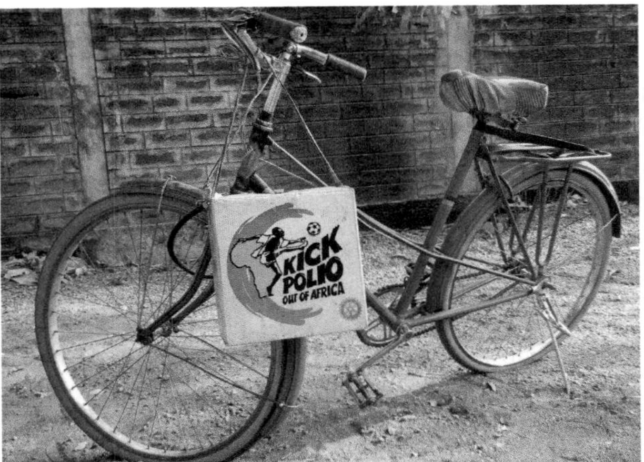

FIGURE 14.15 **The Global Polio Eradication Initiative.** Health workers and volunteers paid individual visits to millions of houses in door-to-door immunization campaigns, which often took place by boat, bicycle, horse, or on foot. Shown here is one of the bicycles used in the central African country of Chad to deliver vaccines to more rural areas of the country. Draped over the handlebars is a vaccine-carrying satchel. *Figure courtesy of the CDC and Minal K. Patel, MD.*

Wild poliovirus — 1988: 125+ countries

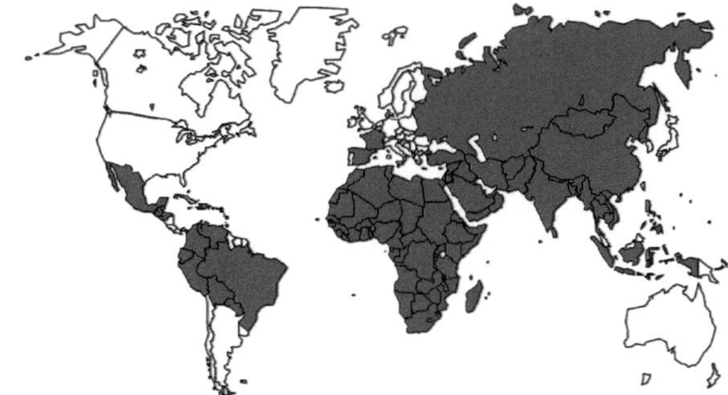

Wild poliovirus — 2006: 4 countries

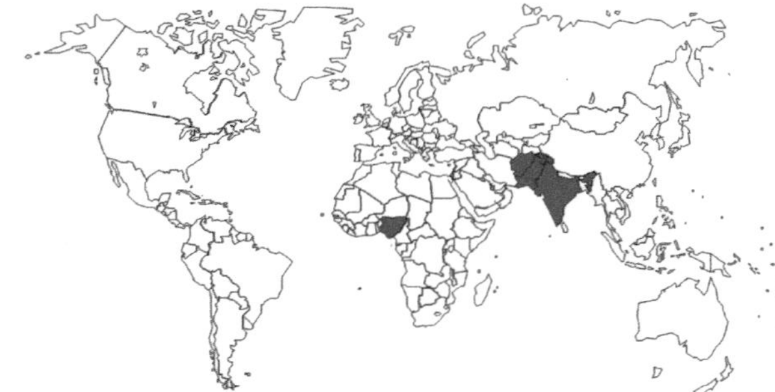

Wild poliovirus — October 2015: 2 countries

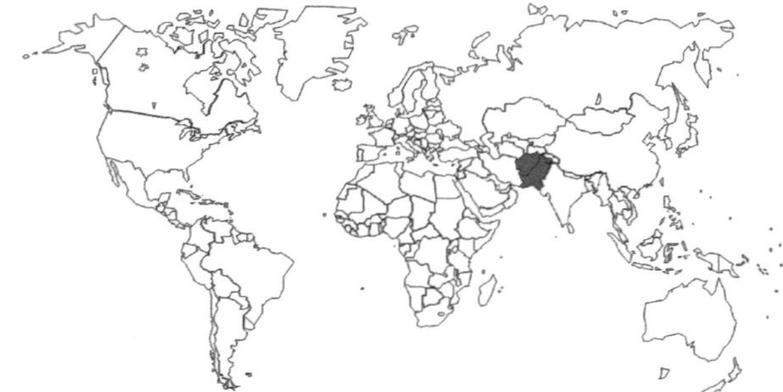

FIGURE 14.16 **The eradication of poliovirus.** The Global Polio Eradication Initiative was announced in 1988, when over 125 countries had endemic wild poliovirus infections. By 2006, this number dropped to four countries—India, Nigeria, Pakistan, and Afghanistan. India and Nigeria logged their last case in 2011 and 2012, respectively. *Maps updated from Centers for Disease Control and Prevention, 2015. Poliomyelitis. In: Hamborsky, J., Kroger, A., Wolfe, C., (Eds.), Epidemiology and Prevention of Vaccine-Preventable Diseases, thirteenth ed. Public Health Foundation, Washington DC, pp. 175–186.*

of polio into polio-free countries. For example, in 2010, polio was endemic in only 4 countries, but 10 bordering countries experienced imported poliovirus infections. This is of particular concern in countries where refugees are fleeing due to war or civil unrest, as the large numbers of unvaccinated individuals could incite outbreaks. Even if the country has been declared polio-free, new generations of children are susceptible if not vaccinated. Vaccine workers have often set up camps at the borders of countries, prepared to administer vaccine doses to children passing from one country to another.

Eradication in Pakistan, Afghanistan, and Nigeria has been stifled due to political unrest, poor health infrastructures, and opposition from militant groups, who are distrustful of aid workers and the vaccine contents, which have been accused of harboring chemicals to sterilize children. Since 2012, Pakistan and Afghanistan militants have killed over 70 polio vaccination workers. Vaccine awareness campaigns have been unsuccessful in convincing people of the purpose of vaccination efforts, so in 2015, the Pakistani government resorted to arresting parents who had been giving chronic refusals to vaccine workers.

The choice of the Sabin OPV was prudent due to its reduced cost and ease of administration, but circulating VDPV strains have become a confounding factor in the eradication of poliovirus. The live virus vaccine is excreted in stool, and in underimmunized areas of the world, VDPV strains can circulate and cause VAPP. Serotype 2 has historically caused the most circulating VDPV: since 2000, it has caused 683 of 783 cases (87.2%), despite that wild poliovirus type 2 ceased circulating in 1999. Taking this into consideration, eradication efforts are switching from the use of the trivalent OPV (with all three serotypes) to the bivalent OPV (containing types 1 and 3), and it has been recommended that routine immunization programs should introduce at least one dose of the IPV, toward the goal of an eventual switch to the inactivated vaccine preparation. Laboratory tests that examine the genomic sequences of the virus can distinguish between wild poliovirus and a VDPV strain.

In the United States, occasional epidemics in the pre-World War I years became seasonal epidemics by the early 1930s. Large epidemics occurred in the 1940s and 1950s, peaking with 57,000 cases in the epidemic of 1952, which caused 3145 deaths and left 21,000 people with mild to permanent paralysis. By 1957, 2 years after the approval of the Salk vaccine, cases of polio had dropped to less than a tenth of their usual numbers (Fig. 14.17A). The Salk and Sabin vaccines were viewed positively by parents who were anxious to prevent paralysis in their children (Fig. 14.17B and C). The last cases of paralytic poliomyelitis in the United States were in 1979, when poliovirus was imported from the Netherlands and infected unvaccinated individuals in Amish communities in several Midwestern states. In the following 20 years, 162 cases of paralytic poliomyelitis were reported; 6 were imported from other countries, 154 were VAPP cases caused by the OPV, and 2 were of indeterminant origin. Poliovirus vaccination continues to occur, because outbreaks remain possible from imported cases until the virus is eliminated from every location in the world.

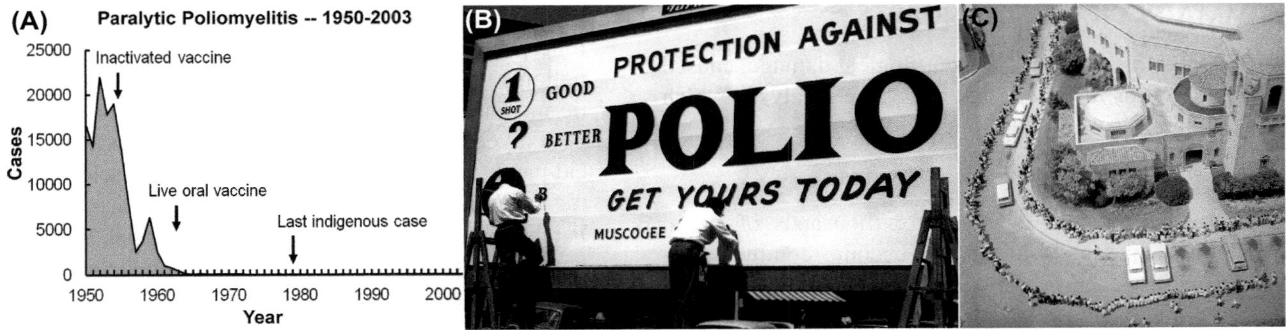

FIGURE 14.17 **Vaccination efforts in the United States.** Cases of paralytic poliomyelitis peaked in 1952, when over 21,000 people had mild to permanent paralysis. Cases dropped precipitously following the approval of the inactivated Salk vaccine in 1955. *Modified from Centers for Disease Control and Prevention, 2015. Poliomyelitis. In: Hamborsky, J., Kroger, A., Wolfe, C., (Eds.), Epidemiology and Prevention of Vaccine-Preventable Diseases, thirteenth ed. Public Health Foundation, Washington DC, pp. 175–186.* (B) A billboard in Columbus, Georgia, used to promote awareness of the polio vaccine. *CDC.* (C) An image from 1962 of a gymnasium in San Antonio, Texas, shows a line of people wrapping around the block, awaiting the polio vaccine. *Courtesy of CDC and Stafford Smith.*

Case Study: Imported Vaccine–Associated Paralytic Poliomyelitis—United States, 2005

Excerpted from the Morbidity and Mortality Weekly Report, 55(4), pp. 97–99.

In March 2005, an Arizona woman aged 22 years contracted paralytic polio while traveling in Central and South America. She arrived in Costa Rica on January 14, 2005, to participate in a university-sponsored study-abroad program. During her stay with a local family, she visited several tourist locations along the Pacific coast in Costa Rica, Panama, Nicaragua, and Guatemala. On March 2, after she returned to the host family's home, she had fever and general malaise. During the next 24 hours, her symptoms worsened, and she began to have headache and neck and back pain. On March 6, she experienced acute leg weakness and was hospitalized locally and soon transferred to a hospital in San Jose, Costa Rica. On March 9, she was transported by air to Phoenix, Arizona, for further evaluation.

Upon admission to a hospital in Phoenix, the patient had lower extremity weakness and respiratory failure requiring intubation. Electrodiagnostic studies displayed reduced compound muscle action potentials, normal sensory nerve action potentials, and widespread denervation, consistent with a severe, asymmetric process involving anterior horn cells or motor axons. Magnetic resonance imaging of the cervical and thoracic spine demonstrated signal abnormality in the anterior cord, indicative of anterior horn cell involvement. Stool specimens were collected on March 20 and sent to the CDC polio reference laboratory. The specimens were positive for Sabin-strain poliovirus types 2 and 3.

During the course of hospitalization, the patient recovered respiratory function, was transferred to a rehabilitation center for physical and occupational therapy, and was eventually discharged home for outpatient therapy. Sixty days after the onset of weakness, she had residual weakness in both legs.

The patient had never been vaccinated with either OPV or IPV because of a religious exemption. The Costa Rican family with whom she lived consisted of a mother, father, and daughter with no young children. The host family's son and daughter-in-law lived next door with two children, aged 2 months and 3 years, who visited the host family frequently. The infant received his first dose of OPV on January 19, 2005, 4 days after the woman arrived to live with the host family. Vaccination records indicated that both children were up to date for all other routine vaccinations. She had no underlying medical or immune-compromising conditions.

Reported by: Landaverde, M., MD, Pan American Health Organization. Salas D., MD, Humberto, M., MD, Ministry of Health Costa Rica. Howard, K., Walker R., MD, St. Joseph's Hospital and Medical Center, Phoenix; Everett, S., MPH, Robyn, S., Yavapai County Health Dept, Prescott; Erhart, L., MPH, Anderson, S., MPH, Goodykoontz, S., Arizona Dept of Health Svcs. Pallansch, M., PhD, Sejvar, J., MD, Div of Viral and Rickettsial Diseases, National Center for Infectious Diseases; Kenyan, K., MPH, Alexander, J., MD, Alexander, L., MPH, Seward, J., MBBS, Epidemiology and Surveillance Div, National Immunization Program, CDC.

SUMMARY OF KEY CONCEPTS

- Poliomyelitis refers to the neurological damage, caused by poliovirus infection, that can cause paralysis, which occurs in less than 1% of infections.

Section 14.1 The Early Years of Poliovirus

- Sporadic polio has occurred for thousands of years, although epidemics have only become common during the last century. This is possibly due to improved sanitation, which would result in reduced endemicity of poliovirus and a population more susceptible to epidemics because children would not be infected while they were protected by their mother's antibodies.
- Landsteiner and Popper discovered poliovirus in 1909. Difficulties in propagating the virus in small mammals stymied research progress. In 1949, Enders, Weller, and Robbins determined how to grow the virus using tissue culture, which paved the way for the mass production of virus for vaccine formulations.
- In 1955, Jonas Salk's laboratory developed a formalin-inactivated virus vaccine against polio, which was known as the inactivated polio vaccine. Albert Sabin later developed the oral polio vaccine containing live attenuated poliovirus. Both are highly effective in producing immunity against all three poliovirus serotypes.
- The attenuated strains in the OPV infrequently revert back to a neurovirulent strain capable of causing VAPP. In locations where poliovirus had been eliminated, VAPP has been the major cause of poliovirus-induced paralysis.
- Over 100 million poliovirus vaccines administered between 1955 and 1963 contained SV40, a simian virus that is capable of transforming cells and causing tumors in hamsters. SV40 DNA has also been found in human tumors, although SV40 is also found in normal human tissues and in people who could not have received contaminated vaccine. Thus far, it has been difficult to show that SV40 is a cause of human cancers.

Section 14.2 Clinical Course of Infection

- Poliovirus is within the *Picornavirales* order, *Picornaviridae* family, and *Enterovirus* genus. It is officially the *Enterovirus C* species. Three poliovirus serotypes exist: type 1, 2, and 3.
- Poliovirus is nonenveloped and acid resistant, primarily transmitted through the fecal–oral route.
- Poliovirus infects cells of the oropharynx and gastrointestinal tract mucosa before replicating within nearby locations, specifically the tonsils and Peyer's patches, respectively. This produces the minor viremia that leads to the infection of distal sites and a major viremia that results in the infection of the central nervous system.
- Infection is inapparent in 72% of infections. Twenty-four percent of infected individuals exhibit abortive poliomyelitis, 1–5% present with nonparalytic aseptic meningitis, and <1% of cases progress to paralytic poliomyelitis in which the individual develops acute flaccid paralysis, the weakening of skeletal muscles due to the damage of the nerves that innervate them.
- Paralytic poliomyelitis is classified into spinal, bulbar, or bulbospinal poliomyelitis, depending upon the site of the nerves that are damaged. Paralysis can be temporary or permanent. Post-polio syndrome is muscle weakness and fatigue that appear 15–40 years after infection.

Section 14.3 Poliovirus Replication

- Poliovirus is a small, nonenveloped icosahedral virus that possess a 7.5 kb +ssRNA genome that is infectious.
- Poliovirus attaches to cells using CD155, an adhesion molecule. The virion is endocytosed through clathrin- and caveolin-independent pathways, and the capsid undergoes a conformation change whereby VP1 forms a pore in the endosomal membrane through which the genome is released.
- Translation begins immediately through ribosomes binding an IRES in the 5′-NTR region of the genome. Poliovirus has a single open reading frame that is cleaved sequentially into intermediate precursors and mature proteins. VPg functions as a primer for genome replication, which occurs in replication complexes.
- VP1, VP3, and VP0 spontaneously assemble into protomers that form the procapsid, which is packaged with the +ssRNA genome. Maturation occurs as VP0 is cleaved into VP2 and VP4 by a host protease. Release is through cell lysis or exocytosis.

Section 14.4 Epidemiology and Worldwide Eradication Efforts

- Permanent eradication of poliovirus through vaccination is theoretically possible because poliovirus infects only humans, has a limited number of serotypes, and does not cause persistent infections. Most infections are asymptomatic, however, and so all children must be vaccinated, and four doses are required for complete immunity.
- The Global Polio Eradication Initiative used the OPV due to its ease of administration and lower price. It focused on promoting high routine immunization coverage, supplemental immunization through "National Immunization Days," effective surveillance, and door-to-door immunization campaigns.
- 350,000 annual polio cases occurred worldwide in 1988, the beginning of the campaign. In 2015, 72 cases of wild poliovirus were reported in only 2 countries: 54 from Pakistan and 20 from Afghanistan.

FLASH CARD VOCABULARY

Poliovirus	Asymmetrical
Poliomyelitis	Spinal poliomyelitis
Foot drop deformity	Bulbar poliomyelitis
Dorsiflexion	Bulbospinal poliomyelitis
Inactivated polio vaccine (IPV)	Post-polio syndrome (PPS)
Oral polio vaccine (OPV)	VPg
Vaccine-derived poliovirus (VDPV)	VP1, VP2, VP3, VP4
Vaccine-associated paralytic polio (VAPP)	VPg unlinkase
Fomites	$3D^{pol}$
Primary (minor) viremia	$2A^{pro}/3CD^{pro}/3C^{pro}$
Secondary (major) viremia	Replication complex (RC)
Abortive poliomyelitis	*Cis*-acting response element (CRE)
Nonparalytic aseptic meningitis	Protomers
Paralytic poliomyelitis	Procapsid
Acute flaccid paralysis	National Immunization Days

CHAPTER REVIEW QUESTIONS

1. Symptomatic polio infections used to be sporadic in nature. Why is it thought that polio began causing epidemics in the 1900s?
2. Why were experiments that infected rabbits or guinea pigs with poliovirus not helpful in characterizing the virus?
3. Why was the discovery of how to propagate poliovirus using tissue culture a necessary step in the creation of a poliovirus vaccine?
4. What are the differences and similarities between the Salk and Sabin vaccines?
5. What is VAPP, and what causes it?
6. SV40 contaminated over 100million doses of polio vaccine between 1955 and 1963, and SV40 has been found in human tumors. Why isn't this enough to show that SV40 is the cause of certain human cancers?
7. For purposes of infection, why is it necessary for the poliovirus capsid to be acid resistant?
8. What are the clinical outcomes of poliovirus infection? List them in order of severity and describe each condition.
9. What causes the three types of paralytic poliomyelitis, and what are their effects?
10. As a virus, why is it economical to encode a single polyprotein instead of several individual proteins?
11. Make a list of the seven stages of viral replication and describe how poliovirus achieves each one.
12. What have been the biggest challenges in the global eradication of polio?

FURTHER READING

Bird, S.W., Kirkegaard, K., 2015. Escape of non-enveloped virus from intact cells. Virology 479–480, 444–449.

Brandenburg, B., Lee, L.Y., Lakadamyali, M., Rust, M.J., Zhuang, X., Hogle, J.M., 2007. Imaging poliovirus entry in live cells. PLoS Biol. 5, 1543–1555.

Cameron, C.E., Suk Oh, H., Moustafa, I.M., 2010. Expanding knowledge of P3 proteins in the poliovirus lifecycle. Future Microbiol. 5, 867–881.

Centers for Disease Control, 2015. Poliomyelitis. In: Hamborsky, J., Kroger, A., Wolfe, C. (Eds.), Epidemiology and Prevention of Vaccine-Preventable Diseases, thirteenth ed. Public Health Foundation, Washington, DC, pp. 297–310.

College of Physicians of Philadelphia, 2014. Polio – Timelines – History of Vaccines. http://www.historyofvaccines.org/content/timelines/polio.

De Jesus, N.H., 2007. Epidemics to eradication: the modern history of poliomyelitis. Virol. J. 4, 70.

Koike, S., Taya, C., Kurata, T., et al., 1991. Transgenic mice susceptible to poliovirus. Proc. Natl. Acad. Sci. U.S.A. 88, 951–955.

Lévêque, N., Semler, B.L., 2015. A 21st century perspective of poliovirus replication. PLoS Pathog. 11, e1004825.

Lyle, J.M., Bullitt, E., Bienz, K., Kirkegaard, K., 2002. Visualization and functional analysis of RNA-dependent RNA polymerase lattices. Science 296, 2218–2222.

Oshinsky, D.M., 2006. Polio: An American Story. Oxford University Press, New York.

Racaniello, V.R., 2006. One hundred years of poliovirus pathogenesis. Virology 344, 9–16.

Skern, T., 2010. 100 years poliovirus: from discovery to eradication. A meeting report. Arch. Virol. 155, 1371–1381.

Trevelyan, B., Smallman-Raynor, M., Cliff, A.D., 2005. The spatial dynamics of poliomyelitis in the United States: from epidemic emergence to vaccine-induced retreat, 1910–1971. Ann. Assoc. Am. Geogr. 95, 269–293.

Virgen-Slane, R., Rozovics, J.M., Fitzgerald, K.D., et al., 2012. An RNA virus hijacks an incognito function of a DNA repair enzyme. Proc. Natl. Acad. Sci. 109, 14634–14639.

World Health Organization, 2015. Global Polio Eradication Initiative. http://polioeradication.org.

Chapter 15

Poxviruses

Poxviruses are a family of complex dsDNA viruses that infect vertebrates and invertebrates alike. The poxviruses acquired their name due to the visible lesions caused by the viruses. On humans, these were referred to in the 10th century as *pocs*, meaning bag or pouch, and later as *pockes*.

Smallpox was one of the most deadly plagues in human history, a devastating disease that killed about a third of those that were infected. The smallpox rash often left survivors blinded or visibly scarred for life, including on the face (Fig. 15.1). **Variola virus** (VARV), the virus that causes smallpox, is thought to have killed more people throughout history than all other infectious diseases combined. In the 18th century, smallpox killed over 400,000 people in Europe *each year*. It caused the death of 300–500 million people in the 20th century alone, and in the 1950s—150 years after the smallpox vaccine was invented—50 million global cases of smallpox were still occurring each year. The virus was the first ever to be eradicated from the earth, the importance of which cannot be overstated. This was a monumental undertaking achieved through massive vaccination and surveillance efforts.

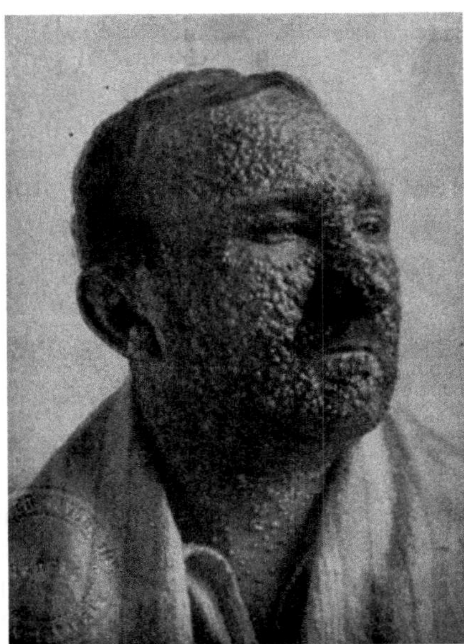

FIGURE 15.1 **Smallpox rash.** This photograph from 1912 illustrates that the smallpox rash could be very extensive, leading to permanent scarring and sometimes blindness. *Image courtesy of the Illinois Department of Public Health.*

15.1 TAXONOMY

The *Poxviridae* family contains 38 viruses that together infect a wide range of hosts, including mammals, birds, reptiles, and insects. The family includes two subfamilies, *Chordopoxvirinae* and *Entomopoxvirinae* (Table 15.1). *Entomopoxvirinae* members infect invertebrates (*entomo* indicates an insect, such as in *entomology*, the study of insects), while the *Chordopoxvirinae* subfamily is composed of the poxviruses that infect vertebrates, including humans *(chordo* is from the Greek for "cord," referring to the spinal cord of vertebrates). There are 10 genera classified within the *Chordopoxvirinae* subfamily. The *Orthopoxvirus* genus includes VARV, the cause of smallpox. It also includes Cowpox virus, Monkeypox virus, Camelpox virus, and Vaccinia virus (VACV), which is currently used in the smallpox vaccine (Fig. 15.2A). All of the viruses within the *Orthopoxvirus* genus generate **cross-reactive** immune responses, meaning that antibody-producing B lymphocytes and cytotoxic T lymphocytes produced during infection with one virus protect against infection by the other members of the genus (Fig. 15.2B). This is the scientific basis behind why Edward Jenner's cowpox inoculations and current-day smallpox vaccines containing VACV protect against VARV.

VARV is the only orthopoxvirus that infects humans exclusively. There are two types of VARV, *Variola major* and *Variola minor*, which have distinct biological properties that result in different clinical severity.

15.2 CLINICAL COURSE OF VARIOLA INFECTION

Smallpox is the clinical condition that is caused by VARV. The word *variola* was first used in 570 AD by Bishop Marius of Avenches, Switzerland, to describe the characteristic smallpox rash. The word is derived from the Latin *varius*, meaning "spotted," or *varus*, meaning "pimple." VARV is transmitted through airborne transmission of large respiratory droplets. It is also possible to transmit the virus through fomites that have been in contact with infectious material from the smallpox rash. Smallpox cases historically peaked in winter and early spring, which is thought to be because the virus is more stable at lower temperatures and humidity. In laboratory experiments,

Essential Human Virology.

TABLE 15.1 Orthopoxviruses

	Species	Abbreviation	Genome size[a] (kb)	Notes
Family	*Poxviridae*			
Subfamily	*Chordopoxvirinae*			
Genus	*Orthpoxvirus*			
Species	*Camelpox virus*	CMLV	206	Narrow host range: infects camels only
	Cowpox virus	CPXV	224	Broad host range, original smallpox vaccine
	Ectromelia virus	ECTV	210	Causes mousepox
	Monkeypox virus	MPXV	206	Causes serious clinical illness in humans
	Raccoonpox virus	RCNV	215	Indigenous to USA
	Taterapox virus	TATV	198	Causes gerbilpox
	Vaccinia virus	VACV	165–200	Current smallpox vaccine, prototype orthopoxvirus
	Variola virus	VARV	185–188	Cause of smallpox, narrow host range (humans)
	Volepox virus	VPXV	Not sequenced	Infects voles, found in USA

[a]*Strains may have different genome sizes.*

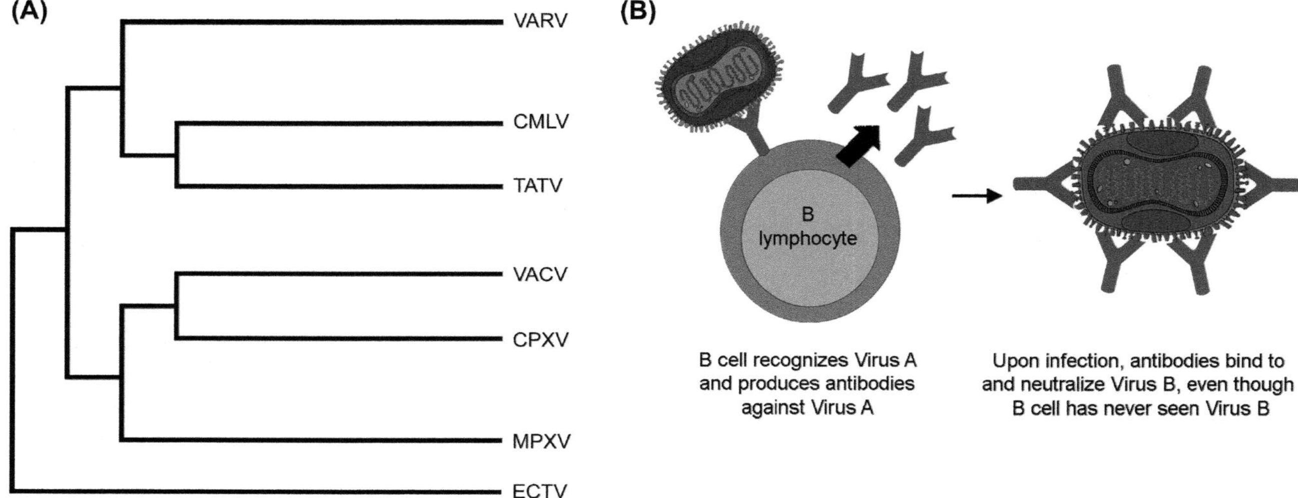

FIGURE 15.2 **Orthopoxvirus characteristics.** (A) Phylogenetic tree showing the genetic relationship between certain orthopoxviruses. See Table 15.1 for abbreviations. *(Simplified from data derived from Babkin et al., 2015. Viruses 7, 1100–1112, and Smithson et al., 2014. PLoS One 9 (3), e91520.)* (B) All members of the *Orthopoxvirus* genus result in the creation of cross-reactive antibodies by B cells that recognize the other orthopoxviruses. Cross-reactive cytotoxic T cells are also generated. This is the scientific basis behind why vaccination with vaccinia virus protects against variola virus infection.

90% of aerosolized VARV is inactivated within 24 h, although virions captured in the crusts of smallpox lesions can remain infectious for years.

The incubation period for variola infection is 10–14 days (range: 7–17 days) (Fig. 15.3). During this asymptomatic period, the virus replicates within the oropharyngeal and respiratory mucosa, spreading to the regional lymph nodes. A primary viremia occurs 3–4 days after infection, spreading the virus to the bone marrow, spleen, and lymphatics. A second viremia develops around day 8–10 of infection and is followed by a prodromal period during which an infected individual appears quite ill, presenting with a high fever (101–104°F/38–40°C) that is most often accompanied by headache, severe back pain, malaise, **prostration** (extreme weakness or exhaustion), and sometimes vomiting and abdominal pain, among other symptoms (Table 15.2).

The fever subsides within 2–4 days, at which point the person feels better. It is at this time that the characteristic

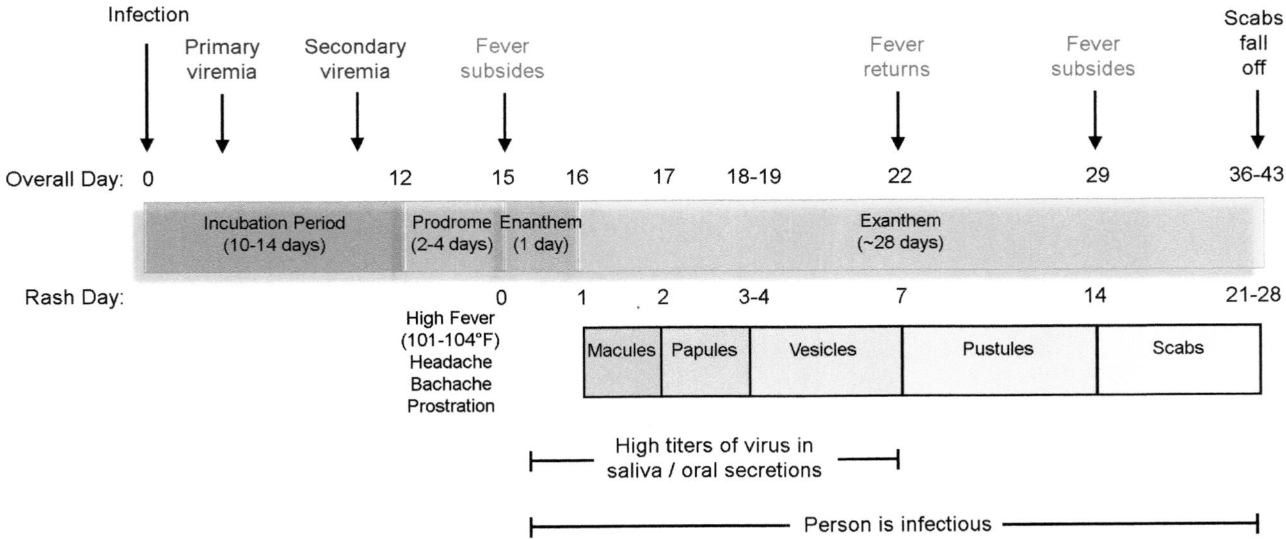

FIGURE 15.3 **Clinical course of smallpox infection.** The average incubation period is 10–14 days following respiratory infection with variola virus. During this time, the virus replicates within the mucosa and regional lymph nodes. A primary viremia around day 3–4 spreads the virus to the bone marrow, spleen, and lymphatics, which causes a secondary viremia around day 8–10. The prodrome lasts 2–4 days, at which point the enanthem appears on the tongue and mucous membranes. These lesions rupture, which releases large amounts of virus into the throat and mouth. The exanthem lasts about a month and progresses through four stages (macules, papules, vesicles, and pustules) before scabbing over. The person is no longer infectious when the crusts finally separate from the skin.

TABLE 15.2 Frequency of Symptoms Observed With Smallpox Infection

Symptom	Frequency (Percentage)
Fever	100
Headache	90
Backache	90
Chills	60
Vomiting	50
Pharyngitis	15
Delirium	15
Abdominal Pain	13

Data from Rao, A.R., 1972. Smallpox. Kothari Book Depot, Bombay, describing an outbreak in India.

rash appears as a result of the replication of the virus within the small blood vessels of the skin. The rash begins as an **enanthem**, a rash of tiny small spots found on the tongue and mucous membranes of the oropharynx and nasopharynx (see Fig. 10.1 for a review of these anatomical locations). These lesions enlarge quickly and rupture, releasing large amounts of virus into secretions within the throat and mouth that facilitate the transmission of the virus to others. An **exanthem**, or external rash, follows within 24h. A few red spots, known as "herald spots," initially appear on the face, particularly on the forehead. Lesions then appear on the proximal portions of the limbs and spread to the trunk and extremities within 1–2 days. The **centrifugal rash** is characterized by lesions that are more prevalent on the extremities and face (Fig. 15.4A), providing a visual clue to differentiate variola infection from other viruses that cause rashes. The disease most likely to be confused with smallpox is chickenpox, caused by varicella zoster virus (see Fig. 13.6 for a comparison of these two rashes). Smallpox lesions are often present on the palms of the hands and soles of the feet, which rarely occurs in individuals with chickenpox.

Patients are most infectious during the first week of the rash, correlating with high amounts of virus in saliva and oral secretions. The variola rash progresses through four stages over the course of 1–2 weeks. Although they may be of differing sizes, the lesions all progress together through the stages within a given area of the body, another clue that differentiates smallpox from chickenpox, where lesions are found in various stages within a single location of the body. **Macules** are red spots that develop into raised bumps, known as **papules**, by the second day of the rash. Together, these known as a **maculopapular** rash. Papules fill with opalescent fluid to become blistery **vesicles** by the third or fourth day of the rash. High titers of virus are found in the vesicular fluid. It is at this point that the fever returns and remains high until the lesions have crusted over. Vesicles are firm to the touch and can possess a central dimple called an **umbilication** (Fig. 15.4B), differentiating them from a

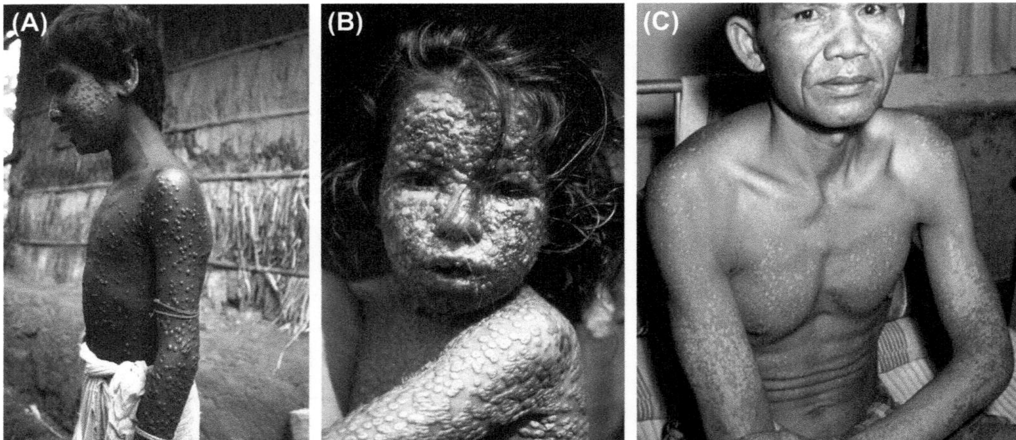

FIGURE 15.4 **Characteristics of the smallpox rash.** (A) The centrifugal rash is characterized by lesions that are more prevalent on the extremities and face, as seen in this photograph of a Bengali boy with variola major. *(Courtesy of CDC/Jean Roy.)* (B) This girl with confluent smallpox has lesions on her arm that possess visible umbilications. *(Courtesy of CDC/James Hicks.)* (C) A depigmented scar remains after crusts separate, as seen on this Indonesian man recovering from smallpox. *(Courtesy of CDC/Dr. J.D. Millar.)*

superficial chickenpox rash. They develop into deep-seated pus-filled **pustules** by day 7. The fluid slowly leaves the pustules, and they crust over into a scab by day 14. When the scab falls off at day 21–28, a depigmented scar is left behind (Fig. 15.4C).

Because the scabs contain infectious virus, patients remain infectious until their scabs separate. However, transmission is more likely from the rupture of pharynx lesions that release large amount of virus into saliva and respiratory secretions. Contact infection through the skin is less common and thought to require that the virus pass through the epidermis and reach the dermal layer in order to establish an infection. Virions in scabs or vesicular fluid display different morphologies when viewed with an electron microscope. The "capsular" or C form is found in dried scabs, while the "mulberry" or M form is associated with vesicular fluid (so-called due to their mulberry-like appearance that resembles blackberries).

In some cases, the smallpox lesions are so numerous that there is no normal skin visible in between them. This is known as **confluent smallpox** (see Fig. 15.4B). It is associated with severe, prolonged disease with a high fever and symptoms that persist even after scabs have formed. It is also associated with a fatality rate of over 50%.

Variola major is the more severe type of VARV. It has an overall fatality rate of around 30%, with death occurring between day 10 and 16 of illness. In contrast, variola minor (also known as "alastrim") causes less severe disease and has a fatality rate of 1% of less. Variola major is categorized into four principal clinical presentations: **ordinary**, **modified**, **flat (malignant)**, and **hemorrhagic smallpox**. Ordinary smallpox occurred in ~90% of the smallpox cases of unvaccinated individuals and is characterized by the symptoms described above. Modified smallpox is a mild, rarely fatal disease that occurred in around 5% of people who were previously vaccinated against smallpox. The person usually had mild symptoms and lesions that progressed quickly. On the other hand, the flat and hemorrhagic forms of smallpox were usually fatal. Flat smallpox, which occurred in about 5–10% of people that contracted smallpox, is characterized by vesicles that are velvety, nonumbilicated, and flush with the skin, rather than raised (Fig. 15.5A). These patients were very ill. Their fever remained high throughout the course of infection, and skin lesions developed very slowly. They also had difficulty eating or drinking due to severe throat pain from pharyngeal lesions. Large areas of skin would slough off with the slightest pressure, leading to severe skin destruction that resulted in burnlike wounds susceptible to secondary bacterial infections. Flat smallpox occurred most frequently in children and has been shown to have >90% fatality rate. Hemorrhagic smallpox, which occurred in 2–3% of cases, was more common in adults and pregnant women. As the name suggests, hemorrhagic smallpox is associated with extensive bleeding into the skin, mucous membranes, and gastrointestinal tract (Fig. 15.5B). Death generally occurred between day 5 and 7 of illness, and hemorrhagic smallpox had a >98% fatality rate.

Individuals that contracted smallpox from a person with flat or hemorrhagic smallpox were no more likely to present with these forms of the disease, indicating that the clinical presentation of variola major does not correlate with a particular strain of virus. Rather, it is thought that the magnitude and type of a person's immune response to the virus correlates with the clinical presentation that ensues. Regardless of the clinical manifestation, the majority of those who survived were permanently scarred. Scarring of the eye could also lead to blindness, which occurred in about 1% of

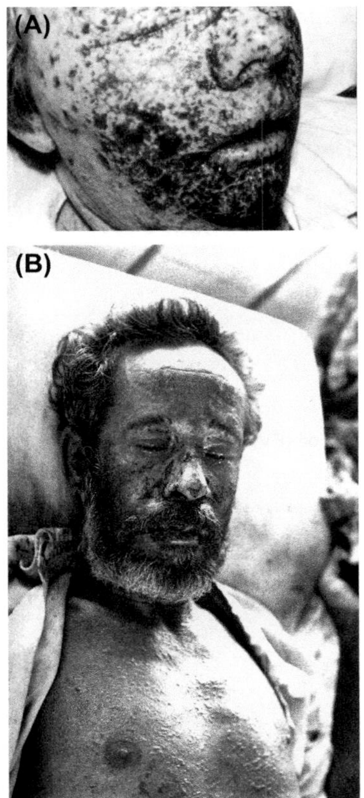

FIGURE 15.5 **Flat and hemorrhagic forms of variola major.** Flat and hemorrhagic forms of variola major are nearly always fatal. (A) Individuals with flat smallpox have velvety lesions that are flush to the skin, as observed with the papules on this woman's face. *(Courtesy of CDC/Dr. Robinson.)* (B) Hemorrhagic smallpox is associated with extensive bleeding into the skin, mucous membranes, and gastrointestinal tract. Hemorrhaging lesions are visible on this man's face and chest. *(Courtesy of CDC/WHO and Stanley O. Foster, M.D., M.P.H.)*

smallpox survivors. VARV does not cause chronic or persistent infections, and immunity against the virus is thought to be long-lived, if not lifelong.

Study Break
What are at least four visible clues to distinguish a smallpox rash from a chickenpox rash?

15.3 POXVIRUS REPLICATION CYCLE

Poxviruses are enveloped, brick-shaped viruses of approximately 350 nm by 250 nm (Fig. 15.6A and B), placing them among the largest human viruses. In fact, they are just barely discernable using a light microscope, although an electron microscope is required to see any structural features of the virion. Poxviruses possess large dsDNA genomes (Table 15.1) that encode ~200 genes, although they differ from most dsDNA viruses in that

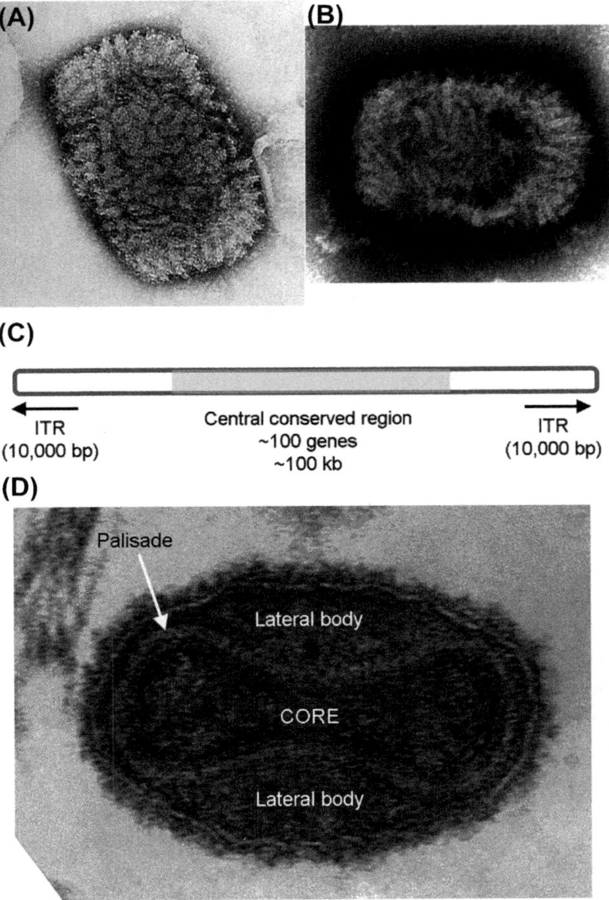

FIGURE 15.6 **Molecular characteristics of orthopoxviruses.** Orthopoxviruses have large, brick-shaped virions. Visible in these electron micrographs of vaccinia virus (A) and variola virus (B), ridges on the external surface of the virion are formed by the outer membrane. *(Courtesy of CDC/Dr. Fred Murphy and Sylvia Whitfield (A) and J. Nakano (B).)* (C) Poxviruses have linear dsDNA genomes that have inverted terminal repeats (ITRs) on each end that create a closed hairpin loop. The central region consists of ~100 genes conserved in all chordopoxviruses. (D) This transmission electron micrograph of variola virus clearly shows the dumbbell-shaped core and adjacent lateral bodies. *(Courtesy of CDC/Dr. Fred Murphy and Sylvia Whitfield.)*

they replicate entirely within the cytoplasm. The two strands of the dsDNA genome contain complementary regions at the ends, called **inverted terminal repeats (ITRs)**, that fold back upon themselves into hairpins, forming a covalently closed, continuous chain of DNA nucleotides (Fig. 15.6C).

The poxvirus virion is complex in nature. The viral dsDNA associates with at least four different proteins that organize the DNA into nucleosome-like structures resembling eukaryotic chromatin. This complex is surrounded by a well-defined, dumbbell-shaped core that is composed of two layers, a thinner inner layer and an outer **palisade** layer of aligned rod-shaped or possibly T-shaped proteins (Fig. 15.6D). (A *palisade* is a fence made of adjacent stakes that was constructed around castles and forts.) Several proteins are also found within the core, including the DNA-dependent RNA polymerase and 20+ enzymes involved in transcription and processing of early mRNA transcripts.

The core is wrapped with a lipid membrane, and the resultant areas between the dumbbell-shaped core and envelope are known as **lateral bodies**. These are visible using standard electron microscopy and assist in the formation of the core. The surface of the envelope appears ridged due to protrusions of the virion around which the membrane forms (see Fig. 15.6A and B).

Poxvirus virions have two infectious forms. The **mature virion** (**MV**, also known as intracellular mature virus) has a dumbbell-shaped core wrapped with a single lipid membrane (Fig. 15.7A), while the **enveloped virion** (**EV**, also known as extracellular enveloped virus) possesses an additional lipid membrane (Fig. 15.7B). Both forms contain the same internal structure and constituents. At least 20 viral proteins are associated with the MV envelope and nine additional proteins with the EV. Because different surface proteins are embedded within the two membranes, the mechanism of entry differs depending upon the form of the virion. MVs are very stable and are thought to be involved in infection *between* hosts, while the fragile extra envelope of EVs result in a virion that is better suited for infection of cells *within* a single host. How MVs and EVs are generated will be discussed later in this chapter.

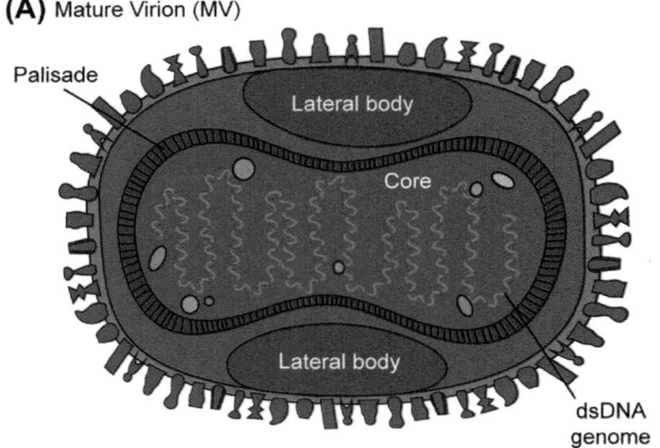

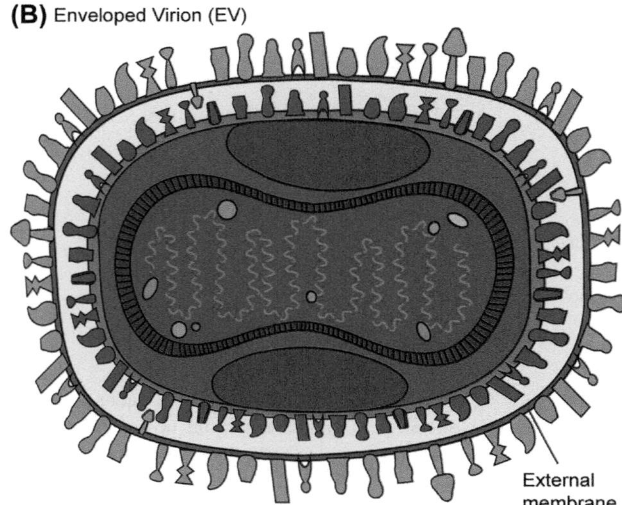

FIGURE 15.7 **The two infectious forms of vaccinia virus.** Poxviruses have two infectious forms. The mature virion (A) possesses a dumbbell-shaped core that encloses the dsDNA genome and several viral proteins. Lateral bodies assist in the formation of the core's shape. Over 20 proteins are associated with the surrounding lipid envelope. The enveloped virion (B) possesses an additional external membrane with a different complement of at least nine viral proteins. Mature virions are very stable and are thought to be involved in dissemination between hosts, while the fragile extra envelope of enveloped virions result in a virion that is better suited for infection within a single host.

15.3.1 Poxvirus Attachment, Penetration, and Uncoating

Research involving live VARV can only be performed within two biological safety level 4 facilities, located either at the Centers for Disease Control and Prevention (CDC) in Atlanta, Georgia, USA, or the State Research Center of Virology and Biotechnology (Vector Institute) in Koltsovo, Russia. Because of this, the related VACV has been the prototype for poxvirus research. Many important advances in molecular biology have been revealed through research with VACV. It is the virus found in the smallpox vaccine, and it is also being used as a live vector virus for HIV vaccines. However, many aspects of poxvirus replication are still being investigated.

The attachment and entry of VACV differ depending upon the infectious form of the virus. The MV possesses at least seven envelope proteins that mediate adsorption of the virus to the cell. Some of these bind to cell surface glycosaminoglycans (Fig. 15.8). For example, viral protein D8 binds to chondroitin sulfate, while A27 and H3 interact with heparan sulfate. Additionally, A26 binds to the extracellular matrix protein laminin, and viral L1 binds to an unidentified cellular receptor. On **keratinocytes**, the major cells of the skin epidermis, VACV also binds directly to a cell surface receptor known as macrophage receptor with collagenous structure (MARCO). None of these proteins alone are essential for attachment. Rather, it is thought that the set of viral proteins bind to their cellular proteins in a concerted manner.

The additional envelope found in EVs presents a problem, because if the EV fused with the plasma or endosomal membrane, the core that would be released into the cytosol would still be surrounded by the MV envelope. Accordingly, the proteins that are found on the outer envelope of the EV, specifically A34 and B5, bind to cell surface molecules and cause the dissolution of the outer envelope (Fig. 15.8). This leaves behind the MV, which is then free to attach to the cell surface as described above.

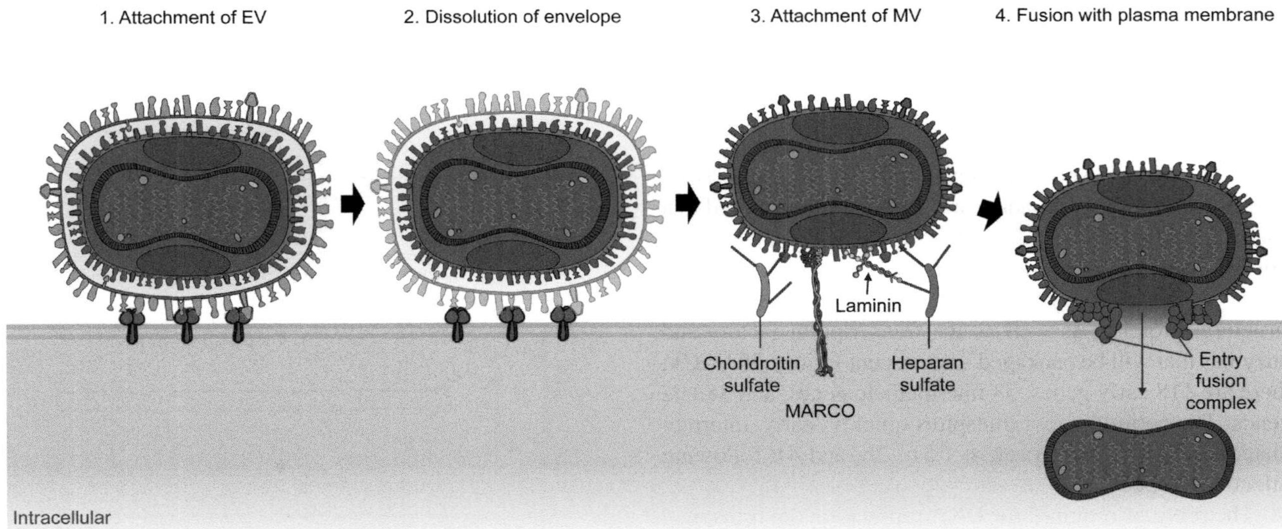

FIGURE 15.8 Entry of vaccinia virus into the cell. The attachment of vaccinia virus differs depending upon the virion type. After binding to cell surface molecules (1), the EV possesses proteins on its outer membrane, including proteins A34 and B5, that cause the dissolution of the outer membrane (2) and release of the MV. The MV possesses at least seven membrane proteins that facilitate attachment. Some of these bind to surface glycosaminoglycans chondroitin sulfate and heparan sulfate, while others bind to proteins associated with the membrane, such as MARCO and laminin (3). The entry fusion complex of viral proteins fuses the viral envelope with the plasma membrane (4), allowing the core to enter the cytosol of the cell.

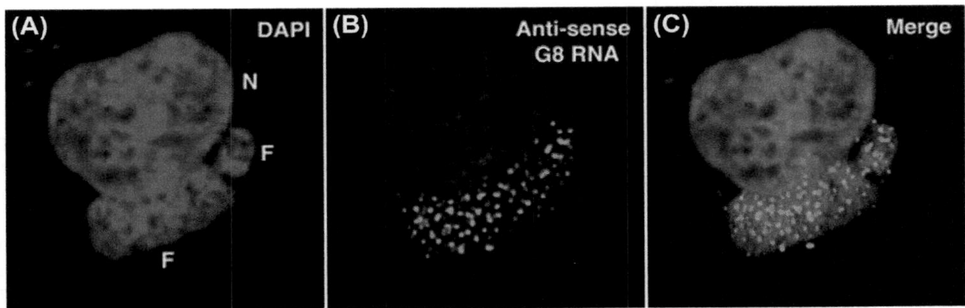

FIGURE 15.9 Poxvirus virus factory. The virus factory, enclosed by membranes derived from the endoplasmic reticulum, is the site of vaccinia transcription, translation, DNA replication, and assembly. This image, acquired by confocal microscopy, used the blue fluorescent dye DAPI to stain DNA within a cell (A), revealing DNA within the nucleus (N) and two virus factories (F). In (B), a short fluorescent green antisense RNA fragment was used to identify mRNA encoding the viral intermediate protein G8. When merged together (C), the images show the location of the virus factory in the cytoplasm adjacent to the nucleus. *Reprinted from Katsafanas, G.C., Moss, B. Colocalization of transcription and translation within cytoplasmic poxvirus factories coordinates viral expression and subjugates host functions. Cell Host Microbe 2 (4), p. 221–228 (Figure 1C), Copyright 2007, with permission from Elsevier.*

Multiple VACV proteins and host cell components mediate the fusion of the MV envelope to the plasma membrane. These 11+ proteins are collectively referred to as the **entry fusion complex (EFC)**, which tightly associates with the viral membrane. The EFC brings the viral envelope toward the plasma membrane and fuses the two, forming a pore through which the core passes to enter the cytosol.

Certain strains of VACV can also enter cells through endocytosis. In fact, the binding of VACV to the cell surface integrin β1 has been shown to signal the cell to endocytose the virion. In this case, the low pH of the endosome is required for fusion of the virus and endosomal membrane.

15.3.2 Viral Protein Expression and Genome Replication

Following fusion of the virion with the plasma or endosomal membrane, the core travels deeper into the cell using microtubules to a discrete location outside the nucleus. Here, a **virus factory** is formed that is enclosed by the membranes of the endoplasmic reticulum (Fig. 15.9). The virus factory is an infection-specific cytoplasmic domain where transcription, DNA replication, and nascent virion assembly take place. A single virion can initiate the creation of a virus factory, and cells can contain multiple virus factories that often fuse into larger virus factories.

Poxviruses do not require entry into the nucleus for replication because all the enzymes required to initiate transcription and processing of viral mRNAs are carried into the cell within the virion core. Transcription begins about 20 min after virus entry, and transcripts are released from a partially uncoated core. The transcriptional system includes a DNA-dependent RNA polymerase, enzymes for adding a 5′-cap and 3′-poly(A) tail, and early gene transcription factors. Transcription occurs in a cascade: early genes are transcribed that encode proteins required for DNA replication and transcription of intermediate genes, whose protein products result in the expression of late genes. Late gene products include structural proteins, as well as the transcription factors and enzymes that will be packaged into nascent virions. In VACV, there are 118 early genes, 53 intermediate genes, and 38 late genes. The entire process transpires quickly: early, intermediate, and late mRNAs peak at 1.5 h, 2 h, and 4 h following infection, respectively.

The proteins required for DNA replication are translated from early mRNAs. In addition to the viral DNA polymerase E9, early gene products also include helicase–primase D5, DNA ligase A50, and several enzymes that increase the pool of nucleotides available within the cell—particularly important since viral replication is occurring outside of the nucleus. Within the virus factories, DNA replication starts about 2 h following infection and immediately after early gene transcription and translation. An unidentified nuclease nicks one strand of the continuous strand of viral DNA, near one of the hairpins (Fig. 15.10). The released DNA acts as a template, and DNA polymerase extends the new strand. Because of the complementarity of this region, the nascent strand folds back upon itself, allowing DNA polymerase to use the strand as its own template. The entire strand is copied, creating a concatemer of two joined genomes. Intermediate and late gene products cleave the concatemer into monomers that will be packaged into new virions.

One hallmark of poxviruses is that they encode a large number of proteins that are involved in evading the immune response. In fact, 30–50% of poxvirus genes modulate the immune response. These proteins interfere with a variety of immune responses by neutralizing interferons and cytokines, inhibiting apoptosis, and interfering with cellular antiviral-signaling pathways, among others. A subset of these are packaged directly into the core for delivery into cells upon infection.

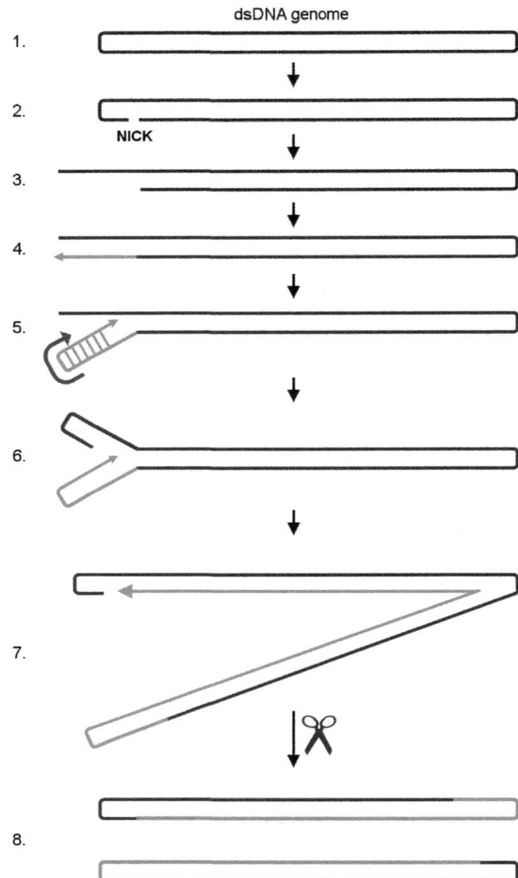

FIGURE 15.10 **Poxvirus DNA replication.** The poxvirus genome is composed of dsDNA that is covalently closed at each end (1). An unidentified nuclease nicks one strand near a hairpin (2). The released DNA acts as a template for synthesis of new DNA (3 and 4). The nascent strand folds back upon itself due to nucleotide complementarity (5). The released strand does the same (6). DNA synthesis continues, using the linearized old strand as a template, until it reaches the end (7). The molecule is cleaved by viral proteins into individual genomes (8) that are then packaged into nascent virions. The continuation of DNA replication through the hairpins can lead to the formation of large concatemers before cleavage into individual genomes occurs.

15.3.3 Assembly, Maturation, and Release

VACV goes through several stages of virion **morphogenesis**, meaning the development of its shape (Fig. 15.11A). The first visible indication of virus assembly is the presence of **crescent membranes** derived from the endoplasmic reticulum. Over nine **viral membrane assembly proteins** (VMAPs, pronounced "vee-maps") are required for the formation of crescent membranes (Fig. 15.11B). Notably, viral scaffold protein D13 associates with the convex surface of the crescent and forms a hexagon-shaped, honeycomb lattice. The D13 lattice begins forming on the *inside* of the endoplasmic reticulum membrane, and the lattice maintains the curvature of the membrane as it continues forming into a sphere. The genomic DNA and core proteins are also recruited to the interior before it is sealed, forming a spherical **immature virion (IV)**. At least 20 proteins are embedded within or associate with the IV membrane.

The immature virion does not become an infectious **mature virion (MV)** until maturation occurs. In the process of maturation, the D13 scaffold is removed and additional viral surface proteins associate with the membrane. Several core proteins are cleaved by viral proteases I7 and G1, and proteins associated

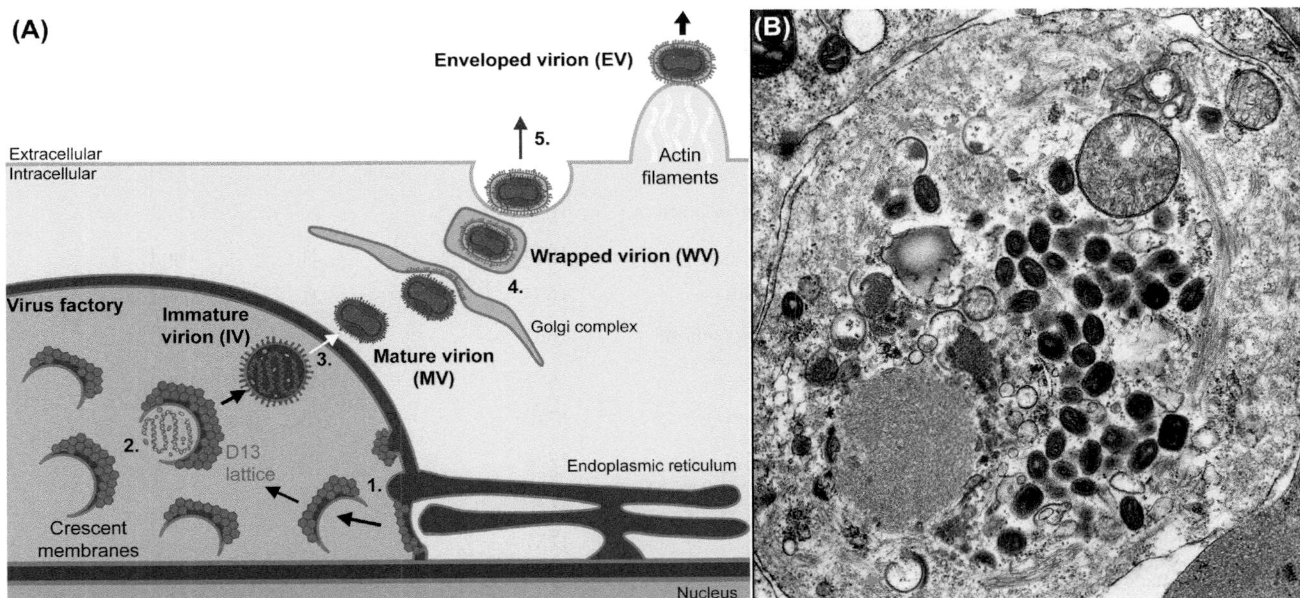

FIGURE 15.11 **Virion morphogenesis and release.** (A) During infection, the remodeling of endoplasmic reticulum membranes creates virus factories that act as the site of viral transcription, translation, genome replication, and assembly. During assembly, viral membrane assembly proteins (VMAPs) form crescent-shaped membranes derived from the endoplasmic reticulum (1). Genomic DNA and core proteins are recruited to the interior (2) and an immature virion is formed. In the process of maturation (3), additional proteins are recruited to the viral membrane while remodeling of the core occurs through the action of viral proteases and other proteins associated with the lateral bodies. This creates the mature virion that is released upon cell lysis. The enveloped virion is formed when the MV passes through the Golgi complex or an endosome (4). Here, the MV obtains a double membrane to become a wrapped virion. The outermost membrane fuses with the plasma membrane, exocytosing the EV (5). EVs can be released farther from the cell when actin tails form beneath the cell surface, creating projections of the plasma membrane. (B) This transmission electron micrograph of a tissue section shows several variola virions in the process of assembly. Note the crescent membranes in various stages of assembly (*orange arrows*) and the dumbbell-shaped virion cores. *Photo courtesy of CDC/Fred Murphy and Sylvia Whitfield.*

with the two lateral bodies assist in the creation of the final dumbbell shape of the core. The result is a brick-shaped MV that is released from the cell upon lysis. Notably, none of the IV or MV proteins are glycosylated, further emphasizing that they do not traffic through the cell's secretory pathway.

An alternate pathway also exists for the egress of VACV virions. Some of the MVs are trafficked on microtubules and obtain a double membrane from the Golgi network or an endosome, becoming **wrapped virions (WVs)**. WVs continue moving along microtubules to the plasma membrane, where the outer membrane of the WV fuses with the plasma membrane. This results in exocytosis of the virion, now known as an **enveloped virion (EV)**. The EV possesses two membranes, the one initially formed from the crescent membrane and the single remaining membrane derived from the Golgi network or endosome. The EV contains 9+ viral proteins that it acquired upon its final wrapping that are absent from the MV. The EV can remain associated with the cell through the action of its outer envelope proteins, or it can be released farther from the cell when actin tails form beneath the cell surface, creating projections that promote the release of the virion toward neighboring cells. EVs are associated with localized, cell-to-cell spread whereas MVs are thought to be involved in infection between hosts.

Study Break

What do the abbreviations IV, MV, WV, and EV stand for? Explain how these terms are connected.

15.4 ERADICATION OF SMALLPOX

Evidence suggests that VARV emerged in Africa between 4000 and 10,000 years ago, likely corresponding to the establishment of large farming settlements in the region. The mummy of Egyptian pharaoh Ramses V, who died in 1156 BCE, displays evidence of smallpox-like lesions on his face and neck (see Fig. 1.1B). Historical texts indicate the disease spread to China and India before the common era, later reaching Europe between the fifth and seventh century AD. Spanish and Portuguese expeditions to the New World in the 16th century introduced smallpox to South and Central America. The disease decimated native populations, leading to the fall of Aztec and Inca empires. Early colonization in North America brought smallpox to the Native Americans. As a respiratory disease, smallpox spread easily and did not discriminate against its victims. Many notable historic figures were infected with smallpox (Table 15.3).

TABLE 15.3 Some Notable Historic Cases of Smallpox

Person	Contribution to History	Year Contracted	Age	Outcome
Marcus Aurelius[a]	Roman emperor	180 AD	58	Died
Cuitláhuac	Aztec emperor	1520	44	Died
William II	Prince of Orange (Netherlands)	1650	24	Died
Ferdinand IV	King of the Romans	1654	21	Died
Emperor Go-Komyo	110th emperor of Japan	1654	21	Died
Shunzhi Emperor	Third emperor of Qing dynasty	1661	22	Died
Queen Mary II	Queen of England	1694	32	Died
Emperor Higashiyama	113th emperor of Japan	1710	34	Died
King Louis I	King of Spain	1724	17	Died
Tsar Peter II	Emperor of Russia	1730	14	Died
George Washington	1st US President	1751	19	Survived
Wolfgang Amadeus Mozart	Composer	1767	11	Survived
King Louis XV	King of France	1774	64	Died
Andrew Jackson	7th US President	1781	14	Survived
Abraham Lincoln	16th US President	1863	54	Survived
Joseph Stalin	Soviet dictator	1885	7	Survived

[a]*Based upon historical accounts of the symptoms at the time of his disease.*

15.4.1 Origin of Smallpox Vaccination

Variolation, the intentional inoculation of an individual with smallpox material, traces back to 16th century China. Variolation used a lancet or needle to introduce pulverized dried smallpox scabs or pustule fluid into the skin of an individual. This resulted in a milder form of the disease with a much lower fatality rate (2–3%), compared to natural smallpox infection. However, a person contracting smallpox through variolation could still transmit it to other individuals, and variolation spread other diseases, such as syphilis and hepatitis. Variolation was widely practiced in Africa, India, and China, and in the absence of a better solution, variolation was adopted in Europe and its colonies in the 1700s.

As described in Chapter 8, "Vaccines, Antivirals, and the Beneficial Uses of Viruses," Edward Jenner (Fig. 15.12) created the first vaccine in 1796, when he tested and promoted the use of cowpox injections to confer immunity against smallpox. Country doctors had previously taken note that milkmaids that became infected with cowpox did not contract smallpox, and during an apprenticeship at age 13, Jenner himself heard a milkmaid say, "I shall never have smallpox for I have had cowpox. I shall never have an ugly pockmarked face." Jenner became a sharp, skilled doctor who was enthusiastic about testing and advancing science. In May of 1796, he inoculated an 8-year-old boy, James Phipps, with the fluid from a fresh cowpox lesion taken from milkmaid Sarah Nelmes (see Fig. 8.1B). Phipps presented with mild symptoms that resolved within 2 weeks. When

FIGURE 15.12 **Portrait of Edward Jenner, M.D.** *Drawn by J. Northcote and engraved by W. Say. Courtesy of the National Library of Medicine.*

exposed to infectious smallpox, Phipps was protected from contracting the disease. Jenner promoted his **vaccine**, a word derived from the Latin word *vacca*, meaning "cow," and sent vaccine material to anyone that requested it. Within 4 years, the process of vaccination was used in most European countries. In the United States, Harvard University professor Benjamin Waterhouse introduced vaccination to New England. He convinced Thomas Jefferson to support vaccination, and Jefferson set up a national vaccination program as a result. Although Jenner was not the first person to attempt the use of cowpox to protect against smallpox, he received deserved credit for its discovery because he was the first to scientifically test the hypothesis and promote its widespread use.

15.4.2 Modern Attempts at Smallpox Eradication

Edward Jenner used the fluid from a fresh cowpox lesion on the hand of milkmaid Sarah Nelmes for his initial vaccine material. However, it was realized in the early 1900s that the vaccine strains being used for vaccination were no longer cowpox virus, at which point the vaccine virus was designated vaccinia virus. The precise origin of VACV is unknown but is thought to be derived through the passaging of either cowpox virus, VARV, or horsepox virus through animals. The virus is assumed to have mutated in the artificially infected host animals into VACV. Evidence of multiple recombination events is also present in the vaccinia genome.

By 1853, smallpox vaccination of children had become mandatory in the United Kingdom. In 1855, Massachusetts became the first state to require vaccination, a law that was upheld by the Supreme Court of the United States. Those against vaccination were active in Europe and the United States, incorrectly promoting that smallpox could not be contracted by healthy people and that vaccine materials derived from cowpox lesions were "poisonous," "revolting," and "bestial" in nature. Despite continued outbreaks over the next century due to low vaccination rates in certain areas, the last naturally occurring case of smallpox in the United States occurred in 1949 with an outbreak in eight people in Hidalgo County, Texas. Several cases that were imported from other countries would occur in the following years.

In the early 1950s, 50 million cases were still occurring worldwide each year. In 1967, the World Health Organization (WHO) received $2.5 million from the Union of Soviet Socialist Republics and the United States to fund the Intensified Global Eradication program through vaccination. At the time, the virus was still causing 10–15 million cases of smallpox in endemic areas located in 31 countries in South America, Africa, and Asia. Two million people died annually from infection.

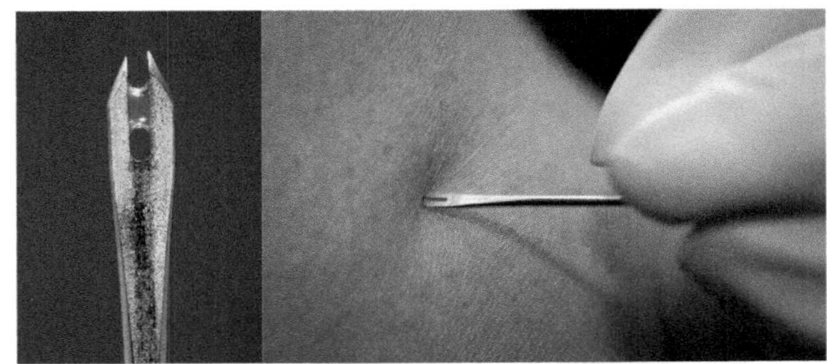

FIGURE 15.13 **Bifurcated needle.** The bifurcated needle held a drop of vaccine material after being dipped into the reconstituted vaccine solution. The needle would then be used to rapidly and superficially puncture the skin of the individual. Those receiving the vaccine for the first time received 3 punctures, while those receiving a booster received 15. Vaccinated individuals were often left with a lifelong scar at the vaccination site. *Courtesy of CDC / James Gathany.*

Eradication of smallpox was theoretically possible due to several fortunate aspects of VARV and the vaccine against it. First, VARV does not have a nonhuman reservoir and does not induce subclinical or latent infections. As a result, no animals or chronic human carriers would contribute to continued circulation of the virus. Additionally, because the virus is spread through droplets, it is uncommon for transmission to occur over long distances, meaning that only close contacts of those infected were most likely to acquire the virus. Second, a **lyophilized** (freeze-dried) preparation of the vaccine had become available that did not require refrigeration and was stable in warm climates. It was reconstituted into a liquid form using a sterile diluent at the vaccination clinic, immediately before use. The **bifurcated needle** was invented in 1965. This narrow steel rod possessed two prongs at one end (Fig. 15.13). The needle was dipped into the vaccine solution, and a drop of vaccine material was held between the two prongs. In the process of **scarification**, the needle was held perpendicular to the site of insertion, usually the upper arm, and several punctures with the needle introduced the vaccine into the skin. The bifurcated needle provided a consistent manner to administer the vaccine and used a quarter of the material that previous injections required. Local replication of VACV at the vaccination site created a lesion that progressed through the rash stages, leaving behind a permanent scar as evidence of vaccination. One final attribute that allowed the eradication of smallpox is that only 1 dose of vaccine was required to induce immunity. Having to revaccinate would have made eradication efforts significantly more difficult.

The WHO eradication plan initially called for vaccination of every individual. However, it was noted early that outbreaks could be controlled by identifying an infected individual and vaccinating his/her immediate contacts (see Chapter 6, "The Immune Response to Viruses," for a review of herd immunity). This led to an intensive "**surveillance and**

FIGURE 15.14 **Smallpox eradication efforts.** This photograph shows villagers helping to push a jeep up a muddy hill, one of many struggles encountered by public health practitioners traveling through Bangladesh in 1975 in an attempt to eradicate smallpox from the country. *Courtesy of CDC/WHO and Stanley O. Foster, MD, MPH. Photograph by Pierre Claquin, MD, BAC.*

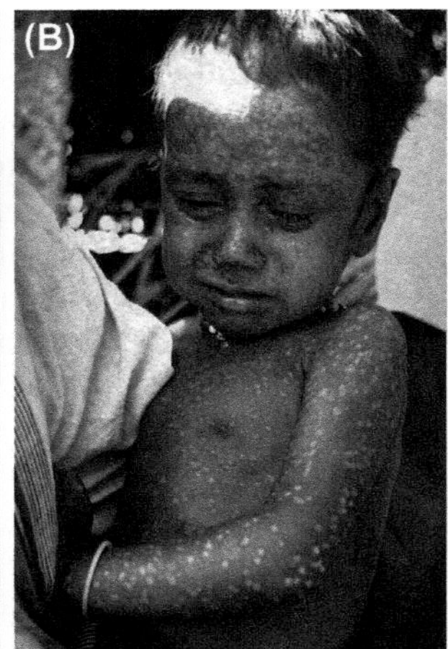

FIGURE 15.15 **Last cases of variola minor and variola major.** (A) Ali Maow Maalin had the last known cases of smallpox in the world, in Somalia in 1977. It was variola minor. The last case of variola major occurred in a child named Rahima Banu, in Bangladesh in 1975 (B). *Couresy of CDC/WHO and Stanley O. Foster, MD, MPH.*

containment" regimen with programs developed in every endemic country that notified health officials of smallpox cases. Once an infected person was discovered, the patient was isolated and his/her contacts were vaccinated. This was referred to as **ring vaccination**, referring to the close contacts of each infected individual that were immunized. This was a monumental undertaking, because every case of smallpox needed to be found (Fig. 15.14). In India, over 100 million homes were visited by 120,000 healthcare workers. When the number of cases started dropping, a reward system was instituted that provided a monetary reward to the individual that first reported a new case of smallpox. For example, the reward for reporting a smallpox case was initially 50 rupees in India but gradually rose to 1000 rupees (over 100 US dollars) for the last of the smallpox cases.

The effort worked. By 1975, smallpox persisted only in the Horn of Africa, within Ethiopia, Kenya, and Somalia. The last case of smallpox in Ethiopia was in August of 1976, and the last case in Kenya was in February of 1977. The last known natural case of smallpox in the world was identified on October 26, 1977, in Somalia (Fig. 15.15A). It was variola minor; the last case of variola major had occurred in Bangladesh in 1975 (Fig. 15.15B). All 211 contacts of this individual were traced, immunized, and kept under surveillance. Following no additional outbreaks for the next 2 years, an international team of experts was assembled by the WHO that certified on December 9, 1979, that the world had successfully eradicated smallpox.

TABLE 15.4 Risks of Smallpox Vaccine Complications

Complication	Rate (per million[a])
Inadvertent autoinoculation	529
Generalized vaccinia	242
Eczema vaccinatum	39
Progressive vaccinia	1.5
Postvaccinial encephalitis	12

[a]*Per million primary vaccinations.*

15.4.3 Cessation of Smallpox Vaccination

In 1980, the World Health Assembly (the governing body of the WHO) recommended that all countries cease routine smallpox vaccinations. The United States had already ceased smallpox vaccinations in 1972, even though smallpox was still prevalent at the time in certain areas of the world. The reason vaccination was discontinued is that the smallpox vaccine possesses some potential complications because of the live VACV used in the vaccine. The overall chance of a complication is relatively high, at around 1 out of every 1200 vaccinated persons (Table 15.4). During a smallpox outbreak, the benefits of vaccination outweigh the risks. However, once smallpox was no longer endemic or commonly imported in a region, the risk of vaccine complications was much higher than the risk of contracting smallpox itself. It is for this reason that it was recommended that smallpox vaccinations cease following eradication of the virus.

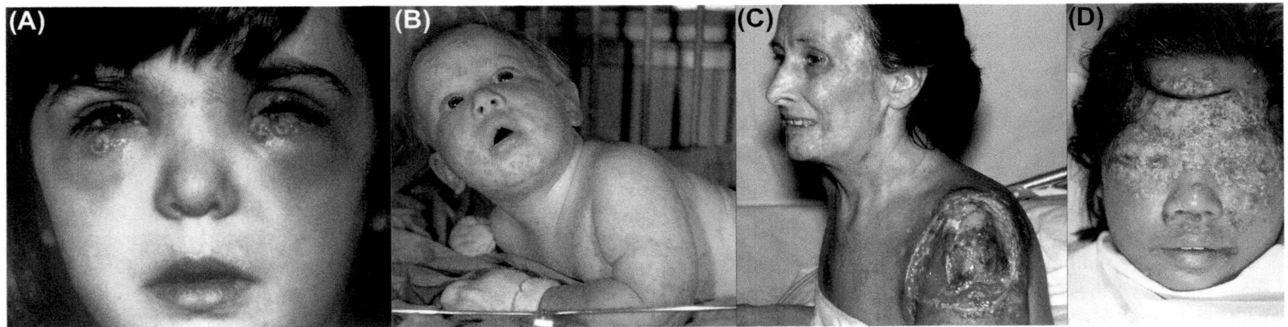

FIGURE 15.16 **Potential complications of the smallpox vaccine.** (A) Inadvertent autoinoculation, the transfer of vaccinia virus from the vaccination site to a different site on the body, was the most common complication of smallpox vaccination. This girl has transferred the virus to her eyes. *(Courtesy CDC/Arthur E. Kaye.)* (B) Generalized vaccinia occurred when vaccinia virus caused a systemic infection, as occurred in this 8-month-old child. *(Courtesy of Allen W. Mathies, MD, California Emergency Preparedness Office, Immunization Branch.)* (C) Progressive vaccinia was a serious complication due to replication of vaccinia virus at the vaccination site and at other sites of the body. It primarily occurred in immunocompromised individuals. *(Courtesy of CDC/California Department of Health Services.)* (D) Unless it was during an outbreak, vaccination of people with known skin conditions was not advised due to the risk of eczema vaccinatum. This 28-year-old woman with a history of atopic dermatitis presented with eczema vaccinatum after coming in contact with vaccinia virus from her recently immunized child. *(Courtesy of Allen W. Mathies, MD, John Leedom, MD and the California Emergency Preparedness Office, Immunization Branch.)*

The most common major complication of vaccination was **inadvertent autoinoculation**, the transfer of VACV from the vaccination site on the arm to another area on a person's body. The majority of inadvertent autoinoculations were to the eye, resulting in vaccinia lesions on the eyelid or inflammation of the cornea, iris, or conjunctiva (Fig. 15.16A). Autoinoculation also frequently occurred to the nose, mouth, genitalia, and rectum. The second most common complication was **generalized vaccinia** in which the vaccinated individual came down with a systemic vaccinia infection with lesions on other parts of the body (Fig. 15.16B). Generalized vaccinia was usually not severe. On the other hand, **postvaccinial encephalitis** resulted in inflammation and swelling of the brain, leading to confusion, seizures, or coma. This is thought to be caused by the immune response to vaccinia, rather than the virus directly infecting the central nervous system. Postvaccinial encephalitis resulted in death 15–25% of the time, with 25% of cases resulting in some degree of neurological damage.

Progressive vaccinia or **vaccinia necrosum** occurred in less than 1 vaccinated individual out of 500,000, primarily in those with immune system defects. In progressive vaccinia, the initial vaccination site forms an expanding, nonhealing ulcer that leads to tissue **necrosis**, the death and destruction of the tissue (Fig. 15.16C). This is thought to occur because the person's immune system is deficient and allows vaccinia to continue replicating unchecked following vaccination, spreading locally and systemically. Once the virus gains entry into the bloodstream, additional ulcers can form at distal sites.

The vaccine was not recommended for individuals with skin conditions, such as eczema or atopic dermatitis, because **eczema vaccinatum** could occur (Fig. 15.16D). Although very rare, eczema vaccinatum is a serious vaccinia rash, often with confluent papules, vesicles, or pustules. Individuals with widespread skin involvement can become very ill. Eczema vaccinatum is thought to occur when vaccinia becomes introduced into areas of the skin that are already compromised by the previous skin condition, allowing the virus to replicate and spread. Eczema vaccinatum can occur from direct vaccinia vaccination or from being in close contact to a recently vaccinated person.

Because of these potential complications, the US ceased vaccinating in 1972 and WHO recommended in 1980 that all countries stop smallpox vaccinations. WHO also requested that all laboratories destroy their remaining stocks of VARV or transfer them to one of the WHO-approved reference laboratories. Although laboratories were thought to have complied with this request, concerns remain that hidden or unknown stocks of VARV could be used for malicious purposes. The idea of using smallpox as a biological weapon is not new; blankets from a smallpox hospital were given to Native Americans in 1763 by British colonists, and circumstantial evidence exists that the British may have tried to use smallpox during the American Revolutionary War. Because of its relative ease of dissemination and drastic outcomes, VARV remains a biological agent of concern. In the United States, over 150 million people—nearly half of the country's current population—have been born since smallpox vaccination was discontinued in 1972. Worldwide, around 2.5 billion individuals have been born since vaccination ceased in 1980. The result is that a large proportion of those alive would be susceptible to infection with VARV. Additionally, the immunity against smallpox in people that were immunized may be declining without continued boosting from natural infection.

Case Study: Multistate Outbreak of Monkeypox—Illinois, Indiana, and Wisconsin, 2003

Excerpted from Morbidity and Mortality Weekly Report, June 13, 2003/52(23); 537–540.

CDC has received reports of patients with a febrile rash illness who had close contact with pet prairie dogs and other animals. The Marshfield Clinic, Marshfield, Wisconsin, identified a virus morphologically consistent with a poxvirus by electron microscopy of skin lesion tissue from a patient, lymph node tissue from the patient's pet prairie dog, and isolates of virus from culture of these tissues. Additional laboratory testing at CDC indicated that the causative agent is a monkeypox virus, a member of the orthopoxvirus group. This report summarizes initial descriptive epidemiologic, clinical, and laboratory data, interim infection-control guidance, and new animal import regulations.

As of June 10, a total of 53 cases had been investigated in Illinois, Indiana, and Wisconsin. Of these, 29 (49%) cases were among males; the median age was 26 years (range: 4–53 years). Data were unavailable for sex and age for 2 and 14 patients, respectively. A total of 14 (26%) patients have been hospitalized, including a child aged <10 years with encephalitis.

Detailed clinical information was available for 30 cases reported in Illinois and Wisconsin. Among these, the earliest reported onset of illness was on May 15, 2003. For the majority of patients, a febrile illness has either preceded or accompanied the onset of a papular rash; respiratory symptoms, lymphadenopathy, and sore throat also were prominent signs and symptoms (Table 15.5). The rash typically progressed through stages of vesiculation, pustulation, umbilication, and encrustation. Early lesions became ulcerated in some patients. Rash distribution and lesions have occurred on the head, trunk, and extremities; many patients had initial and satellite lesions on palms, soles, and extremities. Rashes were generalized in some patients.

All patients have had contact with animals; however, at least two patients also reported contact with another patient's lesions or ocular drainage. A total of 51 patients reported direct or close contact with prairie dogs (*Cynomys* sp.), and 1 patient reported contact with a Gambian giant rat (*Cricetomys* sp.). One patient had contact with a rabbit (Family *Leporidae*) that became ill after exposure to an ill prairie dog at a veterinary clinic. Traceback investigations have been initiated to identify the source of monkeypox virus introduced into the United States and have identified a common distributor where prairie dogs and Gambian giant rats were housed together in Illinois. A search of imported animal records revealed that Gambian giant rats were shipped from Ghana in April to a wildlife importer in Texas and subsequently were sold to the Illinois distributor. The shipment contained approximately 800 small mammals of nine different species that might have been the actual source of introduction of monkeypox.

As of June 9, specimens obtained from 10 patients in Illinois, Indiana, and Wisconsin had been forwarded to CDC for testing; nine patients with skin lesions had DNA sequence signatures specific for monkeypox. No skin lesions were observed in one patient who tested negative by polymerase chain reaction. Skin biopsies were available for five patients; four showed orthopox viral antigens by immunohistochemical testing. Skin lesions from four of the 10 patients were evaluated by negative stain electron microscopy, and pox viral particles were found in three patients. Monkeypox specific DNA signatures also were found in a viral isolate derived from lymphoid tissue of a patient's ill prairie dog.

Reported by: J Melski, MD, K Reed, MD, E Stratman, MD, Marshfield Clinic and Marshfield Laboratories, Marshfield; MB Graham, MD, J Fairley, MD, C Edmiston, PhD, KS Kehl, PhD, Medical College of Wisconsin; SL Foldy, MD, GR Swain, MD, P Biedrzycki, MPH, D Gieryn, Milwaukee Health Dept; K Ernst, MPH, Milwaukee–Waukesha Consortium for Emergency Public Health Preparedness, Milwaukee; D Schier, Oak Creek Health Dept, Oak Creek; C Tomasello, Shorewood/Whitefish Bay Health Dept, Shorewood; J Ove, South Milwaukee Health Dept, South Milwaukee; D Rausch, MS, N Healy-Haney, PhD, Waukesha County Health Dept, Waukesha; N Kreuser, PhD, Wauwatosa Health Dept, Wauwatosa; MV Wegner, MD, JJ Kazmierczak, DVM, C Williams, DVM, DR Croft, MD, HH Bostrom, JP Davis, MD, Wisconsin Dept of Health and Family Svcs; R Ehlenfeldt, DVM, Wisconsin Dept of Agriculture, Trade and Consumer Protection; C Kirk, Wisconsin State Laboratory of Hygiene. M Dworkin, MD, C Conover, MD, Illinois Dept of Public Health. R Teclaw, MD, H Messersmith, MD, Indiana State Dept of Health. Monkeypox Investigation Team; MJ Sotir, PhD, G Huhn, MD, AT Fleischauer, PhD, EIS officers, CDC.

TABLE 15.5 Clinical Features of Persons With Monkeypox—Illinois and Wisconsin, 2003[a]

Clinical Features	No. Cases	Percentage
Rash	25	83
Fever	22	73
Respiratory[b]	16	64
Lymphadenopathy	14	47
Sweats	12	40
Sore throat	10	33
Chills	11	37
Headache	10	33
Nausea and/or vomiting	6	20

[a]*N = 30. As of June 10, 2003.*
[b]*Includes cough, shortness of breath, and nasal congestion. Data were missing for five patients.*

Another interesting consequence of the discontinuation of vaccination is the potential for other poxviruses to infect humans. Monkeypox virus, an orthopoxvirus related to VARV, infects a wide range of mammals, including humans. The smallpox vaccine has been shown to be at least 85% effective in preventing monkeypox due to the same principles of cross-reactivity that lead to protection against VARV. Monkeypox was first isolated from a colony of monkeys kept for research purposes, although rodents are thought to be the most probable natural reservoirs of the virus.

In humans, monkeypox virus causes symptoms that are clinically indistinguishable from smallpox (Fig. 15.17), with the addition of **lymphadenopathy** (swollen lymph nodes). The mortality rate is estimated to be between 1% and 10%, much lower than that of smallpox. However, concerns exist that increased human infections due to the cessation of smallpox vaccination could result in greater opportunity for the virus to evolve into one that could be as virulent as VARV.

Monkeypox is endemic in the Democratic Republic of Congo and has been reported in several other African countries. An outbreak in the United States in 2003 was the only time that the virus has been reported outside of Africa. A total of 47 confirmed and probable cases of monkeypox were reported from six states. The source of the outbreak was found to be pet prairie dogs that became infected after being in close proximity to infected rodents that were imported into the country from Ghana (see Case Study).

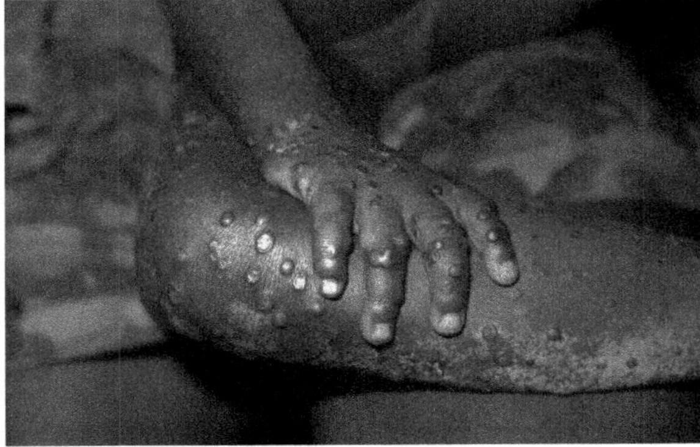

FIGURE 15.17 **Monkeypox.** Monkeypox causes symptoms that are clinically indistinguishable from those of smallpox, including the characteristic rash shown in this photograph of the hand and leg of a 4-year old during an outbreak in Liberia in 1971. *Courtesy of CDC.*

SUMMARY OF KEY CONCEPTS

Section 15.1 Taxonomy

- The *Poxviridae* family contains 38 viruses. It is subdivided into *Entomopoxvirinae* and *Chordopoxvirinae* subfamilies. The *Orthopoxvirus* genus within *Chordopoxvirinae* contains nine viruses that include *Variola virus, Vaccinia virus, Cowpox virus,* and *Monkeypox virus.* Some of these have wide host ranges, while VARV infects humans exclusively.

Section 15.2 Clinical Course of Variola Infection

- Smallpox is caused by VARV. It is transmitted most commonly through respiratory droplets, although contact with infectious material can also transmit the virus.
- The incubation period for variola infection is 10–14 days. The prodrome is characterized by a high fever, headache, severe back pain, malaise, prostration, and sometimes vomiting and abdominal pain.
- The fever subsides within 2–4 days, at which point the rash begins as an enanthem. An exanthem follows within 24 hours. Herald spots appear on the face initially, followed by a centrifugal rash that is more prevalent on the extremities and face. Lesions are often present on the palms of the hands and soles of the feet.
- The lesions of the rash progress together through four stages. Macules are red spots that develop into raised papules, which fill with opalescent fluid to become blistery vesicles. Vesicles are firm to the touch, deep-seated, and can possess a central umbilication. These fill with pus to become pustules, which crust over into scabs that eventually separate from the lesion, leaving a depigmented scar behind. Scabs also contain infectious virions.
- Confluent smallpox occurs when the lesions are so closely associated that no skin is visible between them. It has a fatality rate >50%.
- Variola major has an overall fatality rate of around 30% with death occurring between days 10–16 of infection. Variola minor is less virulent, associated with <1% fatality. Variola major is classified into four principal clinical presentations: ordinary, modified, flat, and hemorrhagic smallpox. Flat and hemorrhagic forms are usually fatal.

Section 15.3 Poxvirus Replication Cycle

- Poxviruses are enveloped, brick-shaped viruses of approximately 350 nm by 250 nm, placing them among the largest human viruses. They possess large dsDNA genomes encoding ~200 genes. ITRs at the end of the genome form a covalently closed, circularized genome. Most of our knowledge of poxvirus molecular biology has been obtained by studying VACV.
- The viral dsDNA associates with at least four different proteins that organize the DNA into nucleosome-like structures, resembling eukaryotic chromatin. The dumbbell-shaped core of the virus is composed of a thinner inner layer and outer palisade layer. Lateral bodies assist in the shaping of the core.
- Poxvirus virions have two infectious forms. The MV is wrapped with a single envelope, while the EV possesses an additional lipid membrane.
- The VACV MV contains at least seven envelope proteins that together mediate adsorption of the virus to the cell. Notably, A27 and H3 interact with heparan sulfate, D8 binds to chondroitin sulfate, A26 binds to the extracellular matrix protein laminin, and L1 binds to an unidentified receptor. The EV contains membrane proteins A34 and B5 that bind to cell surface molecules and dissolve the outer envelope at the surface of the cell.
- The EFC comprises 11+ proteins that mediate the fusion of the MV envelope to the plasma membrane. Binding to cell surface integrin β1 causes endocytosis of the virion, and fusion occurs in the low pH of the endosome.
- The virus core travels to the endoplasmic reticulum using the microtubule network. An enclosed virus factory is created from the membranes of the endoplasmic reticulum. The virus factory is an infection-specific cytoplasmic domain where transcription, DNA replication, and nascent virion assembly takes place.
- Poxviruses do not require entry into the nucleus for replication because they carry all the proteins required to initiate transcription and processing of early mRNAs. The transcriptional system includes a DNA-dependent RNA polymerase, enzymes for adding a 5′-cap and 3′-poly(A) tail, and early gene transcription factors.
- Transcription occurs in a cascade: early genes are transcribed that encode proteins required for DNA replication and transcription of intermediate genes, whose protein products result in the expression of late genes. Late gene products include structural proteins, as well as the transcription factors and enzymes that will be packaged into nascent virions. VACV has 118 early genes, 53 intermediate genes, and 38 late genes.
- The proteins required for DNA replication, which occurs in the virus factory, are translated from early mRNAs. In addition to the viral DNA polymerase E9, early gene products also include helicase–primase D5, DNA ligase A50, and several enzymes that increase the pool of nucleotides available within the cell. Concatemers are created that are cleaved into individual genomes that are packaged into new virions.
- VMAPs facilitate the formation of crescent membranes from endoplasmic reticulum membrane. Viral scaffold protein D13 creates a hexagon-shaped, honeycomb lattice on the convex surface of the crescent and continues the membrane curvature until it encloses the genomic DNA and core proteins. The sealed spherical body is known as an immature virion.

- The IV undergoes maturation to the MV with the removal of the D13 scaffold, addition of extra surface proteins to the viral membrane, and cleavage of core proteins by viral proteases I7 and G1. Proteins associated with the lateral bodies assist in the shaping of the core.
- Some of the MVs obtain a double membrane by passing through the Golgi network or an endosome, becoming a WV. The outer membrane of the WV fuses with the plasma membrane to release the EV.
- The EV can remain associated with the cell through the action of its outer envelope proteins, or it can be released when actin tails form beneath the cell surface, creating projections that promote the release of the virion toward neighboring cells. EVs are associated with localized, cell-to-cell spread whereas MVs are thought to be involved in infection between hosts.

Section 15.4 Eradication of Smallpox

- VARV likely emerged in Africa between 4000 and 10,000 years ago.
- Variolation was the intentional inoculation of an individual with smallpox material. It resulted in a milder form of disease than smallpox itself but still had a 2–3% fatality rate.
- In 1796, English country doctor Edward Jenner found that cowpox inoculation protected against smallpox infection. He widely publicized vaccination and provided vaccine material to anyone that requested it. Vaccination was soon adopted by other European countries and the United States.
- In the early 1950s, 50 million cases of smallpox were still occurring worldwide each year. The WHO championed the Intensified Global Eradication program to eliminate smallpox through vaccination.
- Eradication was possible because VARV only infects humans, does not induce subclinical or latent infections, and close contact is necessary for transmission. A lyophilized preparation allowed for the vaccine to be delivered to areas lacking refrigeration, and the bifurcated needle provided a consistent and reproducible manner to deliver the vaccine.
- Eradication was carried out through a surveillance and containment regimen that included identifying all infected persons and performing ring vaccination of his/her close contacts.
- The last case of variola major occurred in Bangladesh in 1975. The last known natural case of smallpox in the world was variola minor. It occurred in Somalia in October of 1977.
- Following smallpox eradication, the WHO recommended in 1980 that all countries cease vaccinating against smallpox because the risk of complications due to the live VACV preparation was then higher than the risk of contracting smallpox. Possible complications included inadvertent autoinoculation, generalized vaccinia infection, postvaccinial encephalitis, progressive vaccinia/vaccinia necrosum, and eczema vaccinatum.
- Over 2.5 billion people have been born since smallpox vaccination was discontinued. These individuals are therefore susceptible to related orthopoxviruses, such as monkeypox virus, and would be susceptible if VARV were to be used as biological warfare.

FLASH CARD VOCABULARY

Smallpox	Flat smallpox
Cross-reactive	Hemorrhagic smallpox
Variola virus (VARV)	Inverted terminal repeats (ITRs)
Prostration	Palisade
Enanthem	Lateral bodies
Exanthem	Vaccinia virus (VACV)
Centrifugal rash	Keratinocytes
Macules	Entry fusion complex (EFC)
Papules	Virus factory
Maculopapular	Morphogenesis
Vesicles	Crescent membranes
Pustules	Viral membrane assembly proteins (VMAPs)
Umbilication	Immature virion (IV)
Confluent smallpox	Mature virion (MV)
Ordinary smallpox	Wrapped virion (WV)
Modified smallpox	Enveloped virion (EV)
Variolation	Inadvertent autoinoculation
Vaccine	Generalized vaccinia
Lyophilized	Postvaccinial encephalitis
Bifurcated needle	Progressive vaccinia/vaccinia necrosum
Scarification	Necrosis
Surveillance and containment	Eczema vaccinatum
Ring vaccination	Lymphadenopathy

CHAPTER REVIEW QUESTIONS

1. Why is VACV—or any orthopoxvirus—able to be used as a vaccine to protect against VARV infection?
2. Draw out the course of variola infection. Be sure to include the incubation period, prodromal period, and the stages of the rash.
3. What are the differences between macules, papules, vesicles, and pustules?

4. There are four principal clinical presentations of variola major. List them and describe how they are different.
5. Draw out and label the structure of a poxvirus virion.
6. Recap how VACV accomplishes the seven stages of the viral replication cycle.
7. Poxviruses are dsDNA viruses. Why do they not need to enter the nucleus to replicate?
8. How does a poxvirus immature virion become a mature virion?
9. What are the differences between a mature virion and an enveloped virion?
10. Why did Edward Jenner suspect that cowpox infection may protect an individual from contracting smallpox?
11. Why was the eradication of smallpox theoretically possible? Describe at least five factors that made it a scientifically feasible target.
12. The smallpox vaccine is still occasionally used. Can you think of any situations where its use would be reasonable?

FURTHER READING

Babkin, I., Babkina, I., 2015. The origin of the variola virus. Viruses 7 (3), 1100–1112.

Chiu, W.-L., Lin, C.-L., Yang, M.-H., Tzou, D.-L.M., Chang, W., 2007. Vaccinia virus 4c (A26L) protein on intracellular mature virus binds to the extracellular cellular matrix laminin. J. Virol. 81 (5), 2149–2157.

Dixon, C.W., 1962. Smallpox. J. & A. Churchill Ltd., London.

Downie, A.W., 1939. The Immunological relationship of the virus of spontaneous cowpox to vaccinia virus. Br. J. Exp. Pathol. 20 (2), 158–176.

Emerson, G.L., Li, Y., Frace, M.A., Olsen-Rasmussen, M.a., Khristova, M.L., Govil, D., Carroll, D.S., 2009. The phylogenetics and ecology of the orthopoxviruses endemic to North America. PLoS One 4 (10).

Izmailyan, R., Hsao, J.-C., Chung, C.-S., Chen, C.-H., Hsu, P.W.-C., Liao, C.-L., Chang, W., 2012. Integrin 1 mediates vaccinia virus entry through activation of PI3K/Akt signaling. J. Virol. 86 (12), 6677–6687.

Laliberte, J.P., Weisberg, A.S., Moss, B., 2011. The membrane fusion step of vaccinia virus entry is cooperatively mediated by multiple viral proteins and host cell components. PLoS Pathog. 7 (12).

Law, M., Carter, G.C., Roberts, K.L., Hollinshead, M., Smith, G.L., 2006. Ligand-induced and nonfusogenic dissolution of a viral membrane. Proc. Natl. Acad. Sci. U.S.A. 103 (15), 5989–5994.

Liu, L., Cooper, T., Howley, P., Hayball, J., 2014. From Crescent to mature virion: vaccinia virus assembly and maturation. Viruses 6 (10), 3787–3808.

MacLeod, D.T., Nakatsuji, T., Wang, Z., di Nardo, A., Gallo, R.L., 2014. Vaccinia virus binds to the scavenger receptor MARCO on the surface of keratinocytes. J. Invest. Dermatol. 135 (October 2013), 142–150.

Mercer, A.A., Schmidt, A., Weber, O. (Eds.), 2007. Poxviruses. Berkhauser Verlag, Basel.

Moss, B., 2013. Poxvirus DNA replication. Cold Spring Harb. Perspect. Biol. 5, a010199.

Moss, B., 2015. Poxvirus membrane biogenesis. Virology 479-480, 619–626.

Reynolds, M.G., Carroll, D.S., Karem, K.L., 2012. Factors affecting the likelihood of monkeypox's emergence and spread in the post-smallpox era. Curr. Opin. Virol. 2 (3), 335–343.

Riedel, S., 2005. Edward Jenner and the history of smallpox and vaccination. Proc. Bayl. Univ. Med. Cent. 18 (1), 21–25.

Thèves, C., Biagini, P., Crubézy, E., 2014. The rediscovery of smallpox. Clin. Microbiol. Infect. 20 (3), 210–218.

Chapter 16

Emerging and Reemerging Viral Diseases

In late spring of 2014, a man from Bourbon County, Kansas, found an engorged tick on his shoulder while working outside on his property. Several days later, he became ill with nausea, vomiting, and diarrhea. The following day, he developed a fever, chills, headache, myalgia, and arthralgia. He was prescribed antibiotics by his doctor for a presumed tickborne bacterial infection. The following morning, his wife found him with reduced consciousness and took him to the hospital, where he was admitted. There they noticed a maculopapular rash and low white blood cell and platelet counts. He continued to decline, dying 11 days after first becoming ill.

A novel virus was isolated from a specimen of the man's blood (Fig. 16.1). The virus, named Bourbon virus after the county in which it was isolated, is in the *Orthomyxoviridae* family and *Thogotovirus* genus. Like influenza virus, another orthomyxovirus, Bourbon virus possesses a segmented −ssRNA genome. This was the first report of a Thogotovirus originating in the Western Hemisphere.

Although all viruses have a source, some viruses seem to randomly appear or reappear in the human population. An **emerging infectious disease (EID)** is defined as a disease caused by a pathogen that has not before been observed within a population or geographic location. Similarly, a **reemerging infectious disease** is caused by an established pathogen that appears in a new geographical location or was once controlled but begins appearing at a higher incidence. A significant number of emerging and reemerging infectious diseases are viral in nature, and they are becoming more frequent. The 2014–16 epidemic of Ebolavirus in West Africa, as well as the human immunodeficiency virus (HIV) and past epidemics of yellow fever virus, West Nile virus (WNV), and Nipah virus, among others, emphasize that emerging and reemerging viral diseases can be of great concern for global public health.

16.1 FACTORS INVOLVED IN THE EMERGENCE OF VIRAL INFECTIOUS DISEASES

Many aspects influence the emergence or reemergence of viral diseases by changing the exposure of people to the virus. These can broadly be divided into human, environmental/ecological, and viral factors (Table 16.1). Three-quarters of emerging or reemerging infectious diseases are **zoonoses**, infectious diseases of animals that are transmitted to humans. Most zoonoses arise from wildlife, although they can also be derived from domesticated animals or through **vectors** such as mosquitoes and ticks. Other viral diseases, such as measles, are reemerging due to the changing susceptibility of the human population to the virus because of reduced immunization rates.

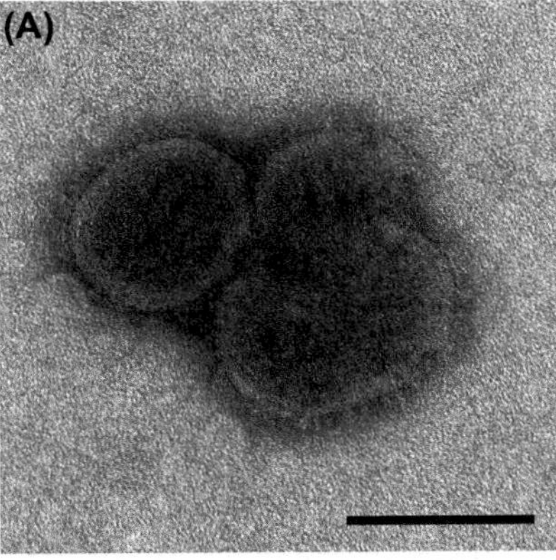

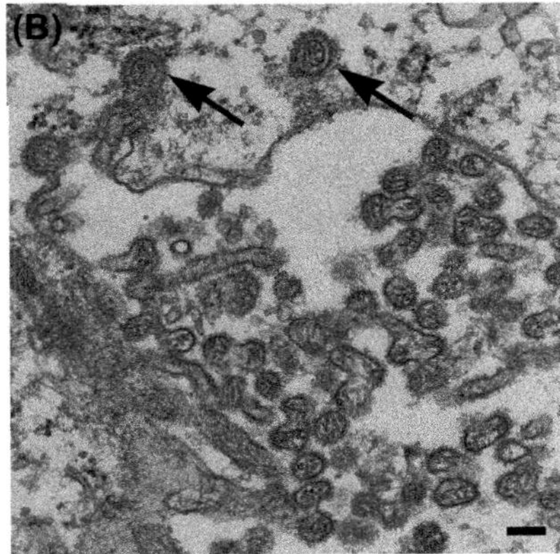

FIGURE 16.1 **Bourbon virus, a novel Thogotovirus.** Bourbon virus was isolated from a man in Bourbon County, Texas, that had died following a possible tick-borne infection. (A) Electron micrograph of spherical Bourbon virus particles with distinct surface projections. (B) Electron micrographs of cells infected with Bourbon virus show numerous extracellular virions. *Arrows* indicate virions that have been endocytosed into a cell. *Images courtesy of Kosoy, O.I., et al., 2015. Emerg. Infect. Dis. 21(5), 760–764.*

TABLE 16.1 Factors Involved in the Emergence/Reemergence of Viral Diseases

Human (including economical, governmental, and societal)	
Population growth	Inadequate public health infrastructures
Urbanization	Lack of access to health care
International travel	Reduced immunization policies
Tourism	Conflicts/wars
Human behaviors	Free trade economies
Human susceptibility to virus	Human invasion of wild animal habitats
Environmental and ecological	
Rainfall accumulation	Wildlife host species richness
Temporary weather patterns	Human encroachment into pristine lands
Long-term climate change	Building of dams, land development
Dew point	Deforestation for farming
Soil moisture	
Viral	
Genomic changes through reassortment, recombination, or mutation	

16.1.1 Human Factors

An abundance of factors related to human behaviors and social systems are pivotal in the emergence or reemergence of viral diseases. Increases in **urbanization**, the process of people living together in towns and cities rather than rural areas, result in denser populations that are subsequently more prone to spreading infectious diseases. This is particularly true in overcrowded urban centers and those that lack adequate sanitary conditions and clean water supplies. **Globalization** is the expansion of economies, populations, and businesses to areas throughout the world. Globalization has facilitated the ability of infectious diseases to spread rapidly between countries through trade and travel. Increased trade between countries can promote the spread of infectious diseases, particularly those found in food or animal products. Illegal activities, such as the trade of bushmeat, also contribute to the spread of infectious diseases. As an example, the U.S. Geological Survey found that bushmeat that had been confiscated at US borders contained novel herpesviruses and retroviruses. The illegal trade of bushmeat is extensive, providing an avenue for zoonoses to be transferred to the human population. In fact, the HIV pandemic may have begun with the acquisition of bushmeat contaminated with simian immunodeficiency viruses.

A notable contribution to globalization is the ability of humans to quickly and easily fly between distant countries. A person traveling from one country to another can spread infectious diseases in this way, as occurred in the 2003 outbreak of **severe acute respiratory syndrome (SARS)**, which was characterized by severe lower respiratory symptoms, pneumonia, and fever. SARS was later determined to be caused by a novel coronavirus, named **SARS-associated coronavirus (SARS-CoV)**. In this case, the virus initially appeared in November of 2002 in Guangdong Province, China, and was spread by travelers within weeks to over 30 countries in Europe, North America, and other areas of China and Asia. In total, 8098 people were infected, and 774 (9.6%) died. In Guangdong Province markets, the virus was isolated from palm civets, a meat- and fruit-eating mammal with a cat-like appearance that was sold for meat. SARS-CoV was also isolated from civet meat from a restaurant that employed a worker that had contracted SARS. Although the palm civets were the most likely source of SARS-CoV in this outbreak, much higher rates of SARS-CoV were found in palm civets from local markets as compared to palm civets on distant farms. This suggested that the palm civets themselves may have contracted SARS-CoV from another animal source while in the market. The virus has also been identified in horseshoe bats, the likely natural reservoir of this zoonosis.

Imported insects and animals can also carry viruses into new environments. West Nile Virus (Fig. 16.2A) is a mosquito-borne flavivirus that was first identified in Uganda in 1937 and later identified in other parts of Africa, as well as Asia, Europe, and the Middle East. Although mostly asymptomatic, the virus causes fever, body aches, joint pain, vomiting, diarrhea, and rash in about 20% of those infected. Less than 1% of total infected individuals develop encephalitis or meningitis, but 10% of those that develop these neurological effects die.

WNV first emerged in North America in October of 1999, when an outbreak occurred in both humans and birds in New York City. It was found to be most closely related to a strain circulating in geese and humans in Israel. Although the initial source of WNV within the United States is unknown, it is thought that imported birds or mosquitoes—either accompanying their natural avian hosts or as stowaways on international flights—were responsible for introducing the virus to North America. From 1999 to 2014, a total of 41,762 US cases of WNV disease were reported to the Centers for Disease Control and Prevention (CDC). This caused 1765 deaths (4.2% of cases).

WNV normally cycles between mosquitoes (*Culex* species, Fig. 16.2B) and birds, although mosquitoes can also transmit the virus to horses and humans, which are both considered "dead end" hosts because they do not develop high enough viral titers in the bloodstream to continue the transmission cycle to mosquitoes. However, WNV has been found to be transmitted through organ transplantation or

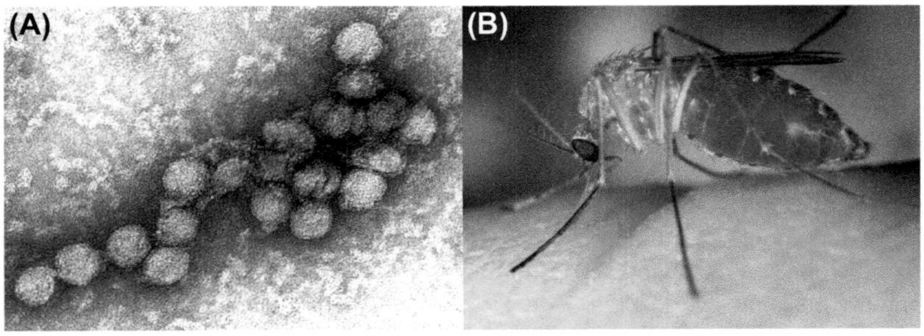

FIGURE 16.2 **West Nile virus.** West Nile virus (A) is a flavivirus that emerged in the United States in 1999. It is transmitted by *Culex* mosquitoes, such as this engorged female *Culex quinquefasciatus* (B), the primary vector of the virus in southern California and the southeastern United States. *Images courtesy of CDC and (A) P.E. Rollin and Cynthia Goldsmith and (B) Games Gathany.*

blood transfusions, as well as from mother to child during pregnancy, delivery, or breastfeeding. As a result, nucleic acid testing of the blood supply for WNV began in 2003.

Increased tourism to exotic and pristine areas also puts humans at risk of coming in contact with novel zoonoses. It also exposes native animal populations to human diseases. An **anthroponosis** or **anthroponotic disease** is an infectious disease transferred from humans to other animals. Due to their genetic similarity to humans, great apes such as chimpanzees and gorillas are particularly susceptible to anthroponoses. Chimps have died from human viruses such as respiratory syncytial virus and metapneumovirus, and primatologist Jane Goodall noted in her book *The Chimpanzees of Gombe* that a local outbreak of polio in humans was accompanied by similar symptoms of flaccid paralysis in nearby chimpanzee populations.

Many sociological factors may also contribute to the emergence/reemergence and spread of viral diseases. Some of these factors include inadequate public health infrastructures, a lack of access to vaccinations or reduced immunization policies (see *In-Depth Look*), and political conflict or wars that displace millions of people, creating overcrowded refugee populations without access to basic health care or clean water supplies.

Human behaviors also influence the spread of emerging/reemerging infectious diseases. The popularity of restaurants and partially prepared foods provides an avenue to disseminate foodborne viruses like hepatitis A virus, while importing fruits, vegetables, and nuts from other countries can lead to the widespread distribution of a virus. Sexual activities, body art (tattoos, piercings), and the use of intravenous drugs can spread viruses such as HIV or hepatitis. Religious or spiritual practices can also affect the transmission of viruses. For instance, the touching and handling of deceased individuals has been shown to spread Ebolaviruses, and outbreaks of polio and measles, among others, often occur in religious communities that avoid vaccinations.

Finally, the susceptibility of the human host also plays a role in the ability of a virus to emerge as a human pathogen. The host must possess receptors for the virus and the intracellular factors required for the virus to replicate. Viruses that are easily combatted by the host immune system will be quickly eliminated, whereas viruses that elicit an internal "cytokine storm" by immune system cells are likely to induce rapid and severe pathology. Immunocompromised individuals, pregnant women, malnourished people, the elderly, and the very young are often more susceptible to infection due to their weaker immune systems.

16.1.2 Environmental and Ecological Factors

Nearly every emerging viral disease can be associated with an environmental or ecological component. The consequences of urbanization and globalization—including but not limited to deforestation, habitat modification, and the use of wild land for farming—impact the emergence and reemergence of potential zoonoses. Increases in the human population have led to the encroachment of humans into animal habitats formerly unoccupied by humans. These circumstances contribute to humans coming in contact with animals with which they would not normally interact, putting them at risk of being exposed to zoonotic viruses that are found in the local wildlife populations. The transfer of zoonotic viruses to humans can occur directly or indirectly. For example, during an outbreak of 276 cases of Nipah virus in Malaysia and Singapore in 1998–99, most infected individuals had contact with sick pigs that were being farmed commercially as livestock and then slaughtered for meat. The pigs contracted Nipah virus from the local bats, which were thought to have spread the virus through urine or by dropping partially eaten fruit, laden with bat saliva, into the pig stalls. The bats may have acquired the fruit from the fruit trees on the farms. Human infection with Nipah virus causes encephalitis, fever, headache, and reduced consciousness. Forty percent of infected individuals died during this outbreak. Interestingly, Nipah virus infection is asymptomatic in bats.

In contrast to the indirect mode of transmission observed in the 1998–99 outbreak in Malaysia and Singapore, transmission of Nipah virus in Bangladesh has occurred directly from bats to humans. The most frequently implicated route

In-Depth Look: Is Measles a Reemerging Disease?

Measles is a respiratory virus (Fig. 16.3A) that causes a fever (as high as 105°F) and a maculopapular rash that covers most of the body (Fig. 16.3B). It is highly contagious, and infection leads to a severe illness that lasts about a week. Infection can cause serious complications, including deafness, encephalitis, blindness, seizures, and pneumonia. Death occurs in 2–3 of every 1000 infected individuals. Worldwide, 20 million people are infected annually, causing over 145,000 deaths each year. It is one of the leading causes of death in young children.

Measles is not a newly emerging disease. The first documentation of infection dates back to the year AD 900 in an account written by a Persian doctor trying to distinguish measles from smallpox. It remains endemic in certain areas of the world, including portions of Europe and Africa. However, vaccination has proven effective in reducing cases of measles and its complications. For example, in the United States, three to four million people contracted measles each year before the vaccine program was initiated in 1963. These infections caused 400–500 deaths, 48,000 hospitalizations, and 4000 cases of encephalitis annually. In 2000, the United States declared that measles had been eliminated from the country, attributable to strong vaccination campaigns.

Unfortunately, measles has been making a comeback in the US in the past decade, imported from foreign countries or from unvaccinated individuals that have visited these countries. Unlike previous occasions when the virus did not meet many susceptible hosts, measles has recently been able to spread locally in the United States because of pockets of individuals that have refused vaccination. Although the vaccine is 97% effective, a small percentage of previously vaccinated individuals have also contracted the disease, further emphasizing the importance of high vaccination rates.

Several large outbreaks have occurred from 2008 to the present (Fig. 16.3C). 2014 saw 668 cases of measles in the United States, the largest number observed since 2000 and more cases than occurred in 1996. As of September 18, 2015, 189 measles cases have occurred in the United States during 2015. Most of these (117) were associated with an outbreak at two Disney theme parks in California. Among 110 infected California residents, 45% were unvaccinated. Twelve were too young to receive the vaccine, but 76% of the unvaccinated individuals declined vaccination due to personal beliefs. The uptick of measles in the United States and other countries in which the virus was previous eradicated emphasize the importance of the measles vaccine and efforts to maintain high immunization rates.

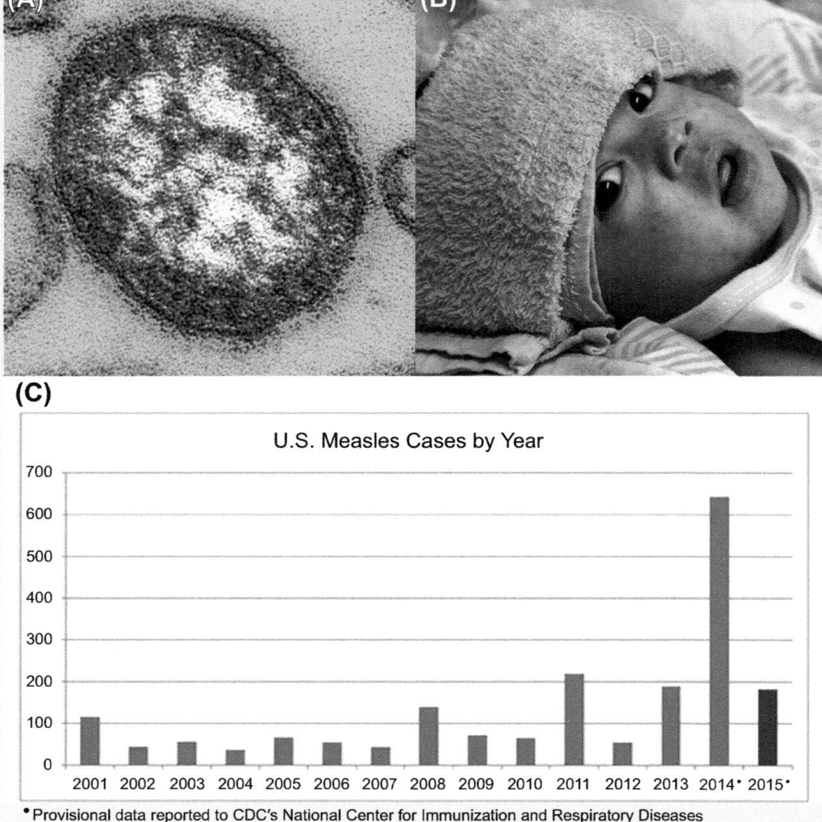

FIGURE 16.3 Reemergence of measles virus. (A) This transmission electron micrograph shows the measles virus, a –ssRNA paramyxovirus. (B) A child in the Philippines capital of Manila in 2014 with the characteristic maculopapular measles rash. (C) Cases of measles have been increasing in the last decade. A record 668 cases of measles were diagnosed in 2014. *Images courtesy of CDC and (A) Cynthia S. Goldsmith and William Bellini, PhD (B) Molly Kurnit, MPH and Jim Goodson, MPH, and (C) National Center for Immunization and Respiratory Diseases, Division of Viral Diseases, http://www.cdc.gov/measles.*

was through the ingestion of fresh date palm sap, which is harvested from date palm trees much in the same way that maple sap is tapped from maple trees to produce maple syrup. Local bat species also visit the pots of date palm sap and lick the sweet sap as it is being collected, thereby contaminating the sap with virus-containing saliva and providing a route of exposure to humans that drink it. Human-to-human transmission of Nipah virus also occurred after the virus entered the human population.

Weather and climate change also play a role in the emergence of novel viral diseases. These can be short-term seasonal and regional changes in weather patterns, or longer-term climate change trends (global warming). For example, increases in rainfall in certain areas lead to growth of the vegetation that supports rodent populations. As a consequence, rodent exposure to humans is increased, concurrent with the frequency of rodentborne viruses. This occurred in the emergence of the Sin Nombre virus, a hantavirus that causes a potentially deadly respiratory syndrome, in the Four Corners region of the Southwestern United States. Preceded by six years of drought, the 1993 spring season was characterized by extremely heavy rains (as a result of an El Niño) that increased 10-fold the population of deer mice (Fig. 16.4A). This resulted in 42 people contracting Sin Nombre virus, a previously unrecognized hantavirus carried by deer mice. Sixty-two percent of the individuals that contracted Sin Nombre virus died of hantavirus pulmonary syndrome.

Increases in rainfall also affect mosquito populations by providing stagnant pools and puddles (Fig. 16.4B) that function as extra breeding grounds for the insects (Fig. 16.4C). The prevalence of dengue virus (DENV) and Rift Valley fever virus, two mosquito-borne viruses, are significantly affected by rainfall. Similarly, WNV is a seasonal epidemic in the United States, appearing in the summer and fall months, the time of the year when rainfall supports mosquito larval habitats.

Depending upon the species, other environmental factors also affect the growth of viral vectors. These include mean temperature, wind speed, dew point, soil moisture, and rain accumulation, among others. Long-term increases in global temperature due to climate change also affect the emergence/reemergence of viral diseases. According to the National Oceanic and Atmospheric Administration, the combined land and ocean surface temperature in 2014 was 1.24°F above the 20th century average, and the 20 warmest years on record have all occurred within the past 20 years. Warmer winter weather allows female mosquitoes to more

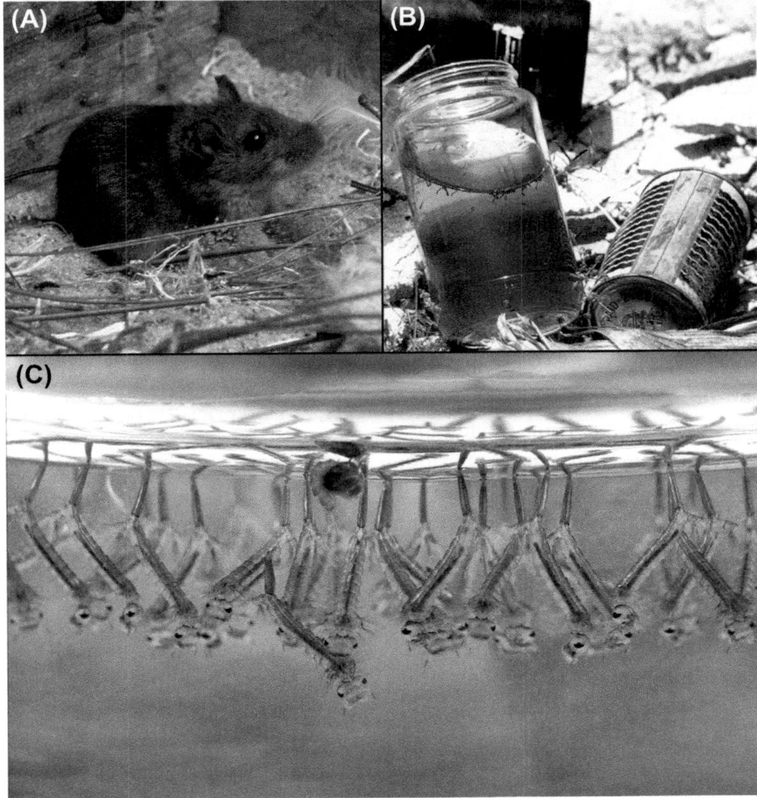

FIGURE 16.4 **Factors in the emergence of viral diseases.** Changes in rainfall levels can lead to an increase in rodent vectors, such as the deer mouse, *Peromyscus maniculatus* (A), the natural reservoir of Sin Nombre virus. (B) Stagnant water provides breeding grounds for mosquito larvae, as seen in this jar of rainwater. (C) This photograph shows *Culex* mosquito larvae found in standing water just under the surface. A prominent breathing siphon enables the larvae to access their air supply. *Photographs courtesy of CDC and (B) Graham Heid and Dr. Harry D. Pratt and (C) James Gathany.*

easily overwinter, surviving to the spring. Warmer daily temperatures in an area also expand the habitat for certain mosquito populations, concurrently expanding the populations that can be infected by the viruses they carry. In the United States, mosquito populations in tropical and subtropical Florida support local infections of DENV and Chikungunya virus, the latter of which first emerged in Florida in 2014. As average temperatures increase in states, so will the frequency of zoonotic viruses into states whose climate did not previously support these mosquito species. Similarly, climate changes affect the distribution of ticks and the migratory routes of birds and other animals, changing the locations that are exposed to the zoonoses they carry. In the case of birds, this includes influenza viruses and WNV, among others.

Study Break

Provide four examples of how human or environmental/ecological factors have affected the emergence or reemergence of a viral disease.

16.1.3 Viral Factors

Although human behaviors and environmental/ecological factors are more commonly associated with outbreaks, pathogen characteristics also are involved in the success of emergent/reemergent infectious diseases. For viruses, the molecular makeup of the viral genome often determines whether the virus will successfully integrate into the new population. This generally occurs through reassortment, recombination, or mutation.

Many emerging viruses are unable to establish themselves because their new hosts function as "dead-end" infections that do not sustain person-to-person spread. As described in Chapter 8, "Influenza," antigenic shift occurs when subtypes of influenza that are circulating in animal populations enter the human population for the first time (see Fig. 6.15B and Section 10.4 of Chapter 8, "Influenza," for a refresher). The properties of the virus determine whether the antigenic shift will result in a pandemic or will be unable to undergo human-to-human transmission. The 1918 H1N1 virus was able to spread throughout the human population, but H5N1 and H7N9 viruses have only been transmitted to humans through direct contact with birds. However, because influenza viruses are segmented, coinfection of a susceptible host with both avian and human viruses could lead to the reassortment of genetic segments of the two viruses and the creation of a novel human influenza virus subtype. Because of these concerns and the high fatality rate of certain avian influenza strains in the human population, over 45 million chickens, turkeys, and ducks were culled in the Midwestern United States in 2015 in an effort to prevent the further spread of H5N2 in the poultry population and its possible jump to humans.

Another means of acquiring genetic changes is through recombination. As described in Chapter 4, "Virus Replication," recombination occurs when the RNA or DNA polymerase that is copying the viral genome transfers to the template of another strain of the virus, thereby creating a hybrid genome from two different strains of the virus. Recombination has as much chance of conferring an evolutionarily disadvantageous or neutral result as an advantageous one, but it does contribute to genetic variability within individual virions that could be selected for if beneficial within the environment.

RNA viruses are the most common cause of emerging diseases in humans, attributable to the high mutation rate in RNA viruses compared to DNA viruses. Unlike the DNA-dependent DNA polymerases of living organisms and DNA viruses, RNA-dependent RNA polymerases (including reverse transcriptases) do not possess proofreading ability. This leads to a decrease in enzyme fidelity, inserting an incorrect nucleotide every 10^5 bases (compared to 1 error per 10^9 bases for DNA polymerases). As a result, RNA viruses have some of the highest mutation rates of all biological entities. This ensures a range of genetic diversity that maintains virulence and promotes potential transmission to new hosts.

16.2 NOTABLE EMERGING/REEMERGING VIRAL DISEASES

Within the past 50 years, dozens of viruses have emerged or reemerged within the human population, several of which have been mentioned above (see Table 16.2 for a selection of emerging viruses). It is important to note that other viruses have likely also emerged but are unnoticed because they are not associated with a notable clinical infection. Because the great majority of emerging viral diseases are zoonoses, below we focus on notable zoonotic viruses transmitted by arthropods and by nonhuman mammals to humans.

16.2.1 Arboviruses

Arboviruses (**ar**thropod-**bo**rne **viruses**) are viruses that are transmitted to humans or other mammals by arthropods, invertebrate animals possessing an exoskeleton. The major arthropod vectors for the transmission of viruses are mosquitoes and ticks. In 1930, only six arboviruses had been identified, one of which—yellow fever virus—caused disease in humans. Currently, the CDC arbovirus catalog lists 537 known arboviruses, approximately a quarter of which cause disease in humans. With only a few exceptions, arboviruses are RNA viruses.

Most arboviruses are maintained in a transmission cycle between an arthropod vector and vertebrate hosts, usually birds or small mammals. The virus is acquired by the vector when it feeds upon an infected individual,

TABLE 16.2 Selected Emerging/Reemerging Viral Diseases

Family	Virus	Natural reservoir	Transmitted to humans by:	Mode of transmission	Disease caused
Adenoviridae	Adenovirus 14	Humans	Humans	Respiratory secretions	Acute respiratory disease
Arenaviridae	Guanarito virus	Rodents	Rodents (*Zygodontomys* sp.)	Urine or feces of infected rodents	Fever, hemorrhagic fever
	Lassa fever	Rodents	Rodents (*Mastomys* species)	Urine or feces of infected rodents	Fever, hemorrhagic fever
	Machupo virus	Rodents	Rodents (*Calomys* species)	Urine or feces of infected rodents	Fever, hemorrhagic fever
Bunyaviridae	Crimean–Congo hemorrhagic fever virus	Ticks	Ticks (*Hyalomma* species)	Tick bite	Fever, encephalitis[a], hemorrhagic fever
	Rift Valley fever virus	Sheep, cattle	Mosquitoes, infected blood/fluids	Mosquito bite, mucosal infection	Encephalitis, hemorrhagic fever
	Sin Nombre virus	Rodents	Rodents	Urine or feces of infected rodents	Hantavirus pulmonary syndrome
Caliciviridae	Norovirus	Humans	Humans	Fecal-oral	Gastroenteritis, diarrhea, vomiting
Coronaviridae	MERS coronavirus	Camels	Camels, humans	Respiratory secretions	Severe acute respiratory illness
	SARS coronavirus	Horseshoe bats	Civets	Respiratory secretions	Severe acute respiratory illness
Filoviridae	Ebola virus	Bats	Bats, primates, humans	Bodily fluids	Hemorrhagic fever, shock
	Marburg virus	Bats	Bats, primates, humans	Bodily fluids	Hemorrhagic fever, shock
Flaviviridae	Dengue virus	Primates, humans	Mosquitos (*Aedes* species)	Mosquito bite	Hemorrhagic fever, shock
	Hepatitis C virus	Humans	Humans	Blood products, sexual activity, vertical	Hepatitis, liver cancer
	Powassan virus	Woodchucks, squirrels, mice	Ticks (*Ixodes* species)	Tick bite	Encephalitis, meningitis
	St. Louis encephalitis virus	Birds	Mosquitoes (*Culex* species)	Mosquito bite	Encephalitis
	West Nile virus	Birds	Mosquitoes (*Culex* sp.), birds	Mosquito bite	Encephalitis, meningitis
	Yellow fever virus	Humans, primates	Mosquitoes (*Aedes* species)	Mosquito bite	Fever, myalgia, hemorrhagic fever
Orthomyxoviridae	Avian influenza viruses	Waterfowl, birds	Poultry	Respiratory secretions	Severe respiratory illness, pneumonia

Continued

TABLE 16.2 Selected Emerging/Reemerging Viral Diseases—cont'd

Family	Virus	Natural reservoir	Transmitted to humans by:	Mode of transmission	Disease caused
Paramyxoviridae	Hendra virus	Fruit bats (*Pteropodidae* sp.)	Horses	Bodily fluids, urine, feces	Respiratory illness, encephalitis
	Measles virus	Humans	Humans	Respiratory secretions	Fever, rash, conjunctivitis
	Nipah virus	Bats	Bats, pigs, humans	Bodily fluids, urine, feces	Encephalitis
Picornaviridae	Poliovirus	Humans	Humans	Fecal–oral	Paralysis
Poxviridae	Monkeypox	Unidentified rodents	Rodents, marsupials, primates	Bodily fluids, respiratory secretions	Fever, rash, encephalitis
Reoviridae	Rotavirus	Humans	Humans	Fecal-oral or respiratory secretions	Gastroenteritis, diarrhea, dehydration
Retroviridae	Human immunodeficiency virus	Chimpanzees	Chimpanzees, humans	Blood products, sexual activity, vertical	Immunodeficiency
Rhabdoviridae	Rabies virus	Bats	Bats, dogs, raccoons, skunks	Animal bite	Encephalomyelitis
Togaviridae	Chikungunya virus	Bats, rodents, primates	Mosquitoes (*Aedes* species)	Mosquito bite	Arthralgia, rash
	Eastern equine encephalitis virus	Birds, rodents	Mosquitoes (*Culex* species)	Mosquito bite	Encephalitis
	Venezuelan equine encephalitis virus	Rodents, horses	Mosquitoes (*Culex* species)	Mosquito bite	Encephalitis
	Western equine encephalitis virus	Birds	Mosquitoes (*Culex* species)	Mosquito bite	Encephalitis

[a]*Inflammation of the brain.*

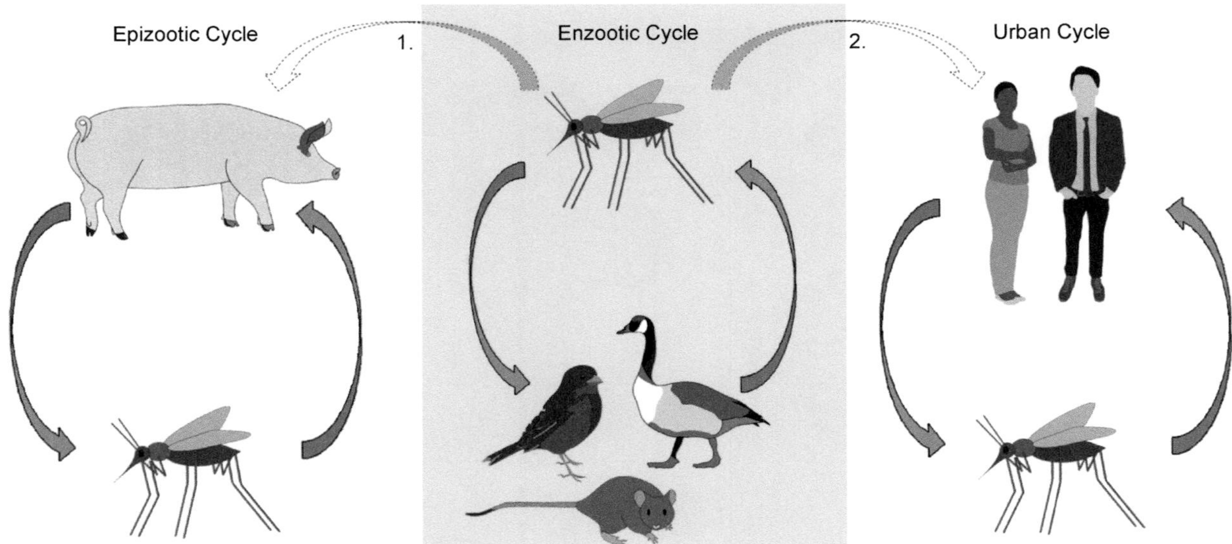

FIGURE 16.5 **Arbovirus transmission models.** An enzootic cycle occurs when the virus is transmitted between its arthropod vector and its natural reservoir, a wild animal. The vector may infect (1) and cause an epidemic within domestic animals, which can further amplify the virus in an epizootic cycle. Humans may also become infected tangentially from enzootic cycles (2). In the urban cycle, the virus cycles directly between the vector and a human host.

taking a blood meal that contains the virus. Only female mosquitoes feed on blood meals (males feed on flower nectar). Within the infected mosquito, the virus breaches the midgut to infect the salivary glands, becoming transmitted within the saliva of the mosquito when it bites a new individual. On the other hand, ticks attach to the skin of the host and create a feeding pool within the epidermis, becoming engorged with blood. As with the mosquito, the virus enters the tick within the blood meal, replicates within the salivary glands, and is transmitted through saliva to the next host. Once infected, the mosquito or tick can remain infectious for the duration of its life, generally 6–8 weeks for female mosquitoes, and years for ticks. Viruses can also be transmitted vertically from an infected female mosquito or tick to the next generation within eggs, although this is not thought to be the primary means of arbovirus persistence within the population.

Arboviruses are transmitted via several models of transmission (Fig. 16.5). The **enzootic cycle**, also known as the jungle cycle or sylvatic cycle (*sylvatic* means "occurring in wild animals"), occurs when the virus cycles between the arthropod vector and the natural reservoir, a wild animal. Some arboviruses may also possess an **epizootic cycle**, indicating the virus has the potential to cause an epidemic within animals that become infected by the vector. This can occur in wild or domestic animals, such as pigs and horses. For example, Japanese encephalitis virus normally circulates between mosquitoes and birds but can be amplified in pigs, and Venezuelan equine encephalitis virus has an enzootic cycle between mosquitoes and rodents but is capable of initiating a rural epizootic cycle in horses. Both of these transmission cycles can result in tangential infection of humans if infected by the arthropod vector. On the other hand, an **urban cycle** involves the direct transmission of the virus between mosquitoes and humans. In this case, the viremia that occurs upon human infection is sufficiently large to transmit the virus when the person is bitten by the mosquito. Urban cycles occur with yellow fever virus and DENV, although both of these viruses possess enzootic transmission cycles as well. Intermediate cycles are also possible in which vectors transmit the virus back and forth between human and nonhuman hosts.

The *Flaviviridae* family contains several well-known viruses, including hepatitis C virus. These viruses are enveloped and possess +ssRNA genomes within an icosahedral capsid. Within *Flaviviridae*, the *Flavivirus* genus contains over 50 different species, the majority of which are transmitted by vectors. It includes several notable human pathogens, including yellow fever virus, DENV, Japanese encephalitis virus, WNV, and St. Louis encephalitis virus (see Table 16.2). DENV is currently the most common arboviral disease in the world, infecting around 400 million individuals and causing over 12,000 deaths each year. Dengue-like epidemics have been recorded in tropical areas of the world since the 1600s, and the virus was first isolated in 1943–44 by Albert Sabin and others from soldiers in the Pacific and Asia during World War II. DENV is considered a reemerging virus due to the expansion of the domesticated *Aedes aegypti* mosquito (Fig. 16.6A), the primary vector, and *Aedes albopictus*, a lesser vector for DENV. The spread of these mosquito species is attributable to the factors mentioned above. Currently, the mosquito species live mainly

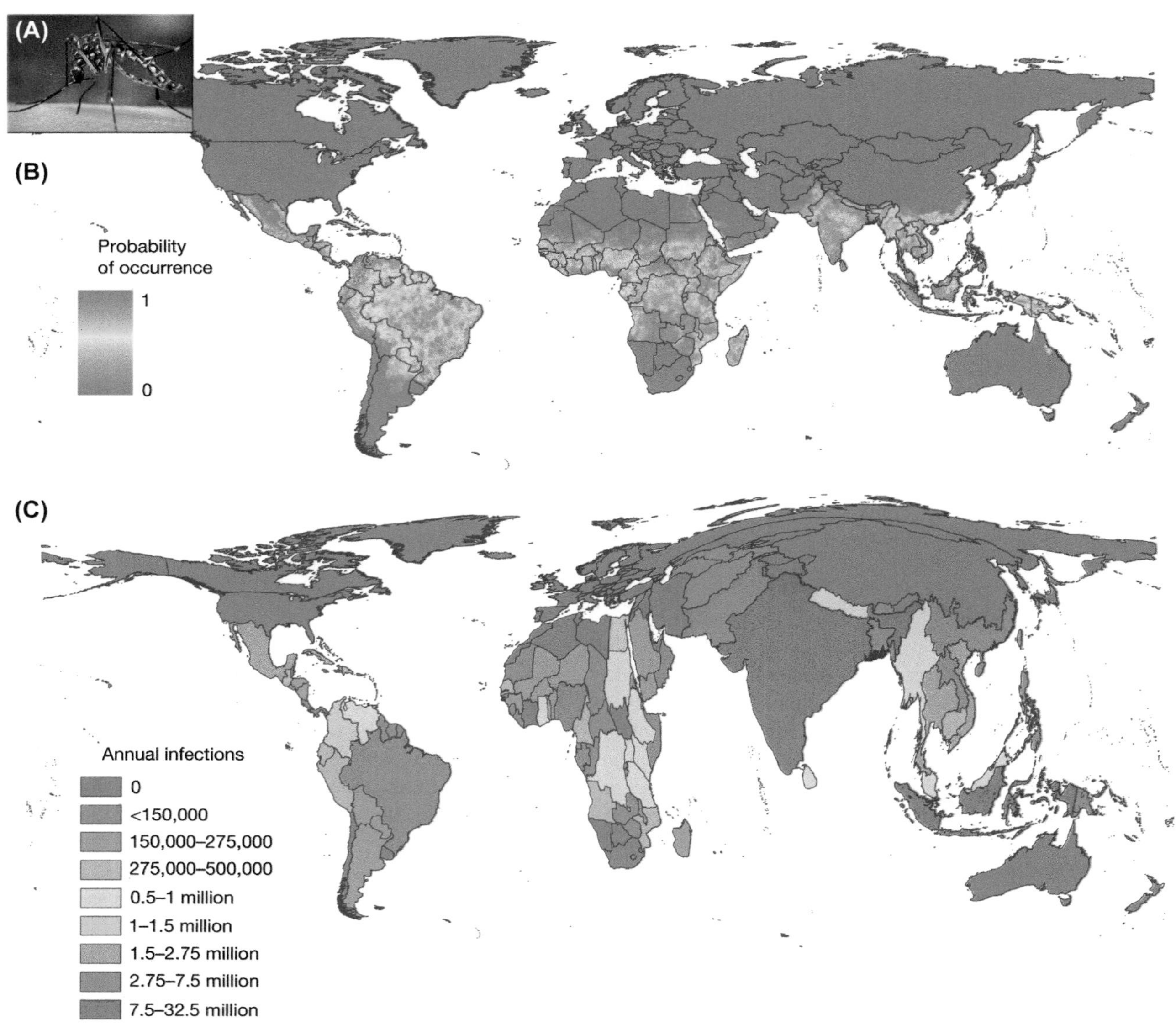

FIGURE 16.6 **Geographical distribution of dengue virus.** (A) *Aedes aegypti* is a domestic, day-biting mosquito that prefers to feed on humans. It is the most common vector of dengue virus. *(Image courtesy of CDC/Prof. Frank Hadley Collins and James Gathany.)* (B) Map showing the probability of dengue occurrence. Areas with a high probability of occurrence are shown in red and areas with a low probability in green. (C) A cartogram that has distorted the geographic area to correspond to the annual number of dengue infections for all ages. *((B) and (C) reprinted by permission of Bhatt, S. et al., 2013. The global distribution and burden of dengue. Nature 496(7446), 504–507. Macmillan Publishers Ltd., copyright 2013.)*

between the 35°S and 35°N latitudes and below 3200 feet in areas where the evening temperature does not fall below 50°F (Fig. 16.6B and C). They are associated with human-populated areas and cities, and as such, DENV is communicated primarily via an urban transmission cycle.

Although they cause similar clinical symptoms, the four serotypes of DENV (DENV-1, DENV-2, DENV-3, and DENV-4) are genetically distinct. As such, immunity against one serotype does not provide long-term protection against another serotype, although some short-term protection may occur. Within 5–7 days of being bit by an infected mosquito, a person develops a viremia that lasts 5–12 days. DENV can be transmitted from human to human through blood transfusions or organ transplants, and perinatal transmission can occur if the mother is infected around the time of delivery. The virus can be transmitted back to a mosquito host if the insect feeds on an infected person during the viremia. Following infection of a mosquito, the virus takes 8–12 days to disseminate to the salivary glands and become transmissible to another human in saliva.

TABLE 16.3 Case Definitions[a] of Symptomatic Dengue Illness

Dengue without warning signs
Fever and two of the following: Nausea, vomiting Rash Aches and pains Leukopenia Positive tourniquet test
Dengue with warning signs
Dengue as defined above with any of the following: Abdominal pain or tenderness Persistent vomiting Clinical fluid accumulation Mucosal bleeding Lethargy, restlessness Liver enlargement >2 cm Laboratory: increase in hematocrit with decrease in platelet count
Severe dengue
Dengue with at least one of the following criteria: Severe plasma leakage leading to: • Shock • Fluid accumulation with respiratory distress Severe bleeding as evaluated by clinician Severe organ involvement: • Liver: AST or ALT ≥1000 • CNS: impaired consciousness • Failure of heart and other organs

[a]*Most recent (2009) World Health Organization case definitions.*

As many as three-quarters of DENV infections are asymptomatic. Symptomatic illness is divided into nonsevere dengue and severe dengue (Table 16.3). Nonsevere dengue is further divided into dengue without warning signs or dengue with warning signs. **Dengue without warning signs** is the less severe form of disease, characterized by fever and at least two of the following symptoms: nausea/vomiting, rash, aches and pains, **leukopenia** (low white blood cell count), or a positive tourniquet test. The tourniquet test is performed by applying a blood pressure cuff on the individual for 5 min at a pressure between the person's systolic and diastolic pressures and then counting the number of **petechiae**, small red spots indicative of capillary **hemorrhages** (bleeds). Dengue is also known as "breakbone fever," eluding to the extreme bone pain that can be experienced.

During **dengue with warning signs**, individuals display any of the above symptoms with the addition of abdominal pain, mucosal bleeding, lethargy, liver enlargement >2 cm, persistent vomiting, or accumulation of fluid in the abdomen or between the pleural linings covering the lungs. They may also have an increase in hematocrit (the volume within the blood that is occupied by red blood cells), indicating that plasma is leaving the bloodstream. These symptoms may progress to **severe dengue**, which is characterized by **hemorrhagic fever**. This is a severe multiorgan syndrome that results from increased permeability of the vascular system, leading to petechiae and bleeding within the internal organs or from orifices like the mouth, nose, and eyes. Severe dengue usually occurs after **defervescence**, the abatement of fever. During this 1–2-day "critical period," plasma leakage can also result in **shock**. The leakage of plasma from major blood vessels causes a loss of blood pressure and subsequent damage to multiple organs due to an inadequate supply of blood and the clotting of blood in small vessels. There are no antivirals approved for use against dengue virus, so treatment is supportive in an attempt to prevent progression to shock. Severe dengue may be fatal for up to 5% of those with the condition, while recovery often takes months in those that survive.

The emergence and spread of arthropod-borne zoonotic viruses are a major concern because many arboviruses cause serious disease—including death—in humans. Besides *Flaviviridae*, arboviruses are also found in the *Bunyaviridae*, *Reoviridae*, and *Togaviridae* families. Major control measures are aimed at eliminating the vector and preventing transmission. For example, mosquito-control measures can reduce vector populations, while public education can lessen the interaction of humans with mosquitoes or ticks by encouraging the use of insect repellents, clothes that cover exposed skin, or bed nets. Currently, yellow fever virus and Japanese encephalitis virus are the only two arboviruses for which human vaccines are available, although vaccines for DENV and WNV are in development.

Study Break
What factors might contribute to the expanding range of dengue virus?

16.2.2 Vertebrate Zoonoses

Although arboviruses are a major cause, wildlife is the primary source of viral zoonoses. The two most species-rich orders are *Rodentia* (rodents) and *Chiroptera* (bats), so it is not surprising that both are host to a variety of high-impact zoonotic viruses. Rodent and bat species are host to 68 and 61 zoonotic viruses, respectively, although bats carry more zoonotic viruses per species. Viruses within the *Arenaviridae* and *Filoviridae* families are transmitted by rodents and bats, respectively. Like dengue virus, both families contain viruses that can cause serious disease, including hemorrhagic fever leading to hypotension, shock, and multiple organ failure.

Word Origin: Arenaviruses

Arenaviruses derive their name from the Latin word *arena*, meaning "sandy." When viewed with an electron microscope, the interior of an arenavirus virion contains sandy or grainy particles (Fig. 16.7). These are ribosomes that were enclosed within the lipid envelope during assembly. The ribosomes are nonfunctional and not required for replication. Like all viruses, arenaviruses rely upon entry into a host cell for replication.

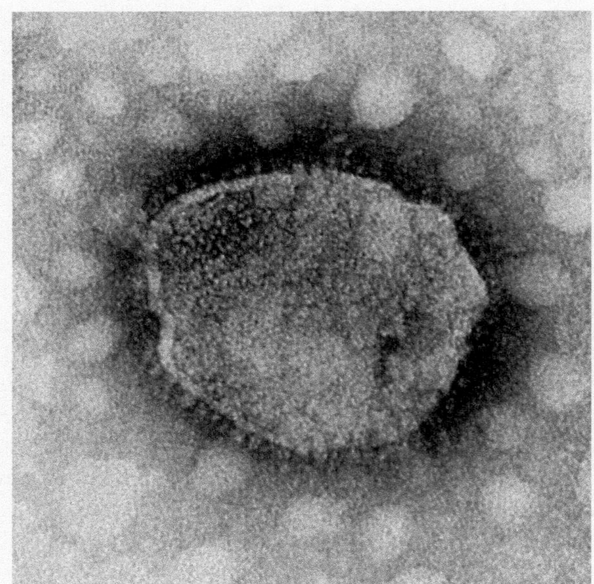

FIGURE 16.7 **Arenaviruses.** Arenaviruses derive their name from the Latin word *arena,* meaning "sandy," in reference to the appearance of the interior of the virion when viewed with an electron microscope. Several arenaviruses are transmitted by rodent vectors and are capable of causing severe disease, including hemorrhagic fevers. *Image courtesy of CDC and E.L. Palmer.*

Arenaviruses are enveloped, ambisense viruses with a bisegmented genome. Viruses within the *Arenaviridae* family are transmitted by rodents and are capable of causing hemorrhagic fever, including Junin virus, Machupo virus, Lassa virus, and Guanarito virus (Table 16.4). The first arenavirus, lymphocytic choriomeningitis virus, was isolated in 1933 and found to be a cause of viral meningitis. Since the 1950s, new arenaviruses have been discovered every few years.

Each of these viruses is limited to the geographical area in which its rodent host is found (Table 16.4). The animals become infected through fighting and bites, although some arenaviruses are transmitted vertically to offspring, thereby maintaining the infected population. The rodents are unaffected by the virus but remain chronically infected, shedding virus into the environment through urine and feces. Humans become infected when abraded skin comes in contact with rodent excretions, or when they ingest contaminated food or inhale aerosolized particles. Some arenaviruses have been shown to be transmitted from person-to-person, including Lassa virus and Machupo virus, through infectious droplets or bodily secretions. Fatality rates for the arenaviruses that cause hemorrhagic fever range from 5% to 35%, although rates up to 50% have been documented in certain outbreaks.

Whereas rodents transmit arenaviruses to humans, bats are thought to be the natural reservoir of viruses within the *Filoviridae* family (*filo* is Latin for "thread," referring to the threadlike appearance of these enveloped helical −ssRNA viruses) (Fig. 16.8). The first filovirus was discovered in 1967 when 31 cases of hemorrhagic fever were documented in German and Yugoslavian laboratory workers who had been handling tissues obtained from African green monkeys imported from Uganda. Of these individuals, seven died (22.5%). The virus was named **Marburg virus**, after the location of one of the outbreaks.

The second filovirus was discovered 9 years later, in 1976. Two outbreaks of a hemorrhagic disease simultaneously struck in Africa, one in northern Zaire (currently the Democratic Republic of the Congo) and one in southern Sudan. The disease was found to be spread through close personal contact and shared syringes in hospitals and clinics. Eighty-eight percent of the 318 infected in Zaire succumbed to the disease, while 53% of those infected in Sudan died.

The virus was soon determined to be a filovirus unique from the Marburg virus. The novel virus was named after a river 60 miles from the site of the Zaire outbreak: the Ebola River. The two outbreaks were found to be caused by different species of ***Ebolavirus***, *Zaire ebolavirus* and *Sudan ebolavirus*. Since that time, three additional species have been identified: *Reston ebolavirus* in 1989, *Taï Forest ebolavirus* in 1994, (formerly Côte d'Ivoire EBOV), and *Bundibugyo ebolavirus* in 2007. The viruses within these five species are known as Ebolavirus (EBOV), Sudan virus (SUDV), Reston virus (RESTV), Taï Forest virus (TAFV), and Bundibugyo virus (BDBV), respectively (Table 16.5). RESTV is the only ebolavirus that has not shown any symptoms in humans, thus far only being seen in monkeys from the Philippines (being held in quarantine facilities in the United States and Italy) and on a pig farm in the Philippines.

Other than accidental laboratory infections or imported cases, all of the human ebolavirus outbreaks have originated in Africa (Fig. 16.9A). In fact, before 1994, all known natural cases of ebolaviruses were in

TABLE 16.4 Arenavirus Infections of Humans

Arenavirus	Possible disease	Natural host	Geographic range	Year isolated
LCMV[a]	Aseptic meningitis	House mouse (*Mus musculus*)	Europe, Americas	1933
Junin virus	Argentinian hemorrhagic fever	Drylands vesper mouse (*Calomys musculinus*)	Argentina	1958
Machupo virus	Bolivian hemorrhagic fever	Large vesper mouse (*Calomys callosus*)	Bolivia	1963
Lassa	Lassa fever	Multimammate rat (*Mastomys natalensis*)	West Africa	1969
Guanarito virus	Venezuelan hemorrhagic fever	Short-tailed cane mouse (*Zygodontomys brevicauda*)	Venezuela	1989
Sabia virus	Brazilian hemorrhagic fever	Unknown	Brazil	1993
Chapare virus	Chapare hemorrhagic fever	Unknown	Bolivia	2008
Lujo virus	Lujo hemorrhagic fever	Unknown	Zambia	2008

[a]*Lymphocytic choriomeningitis virus.*

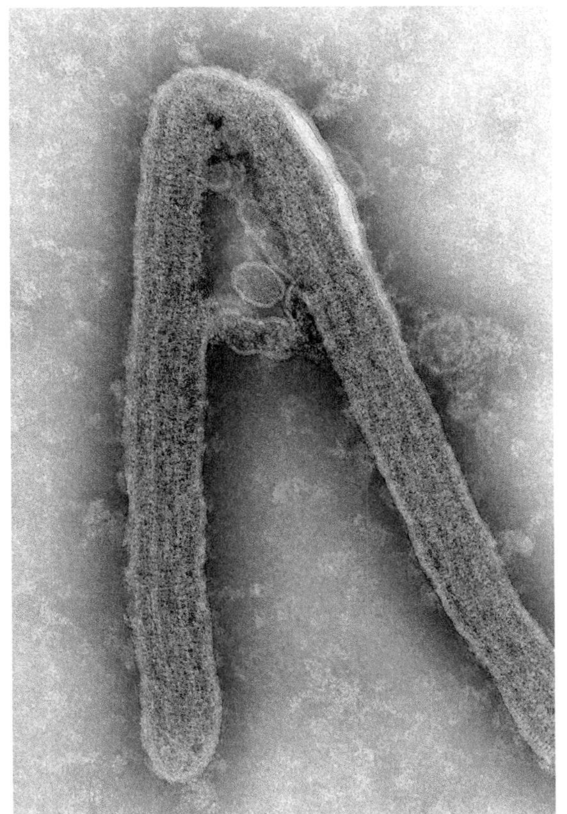

FIGURE 16.8 **Marburg virus.** The first filovirus, Marburg virus, was discovered in 1967 and named after the location of one of the outbreaks (Marburg, Germany). This electron micrograph shows the threadlike appearance that is characteristic of these helical –ssRNA viruses. *Image courtesy of CDC/Frederick Murphy.*

TABLE 16.5 Taxonomy of Ebolaviruses[a]

Order *Mononegavirales*
Family *Filoviridae*
Genus *Ebolavirus*
Species *Taï Forest ebolavirus*
Virus: Taï Forest virus (TAFV)
Species *Reston ebolavirus*
Virus: Reston virus (RESTV)
Species *Sudan ebolavirus*
Virus: Sudan virus (SUDV)
Species *Zaire ebolavirus*
Virus: Ebola virus (EBOV)
Species *Bundibugyo ebolavirus*
Virus: Bundibugyo virus (BDBV)

[a]*International Committee on Taxonomy of Viruses, EC 46, July 2014 (MSL #29).*

five countries in Central Africa: Democratic Republic of Congo (formerly Zaire), South Sudan (previously part of Sudan), Uganda, Republic of the Congo, and Gabon. In 1994, a scientist became ill after conducting a necropsy (animal autopsy) on a wild chimpanzee that had succumbed to a viral hemorrhagic fever in the Taï Forest of the Côte d'Ivoire (Ivory Coast). The virus was identified

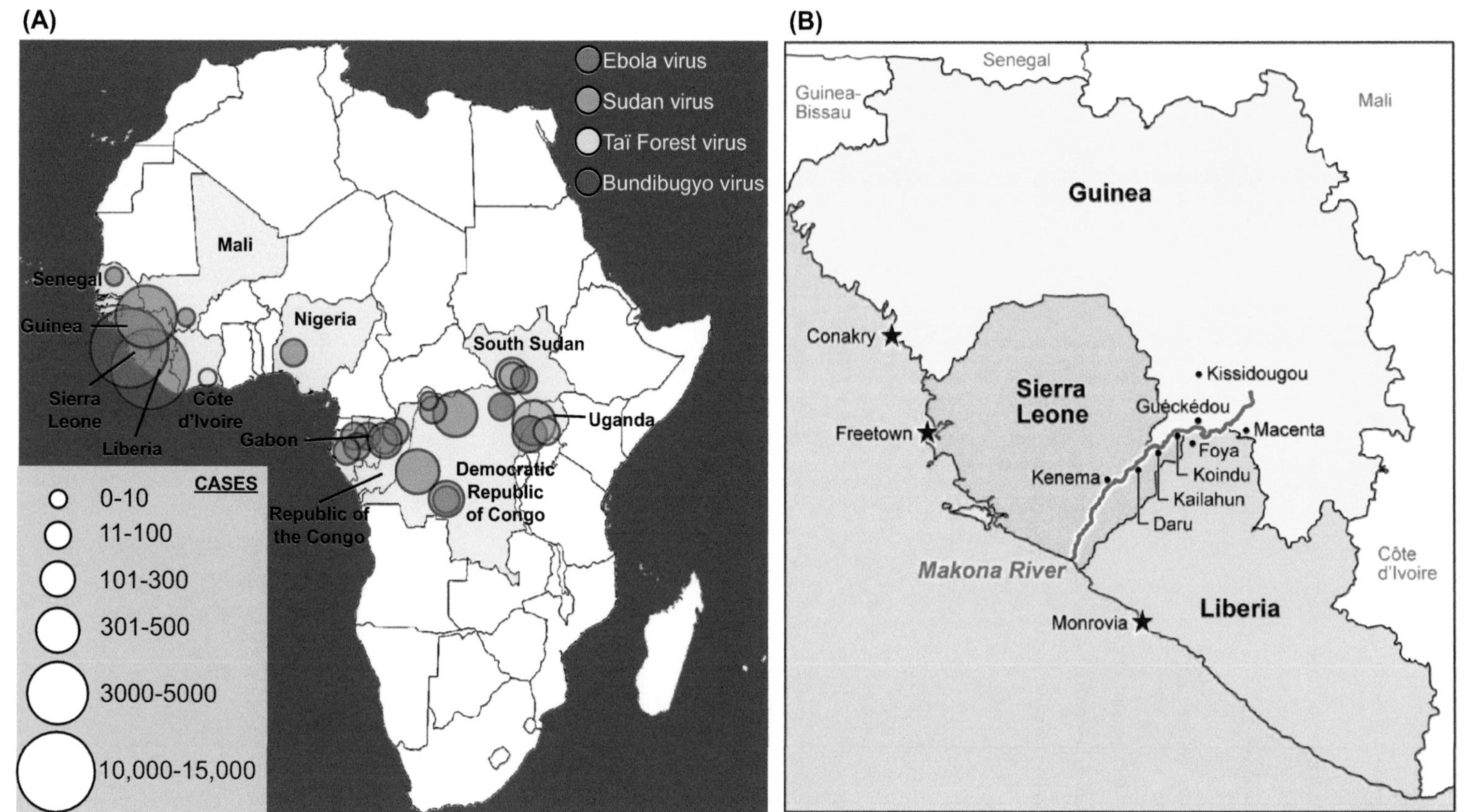

FIGURE 16.9 **Ebolavirus outbreaks.** (A) Other than accidental or laboratory-acquired infections, all human Ebolavirus infections have originated in Africa. Before 1994, when Taï Forest virus was discovered in Côte d'Ivoire (Ivory Coast), all Ebolavirus infections had occurred in five Central African countries: Democratic Republic of Congo, South Sudan, Uganda, Republic of the Congo, and Gabon. The 2014–15 outbreak that originated in Guinea was 10-times larger than all others combined. Note that the circles are representative of the number of cases and not to scale. *(Data derived from CDC (Ebola Virus Disease, http://www.cdc.gov/ebola) and WHO (Ebola virus disease outbreak, http://who.int/ebola).)* (B) Map showing Guinea, Sierra Leone, and Liberia, the three major countries involved in the 2014–15 Ebolavirus outbreak in West Africa. The village of Meliandou is just northeast of Guéckédou in Guinea. *(Updated from Kuhn, J.H., et al., 2014. Viruses 6, 4760–4799.)*

as a new species of ebolavirus, the first and only ever observed outside of Central Africa.

The geographic history of ebolaviruses in Central Africa is part of why world health authorities were caught off guard by an outbreak that began in March of 2014 in the West African country of Guinea (Fig. 16.9B). The etiological agent was determined to be Ebola virus (EBOV, of the *Zaire ebolavirus* species), the ebolavirus with historically the highest fatality rates.

Thus far, the virus is thought to have originated in Meliandou, Guinea, a small village with 31 houses, one school, and one medical center (Fig. 16.10A). On December 26, 2013, a 2-year-old boy developed a fever, vomiting, and black stools; he died 2 days later. It is thought that he may have contracted EBOV while he was playing in a hollow tree (Fig. 16.10B) known to be inhabited by insectivorous free-tailed bats (*Mops condylurus*), a species in which EBOV has previously been reported. It is thought that a **spillover event**, the infection of a human by an infected animal, could have occurred through close contact with infectious animal blood, urine, or bodily fluids. The child's 3-year-old sister died on January 5, 2014, and his mother succumbed 6 days later. By February 1, 2014, an infected member of the boy's family had traveled to and died within a hospital in the capital of Conakry. Without suspecting EBOV, no precautions were taken, and the virus soon spread to other areas of Guinea and into nearby countries of Liberia, Sierra Leone, Nigeria, Senegal, and Mali.

Declared over in January 2016, this was the largest ebolavirus outbreak in history. In total, 28,639 cases with 11,316 deaths (39.5%) occurred. All but 36 cases occurred in Guinea, Sierra Leone, and Liberia. Although Guinea was the origin of the outbreak and sustained high fatality rates, Liberia and Sierra Leone bore the brunt of the toll (Fig. 16.11A). The outbreak was exacerbated by the weak public health systems and lack of resources within these countries (Fig. 16.11B and C). Doctors, laboratories, and governments had never experienced an outbreak and were unprepared to orchestrate an appropriate response. Additionally, in contrast to previous outbreaks in rural Central African villages, the capital cities of Guinea, Sierra Leone, and Liberia were epicenters of the 2014–15 outbreak, providing the human population to allow for rapid transmission. A plethora of other factors contributed to the spread, including mistrust of hospitals, lack of compliance, and denial of the virus.

EBOV is able to spread from person to person, but it is *not* a respiratory virus. It is transmitted through direct contact of broken skin or mucous membranes with infectious blood or bodily fluids, including urine, saliva, sweat, feces, vomit, breast milk, or semen. It can also be transmitted through injection with contaminated syringes. Transmission events occur most often while a person is caring for an infected individual or via direct contact with the body of a deceased individual. The incubation period for EBOV is 2–21 days (average of 8–12 days), and people are not infectious until symptoms develop. Ebola virus disease (EVD) begins with the sudden onset of fever, fatigue, chills, myalgia, headache, and loss of appetite. After about 5 days, this is followed by vomiting, severe watery diarrhea, and abdominal pain. A diffuse maculopapular rash may also develop. Internal and external bleeding is not always present but may occur in some cases, noticeable as bleeding of the gums, blood within the stool, oozing from injection sites, and petechiae. In the 2014–15 outbreak, bleeding was associated with 18% of patients, most often as blood in the stool (6% of patients).

As no antivirals or vaccine exists for EVD, treatment is supportive in an attempt to prevent dehydration, multiorgan

FIGURE 16.10 **The possible origin of the 2014–15 Ebolavirus outbreak in Guinea.** A young boy in the village of Meliandou, Guinea (A), is the earliest infected individual to be identified in the 2014–15 Ebolavirus outbreak. He was known to play around a tree that was inhabited by insectivorous free-tailed bats (B). *Courtesy of Saéz et al., 2015. EMBO Mol. Med. 7, 17–23.*

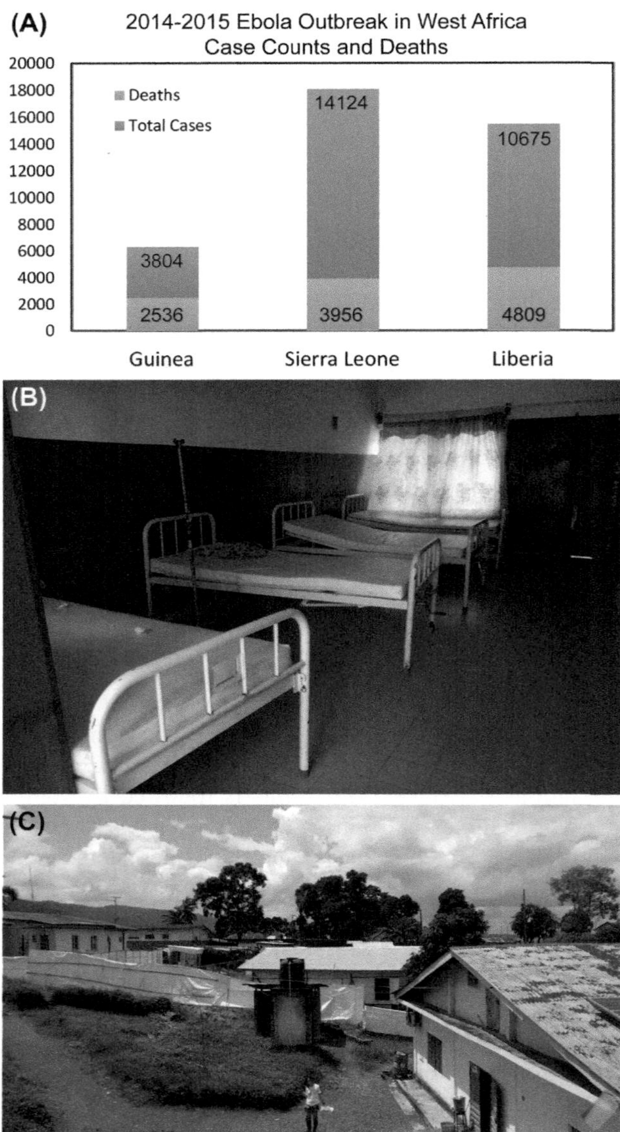

FIGURE 16.11 **The 2014–15 Western Africa Ebolavirus outbreak.** (A) In total, 28,639 cases of Ebola virus disease were diagnosed – primarily in Sierra Leone, Liberia, and Guinea – accounting for 11,316 deaths. *(Data derived from CDC (Ebola Virus Disease, http://www.cdc.gov/ebola) and WHO (Ebola virus disease outbreak, http://who.int/ebola).)* Weak public health systems and lack of resources contributed to the spread of Ebola, as illustrated by photographs of the Infectious Disease ward at Donka Hospital in Guinea's capital of Conakry (B) and one of the region's isolation wards used to accommodate patients ill with Ebola (C). *Photos courtesy of CDC and (A) Dr. Heidi Soeters and (B) Daniel DeNoon.*

failure, shock, hemorrhages, and other complications. Patients with fatal disease die of these conditions typically between day 6 and 16 after the onset of symptoms. In the 2014–15 outbreak in West Africa, the average was 7.5 days. Individuals that survive typically improve around day 6 and are thought to cease being infectious when virus is no longer detectable in the bloodstream or in urine. Interestingly, EBOV has been found in the semen of men that have recovered from EVD long after virus was undetectable in other fluids.

Case Study: Chikungunya Fever Diagnosed Among International Travelers—United States, 2005–06

Excerpted from Morbidity and Mortality Weekly Report, September 29, 2006/55(38);1040–1042.

Chikungunya virus (CHIKV) is an alphavirus indigenous to tropical Africa and Asia, where it is transmitted to humans by the bite of infected mosquitoes, usually of the genus *Aedes*. Chikungunya (CHIK) fever, the disease caused by CHIKV, was first recognized in epidemic form in East Africa during 1952–53. The word "chikungunya" is thought to derive from description in local dialect of the contorted posture of patients afflicted with the severe joint pain associated with this disease. Because CHIK fever epidemics are sustained by human–mosquito–human transmission, the epidemic cycle is similar to those of dengue and urban yellow fever. During 2005–06, 12 cases of CHIK fever were diagnosed serologically and virologically at CDC in travelers who arrived in the United States from areas known to be epidemic or endemic for CHIK fever. This report describes four of these cases.

Case Reports

Minnesota. On May 12, 2005, an adult male resident of Minnesota returned from a 3-month trip to Somalia and Kenya. He had onset of illness hours after arrival in the United States, including fever, headache, malaise, and joint pain mainly in a shoulder and a knee. Serum obtained on May 13 was tested at CDC and determined to be equivocal for CHIKV RNA by reverse-transcription polymerase chain reaction (PCR), consistent with low-level viremia. A recent CHIKV infection was confirmed by demonstration of IgM antibody in this acute-phase serum specimen and neutralizing antibody in convalescent-phase serum (collected 214 days after illness onset). Arthralgias resolved after several weeks.

Louisiana. On January 15, 2006, an adult female resident of India had onset of an illness characterized by fever, joint pain (in the knees, wrists, hands, and feet), and muscle pain (in the thighs and neck). In March 2006, she traveled to Louisiana, where she sought medical attention for persistent joint pain. At CDC, tests of a single serum sample collected on March 30 (74 days after illness onset) were positive for IgM and neutralizing antibodies to CHIKV.

Maryland. An adult female resident of Maryland visited the island of Réunion in the Indian Ocean from October 2005 through mid-March 2006. On February 18, 2006, during an ongoing CHIK fever outbreak in the island, she had onset of fever, joint pain (in the hands and feet), and rash. A local physician clinically diagnosed CHIK fever, but no laboratory tests were conducted. After returning to the United States, the patient sought medical attention for persistent joint pain. At CDC, tests of a single serum sample collected on March 22 (32 days after illness onset) were equivocal for IgM and positive for neutralizing antibody to CHIKV, consistent with a recent CHIKV infection in which IgM antibody was waning. At 5 months after onset, the patient had persistent joint pain (in the hands and feet).

Colorado. An adult male resident of Colorado visited Zimbabwe during April 17–May 29, 2006. On April 29, he had onset of illness with fever, chills, joint pain (in the wrists and ankles), and neck stiffness; a rash appeared a few days later. All symptoms resolved within 2 weeks, except for joint pain, which persisted for approximately 1 month. At CDC, tests of a single serum sample collected on June 12 (44 days after illness onset) were positive for IgM and neutralizing antibody to CHIKV.

Reported by: E Warner, Denver, Colorado. J Garcia-Diaz, MD, Ochsner Clinic Foundation, New Orleans; G Balsamo, DVM, Louisiana Dept of Health and Hospitals. S Shranatan, DO, Johns Hopkins Community Physicians at Hager Park, Hagerstown; A Bergmann, MS, Maryland Dept of Health and Mental Hygiene. L Blauwet, MD, M Sohail, MD, L Baddour, MD, Mayo Clinic College of Medicine, Rochester, Minnesota. C Reed, MD, H Baggett, MD, Div of Global Migration and Quarantine, National Center for Preparedness, Detection, and Control of Infectious Diseases (proposed); G Campbell, MD, T Smith, MD, A Powers, PhD, N Hayes, MD, A Noga, J Lehman, Div of Vector-Borne Infectious Diseases, National Center for Zoonotic, Vector-Borne, and Enteric Diseases (proposed), CDC.

SUMMARY OF KEY CONCEPTS

Section 16.1 Factors Involved in the Emergence of Viral Infectious Diseases

- An emerging infectious disease is a disease caused by a pathogen that has not before been observed within a population or geographic location.
- A reemerging infectious disease is caused by an established pathogen that appears in a new geographical location or was once controlled but begins appearing at a higher incidence.
- 75% of emerging/reemerging diseases are zoonoses, infectious diseases of animals that can be transmitted to humans. Most zoonoses are transmitted by wildlife or arthropod vectors.
- Many human factors are involved in the emergence of infectious diseases. These include urbanization, globalization, travel advances that allow global travel, inadequate public health-care systems, and human behaviors that facilitate transmission. The susceptibility of the human host is also a factor in viral emergence.
- Nearly every emerging viral disease is associated with an environmental/ecological component. The encroachment of humans into pristine environments for farming or land modification puts them in contact with new wildlife and the viruses they carry. Weather and climate, including human-induced changes, also play a role in the emergence of viruses from their natural reservoirs.
- The molecular make-up of the virus determines whether or not an emerging virus will successfully infect and spread within a population. Reassortment of genetic segments, such as with influenza virus, can lead to antigenic shift and the emergence of a novel human virus. Recombination and mutation are two other means of acquiring genetic change.
- RNA viruses are the most common cause of emerging diseases in humans, attributable to the high mutation rate resulting from the low fidelity of their RNA polymerases.

Section 16.2 Notable Emerging and Reemerging Viral Diseases

- The great majority of emerging viral diseases are zoonoses transmitted by arthropods or nonhuman mammals.
- Arboviruses are viruses that are transmitted to humans or other mammals by arthropods, primarily mosquitoes and ticks. The virus is acquired when the arthropod takes a blood meal that contains the virus, which then replicates within the arthropod and is transmitted within infectious saliva to a new host.
- Arboviruses can be transmitted through several transmission models. The virus cycles between the arthropod vector and a wild animal in the enzootic cycle. In the rural epizootic cycle, the virus causes an epidemic within domestic animals when they are infected by the vector. An urban cycle involves the direct transmission of the virus between mosquitoes and humans.
- Flaviviruses include several notable vector-transmitted human pathogens, including yellow fever virus, WNV, Japanese encephalitis virus, St. Louis encephalitis virus, and dengue virus, the most common arboviral disease in the world.
- Dengue virus is transmitted by *Aedes* mosquitoes, whose range has been increasing due to climate change. Three-quarters of dengue infections are asymptomatic, but dengue virus can cause dengue without warning signs, dengue with warning signs, and severe dengue, a serious condition characterized by hemorrhagic fever leading to shock and multiorgan failure.
- *Rodentia* (rodents) and *Chiroptera* (bats) are the two most species-rich orders, and both are host to high-impact zoonotic viruses. Arenaviruses are transmitted by infected rodent urine and feces, and several can lead to hemorrhagic fever.
- Bats are thought to be the natural reservoir of Filoviruses, enveloped helical –ssRNA viruses that are capable of causing hemorrhagic fever. Marburg virus and ebolaviruses are filoviruses. Five species of ebolaviruses have been identified, four of which infect humans and cause high mortality rates.
- Until the 2014–15 EBOV epidemic in West Africa, all ebolavirus outbreaks (with the exception of a researcher infected with TAFV) had occurred in Central Africa.
- The 2014–15 West African EBOV outbreak is thought to have originated with the infection of a 2-year-old boy in Meliandou, Guinea, by a spillover event from an insectivorous free-tailed bat. The virus quickly spread to the capital of Conakry and nearby countries of Liberia, Sierra Leone, Nigeria, Senegal, and Mali.
- The 2014–15 EBOV outbreak was declared over in January 2016. In total, 28,639 cases were documented, accounting for 11,316 deaths (39.5%).
- EBOV is spread from person to person through direct contact of broken skin or mucous membranes with infectious blood or bodily fluids or from contaminated syringes. The incubation period averages 8–12 days, at which point a person develops symptoms and is infectious. EVD begins with the sudden onset of fever, fatigue, chills, myalgia, headache, and loss of appetite. After about 5 days, this is followed by vomiting, severe watery diarrhea, and abdominal pain. Internal and external bleeding occurs in about a fifth of patients.

- No antivirals or vaccine exists for EVD. Treatment is supportive in an attempt to prevent dehydration, multiorgan failure, shock, hemorrhages, and other complications. Patients with fatal disease die of these conditions typically between day 6 and 16 after the onset of symptoms.

FLASH CARD VOCABULARY

Emerging infectious disease (EID)	Rural epizootic cycle
Reemerging infectious disease	Urban cycle
Zoonosis	Dengue without warning signs
Vector	Dengue with warning signs
Urbanization	Severe dengue
Globalization	Hemorrhagic fever
Severe acute respiratory syndrome (SARS)	Defervescence
SARS-CoV	Shock
Anthroponosis/anthroponotic disease	Arenaviruses
Arbovirus	Marburg virus
Enzootic cycle	Ebola virus
Sylvatic	Spillover event

CHAPTER REVIEW QUESTIONS

1. What are the major causes of emerging or reemerging infectious diseases?
2. How does urbanization lead to an environment that is more conducive to the spread of an emerging virus?
3. What aspects of globalization are involved in the spread of EIDs?
4. What kinds of human or societal factors contribute to the emergence or spread of an emerging virus?
5. Why is measles reemerging in the United States and elsewhere in the world?
6. How do deforestation, habitat modification, and the increased use of wild land for farming contribute to the emergence of viral diseases?
7. Give at least two examples of how weather or climate change can increase viral emergence.
8. Why are EIDs often RNA viruses?
9. Explain the process by which a virus is acquired by an arthropod and then transmitted to a new host.
10. Which arboviral mode of transmission is most likely to occur in a rural area where jungle has been cleared for farming purposes? Why?
11. What is hemorrhagic fever? Explain how hemorrhagic fever can lead to death.
12. You are an expert in emerging arboviral diseases and have been asked to suggest control measures to prevent the emergence or spread of new viruses. What measures would you suggest?
13. Explain how rodents transmit arenaviruses to humans. In which sorts of scenarios do you think transmission would be most likely to occur?
14. What factors contributed to the spread of EBOV in West Africa? Describe at least five.

FURTHER READING

Belay, E.D., Monroe, S.S., 2014. Low-incidence, high-consequence pathogens. Emerg. Infect. Dis. 20, 319–321.

Centers for Disease Control and Prevention, 2015. Ebola (Ebola Virus Disease). http://www.cdc.gov/vhf/ebola/.

Choffnes, E.R., Mack, A., Microbial, F., 2015. Emerging Viral Diseases: The One Health Connection Workshop Summary. The National Academies Press, Washington, DC.

Coleman, C.M., Frieman, M.B., 2013. Emergence of the Middle East respiratory syndrome coronavirus. PLoS Pathog. 9, e1003595.

Conway, M.J., Colpitts, T.M., Fikrig, E., 2014. Role of the vector in arbovirus transmission. Annu. Rev. Virol. 1, 71–88.

Engelthaler, D.M., Mosley, D.G., Cheek, J.E., et al., 1999. Climatic and environmental patterns associated with Hantavirus pulmonary syndrome, Four Corners region, United States. Emerg. Infect. Dis. 5, 87–94.

Engering, A., Hogerwerf, L., Slingenbergh, J., 2013. Pathogen–host–environment interplay and disease emergence. Emerg. Microbes Infect. 2, 1–7.

Gurley, E.S., Montgomery, J.M., Hossain, M.J., et al., 2007. Person-to-person transmission of Nipah virus in a Bangladeshi community. Emerg. Infect. Dis. 13, 1031–1037.

Han, B.A., Schmidt, J.P., Bowden, S.E., Drake, J.M., 2015. Rodent reservoirs of future zoonotic diseases. Proc. Natl. Acad. Sci. 112, 201501598.

Holmes, E.C., 2009. The Evolution and Emergence of RNA Viruses. Oxford University Press, Oxford.

Huang, Y.-J., Higgs, S., Horne, K., Vanlandingham, D., 2014. Flavivirus-mosquito interactions. Viruses 6, 4703–4730.

Jones, K.E., Patel, N.G., Levy, M.A., 2008. Global trends in emerging infectious diseases. Nature 451, 990–993.

Kosoy, O.I., Lambert, A.J., Hawkinson, D.J., et al., 2015. Novel thogotovirus associated with febrile illness and death, United States, 2014. Emerg. Infect. Dis. 21, 760–764.

Kuhn, J.H., Becker, S., Ebihara, H., et al., 2010. Proposal for a revised taxonomy of the family Filoviridae: classification, names of taxa and viruses, and virus abbreviations. Arch. Virol. 155, 2083–2103.

Liu, S.-Q., Rayner, S., Zhang, B., 2015. How Ebola has been evolving in West Africa. Trends Microbiol. 23, 387–388.

Luis, A.D., Hayman, D.T.S., O'Shea, T.J., et al., 2013. A comparison of bats and rodents as reservoirs of zoonotic viruses: are bats special? Proc. Biol. Sci. 280, 20122753.

Morens, D.M., Fauci, A.S., 2013. Emerging infectious diseases: threats to human health and global stability. PLoS Pathog. 9, e1003467.

Peters, C.J., 2014. Forty years with emerging viruses. Annu. Rev. Virol. 1, 1–23.

Saéz, A.M., Weiss, S., Nowak, K., et al., 2014. Investigating the zoonotic origin of the West African Ebola epidemic. EMBO Mol. Med. 7, 17–23.

Smith, I., Wang, L.-F., 2013. Bats and their virome: an important source of emerging viruses capable of infecting humans. Curr. Opin. Virol. 3, 84–91.

Weaver, S.C., Barrett, A.D.T., 2004. Transmission cycles, host range, evolution and emergence of arboviral disease. Nat. Rev. Microbiol. 2, 789–801.

White, J.M., Schornberg, K.L., 2012. A new player in the puzzle of filovirus entry. Nat. Rev. Microbiol. 10, 317–322.

World Health Organization, 2009. Dengue: guidelines for diagnosis, treatment, prevention, and control. In: Dengue: Guidelines for Diagnosis, Treatment, Prevention, and Control. WHO Press, p. 160.

World Health Organization, 2015. Ebola Virus Disease Outbreak. http://www.who.int/csr/disease/ebola/en/.

Xu, R.H., He, J.F., Evans, M.R., et al., 2004. Epidemiologic clues to SARS origin in China. Emerg. Infect. Dis. 10, 1030–1037.

Appendix 1

ABBREVIATIONS

3-OS HS 3-O-sulfated heparan sulfate
AAV adeno-associated virus
AAVS1 adeno-associated virus integration site 1
AIDS acquired immune deficiency syndrome
ALT alanine aminotransferase
AMV avian myeloblastosis virus
API active pharmaceutical ingredient
ART antiretroviral therapy
ASGPR asialoglycoprotein receptor
AST aspartate aminotransferase
ATL adult T cell leukemia
ATP adenosine triphosphate
AZT zidovudine
BCR B cell receptor
BSC biological safety cabinet
BSE bovine spongiform encephalopathy
BSL biosafety level
CA capsid protein
cART combination antiretroviral therapy
cccDNA covalently closed circular DNA
CDC Centers for Disease Control and Prevention
CDK cyclin-dependent kinase
cDNA complementary DNA
CFU colony forming unit
CHIKV chikungunya virus
CIN cervical intraepithelial neoplasia
CJD Creutzfeldt–Jakob disease
CPE cytopathic effect
CRE *cis-acting* replication element
cRNA complementary RNA
CTL cytotoxic lymphocyte
DENV dengue virus
DI defective interfering (particles)
DNA deoxyribonucleic acid
DRC Democratic Republic of Congo
dsDNA double-stranded DNA
dsRNA double-stranded RNA
E (gene) early (gene)
EBOV Ebola virus
EBV Epstein–Barr virus
EFC entry fusion complex (poxviruses)
EIA enzyme immunoassay
EID emerging infectious disease
eIF-2α eukaryotic translation initiation factor 2α
ELISA enzyme-linked immunosorbent assay
EV enveloped virion (poxviruses)
EVD Ebola virus disease
FDA Food and Drug Administration
FITC fluorescein isothiocyanate
FMDV foot-and-mouth disease virus
FRET fluorescence resonance energy transfer
GAG glycosaminoglycan
GRID gay-related immune deficiency
HA or H hemagglutinin
HAART highly active antiretroviral therapy
HAV hepatitis A virus
HAVCR1 hepatitis A virus cellular receptor 1
HBcAg hepatitis B core antigen
HBsAg hepatitis B surface antigen
HBV hepatitis B virus
HCMV human cytomegalovirus
HCV hepatitis C virus
HDV hepatitis D virus
HEPA high-efficiency particulate air (filter)
HEV hepatitis E virus
HHV human herpesvirus
HIV human immunodeficiency virus
HPAI highly pathogenic avian influenza
HPS hantavirus pulmonary syndrome
HPV human papillomavirus
HRIG human rabies immunoglobulin
HSV herpes simplex virus
HTLV-I human T-lymphotropic virus type I
HVEM herpesvirus entry mediator
ICAM-1 intercellular adhesion molecule-1
ICTV International Committee on Taxonomy of Viruses
IE (gene) immediate early (gene)
IFA immunofluorescence assay
IFN interferon
Ig immunoglobulin
IHC immunohistochemistry
IN Integrase
IPV inactivated polio vaccine
IRES internal ribosome entry site
ITR inverted terminal repeat
IV immature virion (poxviruses)
KSHV Kaposi's sarcoma-associated herpesvirus
L (gene) late (gene)
LAT latency-associated transcript
LCMV lymphocytic choriomeningitis virus
LDL low-density lipoprotein
LFIA lateral flow immunoassay

LPL lipoprotein lipase
LTNP long-term nonprogressor
LTR long terminal repeat
MA matrix protein
MARCO macrophage receptor with collagenous structure
MCPyV Merkel cell polyomavirus
MERS Middle East respiratory syndrome
MERS-CoV MERS coronavirus
MHC major histocompatibility complex
miRNA microRNA
MMLV Moloney murine leukemia virus
MMR measles-mumps-rubella (vaccine)
MMRV measles-mumps-rubella-varicella (vaccine)
MMWR Morbidity and Mortality Weekly Report
MOI multiplicity of infection
mRNA messenger RNA
MSA metropolitan statistical area
MV mature virion (poxviruses)
NA or N neuraminidase
NAT nucleic acid testing
NC nucleocapsid protein
NK cell natural killer cell
NNRTI nonnucleoside/nucleotide reverse transcriptase inhibitor
NP nucleocapsid protein
NRTI nucleoside/nucleotide reverse transcriptase inhibitor
NS nonstructural
NTCP sodium taurocholate cotransporting polypeptide
NTR nontranslated region
OAS the 2′-5’-oligoadenylate synthetase
OPV oral polio vaccine
ORF open reading frame
PAGE polyacrylamide gel electrophoresis
PAMP pathogen-associated molecular pattern
PBS primer-binding site
PCR polymerase chain reaction
PDA parenteral drug abuser
PEL primary effusion lymphoma
PFU plaque forming unit
pgRNA pregenomic RNA
PKR dsRNA-dependent protein kinase
PML progressive multifocal leukoencephalopathy
PPS postpolio syndrome
PPT polypurine tract
PR protease
pRB retinoblastoma protein
PrP^{C} cellular prion protein
PrP^{Sc} scrapie-causing prion protein
PRR pattern recognition receptor
qPCR quantitative PCR (real-time PCR)
RC replication complex
rcDNA relaxed circular DNA
RdRp RNA-dependent RNA polymerase
rER rough endoplasmic reticulum
RISC RNA-induced silencing complex
RLR RIG-I-like receptor
RNA ribonucleic acid
RNase L ribonuclease L
RNP ribonucleoprotein
rRNA ribosomal RNA
RSV Rous sarcoma virus
RSV respiratory syncytial virus
RT reverse transcriptase
RT-PCR reverse transcriptase polymerase chain reaction
SARS severe acute respiratory syndrome
SARS-CoV SARS coronavirus
SEM scanning electron microscope
siRNA small interfering RNA
SIV simian immunodeficiency virus
ssDNA single-stranded DNA
ssRNA single-stranded RNA
–ssRNA negative-sense RNA
+ssRNA positive-sense RNA
STD sexually transmitted disease
SV40 simian virus 40
T triangulation number
TCR T cell receptor
TEM transmission electron microscope
TIM-1 T cell immunoglobulin and mucin domain 1
TK thymidine kinase
TLR Toll-like receptor
TMV tobacco mosaic virus
tRNA transfer RNA
U3 unique to the 3′ end
U5 unique to the 5′ end
USGS United States Geological Survey
UTR untranslated region
VACV vaccinia virus
VAPP vaccine-associated paralytic polio
VARV variola virus
VDEPT virus-directed enzyme prodrug therapy
VDPV vaccine-derived poliovirus
VLDL very low-density lipoprotein
VMAP viral membrane assembly protein
vmRNA viral mRNA
VPg virion protein, genome-linked
vRNA viral RNA
VZV varicella-zoster virus
WNV West Nile virus
WV wrapped virion (poxviruses)

Appendix 2

GLOSSARY

3′ untranslated region The 3′ portion of an mRNA strand that does not become translated into a protein.

5′ untranslated region The 5′ portion of an mRNA strand that does not become translated into a protein.

Abortive poliomyelitis Poliovirus infection that results in a nonspecific infection that resolves completely.

Accessibility The ability of a virus to come into contact with its host.

Acid-labile A molecule that is not stable in acidic conditions.

Acid-resistant A molecule that is stable under acidic conditions.

Acquired immune deficiency syndrome AIDS; a disease caused by the human immunodeficiency virus that is characterized by the decline of the immune system.

Active pharmaceutical ingredient The molecule within a drug or therapeutic that causes the physiological effect observed by taking the drug.

Acute flaccid paralysis Muscle weakness resulting from neuron damage.

Acute infection An infection in which the pathogen replicates quickly within the host and is cleared by the immune system.

Adaptive immune system The arm of the immune system that uses T and B lymphocytes to generate a pathogen-specific immune response.

Adjuvant A substance used in vaccines that enhances the immune response.

Adsorption The adhesion of a virion to the surface of the cell.

Affinity The strength of binding between one molecule and another.

Agent Used in epidemiology, the infectious pathogen that is necessary to induce a particular disease.

Agglutination The joining together of particles, usually mediated by antibodies.

Alternative splicing The process of forming different mRNAs from a single gene by joining different combinations of exons.

Alveolar macrophages Large phagocytic cells found in the alveoli (air spaces) of the lungs.

Ambisense Type of single-stranded RNA genome that contains both positive and negative sense portions.

Amino acids Biological molecules that are joined together to create peptides and proteins.

Amphipathic A molecule that has both a polar hydrophilic end and a nonpolar hydrophobic end.

Amplicon An amplified fragment of DNA.

Amplification plot Used in real-time PCR, a graph that shows the amplification of DNA compared to the cycle number.

Analytic studies The type of epidemiological study that uses baseline groups to determine how and why a disease occurs/occurred.

Anchorage independence The property of a cell that has lost its requirement to adhere to a substrate.

Angiogenesis The creation of new blood vessels.

Antagonist A molecule that blocks the interaction of a ligand with its specific receptor.

Anterograde transport The transport of virions from a nerve cell body back to the termini of the axon within the tissue.

Antibody/immunoglobulin A protein found in the blood that is produced by B cells as part of the immune response. Each antibody specifically recognized a particular antigen.

Antibody-dependent cell cytotoxicity The ability of natural killer cells and cytotoxic lymphocytes to induce apoptosis in a cell that has already been coated by antibodies.

Anticodon The 3-nucleotide portion of a transfer RNA that recognizes an mRNA codon.

Antigen presentation The presentation of antigen within an MHC molecule to a T cell.

Antigenic drift The accumulation of point mutations within the influenza genome that lead to small changes within the viral proteins.

Antigenic shift A major change in the surface proteins of influenza virus caused by the reassortment of genomic segments from two influenza subtypes.

Antigenic variation Changes within the surface proteins of a pathogen that prevent them from being recognized by previously formed antibodies.

Antigenome The complementary copy of the complete sequence of a DNA or an RNA genome.

Antiparallel Parallel but going in opposite directions.

Antiviral antibodies Antibodies that recognize and bind to a virus.

Antivirals Therapeutics that interfere with viral replication.

Apical The side of a cell that faces the lumen of a body cavity or tube.

Apoptosis Orderly programmed cell death.

Arbovirus A virus that is transmitted between hosts through an arthropod vector, such as a mosquito or tick.

Arthralgia Joint pain.

Assembly The stage of the viral life cycle that brings together the viral genome and proteins to create new virions.

Asymmetrical Occurring on one side of the body (when pertaining to polio paralysis).

Attachment The stage of the viral life cycle in which the virus adheres to the surface of the cell.

Attenuated Weakened.

Autism A spectrum of mental disorders characterized by impaired communication and social interactions.

B cell receptor A signaling receptor on the surface of B lymphocytes that recognizes antigen.

Bacteriophage (Phage) A virus that infects bacteria.

Baltimore Classification System A system used to classify viruses based upon their type of genome and replication strategy.

Base pair Two nucleotides on different strands interacting with each other through a hydrogen bond between nitrogenous bases. Adenine base pairs with thymine or uracil, whereas cytosine base pairs with guanine.

Benign Not cancerous.

Bifurcated needle A needle with two prongs at the end, between which is held a dose of vaccine material. Used for the smallpox vaccination.

Bilirubin A yellowish pigment produced by the breakdown of the heme portion of hemoglobin in the liver.

Bioinformatics The application of math and computer science to extract information from biological data sets.

Biological safety cabinet A piece of equipment that uses HEPA filters to provide a sterile workspace and an environment that protects the user.

Biosafety level A classification system used to define practices and procedures that must be employed when working with a specific pathogen.

Bulbar poliomyelitis Damage to the bulbar region of the brain stem, responsible for autonomic functions.

Bulbospinal poliomyelitis Damage to both bulbar and spinal neurons as a result of poliovirus infection.

Burst size Number of infectious virions released per infected cell.

Bushmeat Hunted meat.

Cancer antigens Antigens that are present only in cancer cells.

Cancer A disease characterized by the uncontrolled growth of cells that proceed to interfere with normal organ functions and invade other tissues.

Capsid A structure containing the viral genome that is composed of repeating viral proteins.

Cap-snatching The process by which influenza viral proteins cleave the 5′ methylguanosine cap from a host mRNA to use it for viral transcripts.

Capsomere A visible morphological unit that composes the capsid (eg, pentamer or hexamer).

Carcinogens Substances that cause cancer.

Carcinoma Cancer that originates in epithelial cells.

Carrier A person that harbors an infectious agent and can transmit it to others but shows no signs of infection.

Case definition A set of uniformly applied criteria to determine whether an individual has a particular disease.

Case-control studies A type of observational study that retroactively assembles a control group to compare to the case patients.

Case-patient A person who meets the case definition for a disease.

Cell culture The process of growing cells in vitro (outside the body) in controlled environmental conditions.

Cell cycle The period during which a cell grows, replicates its DNA, and divides into two daughter cells.

Cell surface receptor The cellular molecule used by a virus to attach to the surface of a cell.

Cell-mediated response The immune response carried out by T lymphocytes.

Central dogma of molecular biology A model that states DNA is transcribed into RNA, which is translated to form proteins.

Centrifugal rash A rash that is distributed more heavily on the face and outer appendages than the trunk of the body.

Centripetal rash A rash that is distributed more heavily on the trunk of the body than on outer appendages.

Chain of infection The process by which an infectious agent leaves its host through a portal of exit, is conveyed by a mode of transmission, and enters a new host through a portal of entry.

Checkpoints Internal controls that prevent a cell from proceeding into the next stage of the cell cycle until the previous stage has been satisfactorily completed.

Chemokines Small secreted proteins that cause cells to traffic to an area of infection.

Chromosome In eukaryotic cells, the structure formed from a piece of genomic DNA wrapped around histone proteins. Different species contain different numbers of chromosomes.

Cirrhosis Scarring of the liver.

Clathrin-coated pit An endocytic vesicle forming from the plasma membrane with the assistance of the clathrin protein, which creates a lattice or cage around the vesicle.

Codon Three adjacent nucleotides within a strand of mRNA that code for a specific amino acid or release factor.

Cognate antigen The antigen that is uniquely recognized by a particular antibody, T cell receptor, or B cell receptor.

Cohort studies A type of observational study that compares two groups of individuals in real time.

Coinfection Infection with two viruses at the same time.

Concatemer Several genome copies linked together.

Confluent smallpox A pattern of smallpox rash in which the lesions are touching with little or no visible skin between them.

Congenital infection An infection of a child acquired during the mother's pregnancy and present at birth.

Conjunctivitis Inflammation of the conjunctiva, the membrane covering the white portion of the eye and the underside of the eyelid.

Contact inhibition The propensity of normal cells to cease dividing when they make contact with an adjacent cell.

Control measures Actions performed to prevent or reduce the spread of an infectious agent.

Convalescent period The period during infection in which the symptoms of the illness subside.

Coreceptor An additional cellular protein that is required for entry of a virus.

Crescent membrane A curved piece of endoplasmic reticulum membrane that is packaged with the poxvirus genome and proteins during assembly to form the immature virion.

Cross-reactivity When antibodies against one virus also recognize and protect against infection with a similar virus.

Cross-sectional studies A type of observational study that gathers data from a random sample of individuals and attempts to draw correlations within the data.

Cyclin-dependent kinases A protein kinase that controls cell cycle progression. It is not activated until its cyclin protein reaches a threshold level.

Cyclins Proteins within the cell that increase or decrease during the cell cycle and regulate the activity of cyclin-dependent kinases.

Cytokine receptor The cellular transmembrane protein receptor that recognizes a cytokine.

Cytokine Small proteins often secreted by cells that induce intracellular signaling within the same, nearby, or distant cells.

Cytology The examination of cells, using microscopy.

Cytopathic effects The visible changes of cells due to viral infection.

Cytoskeleton Formed by proteins, the structural framework of the cell.

Cytotoxic T lymphocyte A type of immune cell that is able to induce apoptosis of virally infected cells.

Daughter cells The term for the resulting cells when a single cell divides in two.

Defective interfering particles Viral particles with incomplete genomes that are released from a cell.

Dendritic cells Cells of the innate immune system that retrieve, process, and present antigen to T lymphocytes.

Deoxyribonucleic acid (DNA) The double-stranded helical molecule composed of nucleotides that contains genetic material of an organism or biological entity.

Dermis The layer of skin located under the epidermis that contains nerves, blood vessels, and glands.

Descriptive studies The type of epidemiological study that reports the who (persons), where (location), and when (time) of a disease occurrence.

Diploid Possessing two of each chromosome.

DNA microarray A test in which thousands of oligonucleotides are added to known locations on a glass or silicon chip and bind to complementary pieces of DNA or RNA within a sample.

DNA polymerase The enzyme that adds new nucleotides onto a growing strand of DNA.

DNA replication The process of creating an exact copy of DNA.

DNA vaccines Experimental vaccines that inject DNA into a person's cells, which transcribe and translate it into a protein.

Dorsiflexion To bend toward the body.

Eclipse period During a one-step growth curve, the period of time required to generate intracellular infectious virions.

Eczema vaccinatum A serious vaccinia infection that occurs in the skin of individuals with previous skin conditions.

Electroporation A technique that applies a brief shock to cells in order to open up pores through which DNA can pass.

Enanthem Internal rash found on a mucous membrane.

Encephalitis Inflammation of the brain.

Endemic The constant presence (or typical prevalence) of a pathogen or condition with a geographical area.

Endocytosis The process by which molecules are brought from the exterior to the interior of a cell.

Endosome A membrane-bound vesicle produced through endocytosis.

Enhancer A region of DNA that contains sequences to which regulatory proteins bind. Enhancers can be a great distance upstream or downstream of the transcription start site.

Enveloped virus A virus that possesses an outer lipid bilayer as its outermost shell. Viral proteins become embedded into the envelope.

Enzyme-linked immunosorbent assay Laboratory test that uses antibodies to determine the amount of antigen or antiviral antibody in a patient sample. The final reagent used in the assay is an enzyme that will change the color of the sample if the antigen/antibody of interest is present.

Epidemic/Outbreak The occurrence of more cases than expected in a given area during a particular period.

Epidemiologic triad model The traditional model of infectious disease causation that involves an external agent, a susceptible host, and an environment that brings the host and agent together so that disease occurs.

Epidemiological variables Factors in the transmission of an infectious disease: what, who, where, when, how/why.

Epidemiology The monitoring and study of the factors related to health conditions and the implementation of control measures to prevent disease occurrence.

Epidermis The outermost layer of the skin composed of five layers of cells.

Epithelium The cells that cover the areas of the body exposed to the environment, either externally or internally.

Epizootic Causing disease within an animal population.

Escape hypothesis The hypothesis that proposes that viruses were once parts of living cells that gained the ability to travel between cells.

Escape mutants Viruses that are not recognized by previously generated host antibodies due to point mutations.

Eukaryote An organism composed of cells that possess a nucleus.

Exanthem A skin rash.

Exocytosis The process of exporting molecules from the inside to the outside of a cell.

Exon The portion of a transcribed pre-mRNA that is not spliced out during mRNA processing.

Experimental studies Planned, controlled epidemiological studies.

Febrile seizures A seizure observed in children, caused by high fever.

Fecal-oral route The route of transmission by which infectious virions are secreted in fecal matter that ends up being ingested.

Fidelity Accuracy (pertaining to DNA replication).

Fluid-mosaic model The current model of plasma membrane architecture that states the plasma membrane is composed of a fluid phospholipid bilayer with freely moving proteins and other biological molecules embedded within it.

Fluorescein isothiocyanate FITC; a fluorescent dye that is commonly coupled to antibodies for fluorescent assays.

Fomites Inanimate objects that may transfer a pathogen from one person to another (eg, towels or utensils).

Fulminant hepatitis Rapid loss of liver function.

Fusion A method of viral entry that involves bringing the virion envelope into close proximity to a cellular membrane (most often the plasma membrane or endosome) so that the two may fuse and create a pore through which the viral genome can be delivered.

Gain-of-function mutations A mutation that results in the constitutive activation of a gene.

Gastrointestinal tract The system of organs involved in the digestion and absorption of food and water.

Gene cloning The placement of a gene of interest (normally without introns) into a plasmid.

Gene therapy The delivery of a normal copy of a gene into a cell to compensate for a nonfunctional defective gene.

Generalized vaccinia The condition that occurs when a person vaccinated against smallpox develops a vaccinia infection.

Genetic code The determined set of rules that dictate which amino acids correspond to specific codons within mRNA.

Genome The complete set of genes of an organism or biological entity.

Genotype The genetic makeup of a cell or organism.

Germ theory The theory that states that infectious diseases are caused by microorganisms.

Glycoprotein A protein to which sugars or carbohydrates have been attached.

Glycosylation The process of attaching sugars or carbohydrates to a protein.

Goblet cells Cells found in mucosal epithelia that produce and secrete mucus.

Golgi complex The organelle of the cell responsible for the final modification and distribution of proteins.

Helix A spiral with a defined, static diameter.

Helper T lymphocyte A type of adaptive immune cell that activates B lymphocytes, macrophages, or other T lymphocytes. There are several classes of helper T lymphocytes that secrete cytokines that fine-tune the immune response.

Helper virus A virus that provides the proteins necessary for a defective/satellite virus to replicate.

Hemagglutination The linking together of red blood cells, mediated by viral proteins or antibodies.

Hematogenous spread Spread through the bloodstream or by blood cells.

Hematopoietic stem cell transplantation (HSCT) The transfer of blood cell precursors from one individual into another.

Hemophiliac An individual that lacks a functional gene for a clotting factor and therefore has reduced clotting capacity in his/her blood.

HEPA filter An air filter that removes 99.97% of particles of the most penetrating size (0.3 μM).

Hepatitis Inflammation of the liver.

Hepatosplenomegaly Enlargement of the liver and spleen.

Herd immunity The protection of unvaccinated individuals due to a high rate of vaccination within the population that eliminates susceptible hosts.

Herpes gladiatorum A skin infection caused by herpes simplex virus-1.

Herpes zoster opthalmicus The reactivation of varicella-zoster virus around the eye.

Herpes zoster A skin rash better known as shingles.

Highly active antiretroviral therapy HAART; a cocktail of several antiviral drugs used against the human immunodeficiency virus.

Highly pathogenic avian influenza virus Influenza viruses of wild birds that cause severe symptoms in humans.

High-throughput sequencing A rapid method of determining a DNA sequence.

Histology The examination of tissues under a microscope.

Homeostasis The maintenance of a steady internal condition despite changes in the external environment.

Homologous Characteristics that are similar because they originated from a common ancestor.

Horizontal transmission Transmission between individuals of the same generation.

Host range The species that can be infected by a particular virus.

Humoral response The immune response carried out by B lymphocytes and antibody.

Hybridization The complementary binding of a probe to a portion of DNA.

Hydrophilic A molecule that associates with water.

Hydrophobic A molecule that avoids water.

Icosahedron A 20-sided figure where each side is an equilateral triangle.

Illness period During an illness, the period when symptoms specific to a particular infection appear.

Immune complexes Clusters formed by the binding of antigen by antibodies.

Immune response The effects caused by the host to defend against a pathogen.

Immunity The state in which a person is protected against becoming infected by a pathogen.

Immunocompetent Possessing a functional immune system.

Immunocompromised Possessing a defective or reduced immune system.

Immunogenic A substance that elicits an immune response.

Immunological memory The ability of lymphocytes to respond faster against a pathogen because the immune system has been previously exposed to the pathogen or parts of the pathogen.

Inactivated virus vaccine A vaccine that uses whole virus that is unable to infect because it has been treated with high heat or a chemical, such as formalin.

Inadvertent autoinoculation The accidental transfer of a virus from the site of vaccination to a distal site, such as the eye or face.

Incidence The frequency of new cases among a population during a specific period.

Inclusion body Visible structures formed by viral material that are observed in the nucleus or cytoplasm of a cell as a result of viral replication.

Incubation period The time before which symptoms of an infection are apparent.

Innate immune system The arm of the immune system that nonspecifically recognizes a pathogen and stimulates the adaptive immune response.

Insertional activation The activation of a gene as a result of retrovirus integration.

Insertional inactivation The inactivation of a gene as a result of retrovirus integration.

Insertional mutagenesis The modification of a gene as a result of retrovirus integration.

Integral protein A protein possessing a portion embedded into a membrane, such as the plasma membrane.

Integration Performed by viruses, the process of joining the viral genome (or a cDNA copy) into the host chromosome.

Internal ribosome entry site A site within a piece of mRNA at which translation initiation factors and a ribosome can bind.

International Committee on Taxonomy of Viruses The international body responsible for the official naming and classifying of viruses.

Intrapartum transmission Transmission of a virus during the process of labor.

Intron The portion of a precursor mRNA that is spliced out from the final translated mRNA.

Isotype A form of an antibody. Different isotypes are produced by B cells during primary or memory responses and can be used to differentiate between a primary or subsequent infection.

Jaundice Yellowing of the eyes and skin.

Kinases Proteins that phosphorylate other proteins.

Latency The state in which a virus becomes dormant (inactive) within the body.

Latent period During a one-step growth curve, the period immediately after initial infection during which no extracellular virus is detectable.

Leukocytes White blood cells.

Ligase A cellular enzyme that joins the sugar phosphate backbones of two adjacent fragments of DNA.

Light microscope A piece of equipment that uses light to illuminate a specimen and glass lenses to magnify it.

Lipoprotein A protein joined with lipid moieties.

Live attenuated virus vaccine The type of vaccine that uses weakened but "live" virus.

Localized infection An infection that remains within the initial site of infection.

Lymphadenopathy Swollen lymph nodes.
Lymphoma Cancer of the lymph nodes.
Lyophilized Freeze-dried.
Lysosome A membrane-bound vesicle that contains enzymes for digesting intracellular molecules.
Macrophages Large phagocytic cells of the immune system.
Macules Flat red bumps.
Maculopapular A rash characterized by macules and papules.
Major histocompatibility complex The molecule in which antigen is presented to a T cell.
Malaise A feeling of discomfort or illness.
Malignant Cancerous. Having the properties that allow cells to replicate uncontrollably and/or spread from the initial tumor site.
Maturation The stage of the viral life cycle in which the virus becomes infectious, usually through structural modifications.
Memory cells T or B lymphocytes that have previously responded to their specific antigen.
Meningitis Inflammation of the meninges, the linings of the brain and spinal cord.
Metabolism The collective set of biochemical reactions that takes place within a cell.
Metastasis The movement of cancerous cells from their initial site to a distant site.
Methylated cap The methylated guanosine that is added by cellular enzymes to the 5′ end of an mRNA transcript.
Metropolitan statistical area Area that contains a core urban population of 50,000 or more.
MicroRNA Small inhibitory pieces of RNA that bind to genes or mRNAs in a complementary fashion.
Microvilli Minute projections from the surface of certain cells.
Mitochondrion The organelle of the cell in which the majority of cellular respiration reactions take place.
Mitosis The division of a cell nucleus to form two identical nuclei.
Mode of transmission How a virus is passed from one susceptible host to another.
Monocistronic An mRNA transcript that encodes one protein.
Mononucleosis A condition caused mainly by Epstein–Barr virus that causes the proliferation of white blood cells.
Morbidity Disease.
Morphogenesis Development of virion structure.
Mortality Death.
Mucosal epithelium Epithelial cells that line the internal, mucus-producing areas of the body (eg, lungs or small intestine).
Multiplicity of Infection The ratio of infectious viral particles to cells used for infection.
Myalgia Muscle pain.
Myeloablative therapy Chemotherapy that kills cancerous and normal cells within the blood and bone marrow.
Myocarditis Inflammation of the heart muscle.
Naïve lymphocyte A T or B lymphocyte that has not previously encountered its cognate antigen.
Naked (nonenveloped) virus A virus whose virions do not possess an external lipid membrane.
Nascent Newly formed.
Natural killer cells Innate immune cells that kill virally infected cells.
Necrosis The disorderly death of a cell through injury.
Negative strand (negative sense) The strand of RNA that is not able to be directly translated by ribosomes.
Neurological Pertaining to nerves and the nervous system.
Neurotropic spread Spread through neurons.
Nomenclature System of applying names.
Nonnucleoside reverse transcriptase inhibitors Drugs that interfere with reverse transcriptase but are not terminal nucleosides.
Nonparalytic aseptic meningitis Inflammation of the brain and spinal cord membranes as a result of poliovirus infection.
Nucleic acid testing Testing a specimen for the presence of specific nucleic acid sequences, such as those of a virus.
Nucleocapsid The viral nucleic acid enclosed by capsid proteins.
Nucleoporins Cellular proteins that compose the nuclear pore complex.
Nucleoside analog A molecule used as an antiviral that terminates transcription when it is incorporated into a growing strand of nucleotides.
Nucleoside/nucleotide The building blocks of nucleic acids, a nucleotide is composed of a sugar attached on one side to a phosphate group and to a nitrogenous base on the other (either adenine, guanine, cytosine, thymine, or uracil). A nucleoside is a nucleotide that lacks the phosphate group. The sugar found in DNA or RNA nucleotides is deoxyribose and ribose, respectively.
Nucleus The central region of a eukaryotic cell that contains the genetic material and is separated from the rest of the cell by a double membrane.
Observational studies Studies that observe participants and collect data but do not have any influence over the exposure of the participant.
Okazaki fragments The shorter pieces of replicated DNA that form the lagging strand during DNA replication.
Oligonucleotides (oligos) Short fragments of nucleic acids.
Oncogene A gene that can cause changes within cells that lead to cancer.
Oncogenesis The development of a tumor/cancer.
Oncolytic virus A virus capable of inducing a tumor/cancer.
Open reading frame A continuous stretch of nucleotides that encodes a protein.
Opportunistic infections Infections that occur when a person's immune system is weakened.
Organelle Specialized structures within a cell.
Oropharynx The part of the throat that is at the back of the mouth.
Pandemic An epidemic that spreads through a large area, such as several countries, a continent, or the entire world.
Papilloma A benign epithelial tumor, such as a wart.
Papules Raised red bumps.
Paralytic poliomyelitis Infection with poliovirus that leads to temporary or permanent paralysis.
Passive immunity The transfer of immune system components, such as antibody, that temporarily provide protection from infection.
Pathogen-associated molecular patterns Molecules that are associated with a specific group of pathogens.
Pattern-recognition receptors Cellular proteins that recognize pathogen-associated molecular patterns.
Penetration The stage of the viral life cycle in which the virus passes into the cell from the extracellular environment.
Percutaneous Through the skin.
Period of communicability The period during a person's infection in which the individual can transmit the pathogen to others.
Peripheral protein Proteins that are not embedded but are associated with a cellular membrane, such as the plasma membrane.
Permissivity Containing all the intracellular factors that support viral replication.
Persistent infection Infections that are not completely cleared by the host immune system.

Phage therapy The use of bacteriophages as an alternative therapy for the treatment of bacterial infections.

Phagocytosis The cellular process of engulfing large particulate matter by forming a plasma membrane vesicle around it.

Phospholipid The major type of molecule that composes the plasma membrane.

Phosphorylation The addition of a phosphate group to a molecule.

Plaque assay An assay used to determine the infectious number of viral particles within a sample.

Plaque forming units A unit of measure that represents the infectious number of viral particles within a sample.

Plasma cell The term for a B cell that is producing antibody.

Plasmid A circular piece of double-stranded DNA that is often used for manipulating or cloning genes.

Pneumonia Inflammation of the lungs.

Poliomyelitis The disease caused by poliovirus.

Poly(A) tail The string of adenine nucleotides added to the end of an mRNA transcript.

Polycistronic An mRNA transcript in which several open reading frames are used.

Polymerase chain reaction A laboratory procedure that rapidly replicates pieces of DNA.

Polyprotein A long chain of amino acids that is cleaved into several separate proteins.

Portal of entry The site at which a virus enters a host.

Portal of exit The site at which a virus leaves a host.

Positive strand (positive sense) A strand of RNA that is able to be translated by ribosomes.

Posttranslational modification The modification of proteins following their translation by ribosomes.

Postvaccinial encephalitis A low-frequency complication of smallpox vaccination that leads to inflammation of the brain due to the immune response against the vaccine material.

Precellular hypothesis The hypothesis that proposes that viruses existed before or alongside cells and may have contributed to the development of life. Also known as the "Virus-First Hypothesis."

Prevalence The total number or proportion of cases among a given population.

Primary infection The initial infection.

Prime-boost strategy A vaccine strategy that uses a DNA vaccine followed by a recombinant vector vaccine.

Primer A short piece of nucleic acid used to initiate the replication of DNA.

Prion A subviral infectious agent caused by the misfolding of a normal cellular protein.

Prodromal period The period of illness characterized by nonspecific symptoms.

Prodrug A drug that must be metabolized by the body in order to become active.

Progressive vaccinia/vaccinia necrosum A low-frequency complication of smallpox vaccination in which the vaccination site forms an unhealing wound.

Prokaryote An organism that lacks a nucleus (eg, bacteria).

Promoter The site within a gene at which transcription factors bind and recruit RNA polymerase to initiate transcription.

Proofreading During DNA replication, the mechanism whereby DNA polymerase removes an incorrectly added nucleotide and substitutes it with the correct nucleotide.

Prostration Extreme physical and/or emotional exhaustion.

Protease An enzyme able to cleave a polypeptide chain.

Protein A biological molecule, composed of amino acids, that is encoded by a gene.

Protooncogenes The normal, functional copy of a gene that, when mutated, may lead to the development of a cancerous cell.

Proviral DNA The genome of a retrovirus that has been reverse transcribed into cDNA and inserted into a cellular chromosome.

Pustules Pus-filled lesions.

Real-time PCR (qPCR) A modification of PCR that allows the user to monitor DNA replication as it is occurring (in "real time").

Reassortment The mixing of genome segments from two viral subtypes to form a novel subtype.

Recombinant DNA DNA from two different sources joined together.

Recombinant protein expression The production of a protein encoded by recombinant DNA.

Recombination The process by which a virus exchanges pieces of its genetic material with another strain of virus, through crossing over or template switching during replication.

Recurrent infection The reactivation of a virus after being latent.

Redundant codons Different codons that correspond to the same amino acid.

Regressive hypothesis The hypothesis that states viruses were once independent intracellular organisms that became unable to replicate independently.

Release The stage of the viral life cycle in which new virions are released from the infected cell.

Replication The stage of the viral life cycle in which the viral genome is copied.

Replication-defective virus Viruses that are able to infect cells but not replicate within them.

Respiratory tract The passages through which air enters and leaves the body.

Retinitis Inflammation of the retina.

Retrograde transport The travel of a virion or virion genome from the synapse of an axon to the cell body.

Retroid virus Any virus that reverse transcribes, whether its genome is RNA or DNA.

Retrovirus A virus that transcribes its RNA genome into DNA upon entry into the cell.

Reverse genetics The creation of a virus by artificially inserting genome segments into a cell to generate functional virions.

Reverse transcriptase The enzyme that creates DNA from an RNA template.

Reverse transcription The process of creating DNA from an RNA template.

Rhinorrhea Runny nose.

Ribonucleic Acid (RNA) A single-stranded biological molecule composed of nucleotides containing ribose.

Ribosome A cellular organelle that is responsible for translating proteins.

Ring vaccination The process of vaccinating all the individuals that came into contact with an infected individual.

RNA splicing The process of removing introns from a piece of precursor mRNA.

Rolling circle replication A method of replicating concatemers of DNA by using a circularized DNA strand as a template for multiple copies.

Roseola infantum A rash of infants caused by HHV-6B or HHV-7.

Rough endoplasmic reticulum An organelle within the cell that is responsible for folding and modifying proteins.

Sarcoma A tumor derived from connective tissues.

Satellite virus A defective virus that does not contain all the viral genes necessary to create nascent virions.

Scanning electron microscope An electron microscope that uses an electron beam to scan the surface of a specimen and derive an image by detecting scattered electrons.

Scarification The creation of shallow cuts within the skin during vaccination.

Segmented genome A genome that is composed of more than one piece of nucleic acid.

Seroconversion The time at which an individual begins making antibody against a particular infectious agent.

Serology Tests involving blood, including those that assay for the presence of antibodies.

Seroprevalence The number or proportion of individuals that produce antibody against a specific infectious agent (and are therefore assumed to have been infected at one time).

Serotype A viral subset classified by differing antigenicity.

Sexually transmitted diseases Infectious agents that are transmitted through sexual contact.

Shedding The release of virus from the host into the environment.

Simian Referring to an ape or monkey.

Slow infection An infection that takes a very long time to produce visible disease.

Small interfering RNA (siRNA) Small pieces of single-stranded RNA that bind to a piece of mRNA using complementary binding, leading to its inactivation and degradation.

Spinal poliomyelitis Damage to motor neurons caused by poliovirus infection.

Splenomegaly Enlargement of the spleen.

Sporadic Occurring randomly or irregularly.

Status epilepticus A seizure that lasts more than 5 min.

Structural unit A repeated set of a single viral protein or multiple viral proteins that form the face of an icosahedral capsid.

Subcutaneous tissue Loose connective tissue beneath the skin containing fat.

Superinfection When an individual infected with one virus is infected with another virus.

Susceptibility The expression by cells of the receptors used for viral entry.

Syncytium A large cell created by the fusion of several cells.

Systemic infection An infection that spreads throughout the body.

T cell receptor The protein receptor found on a T cell that recognizes a specific antigen.

Taxon A category used to classify biological entities.

Taxonomy The classification and naming of biological entities.

Telomerase The enzyme that rebuilds the terminal ends of chromosomes (telomeres).

Transcript A piece of mRNA.

Transcription factors Proteins that assemble on the promoter of a gene and recruit RNA polymerase to initiate the creation of mRNA.

Transcription The process of creating mRNA from a nucleic acid template.

Transfer RNA A small piece of RNA carrying an amino acid that associates with the ribosome.

Transformation When used with bacteria, the process of introducing DNA from an external source. When referring to cancer, the characteristic changes within a cell that accompany oncogenesis.

Translation initiation factors Proteins involved in the initiation phase of eukaryotic translation.

Translation The process of assembling a protein from amino acids using a sequence of mRNA to determine the amino acid order.

Transmission electron microscope A type of electron microscope that passes an electron beam through a specimen and creates an image by detecting the amount of electrons that are transmitted through the specimen.

Transplacental transmission Transfer of a pathogen through the placenta during pregnancy.

Transposable Elements Pieces of DNA that can physically move from one location to another in the genome of a living organism. Also known as transposons.

Triangulation number The number of structural units per face of an icosahedron.

Tropism The specificity of a virus for a particular type of cell or tissue.

Tumor suppressor gene A gene that encodes a protein that regulates the progression of the cell cycle. The loss of tumor suppressor functions can lead to cancer.

Tumorigenic Capable of causing tumors.

Type 1 interferons A family of cytokines that induce antiviral effects within a cell and help to mold the immune response.

Ubiquitination The process of tagging a protein for degradation by attaching ubiquitin molecules to it.

Ulcers A painful lesion characterized by eroding skin.

Umbilication A central depression in a smallpox lesion.

Uncoating The stage of the viral life cycle that releases the viral nucleic acid into the cell.

Vaccination The process of inoculating an individual with inactivated or weakened virus (or pieces of a virus) in order to stimulate their immune system into producing protective immunity.

Variolation The deliberate inoculation of an individual with smallpox scabs in order to induce a lesser infection. Variolation often produced full-blown smallpox.

Vector In epidemiology, an infected species that transmits a pathogen to a human. In molecular biology, the piece of DNA or the virus that is used to transfer genetic material.

Vehicles Nonliving substances or materials that can transmit viruses.

Vertical transmission Transmission from mother to child.

Vesicle In cell biology, a membrane-enclosed structure used for transport. In medicine, a raised lesion filled with fluid.

Villi Finger-shaped formations in the small intestine that increase surface area for absorption.

Viremia Virus in the bloodstream.

Virion An infectious virus particle.

Viroid A small, subviral circular piece of RNA that is transmitted between plant cells and causes damage to the infected plant.

Virotherapy Using viruses for therapeutic purposes.

Virulent Capable of causing disease.

Viruria Virus in the urine.

Virus attachment protein The viral protein used for attachment to a cell surface receptor.

Virus factory An intracellular, membrane-bound structure formed by some viruses that functions as the site of genome replication, transcription, translation, and/or assembly.

Virus A nonliving biological entity that contains an RNA or DNA genome and can be transmitted from cell to cell or host to host.

Zoonosis A pathogen capable of being transmitted from an animal to a human.

Zygote A fertilized egg.

Index

'*Note:* Page numbers followed by "f" indicate figures, "t" indicate tables, and "b" indicate boxes.'

W

Y

9780128009475